Lecture Notes in Computer Science   13052

More information about this subseries at http://www.springer.com/series/7412

Nicolas Tsapatsoulis · Andreas Panayides ·
Theo Theocharides · Andreas Lanitis ·
Constantinos Pattichis · Mario Vento (Eds.)

Computer Analysis of Images and Patterns

19th International Conference, CAIP 2021
Virtual Event, September 28–30, 2021
Proceedings, Part I

Springer

Editors
Nicolas Tsapatsoulis 🆔
Cyprus University of Technology
Limassol, Cyprus

Theo Theocharides 🆔
University of Cyprus
Nicosia, Cyprus

Constantinos Pattichis 🆔
University of Cyprus
Nicosia, Cyprus

CYENS Centre of Excellence
Nicosia, Cyprus

Andreas Panayides 🆔
University of Cyprus
Nicosia, Cyprus

Andreas Lanitis 🆔
Cyprus University of Technology
Limassol, Cyprus

CYENS Centre of Excellence
Nicosia, Cyprus

Mario Vento 🆔
University of Salerno
Salerno, Italy

ISSN 0302-9743   ISSN 1611-3349   (electronic)
Lecture Notes in Computer Science
ISBN 978-3-030-89127-5   ISBN 978-3-030-89128-2   (eBook)
https://doi.org/10.1007/978-3-030-89128-2

LNCS Sublibrary: SL6 – Image Processing, Computer Vision, Pattern Recognition, and Graphics

This Springer imprint is published by the registered company Springer Nature Switzerland AG
The registered company address is: Gewerbestrasse 11, 6330 Cham, Switzerland

Preface

Welcome to the proceedings of the 19th International Conference on Computer Analysis of Images and Patterns (CAIP 2021), which took place during September 28–30, 2021. The CAIP series of biennial international conferences is devoted to all aspects of computer vision, image analysis and processing, pattern recognition, and related fields. Previous conferences were held in Salerno, Ystad, Valletta, York, Seville, Münster, Vienna, and Paris, amongst other places. Due to the ongoing COVID-19 pandemic, CAIP 2021 was held as a virtual conference, but the organizing committee were dedicated to offering the best possible online experience.

The scientific program of the conference consisted of plenary lectures and contributed papers presented in a single track. A total of 129 papers were submitted to CAIP 2021, which were reviewed in a single-blind process with at least two reviewers per paper. A total of 87 papers were accepted. The program featured the presentation of these papers organized into the following 15 sessions:

Session 1: 3D Vision I
Session 2: 3D Vision II
Session 3: Biomedical Image and Pattern Analysis I: Segmentation
Session 4: Biomedical Image and Pattern Analysis II: Segmentation and Classification
Session 5: Biomedical Image and Pattern Analysis III: Disease Diagnosis
Session 6: Deep Learning I: Classification
Session 7: Deep Learning II: Classification
Session 8: Deep Learning III: Image Processing and Analysis
Session 9: Machine Learning for Image and Pattern Analysis
Session 10: Feature Extraction
Session 11: Object Recognition
Session 12: Face and Gesture
Session 13: Guess the Age Contest
Session 14: Biometrics, Cryptography and Security
Session 15: Segmentation and Image Restoration

CAIP 2021 featured a contest on "Guess the Age: Age Estimation From Facial Images with Deep Convolutional Neural Networks" organized by Antonio Greco, University of Salerno, Italy. In addition, the CAIP 2021 program included distinguished plenary and keynote speakers from academia and industry who shared their insights and accomplishments as well as their vision for the future of the field. Moreover, CAIP 2021 included three tutorials as follows:

Tutorial 1: Discovering Patterns in the Road from Genotype to Phenotype
Tutorial 2: Video Summarization for Unpaired Videos
Tutorial 3: Large Scale Video Analytics.

We want to express our deepest appreciation to all the members of the CAIP 2021 organizing committee and Technical Program Committee, as well as all the reviewers, for their dedication and hard work in creating an excellent scientific program. We want to thank all the authors who submitted their papers to CAIP 2021, and all of those who presented and shared their work. Finally, we would like thank all the participants for taking part in the conference.

We hope that you enjoyed this exciting and memorable event, and we look forward to meeting in person at the next CAIP!

October 2021

Constantinos Pattichis
Andreas Lanitis
Nicolas Tsapatsoulis
Andreas Panayides
Theo Theocharides
Mario Vento

Organization

General Chairs

Constantinos S. Pattichis	CYENS and University of Cyprus, Cyprus
Andreas Lanitis	CYENS and Cyprus University of Technology, Cyprus
Nicolai Petkov	University of Groningen, The Netherlands

Program Chairs

Nicolas Tsapatsoulis	Cyprus University of Technology, Cyprus
Andreas Panayides	University of Cyprus, Cyprus
Theo Theocharides	KIOS and University of Cyprus, Cyprus
Mario Vento	University of Salerno, Italy

Local Organizing Committee

Constantinos S. Pattichis	CYENS and University of Cyprus, Cyprus
Constandinos Mavromoustakis	IEEE Cyprus Section and University of Nicosia, Cyprus
Alexis Polycarpou	IET Cyprus Local Network and Frederick University, Cyprus
Toumazis Toumazi	Cyprus Computer Society, Cyprus

Steering Committee

Mario Vento (Chair CAIP 2019)
Gennaro Percanella (Co-chair CAIP 2019)
Michael Felsberg (Chair CAIP 2017)
Nicolai Petkov (Permanent Member)

Awards Chair

Xiaoyi Jiang	University of Münster, Germany

Contests Chairs

Christos Loizou	Cyprus University of Technology, Cyprus
Yannis Avrithis	Inria Rennes-Bretagne Atlantique, France

Tutorials Chairs

Gennaro Percannella	University of Salerno, Italy
Kleanthis Neokleous	CYENS, Cyprus

Workshops Chairs

Nicola Strisciuglio	University of Twente, The Netherlands
Melinos Averkiou	CYENS, Cyprus

Student Activities Chairs

Andreas Aristeidou	University of Cyprus, Cyprus
Sotirios Chatzis	Cyprus University of Technology, Cyprus

Industry Liaison Chairs

Alessandro Artusi	CYENS, Cyprus
Zenonas Theodosiou	CYENS, Cyprus

Publicity Chairs

Alessia Saggese	University of Salerno, Italy
Klimis Ntalianis	University of West Attica, Greece

Publicity Committee

Nikolas Papanikolopoulos	University of Minnesota, USA
Andreas Spanias	Arizona State University, USA
Marios S. Pattichis	University of Mexico, USA
Stefanos Kollias	NTUA, Greece
Andreas Stafylopatis	NTUA, Greece
Xiaoyi Jiang	University of Münster, Germany
Enrique Alegre Gutiérrez	University of Leon, Spain
Alessia Saggese	University of Salerno, Italy

Program Committee

Andreas Lanitis	CYENS and Cyprus University of Technology, Cyprus
Andreas Panayides	University of Cyprus, Cyprus
Constantinos S. Pattichis	CYENS and University of Cyprus, Cyprus
Nicolai Petkov	University of Groningen, Netherlands
Theo Theocharides	KIOS and University of Cyprus, Cyprus
Nicolas Tsapatsoulis	Cyprus University of Technology, Cyprus
Mario Vento	University of Salerno, Italy

Additional Reviewers

Ioannis Anagnostopoulos	University of Thessaly, Greece
Yannis Avrithis	Inria Rennes-Bretagne Atlantique, France
Athos Antoniades	Stremble Ventures Ltd, Cyprus
Zinonas Antoniou	University of Cyprus, Cyprus
Andreas Aristidou	University of Cyprus, Cyprus
Aristos Aristodimou	University of Cyprus, Cyprus
Alessandro Artusi	CYENS, Cyprus
Christodoulou Christodoulos	University of Cyprus, Cyprus
Christoforos Christoforou	University of Cyprus, Cyprus
Adrian-Horia Dediu	SuperData Bucharest, Romania
Constantinos Djouvas	Cyprus University of Technology, Cyprus
Anastasios Doulamis	NTUA, Greece
Basilis Gatos	National Center for Scientific Research "Demokritos", Greece
Enrique Alegre Gutiérrez	University of Leon, Spain
Xiaoyi Jiang	University of Münster, Germany
Minas Karaolis	University of Cyprus, Cyprus
Efthyvoulos Kyriacou	Frederick University, Cyprus
Christos Kyrkou	KIOS Research and Innovation Center, Cyprus
Christos Loizou	Cyprus University of Technology, Cyprus
Alberto Marchisio	Technical University of Viennna, Austria
Costas Neocleous	Cyprus University of Technology, Cyprus
Andreas Neokleous	University of Cyprus, Cyprus
Kleanthis Neokleous	CYENS, Cyprus
Marios Neofytou	University of Cyprus, Cyprus
Athanasios Nikolaidis	Technological Educational Institute of Serres, Greece
Klimis Ntalianis	University of West Attica, Greece
Maria Papaioannou	University of Cyprus, Cyprus
Marios S. Pattichis	University of New Mexico, USA
Gennaro Percannella	University of Salerno, Italy
Ioannis Pratikakis	Democritus University of Thrace, Greece
Benjamin Risse	University of Münster, Germany
Alessia Saggese	University of Salerno, Italy
Zenonas Theodosiou	CYENS, Cyprus
Mario Vento	University of Salerno, Italy

Organized and Sponsored by

Co-organized and Co-sponsored by

Technically Co-sponsored by

Endorsed by

Contents – Part I

Biomedical Image and Pattern Analysis

Machine Learning

Contents – Part II

Object Recognition

Face and Gesture

Guess the Age Contest

Biometrics, Cryptography and Security

Segmentation and Image Restoration

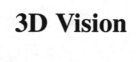

3D Vision

Simultaneous Bi-directional Structured Light Encoding for Practical Uncalibrated Profilometry

Torben Fetzer[1]([⊠]), Gerd Reis[2], and Didier Stricker[1,2]

[1] University of Kaiserslautern, Kaiserslautern, Germany
{torben.fetzer,didier.stricker}@dfki.de
[2] German Research Center for Artificial Intelligence (DFKI), Kaiserslautern, Germany
gerd.reis@dfki.de

Abstract. Profilometry based on structured light is one of the most popular methods for 3D reconstruction. It is widely used when high-precision and dense models, for a variety of different objects, are required. User-friendly procedures encode the scene in horizontal and vertical directions, which allows a unique description of points in the scene. The resulting encoding, can be used to auto-calibrate the devices used. Thus, any consumer or industrial cameras or projectors can be supported and the procedure is not limited to pre-calibrated setups. This approach is extremely flexible, but requires a large number of camera acquisitions of the scene with multiple patterns projected. This paper presents a new approach that encodes the scene simultaneously in horizontal and vertical directions using sinusoidal fringe patterns. This allows to almost halve the number of recorded images, making the approach attractive again for many practical applications with time aspects.

Keywords: Structured light · Reconstruction · Profilometry · Surface encoding

1 Introduction

In the past decades, a multitude of different approaches for the creation of 3D reconstructions have been established. Basically, the procedures can be divided into active and passive procedures, which either directly affect the scene or not. In applications that require highly accurate and dense reconstructions of a wide variety of different objects, structured light approaches have become a preferred choice in industry. These methods usually illuminate the scene with specific patterns, e.g. stripes, pseudo-random-dots, etc. and use their interaction with the scene to estimate depth information. Particularly interesting are fringe projection strategies, where multiple horizontal and vertical sinusoidal patterns are projected onto the scene, in order to encode it in two directions. For cameras, capturing the fringe images, this enables the calculation of dense and robust correspondences (see [3]). This approach also works for un-structured and un-textured surfaces, where most passive methods fail due to the absence of necessary features.

This work was partially funded by the projects MARMORBILD (03VP00293) and VIDETE (01W18002) of the German Federal Ministry of Education and Research (BMBF).

N. Tsapatsoulis et al. (Eds.): CAIP 2021, LNCS 13052, pp. 3–13, 2021.
https://doi.org/10.1007/978-3-030-89128-2_1

Depending on the required accuracy, the frequency of the projected fringes is increased several times in order to successively improve the quality of the encoding. The high number of camera shots required, leads to a considerable expenditure of time in data acquisition, which is the main weakness of the method. To speed up the procedure, often, only one dimension is encoded with structured light and matches are found by intersection with corresponding epipolar lines. Unfortunately, this method is limited to calibrated devices, which means that replacing a camera, lens, changing the settings or moving a device leads to the considerable effort of re-calibration. Furthermore, accuracy often suffers, especially when lens distortions are taken into account. The method presented here, allows auto-calibrations due to the two-dimensional encoding, which makes it significantly more versatile and convenient to use.

According to the state of the art, the phase shifting method is the basis of most structured light methods, due to its texture-invariance. Thereby, sinusoidal fringe patterns with equidistantly shifted phases are projected onto the scene. A superposed phase, which can be calculated from at least three shifted patterns, encodes the scene pixel by pixel in the shifted direction. Doing this, both in horizontal and vertical direction, results in a minimum of 6 captured images. Even more acquisitions are necessary if further refinement steps with higher frequencies are performed. The patterns are sinusoidally modulated in the direction to be encoded and constantly continued in the decoded direction. Therefore, one dimension of the patterns does not carry any information, which is certainly a waste. In the following, more detailed investigations on the phase shifting algorithm and the harmonic addition theorem are carried out. Especially, findings with regard to the resulting amplitude of the phase superposition will encourage us to combine the horizontal and vertical stripe patterns in order to encode both directions simultaneously. The horizontal and vertical phases are then extracted from the combined patterns by a per pixel strategy, making the whole procedure scene-independent.

2   Related Work

Extensive research has been carried out in the field of structured light reconstruction. Various strategies, assuming pre-calibrated setups, have been reviewed and compared by Salvi et al. [9] and more recently by Zanuttigh et al. [14]. The emerging state of the art, based on the phase shifting algorithm, has been reviewed by Servin et al. in [11]. In the mean time, new approaches, like the Fourier-based regularized method of Legarda-Saenz and Espinosa-Romero [6], were developed which, however, could not compete with the state of the art.

Based on the phase shifting method, Mirdehghan et al. [7] recently presented a procedure to generate optimal scene-dependent projection patterns and thus to control the quality of the results. Zhang and Yau presented in [17] a system with two cameras that offers many quality advantages in contrast to standard setups with one camera and one projector. Based on this setup, the devices can be automatically calibrated without any user interaction, as demonstrated in [1] and [2].

Unfortunately, the recording time is the great weakness of the structured light method. One way to shorten the required acquisition time is to distribute several patterns among the different color channels of the cameras and projectors used [5, 15]. These approaches work in theory but suffer from color cross-talk between cameras and projector and a very accurate color calibration is required. In particular, the object color influences the type and strength of the cross-talk. This leads to difficulties in implementing the procedures in practice and even then, one has to expect large quality losses.

To reduce the number of acquisitions required, Guan *et al.* [4] and Sansoni and Redaelli [10] combine patterns of different frequencies into individual patterns. Several fringes are encoded by carrier waves and additively combined. Afterwards they are separated by filter methods. These methods work in theory, but have a poor applicability in practice. The frequencies of the carrier waves must be stable in the image to enable an appropriate extraction. Nevertheless, they made it possible for the first time to provide information in the decoded direction of the patterns. Based on this idea, Yang *et al.* [13] further improved this approach, and created special patterns based on co-prime frequent sine waves that can be more robustly separated by a Garbor filter. Recent advances in real-time measurement with structured light have been detailed and analyzed by Zhang in [16]. Finally, Wang and Yang [12] recently introduced a one-shot approach based on binary stripe patterns, from which the phase can be directly approximated and interpolated. However, the approach assumes the stripes to be continuously visible in the scene, which cannot be guaranteed for general scenarios.

All in all, the problem of combining multiple patterns has already been mentioned, but has not been solved satisfactorily, yet. In particular, the combination of horizontal and vertical patterns has not yet been addressed, before.

3 Mathematical Investigation

In order to develop a new projection pattern, that allows to simultaneously recover the horizontal and vertical phases, we first examine the standard case more closely. This investigation will provide new insights into the amplitude of the superposition of the illuminated scene. These findings will finally enable a subsequent separation of the phase directions from the combined patterns.

3.1 Background: Sinusoidal Phase Shifting Method

Basis of the presented work is the *sinusoidal phase shifting* method. Thereby, patterns are modulated by sine or cosine (convention-dependent) signals and equidistantly shifted at least three times over the periodic domain. The patterns are projected onto the scene and captured by the cameras. Superposition of the resulting images allows to encoded the scene texture-invariant. Horizontal and vertical directions are treated by separate sets of patterns, each modulated by a sine/cosine in the respective direction and continued constantly in the other direction. Horizontal and vertical sets of patterns P_n^H and P_n^V with frequencies F_H and F_V that are shifted $n = 1, ..., N$ times, can be explicitly generated as:

$$P_n^H(i,j) := \cos\left(\frac{2\pi j}{width}F_H + \frac{2\pi(n-1)}{N}\right), \quad P_n^V(i,j) := \cos\left(\frac{2\pi i}{height}F_V + \frac{2\pi(n-1)}{N}\right) \quad (1)$$

Thereby, $i = 1, ..., height$ and $j = 1, ..., width$ denote the image pixels. The first row of Fig. 1 shows an exemplary set of horizontal (a) and vertical (b) patterns with $N = 3$ and $F_H = F_V = 1$. In both cases, the patterns are shown to the left, and the projection of the patterns onto an exemplary scene is shown to the right.

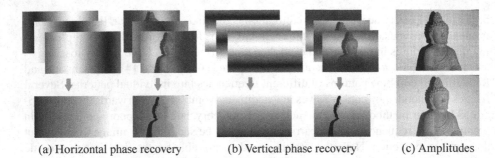

| (a) Horizontal phase recovery | (b) Vertical phase recovery | (c) Amplitudes |

Fig. 1. Example of the phase shifting algorithm for encoding a scene by phase recovery via harmonic addition theorem. The top rows of (a) and (b) show sets of horizontal and vertical fringe patterns and the resulting scene after projection. The bottom rows show the phase images computed by Eq. (3). (c) shows the amplitude of the superposition (Eq. (4)) (top) and the one given by the scaled texture as introduced in Lemma 1 (bottom).

Harmonic Addition Theorem. Structured light approaches with sinusoidal patterns are based on practical application of the *harmonic addition theorem* (see [8]). It states that any superposition of cosines with same phase is again a cosine with the same phase:

$$\sum_{n=1}^{N} I_n \cos(\delta_n) = A \cos(\Phi) \tag{2}$$

$$\text{with} \quad \Phi = \text{atan2}\left(\sum I_n \sin(\delta_n), \sum I_n \cos(\delta_n)\right) \tag{3}$$

$$\text{and} \quad A^2 = \sum_{n=1}^{N}\sum_{m=1}^{N} I_n I_m \cos(\delta_n - \delta_m), \tag{4}$$

where δ_n, δ_m denote the equidistant phase shifts, Φ the phase to be recovered and A the amplitude of the superposition. atan2 denotes the two-dimensional arcustangens function taking into account the quadrants of the input.

Recovering the Phases in the Scene. Projecting patterns from Eq. (1) to a scene I results in images I_n^H and I_n^V for the different phase shifts $n = 1, ..., N$. Applying Eq. (3), the horizontal and vertical phases Φ_H and Φ_V can then be computed by:

$$\Phi_H = \text{atan2}\left(\sum I_n^H \sin(\delta_n), \sum I_n^H \cos(\delta_n)\right)$$

$$\Phi_V = \text{atan2}\left(\sum I_n^V \sin(\delta_n), \sum I_n^V \cos(\delta_n)\right) \tag{5}$$

The second rows of Fig. 1 (a) and (b) show the recovered phases computed for the patterns and the scene. Using this information, the scene points are uniquely encoded by their horizontal and vertical phase values. This allows robust and dense matches between the different camera views and the projector to be achieved. Amplitude A is not needed here and therefore not treated further in literature. For the method that is developed in this work, A plays an important role and is therefore further investigated.

3.2 Amplitude of Superposition

From illuminated images I_n, amplitude A is given by Eq. (4), which can be directly expressed in terms of a captured texture image I of the scene:

Lemma 1. *A captured scene I_n, illuminated by fringe patterns $P_n = sin(x+\delta_n)$, with an arbitrary number of equidistant shifts $n = 1, ..., N \geq 2$, can be superposed to*

$$\sum_{n=1}^{N} I_n \cos(\delta_n) = \frac{N}{4} I \cos(\Phi), \tag{6}$$

where Φ denotes the phase angle and I the fully illuminated scene.

Figure 1 (c) shows the amplitudes of the example scene computed by Eq. (4) (top) and from scaled texture as in Lemma 1 (bottom). Apart from artifacts caused by clipping (due to limited dynamic range of the cameras) and gamma corrections of the devices, these are identical. The lemma can be proved straight forward using properties of trigonometric functions and the harmonic addition theorem.

3.3 Combined Patterns

We are now going to introduce additively combined patterns and setup a mathematical problem with the newly gathered information about the amplitude. Solving this problem enables the recovery of horizontal and vertical phase values simultaneously.

Let the combined patterns P_n^C be defined as

$$P_n^C := \frac{1}{2}(P_n^H + P_n^V) \quad \text{for } n = 1, ..., N. \tag{7}$$

Thereby, two-dimensional sinusoidal patterns result as visualized in Fig. 2 (a, left) and projected to the scene (b, left). The shifting direction naturally becomes the diagonal. The task in the following is to extract the horizontal as well as the vertical phase simultaneously from images of the scene, illuminated by these patterns.

Assuming the optimal case, where cameras and projector linearly respond and do not perform any gamma correction or internal post-processing, a captured scene is proportional to the sum of the separately illuminated scenes I^H and I^V:

$$I^C = P^C \odot \frac{I}{2} = \frac{1}{2}(P^H + P^V) \odot \frac{I}{2} = \frac{1}{2}(I^H + I^V), \tag{8}$$

where $\odot$ denotes pixel-wise multiplication of the patterns and the scene appearance I.

(a) Ideal phases computed from patterns   (b) Scene phases from camera images

Fig. 2. Combined sinusoidal patterns, computed by Eq. (7) and projected onto the scene. The horizontal and vertical phases, to be recovered, are shown to the right.

Problem Formulation. With Lemma 1 and Eq. (8), we directly get the basic properties to set up the problem to be solved, in order to extract the phase information:

$$2\sum_{n=1}^{N} I_n^C \cos(\delta_n) = \frac{N}{4} I \cos(\Phi_H) + \frac{N}{4} I \cos(\Phi_V)$$

$$2\sum_{n=1}^{N} I_n^C \sin(\delta_n) = \frac{N}{4} I \sin(\Phi_H) + \frac{N}{4} I \sin(\Phi_V) \tag{9}$$

This gives us two equations of two phase values, that have to be recovered from the superpositions per pixel. In the following section we treat the problem strictly mathematically, before we apply it again to the real world.

3.4   Mathematical Solution to the Problem

Given the following system of equations:

$$a\cos(x) + a\cos(y) = b$$
$$a\sin(x) + a\sin(y) = c \tag{10}$$

with measured data a, b, c, we want to compute optimal values x, y solving both equations. Using addition theorems of trigonometric functions and dividing leads to:

$$2a\cos(\frac{x}{2} + \frac{y}{2})\cos(\frac{x}{2} - \frac{y}{2}) = b$$
$$2a\sin(\frac{x}{2} + \frac{y}{2})\cos(\frac{x}{2} - \frac{y}{2}) = c \qquad \Rightarrow \quad x + y = 2\arctan\left(\frac{c}{b}\right) \tag{11}$$

Therefore, we can decouple the equations of (11). Both are leading to the same equation:

$$2ab\cos(z) + 2ac\sin(z) = b^2 + c^2 \quad z \in \{x, y\} \tag{12}$$

Using harmonic addition theorem, four explicit solutions for this equation can be derived, where the two feasible ones are given by

$$x/y = 2\arctan\left(\frac{2ac \pm \sqrt{(b^2 + c^2)(4a^2 - b^2 - c^2)}}{b^2 + 2ab + c^2}\right). \tag{13}$$

4   Application to Real World

The left column of Fig. 3 shows the results of Eq. (13) applied to the system (9) that models the real process. If there is a significant influence of ambient light, it may be necessary to subtract an ambient image from the captured images.

With the proposed procedure, phase values Φ_H and Φ_V can be recovered robustly. Unfortunately, due to the symmetric additive superposition of the phases in the whole

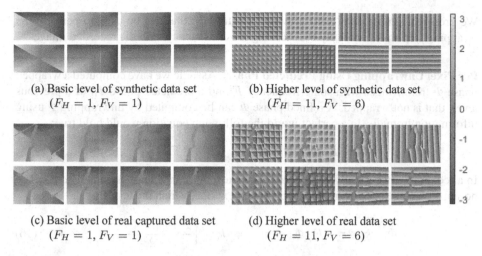

(a) Basic level of synthetic data set
$(F_H = 1, F_V = 1)$

(b) Higher level of synthetic data set
$(F_H = 11, F_V = 6)$

(c) Basic level of real captured data set
$(F_H = 1, F_V = 1)$

(d) Higher level of real data set
$(F_H = 11, F_V = 6)$

Fig. 3. Results of the algorithm applied to synthetic and real data for different frequencies. For each set the two rows show the horizontal and the vertical phase. The left two columns show the results of formula (13) before and after the *comparison step*. The third column shows the phase after the *swapping step*. The ground truth is depicted in the right column. The colorbar indicates the color coding in the range of $[-\pi, \pi]$.

procedure we do not have any information about which of the two phase values corresponds to the horizontal and which to the vertical one. These interchanges can occur at pixel level due to the pixel-wise approach. However, due to the natural continuity of the phases, these interchanges usually occur fragmentary. This can be seen in the first columns of the different examples of Fig. 3 for different frequencies, for both the synthetic and the real case. Note that application directly to the patterns is meaningful for any synthetic scene, through the scene-independent pixel-wise approach.

The first errors can be corrected by simple comparison (*comparison step*), which sorts the values to fragments, lifting the swaps from pixel to region level:

$$\Phi_H = \max\{\Phi_H, \Phi_V\}, \quad \Phi_V = \min\{\Phi_H, \Phi_V\} \tag{14}$$

The second columns of Fig. 3 (a, b, c, d) show the effect to the respective scenes.

4.1 Swapping Step

After this step, still many values are swapped (see Fig. 3). A gradient based strategy could be used to sort them, which would not be per-pixel and therefore scene dependent.

A common procedure, to obtain reconstructions of high accuracy, is the projection of several levels of fringe images with increasing frequencies. It is assumed that separate horizontal and vertical fringe images recorded at frequency 1 were projected in the first level, so that basic phases are available. In order to get a more applicable swapping procedure, that is per-pixel and stable in difficult situations (e.g. discontinuities in the phase) this can be done during the phase unwrapping of the higher level phases.

We present a simple pixel-wise unwrapping strategy, that can be used to perform the swapping step simultaneously by checking for consistency with the recorded images.

Per Pixel Unwrapping Using Predicted Phase. Assume we have computed a wrapped phase $\hat{\Phi}$ from fringes with some frequency F and a predicted phase Φ_0 of a previous level, that is not wrapped. A refined phase Φ can be computed by unwrapping $\hat{\Phi}$ using information from Φ_0. In a perfect world the following equation would hold true:

$$F \cdot \Phi = \hat{\Phi} + k \cdot 2\pi, \qquad k \in \mathbb{Z}_{\geq 0} \tag{15}$$

In a scenario with carefully increased frequencies, one can at least assume that there are no jumps larger than π from one level to the next one, which means

$$\hat{\Phi} + k \cdot 2\pi - F \cdot \Phi_0 \leq \pi \quad \Rightarrow \quad k = \left\lfloor \frac{F \cdot \Phi_0 - \hat{\Phi} + \pi}{2\pi} \right\rfloor \tag{16}$$

with floor rounding $\lfloor \cdot \rfloor$. Therefore, $\hat{\Phi}$ can be explicitly unwrapped to Φ by:

$$\Phi = \frac{\hat{\Phi}}{F} + \frac{2\pi}{F} \left\lfloor \frac{F \cdot \Phi_0 - \hat{\Phi} + \pi}{2\pi} \right\rfloor \tag{17}$$

Per Pixel Swapping. Using the per pixel unwrapping, we can perform the unwrapping step as well on the phase combination (Φ_H, Φ_V) as on the swapped one (Φ_V, Φ_H). The consistency towards captured fringe images can be described by a suitable error like:

$$E_n(\Phi_H, \Phi_V) = \left| \left(\cos(F^H \Phi_H + \delta_n) + \cos(F^V \Phi_V + \delta_n) + 2 \right) I - 4 I_n^C \right|.$$

Since we can assume the refined phases to improve after every unwrapping step, the accumulated consistency error of all captured images should decrease. Therefore, the combination with lower consistency error can be chosen to complete the swapping.

Table 1. Median errors for different levels of the proposed method with combined patterns and separate patterns, applied to the example data.

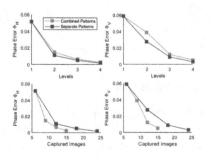

Fig. 4. Behavior of the median phase error with respect to the images that have to be captured.

Level	1	2	3	4
Combined patterns				
Median error of Φ_H	0.0514	0.0149	0.0066	0.0028
Median error of Φ_V	0.0587	0.0390	0.0124	0.0055
Captured images	6	9	12	15
Separate patterns				
Median error of Φ_H	0.0514	0.0110	0.0052	0.0019
Median error of Φ_V	0.0587	0.0278	0.0093	0.0034
Captured images	6	12	18	24

5   Evaluation

In order to evaluate the behavior of the procedure, we have drawn the median pixel error of the calculated phases to the ground truth after several levels in Table 1, as well for the proposed approach with combined patterns as for the separate approach. As expected, the error of the combined phases in each level is slightly higher than the separate procedure with separately computed horizontal and vertical phases. However, less recordings were necessary. Considering the accuracy in relation to the image captures used, even with a two-stage procedure and the 12 shots usually required for this, the combined procedure can take another level and double the accuracy of the calculated phases (see Fig. 4). Figure 5 shows the final phases, computed by the proposed approach, in comparison to the ground truth. It should be noted that the gamma correction of the devices used (especially the projector) strongly influences the quality of later reconstructions, since it violates the assumed linearity condition from Eq. (8). We have applied inverse gamma correction to the projected patterns with a roughly determined gamma value to compensate for this and can therefore demonstrate real results. Nevertheless, the use of industrial projectors without gamma correction or precise gamma calibration of the consumer device used, would significantly improve the quality of the reconstructions. Finally, the calibration results of two other setups and scenes, directly computed from the received point correspondences are visualized in Fig. 5 (d).

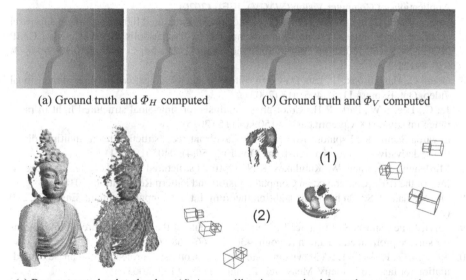

(a) Ground truth and Φ_H computed   (b) Ground truth and Φ_V computed

(1)

(2)

(c) Reconstructed point clouds   (d) Auto-calibrations received from the correspondences

Fig. 5. Ground truth phases of the exemplary scene (a, b, left) in comparison to the recovered phases (a, b, right). The triangulated point clouds from ground truth (left) and simultaneously recovered phase (right) are shown in (c). Auto-calibrations from point correspondences of two additional scenes are visualized in (d).

6 Conclusions

A new method has been introduced, which allows to perform sinusoidal structured light encoding in horizontal and vertical directions, at the same time. Thereby, the recording time is effectively halved. This procedure especially allows to auto-calibrate arbitrary setups directly from the achieved point correspondences. Extensive mathematical investigations were carried out, which yield important new findings in the field of applied harmonic addition theorem. Overall, a method was developed, which can determine the horizontal and vertical phase values from the combined captured patterns pixel-wise, making it scene-independent and therefore applicable to a wide variety of scenarios. The applicability to real scenes besides artificial ones was demonstrated as well. The results are highly interesting both mathematically and from a computer vision point of view, as they open up new possibilities in the field.

References

1. Fetzer, T., Reis, G., Stricker, D.: Robust auto-calibration for practical scanning setups from epipolar and trifocal relations. In: 2019 16th International Conference on Machine Vision Applications (MVA), pp. 1–6. IEEE (2019)
2. Fetzer, T., Reis, G., Stricker, D.: Stable intrinsic auto-calibration from fundamental matrices of devices with uncorrelated camera parameters. In: 2020 IEEE Winter Conference on Applications of Computer Vision (WACV). IEEE (2020)
3. Fetzer, T., Reis, G., Stricker, D.: Fast projector-driven structured light matching in sub-pixel accuracy using bilinear interpolation assumption. In: Tsapatsoulis, N., et al. (eds.) CAIP 2021, LNCS, vol. 13052, pp. 26–36. Springer, Cham (2021). https://doi.org/10.1007/978-3-030-89128-2_3
4. Guan, C., Hassebrook, L., Lau, D.: Composite structured light pattern for three-dimensional video. Opt. Express **11**(5), 406–417 (2003)
5. Je, C., Lee, S.W., Park, R.H.: Color-phase analysis for sinusoidal structured light in rapid range imaging. arXiv preprint arXiv:1509.04115 (2015)
6. Legarda-Saenz, R., Espinosa-Romero, A.: Wavefront reconstruction using multiple directional derivatives and fourier transform. Opt. Eng. **50**(4), 040501 (2011)
7. Mirdehghan, P., Chen, W., Kutulakos, K.N.: Optimal structured light à la carte. In: Proceedings of the IEEE Conference on Computer Vision and Pattern Recognition (2018)
8. Oo, N., Gan, W.S.: On harmonic addition theorem. Int. J. Comput. Commun. Eng. **1**(3), 200 (2012)
9. Salvi, J., Fernandez, S., Pribanic, T., Llado, X.: A state of the art in structured light patterns for surface profilometry. Pattern Recogn. **43**(8), 2666–2680 (2010)
10. Sansoni, G., Redaelli, E.: A 3D vision system based on one-shot projection and phase demodulation for fast profilometry. Meas. Sci. Technol. **16**(5), 1109 (2005)
11. Servin, M., Estrada, J., Quiroga, J.A.: The general theory of phase shifting algorithms. Opt. Express **17**(24), 21867–21881 (2009)
12. Wang, Z., Yang, Y.: Single-shot three-dimensional reconstruction based on structured light line pattern. Opt. Lasers Eng. **106**, 10–16 (2018)
13. Yang, L., Li, F., Xiong, Z., Shi, G., Niu, Y., Li, R.: Single-shot dense depth sensing with frequency-division multiplexing fringe projection. J, Vis. Commun. Image Represent. **46**, 139–149 (2017)

14. Zanuttigh, P., Marin, G., Dal Mutto, C., Dominio, F., Minto, L., Cortelazzo, G.M.: Time-of-Flight and Structured Light Depth Cameras. Technology and Applications. Springer, Cham (2016). https://doi.org/10.1007/978-3-319-30973-6
15. Zhang, L., Curless, B., Seitz, S.M.: Rapid shape acquisition using color structured light and multi-pass dynamic programming. In: Proceedings of the First International Symposium on 3D Data Processing Visualization and Transmission, pp. 24–36. IEEE (2002)
16. Zhang, S.: Recent progresses on real-time 3D shape measurement using digital fringe projection techniques. Opt. Lasers Eng. **48**(2), 149–158 (2010)
17. Zhang, S., Yau, S.T.: Three-dimensional shape measurement using a structured light system with dual cameras. Opt. Eng. **47**(1), 013604 (2008)

Joint Global ICP for Improved Automatic Alignment of Full Turn Object Scans

Torben Fetzer[1]([⊠]), Gerd Reis[2], and Didier Stricker[1,2]

[1] University of Kaiserslautern, Kaiserslautern, Germany
{torben.fetzer,didier.stricker}@dfki.de
[2] German Research Center for Artificial Intelligence (DFKI),
Kaiserslautern, Germany
gerd.reis@dfki.de

Abstract. Point cloud registration is an important task in computer vision, computer graphics, robotics, odometry and many other disciplines. The problem has been studied for a long time and many different approaches have been established. In the case of existing rough initializations, the ICP approach is still widely used as the state of the art. Often only the pairwise problem is treated. In case of many applications, especially in 3D reconstruction, closed rotations of sequences of partial reconstructions have to be registered. We show that there are considerable advantages if ICP iterations are performed jointly instead of the usual pairwise approach (Pulli's approach). Without the need for increased computational effort, lower alignment errors are achieved, drift is avoided and calibration errors are uniformly distributed over all scans. The joint approach is further extended into a global version, which not only considers one-sided adjacent scans, but updates symmetrically in both directions. The result is an approach that leads to a much smoother and more stable convergence, which moreover enables a stable stopping criterion to be applied. This makes the procedure fully automatic and therefore superior to most other methods, that often tremble close to the optimum and have to be terminated manually. We present a complete procedure, which in addition addresses the issue of automatic outlier detection in order to solve the investigated problem data independently without any user interaction.

Keywords: Point cloud registration · Automatic alignment · Drift reduction

1 Introduction

The task of point cloud registration is to align two point sets so that they resemble each other as closely as possible in as many regions as possible. To make

This work was partially funded by the projects MARMORBILD (03VP00293) and VIDETE (01W18002) of the German Federal Ministry of Education and Research (BMBF).

this problem well-defined, it must be assumed that the point clouds represent the same scene or at least that sufficiently large parts of the point clouds represent overlapping parts of the scene. Otherwise, no matching areas can be identified and the problem cannot be solved.

A distinction is made between rigid and non-rigid registration. For the rigid case, two point clouds are aligned only by rotation and translation (in some cases also scaling). The appearance and proportions are fully preserved. In contrast, in the non-rigid registration, deformable objects are aligned by non-linear transformations.

In classical computer vision and robotics, rigid alignment is by far the most common case and has been extensively studied. For this purpose, methods have been established, which simultaneously detect point correspondences and align them iteratively. In particular the method *Iterative Closest Point* (ICP) [3,18] has to be named, which has been successfully applied for decades and that will also form the basis of our procedure.

A special case, which occurs in many practical applications, is given by a sequence of point clouds, which partially overlap pairwise and whose last point cloud closes up with the first one. In this case, it is no longer a matter of several pairwise registration problems but a global overdetermined registration problem. This is because each point cloud has two neighbors (last and next one) with whom it must be aligned. In the case of real data, such as the partial reconstructions of a 3D scanner, pairwise sequential alignment would usually lead to a drift, i.e. a large gap or too much overlap between the last and first position. This drift occurs when the partial alignment errors and possible calibration errors in the partial reconstructions add up to a large error. To avoid such drift, it is common to apply *Pulli's procedure* [11], which involves aligning and merging opposite pairs of adjacent point clouds. The resulting merged larger point clouds are then further treated together. In this way, the error is not concentrated between two scans and the drift is distributed over a larger number of scans.

In this work we will show that there are nevertheless better ways with much better properties to solve the global alignment problem in a stable way and to actually distribute the drift evenly without higher computational effort. We will show that a joint iteration of the pairwise registrations distributes the drift uniformly and achieves lower alignment errors than the state of the art. Furthermore, we present how the standard procedure can be extended into a global procedure by symmetrically registering each scan with the next and last scan in the sequence. This results in a global approach that leads to a much smoother convergence, which allows the reliable use of automatic stopping criteria. In contrast, standard procedures usually begin to tremble near the minimum, which often requires a manual termination of the iterations. Finally, a practical approach to the automatic detection of outliers is presented. This is to provide a complete stable solution to the problem without any user interaction. To allow maximum reproducibility, the entire procedure is attached as pseudo-code at the end of the paper.

2   Related Work

The problem of point cloud registration has been well studied for several decades. Explicit methods for rigidly aligning given point correspondences from two data sets have already been developed in the last century [2,15]. They are based on the singular value decomposition and are still the basis of the modern state of the art. In [1] these methods are robustified by additional weights, based on the certainty of the correspondences and in [5] they are extensively evaluated and compared with other approaches.

For many applications, there is more than two views to be aligned. In order to treat multiple point clouds jointly, an extension of the orthogonal procrustes problem has been introduced in [13]. In [14] the orthogonal constraints are relaxed and the jointly obtained solutions are projected to the space of the orthogonal matrices afterwards. [10] transferred the problem into a simple semidefinite programming in order to ease solving. These methods are no longer explicit and require higher computational effort. In the context of point cloud alignment performed in the upcoming task, the given correspondences are erroneous approximations and change from iteration to iteration. Therefore, a higher accuracy to the costs of additional internal iterations is not reasonable.

Usually, no exact point correspondences are available. A famous principle proven in practice is *Iterative Closest Point* [3,11,18], which iteratively selects the closest points of the data sets as correspondences and calculates infinitesimal updates accordingly. There are also variants that take the normals of the point clouds into account [6,8] and thus improve the alignment for badly sampled and very smooth objects.

In order to accelerate convergence of the methods, the point clouds are either adeptly sampled [6,12] and outliers efficiently detected and rejected [4,12,17] or the iteration updates extrapolated like in [16]. There are also methods to accelerate by a multi-resolution approach [7] or recently by Anderson acceleration [9].

3   Background: Rigid Point Cloud Alignment

The most common algorithm for rigidly aligning point clouds is *Iterative Closest Point*. Thereby, the closest points of two point sets are chosen as correspondences and optimally aligned with each other. Afterwards, new correspondences are chosen, based on the improved alignment. Iteratively, the alignment of the point clouds is improved. For given point correspondences there is a closed form of the optimal rotation matrix and the translation vector for the pairwise case (*Procrustes Analysis*). Since this is also the basis of the method presented in the following and in order to make the paper independently, we will briefly present the procedure for the pairwise case and then show how the method is applied to a full turn according to the current state of the art.

3.1 Orthogonal Procrustes Problem

Assume two sets of point clouds $P = \{\mathbf{p}_0, ..., \mathbf{p}_{N-1}\}$ and $P' = \{\mathbf{p}'_0, ..., \mathbf{p}'_{N-1}\}$ consisting of matching point pairs $\mathbf{p}_n \leftrightarrow \mathbf{p}'_n$, for $n = 0, ..., N - 1$ are given. The task is to find an optimal rotation matrix $\mathbf{R}$ and translation vector $\mathbf{t}$ in order to align points $\mathbf{p}_n$ from P by $\mathbf{R}\mathbf{p}_n + \mathbf{t}$ to points $\mathbf{p}'_n$ from P'. Therefore, the sum of Euclidean distances between all point pairs is minimized:

$$\underset{\mathbf{R},\, \mathbf{t}}{\text{argmin}} \sum_{n=0}^{N-1} \|\mathbf{p}'_n - \mathbf{R}\mathbf{p}_n - \mathbf{t}\|_2^2 \tag{1}$$

Setting the derivative of (1) with respect to translation vector $\mathbf{t}$ equal to zero leads to the minimizing $\mathbf{t}$ of the energy:

$$\mathbf{t} = \frac{1}{N} \sum_{n=0}^{N-1} \mathbf{p}'_n - \mathbf{R}\frac{1}{N} \sum_{n=0}^{N-1} \mathbf{p}_n = \mu_{P'} - \mathbf{R}\mu_P \tag{2}$$

Thereby μ_P and $\mu_{P'}$ denote the centroids of the point clouds computed by the mean of the point sets. Inserting Eq. (2) into the problem formulation (1) decouples the problem. It is equivalent to aligning point clouds with zero centroids by optimal rotation only:

$$\underset{\mathbf{R}}{\text{argmin}} \sum_{n=0}^{N-1} \|\mathbf{q}'_n - \mathbf{R}\mathbf{q}_n\|_2^2, \quad \begin{array}{l} \text{with} \quad \mathbf{q}_n = \mathbf{p}_n - \mu_P \\ \text{and} \quad \mathbf{q}'_n = \mathbf{p}'_n - \mu_{P'} \end{array} \tag{3}$$

Calculating the norm explicitly and replacing the remaining scalar product by the trace formulation leads to the following formulation of the problem, that can be solved in terms of the singular value decomposition of matrix $\mathbf{H}$. The validity of this optimizer can be shown by application of *Cauchy-Schwartz Inequality*.

$$\underset{\mathbf{R}}{\text{argmax}}\ \text{Tr}\Big(\mathbf{R} \sum_{n=0}^{N-1} \mathbf{q}_n \mathbf{q}'^{\mathsf{T}}_n\Big) = \underset{\mathbf{R}}{\text{argmax}}\ \text{Tr}(\mathbf{R}\mathbf{H}) \ \rightarrow \mathbf{R} = \mathbf{V}\mathbf{U}^{\mathsf{T}}, \text{ with } \mathbf{H} = \mathbf{U}\boldsymbol{\Lambda}\mathbf{V}^{\mathsf{T}}$$

$$\tag{4}$$

Weighted Case. When working with real data it is usual to apply certainty weights $w_n \geq 0$ with $\sum_{n=0}^{N-1} w_n = 1$ with respect to the point pairs to the alignment error (1) in order to robustify the approach. The problem is solved similarly to the unweighted case using weighted versions of the centroids and matrix $\mathbf{H}$ (see Algorithm 1).

3.2 Iterative Closest Point (ICP)

Usually, no point correspondences are available between two point clouds. In the procedure of *ICP*, these are approximately chosen in each iteration as the nearest points between the data sets and infinitesimal updates are calculated by *Orthogonal Procrustes Analysis*. Since it is a least-squares formulation, it is important to assess the quality of the correspondence. For this purpose, outliers, i.e. points that obviously have no matches in the other data set, are rejected. All other points are weighted according to their quality, which is often done by the point-to-point distance. Sampling rates and methods can also have a strong influence on performance and should not be disregarded.

Initialization. For this procedure to work, an initial alignment is urgently required. This prevents the procedure from getting stuck in a local minimum. Based on feature points in the object, be it from texture or geometry, usually a few matches can be found which allow a rough alignment of the point clouds as initialization. Based on this initial registration, the ICP algorithm has proven over a long period of time to be a good choice for refining the alignment.

3.3   Full Turn Registration: Pulli's Approach

In a variety of practical applications, full turns of overlapping partial reconstructions are captured as depicted in Fig. 1. Usually, the last scan overlaps with the first one and therefore completes the reconstruction process. In sequential pairwise registrations of the scans, a drift error between the last and the first position often occurs. To avoid or at least reduce this drift error, Pulli's approach [11] has always been the undisputed state of the art. One after the other, scans are registered and merged with their neighbors. These merged point clouds are then registered again until the whole object is composed. In fact, the error is distributed more evenly than in the naive approach and is not added up to a single gap, but it is far from uniform. While the first registration procedures only contain the local alignment errors, the last step combines the alignment errors of several sub-alignments and possible calibration errors.

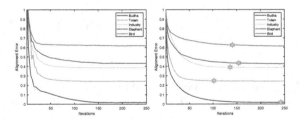

Fig. 1. Partial scans of a full turn and the aligned point cloud (middle).

Fig. 2. Alignment errors of the proposed joint sequential (left) and global (right) ICP variants. Red stars mark automatic stopping points of the global method. (Color figure online)

4   Joint Rigid Point Cloud Alignment

In the following, the alignment problem of a full rotation of scans is formulated as common optimization problem. We assume that two successive scans have at least some overlap and that the last scan closes up to the first one, thus well defining the problem.

Joint Minimization Problem. Let be given a full turn consisting of S scans $\{S_0, ..., S_{S-1}\}$, where the last position S_{S-1} is assumed to be overlapping with the first one S_0. Between two subsequent scans, say scan s and scan $s + 1$, we assume to have N point matches each, given by $\mathbf{p}_n^{(s,s+1)} \leftrightarrow \mathbf{p}_n^{(s+1,s)}$, for $n = 0, ..., N - 1$. The objective error function, that has to be minimized is then given by

$$\underset{\mathbf{R}^{(s)},\, \mathbf{t}^{(s)}}{\operatorname{argmin}} \sum_{s=0}^{S-1} \sum_{n=0}^{N-1} \|\mathbf{R}^{(s+1)}\mathbf{p}_n^{(s+1,s)} + \mathbf{t}^{(s+1)} - \mathbf{R}^{(s)}\mathbf{p}_n^{(s,s+1)} - \mathbf{t}^{(s)}\|_2^2. \quad (5)$$

Note that we assume a periodic arrangement, so that the scans' indices are treated modulo S, which means $S \equiv 0$. Setting the partial derivative with respect to any translation vector $\mathbf{t}^{(s)}$ equal to zero, we get:

$$2\mathbf{t}^{(s)} - \mathbf{t}^{(s-1)} - \mathbf{t}^{(s+1)} = \mathbf{R}^{(s-1)}\mu_{s-1,s} + \mathbf{R}^{(s+1)}\mu_{s+1,s} - \mathbf{R}^{(s)}(\mu_{s,s-1} + \mu_{s,s+1}) \quad (6)$$

which is sufficiently fulfilled for

$$\mathbf{t}^{(s+1)} - \mathbf{t}^{(s)} = \mathbf{R}^{(s)}\mu_{s,s+1} - \mathbf{R}^{(s+1)}\mu_{s+1,s}. \quad (7)$$

Therefore, the objective function (5) can be decoupled to:

$$\underset{\mathbf{R}^{(s)}}{\operatorname{argmin}} \sum_{s=0}^{S-1} \sum_{n=0}^{N-1} \|\mathbf{R}^{(s+1)}(\mathbf{p}_n^{(s+1,s)} - \mu_{s+1,s}) - \mathbf{R}^{(s)}(\mathbf{p}_n^{(s,s+1)} - \mu_{s,s+1})\|_2^2 \quad (8)$$

$$= \underset{\mathbf{R}^{(s)}}{\operatorname{argmin}} \sum_{s=0}^{S-1} \sum_{n=0}^{N-1} \|\mathbf{R}^{(s+1)}\mathbf{q}_n^{(s+1,s)} - \mathbf{R}^{(s)}\mathbf{q}_n^{(s,s+1)}\|_2^2 \quad (9)$$

Joint Sequential ICP. Solving the terms of the joint minimization problem (9) sequentially for one s after the other by simply applying the standard strategy (4) leads to the pairwise approach with joint iterations. Iteratively, closest points between each neighboring pair are chosen and alignment updates by *Orthogonal Procrustes Problem* (4) are applied to each pair. This procedure already avoids drift and the errors are uniformly distributed without additional computational effort.

Joint Global ICP. In order to derive a global formulation, that does not only take pairwise point clouds into account, but also treats the global arrangement information, we are minimizing functional (9) with respect to each $\mathbf{R}^{(s)}$ while fixing the others. This is equivalent to solving the following optimization problem:

$$
\begin{aligned}
\underset{\mathbf{R}^{(s)}}{\operatorname{argmax}} \ & \sum_{s=0}^{S-1} \operatorname{Tr}(\mathbf{R}^{(s)} \mathbf{H}_{s,s+1} \mathbf{R}^{(s+1)^{\mathsf{T}}}) \\
=\underset{\mathbf{R}^{(s)}}{\operatorname{argmax}} \ & \operatorname{Tr}(\mathbf{R}^{(s)} \underbrace{(\mathbf{H}_{s,s+1} \mathbf{R}^{(s+1)^{\mathsf{T}}} + \mathbf{H}_{s,s-1}^{\mathsf{T}} \mathbf{R}^{(s-1)^{\mathsf{T}}})}_{\mathbf{H}_s})
\end{aligned}
\tag{10}
$$

This is a form of a symmetric alignment update of scan $\mathcal{S}_S$ towards last and next adjacent scans $\mathcal{S}_{S-1}$ and $\mathcal{S}_{S+1}$. The problem can be solved similar to (4) using singular value decomposition and does not require special treatment.

Efficient Point Matching. To efficiently find the nearest points between two point clouds, the use of k-d-trees has been established for a long time. Building them means a not inconsiderable effort, but once they are created, the nearest points can be found in logarithmic time. A special feature is, that for each point cloud of a scan only one tree has to be set up, which can be further used after transformation by applying the inverse transformation to the input points, as shown in Algorithm 1. Especially in the iterative application to large point sets, this means an enormous time saving.

5   Outlier Rejection

In order to achieve an automatic procedure that can be applied to a possibly large number of configurations, outliers must be reliably detected in every set of correspondences. Standard procedures, such as rejecting the 10% of correspondence with largest point-to-point distances in each iteration, are widely used, but rely on a well-chosen value. In order to be independent of a fixed value, we have carried out investigations on a large number of data sets.

The task is to separate a set of N point correspondences into two subsets. The separation should divide the outliers as well as possible from the eligible correspondences. Let $D = \{d_0, ..., d_{N-1}\}$ be the set of point-to-point distances of respective correspondences, sorted in a descending order ($d_0 \geq d_1 \geq ... \geq d_{N-1}$). Tests on approximately 25000 different point sets and configurations have shown that a good partition

$$
D = D_{\text{outliers}} \cup D_{\text{inliers}} = \{d_0, ..., d_t\} \cup \{d_{t+1}, ..., d_{N-1}\}
\tag{11}
$$

is achieved at a split point $t \in \{0, ..., N-1\}$ if the *coefficients of variation* of both subsets is equal or as close as possible. Therefore, t can be successively increased until the following equation holds approximately true:

$$
\frac{\frac{1}{t+1} \sum_{n=0}^{t} d_n^2}{\left(\frac{1}{t+1} \sum_{n=0}^{t} d_n\right)^2} = \frac{\frac{1}{N-t} \sum_{n=t+1}^{N-1-1} d_n^2}{\left(\frac{1}{N-t-1} \sum_{n=t+1}^{N-1} d_n\right)^2}
\tag{12}
$$

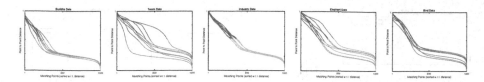

Fig. 3. Outlier rejection strategy (12) applied to the point clouds shown in Fig. 4 bottom right. The lines result from point-to-point distances of matches sorted in descending order. The red segments visualize the detected subset of outliers. (Color figure online)

Fig. 3 shows the behavior of the rejection strategy for the point sets that are evaluated in the upcoming section (see Fig. 4, bottom right). Each reconstruction consists of eight partial scans as it usually comes up from 3D scanners. Between each adjacent pair of scans, matches are computed and outliers are detected by the proposed strategy. Each of the subplots in Fig. 3 shows the 8 curves which result from sorting 1000 matches between each of the 8 point pairs. The red segments visualize the detected set of outliers. For better visualization the plots are given in a logarithmic scale and normalized.

6   Evaluation

For perfect artificial data or uniformly added noise, all alignment strategies work satisfactorily. The situation is different for the real use case of recorded data. In the following we evaluate the considered ICP methods for registration of full turns on a number of sample data sets as they appear from typical 3D scanners. For five independent objects (*Buddha, Totem, Industry, Elephant, Bird*), full rotations of eight partial reconstructions each were created. In order to fully align them, the registration methods must be able to deal with both, local alignment errors of the partial point clouds and calibration errors that can have an impact on the overall fit. The standard procedures were compared to the presented joint ICP variants. Figure 4 bottom right shows the resulting aligned point clouds of the *Joint Global ICP* approach to represent the objects under investigation.

The plots in Fig. 4 show the convergence behaviour of the alignment strategies for increasing numbers of iterations. Both methods that were proposed converge to significantly lower errors than the trivial sequential pairwise alignment (black line) and Pulli's drift preventing procedure (red line) which reflects the current state of the art. Although both, the *Joint Sequential ICP* and the *Joint Global ICP* converge to the same optimum, the alignment error of the sequential variant occasionally alternates depending on the data (see *Totem*). In contrast, the global approach converges completely smoothly and evenly, which leads to a more stable convergence in general.

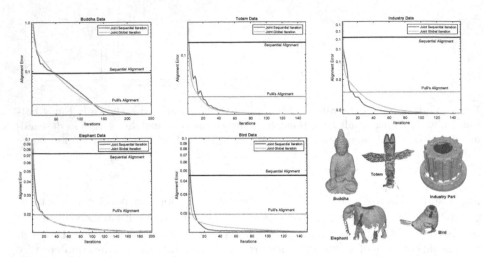

Fig. 4. Convergence behavior for five independent data sets. Alignment error of the naive pairwise approach is given by the black line (contains drift error). State of the art is given by Pulli's approach (red line). Jointly iterating approaches converge to much lower errors. While the sequential procedure (blue) may alternate depending on the data, the global approach (green) converges smoothly (Color figure online).

6.1   Stopping Criterion

The smooth convergence behaviour of the presented global ICP variant provides a considerable advantage over all previous ICP methods. Most of them alternate during the procedure, due to iteratively updated point correspondences, which increases the chance of getting stuck in local minima.

Moreover, the error often starts to alternate around the minimum. Most papers write "we iterate until the alignment error does not reasonable improve any more" without further information. Standard stopping criteria for convergence do not hold in most situations, since the differences between two subsequent iterations may not fall under a given threshold. This is the reason why in many practical implementations the alignment does continue and start to tremble until it is manually stopped.

Figure 2 shows the behavior of the weighted alignment errors for the investigated data sets. Left plot shows the behavior for the *Joint Sequential ICP* and right plot for *Joint Global ICP*. A simple smooth stopping strategy like checking for improvements of the alignment within the last few iterations enables the stable automatic termination of the procedure after a reasonable number of iterations. The stopping points are visualized by the stars in Fig. 2.

Algorithm 1: Joint Global ICP for Full Turn Alignment

Input: Initially aligned partial point clouds $P_0, ..., P_{S-1}$ of scans $\mathcal{S}_1, ..., \mathcal{S}_{S-1}$.

Initialize parameters $\mathbf{R}^{(s)} = \mathbf{I}$, $\mathbf{t}^{(s)} = \mathbf{0}$ for all partial scans $\mathcal{S}_s$, $s = 0, ..., S-1$.

Setup a k-d-tree $\mathcal{T}_s$ for each point cloud P_s.

Sample point clouds P_s to a size of N elements.

for $i = 0, 1, 2, ...$ **do**

    for $s = 0, ..., S-1$ **do**

        Search in k-d-trees of adjacent scans for correspondences:
$$\mathcal{T}_{s-1}\left(\mathbf{R}^{(s-1)^\mathsf{T}}(\mathbf{R}^{(s)}P_s + \mathbf{t}^{(s)} - \mathbf{t}^{(s-1)})\right) \to P^{(s,s-1)}$$
$$\mathcal{T}_{s+1}\left(\mathbf{R}^{(s+1)^\mathsf{T}}(\mathbf{R}^{(s)}P_s + \mathbf{t}^{(s)} - \mathbf{t}^{(s+1)})\right) \to P^{(s,s+1)}$$

        Reject outliers as introduced in Sec. 5.

        Weight correspondences with respect to point distances.

        Subtract centroids from point sets:
$$\mu_{s,s-1} = \sum_{n=0}^{N-1} w_n \mathbf{p}_n^{s,s-1} \quad \to Q^{(s,s-1)} = P^{(s,s-1)} - \mu_{s,s-1}$$
$$\mu_{s,s+1} = \sum_{n=0}^{N-1} w_n \mathbf{p}_n^{s,s+1} \quad \to Q^{(s,s+1)} = P^{(s,s+1)} - \mu_{s,s+1}$$

    end

    for $s = 0, ..., S-1$ **do**

        Set up symmetric system matrices:
$$\mathbf{H}_{s,s-1} = \sum_{n=0}^{N-1} w_n \mathbf{q}_n^{(s,s-1)} \mathbf{q}_n^{(s-1,s)^\mathsf{T}}, \quad \mathbf{H}_{s,s+1} = \sum_{n=0}^{N-1} w_n \mathbf{q}_n^{(s,s+1)} \mathbf{q}_n^{(s+1,s)^\mathsf{T}}$$
$$\to \mathbf{H}_s = \mathbf{H}_{s,s+1}\mathbf{R}^{(s+1)^\mathsf{T}} + \mathbf{H}_{s,s-1}^\mathsf{T}\mathbf{R}^{(s-1)^\mathsf{T}}$$

        Compute SVD $\mathbf{H}_s = \mathbf{U}_s \mathbf{\Lambda}_s \mathbf{V}_s^\mathsf{T}$ and compose updated rotation $\mathbf{R}^{(s)} = \mathbf{V}_s \mathbf{U}_s^\mathsf{T}$.

    end

    for $s = 0, ..., S-1$ **do**

        Update translation vectors $\mathbf{t}^{(s)}$ by Eq. 6.

    end

    Compute weighted alignment error and check for improvement withing the last say
    10 iterations. If no improvement break.

end

Output: Optimal transformations $\mathbf{R}^{(s)}$, $\mathbf{t}^{(s)}$.

---

7 Conclusion

In this paper we presented a procedure that aligns complete closed turns of partial point clouds jointly in a global manner. Not only pairwise adjacent point clouds are considered but also corrected symmetrically to all neighbors. The usual, widely spread ICP procedure can be applied in a slightly adapted way. By iterating the subproblems jointly, alignment and calibration errors are evenly distributed over all scans and drift is prevented. The global approach leads to a smooth convergence behaviour, which enables the credible application of automatic stopping criteria. Together with an introduced outlier rejection strategy, this results in an extremely stable automatic procedure that generates better results than the previous state of the art. This is moreover achieved without any user interaction or additional computational effort. To ease reproduction, the procedure is finally attached as pseudo-code.

References

1. Akca, D.: Generalized procrustes analysis and its applications in photogrammetry. Technical report, ETH Zurich (2003)
2. Arun, K.S., Huang, T.S., Blostein, S.D.: Least-squares fitting of two 3-D point sets. IEEE Trans. Pattern Anal. Mach. Intell. **5**, 698–700 (1987)
3. Besl, P.J., McKay, N.D.: Method for registration of 3-D shapes. In: Sensor Fusion IV: Control Paradigms and Data Structures, vol. 1611, pp. 586–606. International Society for Optics and Photonics (1992)
4. Dorai, C., Wang, G., Jain, A.K., Mercer, C.: Registration and integration of multiple object views for 3D model construction. IEEE Trans. Pattern Anal. Mach. Intell. **20**(1), 83–89 (1998)
5. Eggert, D.W., Lorusso, A., Fisher, R.B.: Estimating 3-D rigid body transformations: a comparison of four major algorithms. Mach. Vis. Appl. **9**, 272–290 (1997). https://doi.org/10.1007/s001380050048
6. Gelfand, N., Ikemoto, L., Rusinkiewicz, S., Levoy, M.: Geometrically stable sampling for the ICP algorithm. In: Fourth International Conference on 3-D Digital Imaging and Modeling, 2003, 3DIM 2003, pp. 260–267. IEEE (2003)
7. Jost, T., Hügli, H.: Fast ICP algorithms for shape registration. In: Van Gool, L. (ed.) DAGM 2002. LNCS, vol. 2449, pp. 91–99. Springer, Heidelberg (2002). https://doi.org/10.1007/3-540-45783-6_12
8. Masuda, T., Sakaue, K., Yokoya, N.: Registration and integration of multiple range images for 3-D model construction. In: Proceedings of 13th International Conference on Pattern Recognition, vol. 1, pp. 879–883. IEEE (1996)
9. Pavlov, A.L., Ovchinnikov, G.W., Derbyshev, D.Y., Tsetserukou, D., Oseledets, I.V.: AA-ICP: iterative closest point with anderson acceleration. In: IEEE International Conference on Robotics and Automation (ICRA), pp. 1–6 (2018)
10. Pizarro, D., Bartoli, A.: Global optimization for optimal generalized procrustes analysis. In: CVPR. IEEE (2011)
11. Pulli, K.: Multiview registration for large data sets. In: Second International Conference on 3-D Digital Imaging and Modeling (Cat. No. PR00062), pp. 160–168. IEEE (1999)

12. Rusinkiewicz, S., Levoy, M.: Efficient variants of the ICP algorithm. In: Proceedings Third International Conference on 3-D Digital Imaging and Modeling. IEEE (2001)
13. Ten Berge, J.M.: Orthogonal procrustes rotation for two or more matrices. Psychometrika **42**(2), 267–276 (1977)
14. Trendafilov, N.T., Lippert, R.A.: The multimode procrustes problem. Linear Algebra Appl. **349**, 245–264 (2002)
15. Umeyama, S.: Least-squares estimation of transformation parameters between two point patterns. IEEE Trans. Pattern Anal. Mach. Intell. **13**, 376–380 (1991)
16. Yaniv, Z.: Rigid Registration: The Iterative Closest Point Algorithm. School of Engineering and Computer Science. The Hebrew University, Israel (2001)
17. Zhang, Z.: Iterative point matching for registration of free-form curves and surfaces. Int. J. Comput. Vis. **13**(2), 119–152 (1994)
18. Zinßer, T., Schmidt, J., Niemann, H.: Point set registration with integrated scale estimation. In: International Conference on Pattern Recognition and Image Processing (2005)

Fast Projector-Driven Structured Light Matching in Sub-pixel Accuracy Using Bilinear Interpolation Assumption

Torben Fetzer[1(✉)], Gerd Reis[2], and Didier Stricker[1,2]

[1] University of Kaiserslautern, Kaiserslautern, Germany
{torben.fetzer,didier.stricker}@dfki.de
[2] German Research Center for Artificial Intelligence (DFKI),
Kaiserslautern, Germany
gerd.reis@dfki.de

Abstract. In practical applications where high-precision reconstructions are required, whether for quality control or damage assessment, structured light reconstruction is often the method of choice. It allows to achieve dense point correspondences over the entire scene independently of any object texture. The optimal matches between images with respect to an encoded surface point are usually not on pixel but on sub-pixel level. Common matching techniques that look for pixel-to-pixel correspondences between camera and projector often lead to noisy results that must be subsequently smoothed. The method presented here allows to find optimal sub-pixel positions for each projector pixel in a single pass and thus requires minimal computational effort. For this purpose, the quadrilateral regions containing the sub-pixels are extracted. The convexity of these quads and their consistency in terms of topological properties can be guaranteed during runtime. Subsequently, an explicit formulation of the optimal sub-pixel position within each quad is derived, using bilinear interpolation, and the permanent existence of a valid solution is proven. In this way, an easy-to-use procedure arises that matches any number of cameras in a structured light setup with high accuracy and low complexity. Due to the ensured topological properties, exceptionally smooth, highly precise, uniformly sampled matches with almost no outliers are achieved. The point correspondences obtained do not only have an enormously positive effect on the accuracy of reconstructed point clouds and resulting meshes, but are also extremely valuable for auto-calibrations calculated from them.

Keywords: Structured light · Matching · Sub-pixel · Linear · Consistent

1 Introduction

Structured light enables the determination of precise and dense point correspondences between a camera and a projector view. No features are required, making it applicable to a wide range of different object types. Often, sinusoidal patterns are projected to encode the scene in two directions. With the help of the deformed patterns, horizontal

---

This work was partially funded by the projects MARMORBILD (03VP00293) and VIDETE (01W18002) of the German Federal Ministry of Education and Research (BMBF).

© Springer Nature Switzerland AG 2021
N. Tsapatsoulis et al. (Eds.): CAIP 2021, LNCS 13052, pp. 26–36, 2021.
https://doi.org/10.1007/978-3-030-89128-2_3

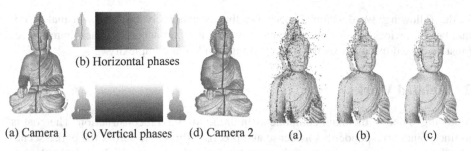

(a) Camera 1   (c) Vertical phases   (d) Camera 2   (a)   (b)   (c)

Fig. 1. Illustration of projector-driven matching of two cameras and a projector. Red lines visualize the encoding of a point by its phases. (Color figure online)

Fig. 2. Resulting point clouds of FPDM with (b, c) and without (a) *TCC* and in addition with *ECC* (c).

and vertical phase images are calculated for each camera view, that theoretically lead to a direct correspondence between each projector pixel and its position in the camera image. From these point correspondences, cameras and projector can be calibrated and a dense point cloud can be triangulated using the obtained camera matrices.

In theory, a setup consisting of a projector as active device, holding the perfect phase, and a camera is sufficient for depth estimation. However, in many practical arrangements, several cameras, at least two, are used in addition to the projector. This is due to a much cleaner projective behavior of high quality cameras compared to most projectors. Since higher quality lenses are available, usually less distortions are caused. Also, most industrial cameras allow gamma correction to be disabled, which has a significant impact on assumptions in computer vision applications. Since this is not possible with affordable projectors, it is of considerable advantage to triangulate the point cloud with the camera information only. To cover the general case of any number of cameras, in this paper the situation with two cameras and one projector is considered. Thus, the procedure can be trivially extended to an arbitrary number of cameras.

The idea of projector-driven matching is to find suitable correspondences in the camera images for each projector pixel. In this way, the camera positions are transitively matched via the projector pixels. Figure 1 illustrates this procedure. (a) and (d) show the texture images of the two camera views. (b) and (c) show the corresponding horizontal and vertical phase images of the cameras and in the center of the projector. The red lines illustrate the unique encoding process of a pixel through the two phases. In the exact execution of dense matching a number of difficulties arises:

- Phase images are discrete samples of continuous functions. Therefore, there is usually no exact pixel-to-pixel mapping. Instead, it is very likely that matches lie between certain camera pixels.
- The topology of the pixels remains locally preserved during the projection process. Thus, certain conditions can be defined which must be fulfilled by the phases and met during the matching process in order to avoid noisy results.
- Matching is only a sub-step in 3D reconstruction and auto-calibration and should therefore be fast. The trivial procedure of searching optimal matches between all images is not practical at all. The procedure would be of quadratic complexity in terms of resolution, and with increasing camera resolutions this is very poor.

In the following, we develop a procedure that is extremely fast and can match any number of devices stably and consistently with sub-pixel accuracy. Each image pixel must be passed through exactly once resulting in a linear complexity.

2 Related Work

Matching is one of the main components in the field of 3D reconstruction. The goal is to find point correspondences as dense and precise as possible across the entire scene. Standard approaches search for suitable candidates along the epipolar lines and evaluate them according to their neighbors using suitable region descriptors [7]. This is a common approach, but requires a calibrated setup and can fail in many cases, as in uniform areas of the scene. If it is, in contrast, possible to create very precise matches without prior calibration information (e.g. [6]), modern auto-calibration methods, such as [4,5] allow to perform an exact calibration of the system directly from these matches, which makes such a computer vision system much more user-friendly and flexible. It also makes it suitable for a variety of other applications where pre-calibration is not possible, extremely tedious or problematic, since the setup may de-calibrate over time.

Common matching procedures without pre-calibration are based on SIFT features [8], which provide robust matches if sufficient object texture is available. Also, there are methods that do not only include appearance but also object geometry into the search [9]. However, both of them most likely fail in the case of very smooth uniform objects, which limits the applicability. To reconstruct untextured objects the structured light approach is a common tool. In [11] the authors use structured light information to handle large disparities in binocular matching. Similarly, in [12] the wrapped phase is used to refine the stereo matches. [14] shows how to get accurate dense matches using only the reconstructed phase. [3] introduces a sub-pixel matching for unsynchronized structured light, while for each match an energy is minimized by gradient descent. Matching based on peak calculation [1,15] also achieves sub-pixel accuracy but requires higher computational effort than the method presented.

In [2] a deep learning approach for structured light matching was proposed recently. It uses a Siamese network trained on a synthetic data set that expects rectified images, which is not suitable for arbitrary uncalibrated systems and auto-calibration. Also in [10] structured light and deep learning are combined to achieve good and exact matches. In [13] matching is even skipped and depth is directly calculated using deep learning.

3 Fast Projector Driven Matching (FPDM)

The task of fast projector-driven matching is to find corresponding positions in the camera phases for each integer projector pixel. Since this is usually not again an integer position, it must be estimated with sub-pixel accuracy. Figure 3 (a) illustrates this for the projector in the middle and camera images left and right. Everything at the sub-pixel level can only be described by the pixels in its environment, since no smaller information is available in an image. In order to compute the sub-pixel matches, it is therefore necessary to find integer camera pixels that span a quadrilateral (Fig. 3, green pixels) that encloses the optimal sub-pixel match as closely as possible. From this quad, the sub-pixel match can then be interpolated in a subsequent step. The quadrilateral does not necessarily have to be square or rectangular, but should at least be convex. This constraint is fulfilled in the general case and only violated at regions with depth discontinuities. It ensures that the enclosed area can be described smoothly through its corners. In addition, there are certain consistency characteristics that should be met.

3.1 Matching Integer Pixel Quads

In a first step, best possible convex quads enclosing the sub-pixel match for each projector pixel are found in each camera image. Each camera pixel should only be processed once in order to maintain linear complexity. Therefore, for each projector pixel, we store the four corner points whose quadrilateral contains the optimal camera correspondence. An array four times the size of the projector resolution is needed as a buffer. Note that the projector image can be selected in any resolution as it is completely imaginary. The resulting density of the point cloud can be precisely controlled in this way. Practice has shown that the projector resolution should be selected in approximately the same order as the camera resolutions, since usually both cover about the same area of the scene.

In the following we assume horizontal and vertical phase images Φ_H and Φ_V of a camera with values in the interval $[0, 1]$. The phases run from left to right and from bottom to top according to the common coordinate axes. Similarly, the optimal projector phases run from 0 to 1 at a selected resolution (w_P, h_P). This is depicted in Fig. 1.

For each camera pixel (x, y), the theoretical corresponding position in the projector image is uniquely given by the vertical and horizontal phase values $\Phi_H(x, y)$ and $\Phi_V(x, y)$. Therefore, a camera pixel (x, y) would theoretically match projector pixel

$$(\hat{x}_P, \hat{y}_P) = \big(\Phi_H(x, y)w_P, \Phi_V(x, y)h_P\big), \tag{1}$$

which is not an integer value, as sought. Nevertheless, it is an approximate position and likely a lower and upper corner of the next integer projector pixels, which is the basic idea of the presented fast (linear) method.

For each integer projector pixel (x_P, y_P) the vertices of the spanned matching quad in the camera image are noted as indicated in Fig. 4 (right). So (x_{00}, y_{00}), (x_{10}, y_{10}), (x_{01}, y_{01}) and (x_{11}, y_{11}) denote the pixels of the quad around sub-pixel match $(\hat{x}, \hat{y})$ with respect to (x_P, y_P). Using the notations $\lfloor \cdot \rfloor$ and $\lceil \cdot \rceil$ for *floor* and *ceil* integer rounding, a camera pixel (x, y) would be a feasible corner point of four adjacent quadrilaterals containing sub-pixel camera matches with respect to four projector pixels. Thereby, it would be exactly one bottom left, one bottom right, one top left and one top right corner of the four corresponding quadrilaterals. The buffers for the projector pixels are filled by traversing the image and assigning each image pixel as:

$$(x, y) \longrightarrow \begin{pmatrix} \lceil \hat{x}_P \rceil_{00}, \lceil \hat{y}_P \rceil_{00} \\ \lfloor \hat{x}_P \rfloor_{10}, \lceil \hat{y}_P \rceil_{10} \\ \lceil \hat{x}_P \rceil_{01}, \lfloor \hat{y}_P \rfloor_{01} \\ \lfloor \hat{x}_P \rfloor_{11}, \lfloor \hat{y}_P \rfloor_{11} \end{pmatrix} \tag{2}$$

Since phases in arbitrary real scenes are usually sampled non-uniformly, it may be possible that several camera pixels are feasible corner points of a specific quadrilateral.

Using the example of a lower left corner point, the quality of the corner point can be calculated by its distance to the optimal sub-pixel value:

$$E = \left| \hat{x}_P - \lceil \hat{x}_P \rceil_{00} \right| + \left| \hat{y}_P - \lceil \hat{y}_P \rceil_{00} \right| \tag{3}$$

If a corner point is already occupied when running through the image, it is only replaced if this error is less than that of the previously stored pixel. This ensures that the enclosing quadrilateral becomes minimal.

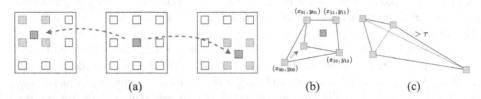

| (a) | (b) | (c) |

Fig. 3. (a) Visualization of optimal sub-pixel matches (red) between projector (center) and two cameras (left and right). Example of a corner point update (b) (Consistency properties stay fulfilled). Quad, that would be removed by diagonal check (b). (Color figure online)

3.2   Topological Consistency Check (TCC)

An important property of a projection is that the topology of the projected points remains consistent. Therefore, also surface points that have been encoded using structured light must remain consistent in the corresponding phase images. Some tests are introduced, that enforce the topology preservation property. Most importantly, they remain valid for non-minimal quads, allowing their application on non-final temporal stores of corner points. This way, incorrect and noisy phase values are excluded from matching, resulting in smoother and more accurate matches with way less outliers.

Before saving any image pixel to a corner point (x_{00}, y_{00}), (x_{10}, y_{10}), (x_{01}, y_{01}) or (x_{11}, y_{11}) with respect to a projector pixel (x_P, y_P), it is ensured that a lower left pixel in the camera phase is also a lower left pixel in the projector phase and so on. In this way, many faulty matches can be detected and avoided. Moreover, it ensures that the resulting quads are convex. The following simple checks have to be fulfilled:

$$
\begin{array}{ccc}
(x_{01}, y_{01}) & \xrightarrow{\ x_{01} \leq x_{11}\ } & (x_{11}, y_{11}) \\[2pt]
\scriptstyle y_{00} \leq y_{01} \big\uparrow & & \big\uparrow \scriptstyle y_{10} \leq y_{11} \\[2pt]
(x_{00}, y_{00}) & \xrightarrow{\ x_{00} \leq x_{10}\ } & (x_{10}, y_{10})
\end{array}
\tag{4}
$$

The tests are applied to the pixels during the storing process while looping through the images. Naturally, therefore, during the storing process, one vertex is checked for consistency with respect to other vertices that are not final and that may be part of non-minimal representations of a quad around a sub-pixel match. As already mentioned these tests are also valid for non-minimal quads as long as they do not represent severe outliers, which moreover would simply lead to finding no match for the projector rather than an outlier. Figure 3 (b) illustrates an update of a corner point to a closer representation. It is easy to see that the convexity properties are fulfilled throughout by all points, while converging to the minimal representation.

Diagonal Check for Weak Quads. Theoretically, the quadrilaterals can take a wide variety of shapes and still satisfy the desired topology and convexity. But the more unusual the shape, the worse its content is determined by bilinear interpolation. An additional optional test avoids unnatural quads by checking the diagonal values:

$$|x_{00} - x_{11}| + |y_{00} - y_{11}| < \tau, \qquad |x_{10} - x_{01}| + |y_{10} - y_{01}| < \tau \qquad (5)$$

The quads should not be of arbitrary size just because they might theoretically be feasible. Usually they will not provide an accurate measurement if the corners are above a certain distance, which can be generously set to $\tau = 5$ pixels for most applications.

For illustration, Fig. 3(c) shows an example of an unfavorable quad that would be removed. Note that this check should only be done after the entire quad matching, otherwise some quads may be removed due to non-minimal representations that may have improved over time.

Epipolar Consistency Check (ECC). In many practical scenarios a rough calibration of the setup is already available. This can be extremely advantageous and easily involved into the scheme. In this case a camera point should only be mapped to a corner if the symmetric epipolar error is below a certain threshold.

In order to illustrate the effect of the checks on calculated matches, Fig. 2 shows the resulting point cloud of *FPDM* without (a) and with *TCC* (b). There are significantly fewer outliers, resulting in less flying points. (c) shows how the matches can be further improved by *ECC* by avoiding faulty assignments, especially in discontinuities of the scene. False matches can also occur due to incorrect but permissible phase regions, caused by (inter-)reflections.

4   Bilinear Sub-pixel Matching

After the quad matching, for each permissible projector pixel a consistent convex quadrilateral is given per camera image. Under certain assumptions it is possible to determine the sub-pixel position of the optimal match from the corners of the quad and their phase values. The optimal sub-pixel position is calculated in the unit patch using bilinear interpolation assumption and then mapped to the convex region as shown in Fig. 4.

4.1   Sub-pixel Position in Unit Patch

Given a unitary patch, with horizontal phase values ϕ_{H00}, ϕ_{H10}, ϕ_{H01} and ϕ_{H11} of the corner points as depicted in Fig. 4, the bilinear interpolated value $\phi_H(\tilde{x}, \tilde{y})$ for any position $(\tilde{x}, \tilde{y}) \in [0, 1]^2$ is given by

$$\phi_H(\tilde{x}, \tilde{y}) = a_0 + a_1\tilde{x} + a_2\tilde{y} + a_3\tilde{x}\tilde{y}, \qquad
\begin{aligned}
a_0 &= \phi_{H00} \\
a_1 &= \phi_{H10} - \phi_{H00} \\
a_2 &= \phi_{H01} - \phi_{H00} \\
a_3 &= \phi_{H11} + \phi_{H00} - \phi_{H10} - \phi_{H01}
\end{aligned} \qquad (6)$$

and analoguesly for the vertical phase by $\phi_V(\tilde{x}, \tilde{y}) = b_0 + b_1\tilde{x} + b_2\tilde{y} + b_3\tilde{x}\tilde{y}$. Figure 5 (a) illustrates how two bilinearly interpolated phases on the unit patch can look like. Task is to find the optimal sub-pixel position inside the patch meeting the phase values $(\hat{\phi}_H, \hat{\phi}_V)$. The patch that interpolates the horizontal phase values defines a two-dimensional curve on which the value $\hat{\phi}_H$ is assumed. The same applies to the patch of the vertical phase values, which describes a curve for $\hat{\phi}_V$. Such curves are visualized by red lines in Fig. 5(a) and in a top view (b). The intersection of the curves within the patch is the optimal position of the sought sub-pixel match and marked by green dots.

In order to find optimal positions $\tilde{x} \in [0, 1]$ and $\tilde{y} \in [0, 1]$ at which the bilinear interpolated patches meet the sought values $\hat{\phi}_H$ and $\hat{\phi}_V$ we explicitly solve:

$$\begin{aligned}
\hat{\phi}_H &= a_0 + a_1\tilde{x} + a_2\tilde{y} + a_3\tilde{x}\tilde{y} \\
\hat{\phi}_V &= b_0 + b_1\tilde{x} + b_2\tilde{y} + b_3\tilde{x}\tilde{y}
\end{aligned} \;\longrightarrow\; \tilde{x} =
\begin{cases}
-\frac{v}{2u} \pm \sqrt{\left(\frac{v}{2u}\right)^2 - \frac{w}{u}}, & u \neq 0 \\
-\frac{w}{v}, & u = 0
\end{cases} \quad (7)$$

$$\text{with}\quad
\begin{aligned}
u &= b_1 a_3 - b_3 a_1 \\
v &= b_1 a_2 + (b_0 - \hat{\phi}_V)a_3 - b_2 a_1 - b_3(a_0 - \hat{\phi}_H) \\
w &= (b_0 - \hat{\phi}_V)a_2 - b_2(a_0 - \hat{\phi}_H)
\end{aligned}$$

The vertical position $\tilde{y}$ can then be computed from Eq. (7). Note that the properties of the quads received in Sect. 3 ensure the existence of intersection within each patch.

Existence of Solution. The interpolated value $\hat{\phi}_H$ is by construction achieved inside the patch and moreover the following holds true due to the consistency checks:

$$\phi_{H00}, \phi_{H01} \leq \hat{\phi}_H \leq \phi_{H10}, \phi_{H11} \qquad (8)$$

Of course for any convex combination with $\tilde{y} \in [0, 1]$, we also have:

$$(1 - \tilde{y})\phi_{H00} + \tilde{y}\phi_{H01} \leq \hat{\phi}_H \leq (1 - \tilde{y})\phi_{H10} + \tilde{y}\phi_{H11} \qquad (9)$$

Therefore the curve, defined by the first equation of (7), that maps feasible positions $\tilde{x}$ to any value $\tilde{y} \in [0, 1]$ has the following property:

$$\tilde{x} = \frac{\hat{\phi}_H - a_0 - a_2\tilde{y}}{a_1 + a_3\tilde{y}} = \frac{\overbrace{\hat{\phi}_H - (1 - \tilde{y})\phi_{H00} - \tilde{y}\phi_{H01}}^{=(i)\geq 0\ (9),}}{\underbrace{(1 - \tilde{y})\underbrace{(\phi_{H10} - \phi_{H00})}_{>0\ (8)} + \tilde{y}\underbrace{(\phi_{H11} - \phi_{H01})}_{>0\ (8)}}_{=(ii)>0\ (9)}} \geq 0 \qquad (10)$$

Thereby, we neglect the situation in which all corner points carry the same value. In this case division by zero would not be defined. Nevertheless, in this case an optimal integer pixel match exists and interpolation is not necessary.

Additionally, the denominator (ii) in this fraction is greater or equal than the numerator (i), which limits the fraction to 1:

$$(ii) - (i) = (1 - \tilde{y})\phi_{H10} + \tilde{y}\phi_{H11} - \hat{\phi}_H \overset{(9)}{\geq} 0 \tag{11}$$

Similar for the vertical phase we obtain for the curves of Eq. (7) the properties:

$$\tilde{x} = \frac{\hat{\phi}_H - a_0 - a_2\tilde{y}}{a_1 + a_3\tilde{y}} \in [0,1] \text{ for } \tilde{y} \in [0,1], \ \tilde{y} = \frac{\hat{\phi}_V - b_0 - b_1\tilde{x}}{b_2 + b_3\tilde{x}} \in [0,1] \text{ for } \tilde{x} \in [0,1]$$

Therefore, the curves are defined for all $\tilde{y}, \tilde{x} \in [0,1]$ and map to values $\tilde{x}, \tilde{y} \in [0,1]$. This proves that the continuous curves describe continuous connections between the top and bottom of the patch (for $\tilde{x} \in [0,1]$) and from left side to right side (for $\tilde{y} \in [0,1]$) that both run inside the patch. These curves must therefore intersect at least once within the patch. This guarantees a solution, which can be explicitly computed by solving the resulting quadratic Eq. (7) and choosing the feasible one inside the patch. □

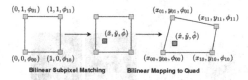

Fig. 4. Procedure of bilinear sub-pixel matching. The position is computed in the unit patch and mapped to the convex quad.

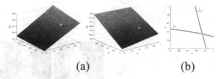

Fig. 5. Unit patch with interpol. phases (a). Red curves are possible solutions, the intersection (green) solves the problem. (Color figure online)

4.2 Mapping to Convex Quad

In general, the corner points around a sub-pixel match will not span a square region. However, for convex quads, the method can be applied by assuming an additional bilinear interpolation scheme. With corresponding corner points in the image given by

$$(x_{01}, y_{01}) \leftrightarrow (0,1) \qquad (x_{11}, y_{11}) \leftrightarrow (1,1)$$
$$(x_{00}, y_{00}) \leftrightarrow (0,0) \qquad (x_{10}, y_{10}) \leftrightarrow (1,0) \tag{12}$$

a point $(\tilde{x}, \tilde{y}) \in [0,1]^2$ in the unit square can be mapped to the convex quadrilateral by:

$$\begin{pmatrix} \hat{x} \\ \hat{y} \end{pmatrix} = \begin{pmatrix} x_{00} & x_{10} & x_{01} & x_{11} \\ y_{00} & y_{10} & y_{01} & y_{11} \end{pmatrix} \begin{pmatrix} 1 & -1 & -1 & 1 \\ 0 & 1 & 0 & -1 \\ 0 & 0 & 1 & -1 \\ 0 & 0 & 0 & 1 \end{pmatrix} \begin{pmatrix} 1 \\ \tilde{x} \\ \tilde{y} \\ \tilde{x}\tilde{y} \end{pmatrix} \tag{13}$$

5   Results

Figure 6 shows the reconstructed point clouds of different scenes as a qualitative illustration of the reconstructions obtained. For each scene, the left reconstruction shows the result of the matches obtained with best-pixel correspondences. The right point cloud shows the result of the *Fast Projector Driven Consistent Sub-Pixel Matching (FPCSM)* presented in this work. For (a–d) the images on the far right show in addition the back-projections of the points onto the projector image, with in- and outliers marked in green and red. The reconstructions are significantly smoother and contain almost no outliers. Of particular note is the *Monkey* data, which was taken from a highly specular metallic brushed monkey statue, which clearly shows the influence and improvements of the consistency checks. Of course there are methods to smooth out noisy results and to remove flying points in post processings, but the method presented here removes outliers during the matching process without any additional computational effort. Also, in contrast to smoothing, erroneous measures are removed and not smeared over the entire point cloud. Especially, if the correspondences are used for auto-calibration procedures this can be a huge advantage. Figure 7 (b) and (c) shows the enlargement of two regions in the reconstructed Buddha statue (a). Due to the optimal sub-pixel matching, the surface is much more uniformly sampled and less noisy. Especially for subsequent meshing and highly precise depth measurement this may have a significant influence.

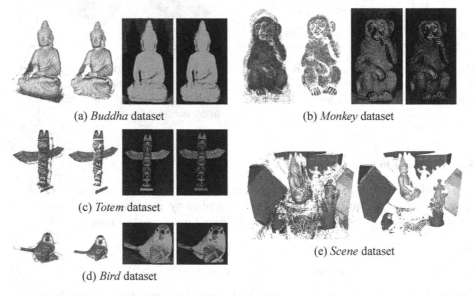

(a) *Buddha* dataset   (b) *Monkey* dataset

(c) *Totem* dataset

(d) *Bird* dataset   (e) *Scene* dataset

Fig. 6. Results of *FPCSM* (right) applied to exemplary scenes in comparison to point clouds obtained by standard best-pixel matching (left). Each set shows the point clouds and for (a–d) their backprojection to the projector image with labeled in- and outliers.

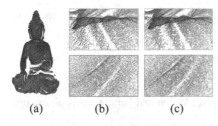

(a)   (b)   (c)

Fig. 7. Reconstr. point clouds using best-pixel matches (b) and FPCSM (c).

Table 1. Median backprojection errors for best-pixel matching and FPCSM.

Set	Best-pixel matches		FPCSM matches	
	Cam 1	Cam 2	Cam 1	Cam 2
Buddha	0.35447	0.35486	0.25326	0.25401
Bird	0.37116	0.37248	0.26199	0.26130
Totem	0.32996	0.33831	0.25896	0.26139
Monkey	0.37887	0.37552	0.27757	0.27818
Scene	0.26731	0.27925	0.17088	0.17866

Finally, Table 1 shows the reduction of the median backprojection errors on the camera images from which they were triangulated. The median error was chosen to avoid overweighting extreme outliers of the standard approach without consistency checks.

6   Conclusions

In this work a matching strategy has been presented which generates high-precision correspondences for structured light systems with any number of cameras. The matches are estimated in sub-pixel accuracy. Therefore, an explicit formula has been derived, which provides matches under the assumption of bilinearly interpolated patches. The existence of such matches has been mathematically investigated and proven. An important contribution is that this is achieved with linear complexity, while simultaneously ensuring topological consistency over the views. This results in high quality matches with nearly no outliers, that are uniformly sampled over the scene. Overall, a method has been developed which reaches extremely high accuracy with extremely low (linear) computational effort, that may be applicable to many active reconstruction applications.

References

1. Donate, A., Liu, X., Collins, E.G.: Efficient path-based stereo matching with subpixel accuracy. IEEE Trans. Syst. Man Cybern. Part B (Cybernetics) **41**(1), 183–195 (2010)
2. Du, Q., Liu, R., Guan, B., Pan, Y., Sun, S.: Stereo-matching network for structured light. IEEE Signal Process. Lett. **26**(1), 164–168 (2018)
3. El Asmi, C., Roy, S.: Subpixel unsynchronized unstructured light. In: VISIGRAPP (5: VISAPP) (2019)
4. Fetzer, T., Reis, G., Stricker, D.: Robust auto-calibration for practical scanning setups from epipolar and trifocal relations. In: 2019 16th International Conference on Machine Vision Applications (MVA), pp. 1–6. IEEE (2019)
5. Fetzer, T., Reis, G., Stricker, D.: Stable intrinsic auto-calibration from fundamental matrices of devices with uncorrelated camera parameters. In: The IEEE Winter Conference on Applications of Computer Vision, pp. 221–230 (2020)
6. Fetzer, T., Reis, G., Stricker, D.: Simultaneous bi-directional structured light encoding for practical uncalibrated profilometry. In: Tsapatsoulis, N., et al. (eds.): CAIP 2021, LNCS, vol. 13052, pp. 3–13. Springer, Cham (2021). https://doi.org/10.1007/978-3-030-89128-2_1

7. Hartley, R., Zisserman, A.: Multiple View Geometry in Computer Vision. Cambridge University Press, New York (2003)
8. Hu, Y.: Research on a three-dimensional reconstruction method based on the feature matching algorithm of a scale-invariant feature transform. Math. Comput. Model. **54**(3–4), 919–923 (2011)
9. Isack, H., Boykov, Y.: Energy based multi-model fitting & matching for 3D reconstruction. In: Proceedings of the IEEE Conference on Computer Vision and Pattern Recognition, pp. 1146–1153 (2014)
10. Li, F., Li, Q., Zhang, T., Niu, Y., Shi, G.: Depth acquisition with the combination of structured light and deep learning stereo matching. Signal Process. Image Commun. **75**, 111–117 (2019)
11. Ma, S., Shen, Y., Qian, J., Chen, H., Hao, Z., Yang, L.: Binocular structured light stereo matching approach for dense facial disparity map. In: Wang, D., Reynolds, M. (eds.) AI 2011. LNCS (LNAI), vol. 7106, pp. 550–559. Springer, Heidelberg (2011). https://doi.org/10.1007/978-3-642-25832-9_56
12. Pribanic, T., Obradovic, N., Salvi, J.: Stereo computation combining structured light and passive stereo matching. Opt. Commun. **285**(6), 1017–1022 (2012)
13. Ryan Fanello, S., et al.: Hyperdepth: learning depth from structured light without matching. In: Proceedings of the IEEE Conference on Computer Vision and Pattern Recognition, pp. 5441–5450 (2016)
14. Scharstein, D., Szeliski, R.: High-accuracy stereo depth maps using structured light. In: 2003 IEEE Computer Society Conference on Computer Vision and Pattern Recognition, 2003, Proceedings, vol. 1, p. I. IEEE (2003)
15. Xie, J., Mo, F., Yang, C., Lia, P., Tian, S.: A novel sub-pixel matching algorithm based on phase correlation using peak calculation. Int. Arch. Photogram. Remote Sens. Spat. Inf. Sci. **1**, 253–257 (2016)

Pyramidal Layered Scene Inference with Image Outpainting for Monocular View Synthesis

Marcos R. Souza[1], Jhonatas S. Conceição[1], Jose L. Flores-Campana[1],
Luis G. L. Decker[1], Diogo C. Luvizon[2], Gustavo Sutter P. Carvalho[2],
Helena A. Maia[1], and Helio Pedrini[1($\boxtimes$)]

[1] Institute of Computing, University of Campinas, Campinas, SP 13083-852, Brazil
helio@ic.unicamp.br
[2] AI R&D Lab, Samsung R&D Institute Brazil, Campinas, SP 13097-160, Brazil

Abstract. Generating novel views from a single input is a challenging task that requires the prediction of occluded and non-visible content. Nevertheless, it is an interesting and active area of research due to its several applications such as entertainment. In this work, we propose an end-to-end architecture for monocular view synthesis based on the layered scene inference (LSI) method. The LSI uses layered depth images that can represent complex scenes with a reduced number of layers. To improve the LSI predictions, we develop two new strategies: (i) a pyramidal architecture that learns LDI predictions for different resolutions of the input and (ii) an image outpainting for filling the missing information at the LDI borders. We evaluate our method on the KITTI dataset, and show that the proposed versions outperform the baseline.

Keywords: Monocular view synthesis · Layered depth image · Pyramidal network · Image outpainting

1 Introduction

The monocular view synthesis aims to produce images from different viewpoints of the scene using a single image as input. Either implicitly or explicitly, this task involves interpreting complex structures in the scene through texture and depth, and filling the content that is not visible in the original viewpoint.

This problem can benefit several other tasks, such as augmented reality systems. A very interesting application is the generation of parallax motion effect [8,12,21], in which a sequence of new views can be created from a single source view. When we see the entire sequence, the objects close to the observer must have a higher perceived velocity than the farther deep objects.

This work was funded by Samsung Eletrônica da Amazônia Ltda., through the project "Parallax Effect", within the scope of the Informatics Law No. 8248/91.

© Springer Nature Switzerland AG 2021
N. Tsapatsoulis et al. (Eds.): CAIP 2021, LNCS 13052, pp. 37–46, 2021.
https://doi.org/10.1007/978-3-030-89128-2_4

Recent work has used deep architectures to predict different representations such as 3D meshes [9] or point clouds [20]. Two other approaches have been widely explored to represent 3D scenes in complex images: multiplane image (MPI) [4,10,15,17,22] and layered depth images (LDI) [2,3,14,18]. According to Shih et al. [14], MPI may produce artifacts on sloped surfaces and contain redundant information across its layers. Moreover, the high number of layers typically used in MPIs leads to a high computational cost. For this reason, here we explore LDI for a compact representation of the scene.

LDI represents a 3D scene by a layered representation. It was originally proposed for image-based rendering [13] and defined as "a view of the scene from a single input camera view, but with multiple pixels along each line of sight". As observed by the authors, the size of the LDI grows linearly with the observed depth complexity in the scene. Each layer of the LDI consists of a 4-channel image, in which the first three channels are the RGB information, and the last one is the corresponding disparity map. The first layer represents the visible content from the original viewpoint and, so its RGB channels correspond to the input image. From the second layer onwards, the LDI stores the information that was occluded in the first layer.

Some recent work aims to calculate new views from calibrated stereo images [4,15,22]. An even challenging approach is the prediction of new views from a single view [3,10,12,17,18]. Most of these works use deep networks to solve intermediate tasks, but do not build a deep end-to-end model for the final task (new view generation). An exception is the layered scene inference (LSI) method [18], in which LDIs are calculated by an end-to-end network from either monocular or stereo-based datasets. This type of strategy has the advantage that intermediate tasks can benefit from the final supervision, which allows the network to find the representation that best minimizes the final loss function.

In this work, we propose a set of incremental strategies based on LSI [18] to improve the LDI prediction through deep end-to-end networks. More specifically, we focus on (a) reducing the distortions present at the rendition boundaries and (b) improving the overall rendering quality (similarity with the ground-truth). To address these issues, we propose and analyze the use of (i) an outpainting method to extrapolate the boundaries of the LDI and (ii) a pyramidal architecture, which learns how to compute LDIs at multiple resolutions.

2   Proposed Method

In this section, we present our method and compare it with the baseline LSI [18]. Both methods are illustrated in Fig. 1. The baseline uses an encoder-decoder architecture, the DispNet [11], to predict LDI from an RGB source image. Then a differentiable rendering is applied to compute the target image.

LSI [18] was trained with a set of N inputs $(I_s^n, I_t^n, K_s^n, K_t^n, R^n, t^n)_{n=1}^{N}$. Images I_s^n and I_t^n are respectively the source image and an arbitrarily sampled target, both from the same scene. Matrices K_s^n and K_t^n define the intrinsic parameters of the cameras that captured the two images. Finally, the matrices R^n and t^n describe the relative rotation and translation between both cameras.

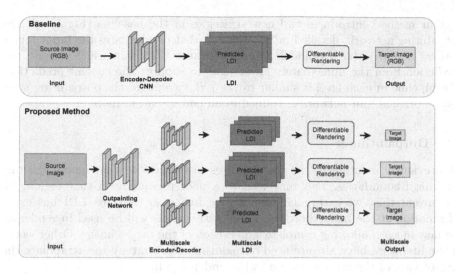

Fig. 1. Overview of the baseline and our proposed method.

An LDI representation with L layers is created from each source image I_s^n. It can be described as $(I^l, D^l)_{l=1}^{L}$, where I^l e D^l are images that represent respectively the texture and disparity of each pixel from I_s^n for the l-th layer. The disparity maps must satisfy the condition $D^l(p) \leq D^{l+1}(p)$, that is, the disparity of the same pixel cannot be greater in a deeper layer. The reason for this condition is that, if the texture of an object took many layers to become visible, it is farther away from the camera or covered by many objects. Therefore, it should have a smaller disparity (greater depth).

From the input camera parameters and a predicted LDI, the LSI can project the source on the target and compute the rendered texture using a differentiable soft z-buffering. Then, the network is supervised by a set of five loss functions: (i) the view synthesis loss $\mathcal{L}_{vs}$ that compares the rendered and ground-truth targets; (ii) the 'min-view synthesis' loss $\mathcal{L}_{m\text{-}vs}$ used to improve the background layers contribution; (iii) the source consistency loss $\mathcal{L}_{sc}$ given by the L_2 norm weighted by the disparity map; (iv) the depth monotonicity loss $\mathcal{L}_{inc}$ that ensures a non-increasing disparity; and (v) the smoothness loss $\mathcal{L}_{sm}$ given by the L_1 norm of the second-order spatial derivatives of the predicted disparity maps. The loss functions $\mathcal{L}_{vs}$ and $\mathcal{L}_{m\text{-}vs}$ are based on a mask M defined to ignore the image boundaries, since these pixels may be out of the field of view in the source image and, therefore, the network does not have the required information for a reliable prediction of them. This aspect will be further discussed in Subsect. 2.1.

From these five loss functions, the final objective function $\mathcal{L}_{final}$ is defined as

$$\mathcal{L}_{final} = \lambda_{vs}\mathcal{L}_{vs} + \lambda_{m\text{-}vs}\mathcal{L}_{m\text{-}vs} + \lambda_{sc}\mathcal{L}_{sc} + \lambda_{inc}\mathcal{L}_{inc} + \lambda_{sm}\mathcal{L}_{sm}, \tag{1}$$

where λ_x defines the weight of each loss function $\mathcal{L}_x$.

Our method introduces two new strategies in the baseline (Fig. 1): (i) the outpainting network, detailed in Subsect. 2.1, that extrapolates the input image borders and (ii) the pyramidal approach (Subsect. 2.2), in which we use the prediction from the immediately lower scale to support the current prediction. The pipeline in each level is similar to the LSI, except for the outpainting step. As the original, our network is trained in an end-to-end fashion.

2.1   Outpainting

As previously mentioned, the baseline uses loss functions that ignore the target image boundaries. This may lead to a poor prediction in that region. To circumvent this, we extrapolate the source image so that the LDI has extra information at the image borders. These extra pixels will be used in rendering the new images, allowing complete supervision of the target image. Other works in the literature have already used outpainting as a strategy to extrapolate the field of view of the representation to be rendered [14].

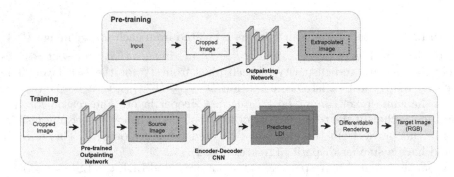

Fig. 2. Overview of the outpainting strategy.

For the outpainting, we use an encoder-decoder network recently proposed [19], which is trained with pixel-wise and adversarial losses. Initially, the outpainting network is pre-trained separately, instead of training the model from scratch along with the LDI predictor. After the pre-training, we use the weights of the outpainting network in the equivalent layers of ours, which is then trained keeping the outpainting layers frozen. This process is illustrated in Fig. 2.

The leftmost image in the pre-training figure is the original one. During the pre-training, we crop their boundaries removing some pixels and the network extrapolates the cropped image, producing a new image with the same resolution as the original one. In the training, the input images are also cropped and extrapolated by the pre-trained model. This new boundary added to the source gives extra information to render the target. Thus, although the target still has half the input size (i.e., cropped image size), it is built from a broader field of view thanks to the extrapolation. For this reason, we modify the loss functions

$\mathcal{L}_{\text{vs}}$ and $\mathcal{L}_{\text{m-vs}}$ to consider the entire images in the supervision, which are now described as

$$\mathcal{L}_{\text{vs}} = \|I_t - \bar{I}_t\|_1, \qquad \mathcal{L}_{\text{m-vs}} = \sum_{p_t} \min_l \|I_t(p_t) - \bar{I}_t^l(p_t)\|_1, \qquad (2)$$

where I_t and $\bar{I}_t$ are the original and predicted target image, and $\bar{I}_t^l$ is the predicted image using only the l-th layer. To train the outpainting network, we use the loss $\mathcal{L}_{\text{sc}}$ defined in Eq. 3, where I_s is the ground-truth image for outpainting, $D^o()$ is the discriminator loss for the extrapolated image I_s^o, and the λ^o is a constant weight. In the LSI network, we use the entire images as inputs without the cropping step.

$$\mathcal{L}_{\text{sc}}^o = \sum_{p_s} \|I_s(p_s) - I_s^o(p_s)\|_2 + \lambda^o D^o(I_s, I_s^o). \qquad (3)$$

2.2   Pyramidal Network Architecture

Our second proposal is the use of a pyramidal architecture. Pyramidal architectures [6] have been successfully used in diverse types of related problems, such as depth estimation [1] and optical flow estimation [16]. This architecture predicts a low resolution LDI and uses this prediction to compute the LDI at the next highest resolution. Figure 3 presents a diagram of this strategy. The architecture adopted for this task was inspired by the pyramidal PWC-Net [16] proposed for optical flow prediction.

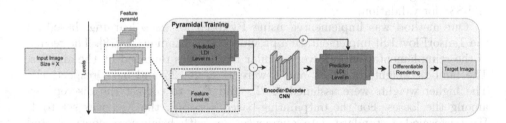

Fig. 3. Representation of the pyramidal architecture.

As seen in the figure, we built a pyramid of features extracted from the input image. It uses a convolutional network with learnable weights, that are learned in the end-to-end training. According to Sun et al. [16], the main advantage of the learnable feature pyramid over an image pyramid is that the former is robust to shadows and lighting changes.

The network predicts an LDI representation for each level using the respective feature as input. To feed the encoder-decoder CNN, this feature is concatenated with the last LDI. The last LDI is bilinearly interpolated to match the feature size. A residual connection combines every representation into a single one. Therefore, the network learns the complement of the LDI at each level. The only exception is the first level of the pyramid, in which we do not use either the concatenation or the residual connection, since no LDI representation is yet available. The prediction of the LDI in each level uses a coarse-to-fine strategy. After computing the coarse LDI (as the complement of the last LDI), it feeds another CNN in order to predict the fine LDI.

The architecture illustrated in Fig. 3 is an end-to-end trained network, including the layers responsible for the construction of the feature pyramid and the encoder-decoder CNN. The new final loss function comprising all pyramid levels is described as

$$\mathcal{L}_{\text{final}} = \frac{\sum_m^M \lambda_{\text{vs}}\mathcal{L}_{\text{vs}}^m + \lambda_{\text{m-vs}}\mathcal{L}_{\text{m-vs}}^m + \lambda_{\text{sc}}\mathcal{L}_{\text{sc}}^m + \lambda_{\text{inc}}\mathcal{L}_{\text{inc}}^m + \lambda_{\text{sm}}\mathcal{L}_{\text{sm}}^m}{M} \tag{4}$$

where M is the number of pyramid levels and $\mathcal{L}_x^m$ is the loss components calculated specifically for the m-th pyramid level.

3   Results

In this section, we show and discuss our results using the raw KITTI [5], which contains videos captured by a recording platform equipped with stereo cameras and other sensors that provide diverse information such as geographic coordinates. The reader may refer to the dataset paper for further details about the cameras and recording setup. As in the LSI [18], we randomly pick the right or left image as the source and the other one as the target. We use 33 sequences and multiple image pairs from each sequence, resulting in 22600 samples for training and 888 for validation.

Our method was implemented using PyTorch for deep learning, based on the TensorFlow LSI implementation provided by Tulsiani et al. [18]. The results of the reimplementation were very similar to the ones reported by the authors. The lambda weights of Eqs. 1 and 4 were 1, 1, 25, 25 and 0.65, respectively. The higher weights were assigned to $\mathcal{L}_{\text{sc}}$ and $\mathcal{L}_{\text{inc}}$ to correct the discrepancy among the losses. For the outpainting-based versions, the λ_{sm} was set to 1. We performed a standard data augmentation, with brightness, contrast, and saturation adjustment. The images were normalized in the range of $[0, 1]$ and resized to 512×256 pixels. The entire network was trained for 40 epochs. The initial learning rate was set to $1e{-}3$ for LSI and $2e{-}5$ for the pyramidal network. The learning rate scheduler multiplies it by 0.1 at every 20 epochs.

The PyTorch implementation of the outpainting network was provided by Hoorick [7]. The outpainting pre-training was done using the same training samples as the entire network. The training loss was given in Eq. 3. It was trained for 130 epochs. The value of λ^o was set to 0.001, 0.005, 0.015 and 0.040, in the epochs 1, 10, 30 and 60, respectively

Table 1 presents the results of the different versions of our method compared to the baseline using the pixel-wise L_1 error and the Structural Similarity (SSIM) metric. In addition to the traditional L_1 error, we also show I-L_1 and

O-L_1, which consider only inner (I-L_1) or outer (O-L_1) pixels to compute the metric. For them, we considered a border of 10% in each side, based on the mask M used in the original LSI. Similarly, we also compare the methods using inner- and outer-SSIM. In the table, LSI, LSI+IO and PyLSI refer respectively to the PyTorch baseline, the LSI with the image outpainting network and the pyramidal LSI. Besides the sequential combination of outpainting and pyramidal strategy presented in Fig. 1 referred to as SeqLSI, we also tested a parallel version (ParLSI) with an offline late fusion. This fusion blends the rendering of two methods (Figs. 2 and 3). The inner and outer pixels of the final rendering are directly computed from the rendering of the pyramidal and the outpainting networks, respectively. To avoid discontinuities, we used a blending mask smoothed by a Gaussian filter.

Table 1. L_1 ↓ error and SSIM ↑ on the validation set of the KITTI dataset.

1-layer LDI				2-layer LDI			
Version	L_1	I-L_1	O-L_1	Version	L_1	I-L_1	O-L_1
LSI [18]	0.0613	0.0539	0.0746	LSI [18]	0.0609	0.0552	0.0710
LSI+IO	0.0533	0.0512	**0.0571**	LSI+IO	0.0670	0.0655	0.0696
PyLSI	0.0482	**0.0396**	0.0643	PyLSI	**0.0597**	**0.0549**	0.0686
SeqLSI	0.0547	0.0547	0.0588	SeqLSI	0.0599	0.0582	**0.0629**
ParLSI	**0.0458**	0.0421	0.0604	ParLSI	0.0606	0.0596	0.0646
Version	SSIM	I-SSIM	O-SSIM	Version	SSIM	I-SSIM	O-SSIM
LSI [18]	0.7289	0.7584	0.5507	LSI [18]	0.7223	0.7499	0.5489
LSI+IO	0.7208	0.7388	0.6217	LSI+IO	0.7034	0.7208	0.6168
PyLSI	0.7567	0.7724	0.5946	PyLSI	0.7424	0.7575	0.5810
SeqLSI	0.7037	0.7195	0.6145	SeqLSI	0.7528	0.7711	**0.6556**
ParLSI	**0.7798**	**0.8047**	**0.6309**	ParLSI	**0.7687**	**0.7931**	0.6350

We can see from Table 1 that the PyLSI outperformed the baseline LSI in all cases. This is more evident in the 1-layer LDI, where the I-L_1, for instance, reaches 0.0396, a 26% drop compared to the LSI. Although the 2-layer PyLSI has also improved the LSI, it was not so intense as in the 1-layer, leading to a decrease of only 0.5% in I-L_1. As expected, the LSI+IO achieved good results in outer pixels, where the LSI has its worst results. In the 1-layer LDI, the O-L_1 was closer to the other metrics and presented a drop of about 23% compared to the LSI. The O-SSIM was increased by around 13%.

The numerical results of the 1-layer SeqLSI were not as good as those achieved by the isolated versions in inner pixels (PyLSI) and outer pixels (LSI+IO), although it did outperform the baseline in most cases. However, the 2-layer version obtained better SSIM values in all regions of the image. On the other hand,

1-layer ParLSI achieved the best results for L_1 and all SSIM values, whereas the 2-layer version obtained the best results on SSIM.

Figure 4 presents a comparison between the results of LSI and PyLSI. Sub-figs. 4a and 4b show the new views of LSI and PyLSI, respectively, stacked with the target image on the RGB channels. On these stacked images, we expect values close to gray when the prediction is good and close to green or purple when it is not. Finally, Subfigs. 4c and 4d present a zoomed region of the disparity and stacked images of the LSI and the PyLSI, respectively.

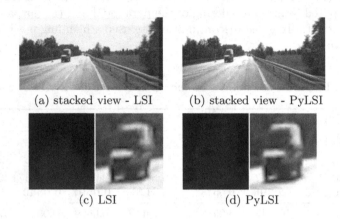

(a) stacked view - LSI　　　　(b) stacked view - PyLSI

(c) LSI　　　　　　　　(d) PyLSI

Fig. 4. Comparison between LSI ($L_1 = 0.0683$) and PyLSI results ($L_1 = 0.0328$).

On an overview of the stacked images, we can see that the PyLSI version has values closer to gray, even in more homogeneous regions, such as the sky. This shows that the intensities and colors of the PyLSI prediction are closer to the ground-truth than the LSI prediction. In addition, as we can see in the zoomed images, the PyLSI disparity in the car is better defined, which leads to better alignment at the prediction.

Figure 5 presents another example for all versions. Comparing the views rendered by LSI and PyLSI, we can see an improvement in the image boundaries. However, as there is no extrapolation and supervision on these regions, they are not reliable and lead to distortions in the objects (highlighted). The outpainting improves these distortions but generates some artifacts in the three versions that include it (LSI+IO, SeqLSI and ParLSI). These artifacts could be removed using a more recent and advanced outpainting network. Despite the artifact, the best visual result is achieved by SeqLSI.

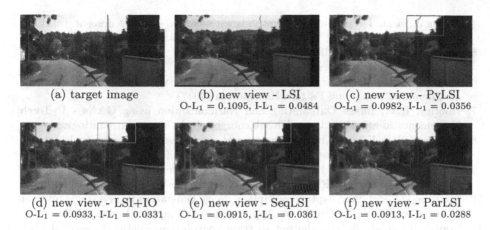

(a) target image $\qquad$ (b) new view - LSI $\qquad$ (c) new view - PyLSI
$\qquad\qquad\qquad$ O-L_1 = 0.1095, I-L_1 = 0.0484 $\quad$ O-L_1 = 0.0982, I-L_1 = 0.0356

(d) new view - LSI+IO $\quad$ (e) new view - SeqLSI $\quad$ (f) new view - ParLSI
O-L_1 = 0.0933, I-L_1 = 0.0331 $\quad$ O-L_1 = 0.0915, I-L_1 = 0.0361 $\quad$ O-L_1 = 0.0913, I-L_1 = 0.0288

Fig. 5. Example of results for all versions.

4   Conclusions

In this work, we presented a new method for monocular view synthesis based
on LSI. Our method is composed of an initial outpainting step and a pyramidal
strategy. The main goal of the outpainting network was to increase the original
field of view and improve the prediction in the boundaries of the new view. The
pyramidal strategy was employed to improve the overall quality by generating
a new view for different resolutions of the input. We tested the two strategies
separately (LSI+IO and PyLSI), and combined them sequentially (SeqLSI) and
in parallel (ParLSI). We compared them with the baseline using the pixel-wise
L_1 and SSIM metrics. For both, we considered three versions (traditional, inner
and outer averages) in order to evaluate the effectiveness of the methods in
different parts of the images. Our results suggested that the pyramidal strategy
achieved superior results with the inner metrics, whereas the outpainting tended
to perform better at the boundaries. ParLSI achieved the best results in most
cases, especially with SSIM, but, in our visual inspection, the SeqLSI generated
images with superior quality. We observed some artifacts in the outpainting-
based versions that can be avoided using a more advanced network in the future.

References

1. Chen, X., Chen, X., Zha, Z.J.: Structure-aware residual pyramid network for
   monocular depth estimation. In: International Joint Conference on Artificial Intelligence (2019)
2. Dhamo, H., Navab, N., Tombari, F.: Object-driven multi-layer scene decomposition
   from a single image. In: IEEE/CVF International Conference on Computer Vision
   (2019)
3. Dhamo, H., Tateno, K., Laina, I., Navab, N., Tombari, F.: Peeking behind objects:
   layered depth prediction from a single image. Pattern Recognit. Lett. (2019)

4. Flynn, J., et al.: DeepView: view synthesis with learned gradient descent. In: IEEE Conference on Computer Vision and Pattern Recognition (2019)
5. Geiger, A., Lenz, P., Stiller, C., Urtasun, R.: Vision meets robotics: the KITTI dataset. Int. J. Robot. Res. (2013)
6. Han, D., Kim, J., Kim, J.: Deep pyramidal residual networks. In: IEEE Conference on Computer Vision and Pattern Recognition (2017)
7. Hoorick, B.V.: Image Outpainting and Harmonization using GANs - PyTorch Implementation (2020). https://github.com/basilevh/image-outpainting
8. Layton, O.W., Fajen, B.R.: Computational mechanisms for perceptual stability using disparity and motion parallax. J. Neurosci. (2020)
9. Liu, S., Li, T., Chen, W., Li, H.: Soft rasterizer: a differentiable renderer for image-based 3D reasoning. In: IEEE International Conference on Computer Vision (2019)
10. Luvizon, D.C., et al.: Adaptive multiplane image generation from a single internet picture. In: Winter Conference on Applications of Computer Vision (2021)
11. Mayer, N., et al.: A large dataset to train convolutional networks for disparity, optical flow, and scene flow estimation. In: IEEE Conference on Computer Vision and Pattern Recognition (2016)
12. Pinto, A., et al.: Parallax motion effect generation through instance segmentation and depth estimation. In: International Conference on Image Processing. IEEE (2020)
13. Shade, J., Gortler, S., He, L.w., Szeliski, R.: Layered depth images. In: 25th Annual Conference on Computer Graphics and Interactive Techniques (1998)
14. Shih, M.L., Su, S.Y., Kopf, J., Huang, J.B.: 3D photography using context-aware layered depth inpainting. In: IEEE Conference on Computer Vision and Pattern Recognition (2020)
15. Srinivasan, P.P., Tucker, R., Barron, J.T., Ramamoorthi, R., Ng, R., Snavely, N.: Pushing the boundaries of view extrapolation with multiplane images. In: IEEE/CVF Conference on Computer Vision and Pattern Recognition (2019)
16. Sun, D., Yang, X., Liu, M.Y., Kautz, J.: PWC-Net: CNNs for optical flow using pyramid, warping, and cost volume. In: IEEE Conference on Computer Vision and Pattern Recognition (2018)
17. Tucker, R., Snavely, N.: Single-view view synthesis with multiplane images. In: IEEE Conference on Computer Vision and Pattern Recognition (2020)
18. Tulsiani, S., Tucker, R., Snavely, N.: Layer-structured 3D scene inference via view synthesis. In: Ferrari, V., Hebert, M., Sminchisescu, C., Weiss, Y. (eds.) ECCV 2018. LNCS, vol. 11211, pp. 311–327. Springer, Cham (2018). https://doi.org/10.1007/978-3-030-01234-2_19
19. Van Hoorick, B.: Image outpainting and harmonization using generative adversarial networks. arXiv preprint arXiv:1912.10960 (2019)
20. Wiles, O., Gkioxari, G., Szeliski, R., Johnson, J.: SynSin: end-to-end view synthesis from a single image. In: Conference on Computer Vision and Pattern Recognition (2020)
21. Zhang, M., Zhang, Y., Piao, Y., Liu, J., Ji, X., Zhang, Y.: Parallax based motion estimation in integral imaging. In: Digital Holography and Three-Dimensional Imaging. Optical Society of America (2019)
22. Zhou, T., Tucker, R., Flynn, J., Fyffe, G., Snavely, N.: Stereo magnification: learning view synthesis using multiplane images. ACM Trans. Graph. (2018)

Out of the Box: Embodied Navigation in the Real World

Roberto Bigazzi[✉], Federico Landi, Marcella Cornia, Silvia Cascianelli,
Lorenzo Baraldi, and Rita Cucchiara

University of Modena and Reggio Emilia, Modena, Italy
{roberto.bigazzi,federico.landi,marcella.cornia,silvia.cascianelli,
lorenzo.baraldi,rita.cucchiara}@unimore.it

Abstract. The research field of Embodied AI has witnessed substantial
progress in visual navigation and exploration thanks to powerful simulating platforms and the availability of 3D data of indoor and photorealistic
environments. These two factors have opened the doors to a new generation of intelligent agents capable of achieving nearly perfect PointGoal
Navigation. However, such architectures are commonly trained with millions, if not billions, of frames and tested in simulation. Together with
great enthusiasm, these results yield a question: how many researchers
will effectively benefit from these advances? In this work, we detail how
to transfer the knowledge acquired in simulation into the real world. To
that end, we describe the architectural discrepancies that damage the
Sim2Real adaptation ability of models trained on the Habitat simulator
and propose a novel solution tailored towards the deployment in real-
world scenarios. We then deploy our models on a LoCoBot, a Low-Cost
Robot equipped with a single Intel RealSense camera. Different from
previous work, our testing scene is unavailable to the agent in simulation. The environment is also inaccessible to the agent beforehand, so
it cannot count on scene-specific semantic priors. In this way, we reproduce a setting in which a research group (potentially from other fields)
needs to employ the agent visual navigation capabilities as-a-Service. Our
experiments indicate that it is possible to achieve satisfying results when
deploying the obtained model in the real world. Our code and models
are available at https://github.com/aimagelab/LoCoNav.

Keywords: Embodied AI · Sim2Real · Visual navigation

1 Introduction

Embodied AI has recently attracted a lot of attention from the vision and learning communities. This ambitious research field strives for the creation of intelligent agents that can interact with the surrounding environment. Smart interactions, however, require fine-grained perception and effective planning abilities.
For this reason, current research focuses on the creation of rich and complex
architectures that are trained in simulation with a large amount of data. Thanks

© Springer Nature Switzerland AG 2021
N. Tsapatsoulis et al. (Eds.): CAIP 2021, LNCS 13052, pp. 47–57, 2021.
https://doi.org/10.1007/978-3-030-89128-2_5

to powerful simulating platforms [7, 15, 19], the Embodied AI community could achieve nearly perfect results on the PointGoal Navigation task (PointNav) [18]. However, current research is still in the first mile of the race for the creation of intelligent and autonomous agents. Naturally, the next milestones involve bridging the gap between simulated platforms (in which the training takes place) and the real world [8]. In this work, we aim to design a robot that can navigate in unknown, real-world environments [2].

We ask ourselves a simple research question: *can the agent transfer the skills acquired in simulation to a more realistic setting?* To answer this question, we devise a new experimental setup in which models learned in simulation are deployed on a LoCoBot [10]. Previous work on Sim2Real adaptability from the Habitat simulator [15] has focused on a setting where the real-world environment was matched with a corresponding simulated environment to test the Sim2Real metric gap. To that end, Kadian *et al.* [8] carry on a 3D acquisition of the environment specifically built for robotic experiments. Here, we assume a setting in which the final user cannot count on the technology/expertise required to make a 3D scan. This experimental setup is more challenging for the agent, as it cannot count on semantic priors on the environment acquired in simulation. Moreover, while [8] employs large boxes as obstacles, our testing scene contains real-life objects with complicated shapes such as desks, office chairs, and doors.

Our agent builds on a recent model proposed by Ramakrishnan *et al.* [12] for the PointNav task. As a first step, we research the optimal setup to train the agent in simulation. We find out that default options (tailored for simulated tasks) are not optimal for real-world deployment: for instance, the simulated agents often exploit imperfections in the simulator physics to slide along the walls. As a consequence, deployed agents tend to get stuck when trying to replicate the same sliding dynamic. By enforcing a more strict interaction with the environment, it is possible to avoid such shortcomings in the locomotor policy. Secondly, we employ the software library PyRobot [11] to create a transparent interface with the LoCoBot: thanks to PyRobot, the code used in simulation can be seamlessly deployed on the real-world agent by changing only a few lines of code. Finally, we test the navigation capabilities of the trained model on a real scene: we create a set of navigation episodes in which goals are defined using relative coordinates. While previous tests were mainly made in robot-friendly scenarios (often consisting of a single room), we test our model, which we call LoCoNav, in a more realistic environment with multiple rooms and typical office furniture (Fig. 1). Thanks to our experiments, we show that models trained in simulation can adapt to real unseen environments. By making our code and models publicly available, we hope to motivate further research on Sim2Real adaptability and deployment in the real world of agents trained on the Habitat simulator.

2   Related Work

There is a broad area of recent research that focuses on designing autonomous agents with different abilities. Among these, a vast line of work concentrates on embodied exploration and navigation [3, 5, 9, 12, 13]. In this setting, the agent's

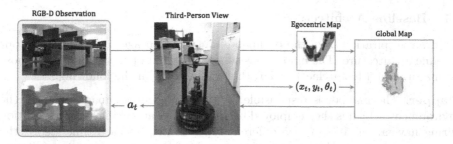

Fig. 1. We deploy a state-of-art navigation architecture on a LoCoBot and test it in a realistic, office-like environment. Our model exploits egocentric and global occupancy maps to plan a route towards the goal.

goal is to explore a new environment in the shortest amount of time. Architectures trained for this task usually employ reinforcement learning to maximize coverage (the area seen during a single episode) [3], surprisal [1], or a reward based on the novelty of explored areas [13]. Usually, this is done by creating internal map representations to keep track of the exploration progress and at the same time help the agent plan for future destinations [3,5,12]. The main advantage of these approaches is their ability to adapt to downstream tasks, such as PointGoal [12] or ObjectGoal [4] navigation. In PointGoal navigation, the target destination is specified using relative coordinates $w.r.t.$ the agent's initial position and heading [15]. Using simulation and impressive computational power, Wijmans $et\ al.$ [18] achieve nearly perfect results. However, their model is trained using 2.5 billion frames and requires experience acquired over more than half a year of GPU time. Unfortunately, models tend to learn simulator-specific tricks to circumvent navigation difficulties [8]. Since such shortcuts do not work in the real world, there is a significant Sim2Real performance gap.

Recent work has studied how to deploy models trained on simulation to the real world [7,8,14]. In their work, Kadian $et\ al.$ [8] make a 3D acquisition of a real-world scene and study the Sim2Real gap for various setups and metrics. However, their environment is very simple as obstacles are large boxes, the floor has an even and regular surface in order to facilitate the actuation system, and there are no doors or other navigation bottlenecks. In this work, instead, we focus on a more realistic type of environment: obstacles are represented by common office furniture such as desks, chairs, cupboards; the floor is uneven as there are gaps between floor tiles that make actuation noisy and very position-dependent, and there are multiple rooms that must be accessed through doorways.

3   Real-World Navigation with Habitat

In this section, we describe our out-of-the-box navigation robot. First, we describe the baseline architecture and its training procedure that takes place in the Habitat simulator [15]. Then we present our LoCoNav agent, which builds upon the baseline and implements various modules to enable real-world navigation.

3.1   Baseline Architecture

We draw inspiration from the occupancy anticipation agent [12] to design our baseline architecture. The model consists of three main parts: a mapper, a pose estimator, and a hierarchical policy, that we describe in the following.

Mapper. The mapper is responsible for producing an occupancy map of the environment, which is then employed by the agent as an auxiliary representation during navigation. We use two different types of map at each time step t: the agent-centric map v_t that depicts the portion of the environment immediately in front of the agent, and the global map m_t that captures the area of the environment already visited by the agent. The global map of the environment m_t is blank at $t = 0$ and it is built in an incremental way. Each map has two channels, identifying the free/occupied and the explored/unexplored space, respectively; each pixel contains the state of a 5 cm × 5 cm area. The mapper module takes as input the RGB and depth observations (o_t^r, o_t^d) at time t and produces the agent-centric map $v_t \in [0, 1]^{2 \times V \times V}$. The RGB observation is encoded to a feature representation $\bar{o}_t^r$ with the first two layers of a pretrained ResNet-18 followed by a three-layered CNN. Instead, the depth observation is used to create a point-cloud and reprojected to form a preliminary map $\bar{o}_t^d$. The resulting agent-centric map v_t is computed by combining $\bar{o}_t^r$ and $\bar{o}_t^d$ with a U-Net. Then, v_t is registered to the global map $m_t \in [0, 1]^{2 \times W \times W}$, with $W > V$, using the agent's position and heading in the environment (x_t, y_t, θ_t).

Pose Estimator. While the agent navigates towards the goal, the interactions with the environment are subject to noise and errors, so that, for instance, the action *go forward* 25 cm might not result in a real displacement of 25 cm. That could happen for a variety of reasons: bumping into an obstacle, slipping on the terrain, or simple actuation noise. The pose estimator is responsible of avoiding such positioning mistakes and keeps track of the agent pose in the environment at each time step t. This module computes the relative displacement $(\Delta x_t, \Delta y_t, \Delta \theta_t)$ caused by the action selected by the agent at time t. It takes as input the RGB-D observations (o_t^r, o_t^d) and (o_{t-1}^r, o_{t-1}^d) retrieved at time t and $t - 1$, and the egocentric maps v_t and v_{t-1}. Each modality is considered separately to obtain a first estimate of the displacement:

$$g_i = W_1 \max(W_2 \star + b_2, 0) + b_1, \tag{1}$$

The final output of the pose estimator is the weighted sum of the three displacement vectors g_i:

$$(\Delta x_t, \Delta y_t, \Delta \theta_t) = \sum_{i=0}^{2} \alpha_i \cdot g_i, \qquad \alpha_i = \mathrm{softmax}(\mathrm{MLP}_i([\bar{o}_t^r, \bar{o}_t^d, \bar{v}_t])), \tag{2}$$

where MLP is a three-layered fully-connected network, $(\bar{o}_t^r, \bar{o}_t^d, \bar{v}_t)$ are the inputs encoded by a CNN and $[\cdot, \cdot, \cdot]$ denotes tensor concatenation. The estimated pose of the agent at time t is given by $(x_t, y_t, \theta_t) = (x_{t-1}, y_{t-1}, \theta_{t-1}) + (\Delta x_t, \Delta y_t, \Delta \theta_t)$.

Hierarchical Policy. Following a current trend in Embodied AI [3,5,12], we employ a hierarchical policy in our baseline navigator. The highest-level component of our policy is the global policy. The global policy selects a long-term goal on the global map, that we call global goal. The input of the global policy at time t is a 4-channel enriched global map $m_t^+ \in [0,1]^{4 \times W \times W}$ obtained as the concatenation of the global map m_t with a spatial representation of visited states and a one-hot representation of the agent position at time t. Finally, we compute an 8-channel input of shape $G \times G$ for the global policy. To that end, we concatenate a cropped and a max-pooled version of m_t^+. The global policy outputs a probability distribution over the $G \times G$ action space. The global goal is sampled from this distribution and is then converted to (x, y) coordinates on the global map. A new global goal is sampled every N time steps during training and is set to the navigation goal during deployment and test. The middle-level component of our hierarchical policy is the planner. After the global goal is set, an A* planner decodes the next local goal within 0.25 m from the agent and on the trajectory towards the global goal. A new local goal is sampled if at least one of the following three conditions verifies: a new global goal is sampled by the global policy, the previous local goal is reached, or the local goal is known to be in an occupied area. Finally, the local policy performs the low-level navigation and decodes the series of actions to perform. The actions available to the agents are *go forwards* 25 cm and *turn* 15°. The local policy samples an atomic action a_t at each time step t.

3.2   Training in Simulation

The baseline architecture described in the previous lines is trained in simulation using Habitat [15] and 3D scans from the Gibson dataset of spaces [19]. The mapper is trained with a binary cross-entropy loss using the ground-truth occupancy maps of the environment, obtained as described in [12]. The navigation policy is trained using reinforcement learning. We choose PPO [16] as training algorithm. The global policy receives a reward signal equal to the increase in terms of anticipated map accuracy [12]:

$$R_t^{glob} = \text{Accuracy}(m_t, \hat{m}) - \text{Accuracy}(m_{t-1}, \hat{m}), \tag{3}$$

where m_t and m_{t-1} represent the global occupancy maps computed at time t and $t-1$ respectively, and $\hat{m} \in [0,1]^{2 \times W \times W}$ is the ground-truth global map. The map accuracy is defined as:

$$\text{Accuracy}(m, \hat{m}) = \sum_{i=1}^{W^2} \sum_{j=1}^{2} \mathbb{1}[m_{ij} = \hat{m}_{ij}], \tag{4}$$

where $\mathbb{1}[\cdot]$ is an indicator function that returns one if the condition $[\cdot]$ is true and zero otherwise. The local policy is trained using a reward that encourages the decrease in the euclidean distance between the agent and the local goal while penalizing collisions with obstacles:

$$r_t^{local} = d_t - d_{t-1} - \alpha * bump_t, \tag{5}$$

where d_t and d_{t-1} are the euclidean distances to the local goal at times t and $t-1$, $bump_t \in \{0, 1\}$ identifies a collision at time t and α regulates the contributions of the collision penalty. The training procedure described in this section exploits the experience collected throughout 6.5 million exploration frames.

3.3   LoCoNav: Adapting for Real World

The baseline architecture described above is trained in simulation and achieves state-of-art results on embodied exploration and navigation [12]. The reality, however, poses some major challenges that need to be addressed to achieve good real-world performances. For instance, uneven ground might give rise to errors and noise in the actuation phase. To overcome this and other discrepancies between simulated and real environments, we design LoCoNav: an agent that leverages the availability of powerful simulating platforms during training but is tailored for real-world use. In this section, we describe the main characteristics of the LoCoNav design. We deploy our architecture on a LoCoBot [10] and use PyRobot [11] for seamless code integration.

Prevent your Agent from Learning Tricks. All simulations are imperfect. One of the main objectives when training an agent for real-world use in simulation is to prevent it from learning simulator-specific tricks instead of the basic navigation skills. During training, we observed that the agent tends to hit the obstacles instead of avoiding them. This behavior is given by the fact that the simulator allows the agent to slide towards its direction even if it is in contact with an obstacle as if there were no friction at all. Unfortunately, this ideal situation does not fit the real world, as the agent needs to actively rotate and head towards a free direction every time it bumps into an obstacle. To replicate the realistic *sticky* behavior of surfaces, we check the $bump_t$ flag before every step. If a collision is detected, we prevent the agent from moving forward. As a result, our final agent is more cautious about any form of collision.

Sensor and Actuation Noise. Another important discrepancy between simulation and real-world is the difference in the sensor and actuation systems. Luckily, the Habitat simulator allows for great customization of input-output dynamics, thus being very convenient for our goal. In order to train a model that is more resilient to the camera noise, we apply a Gaussian Noise Model on the RGB observations and a Redwood Noise Model [6] on the depth observations. Unfortunately, the LoCoBot RealSense camera still presents various artifacts and regions with missing depth values. For that reason, we need to restore the observation retrieved from the depth camera before using it in our architecture. To that end, we apply the hole filling algorithm described in [17], followed by the application of a median filter.

Table 1. List of hyperparameters changes for Sim2Real transfer.

	Height	RGB FoV	Depth FoV	Depth range	Obst. height thresh.
Default for simulation	1.25	H: 90, V: 90	H: 90, V: 90	[0.0, 10.0]	[0.2, 1.5]
LoCoNav (ours)	0.60	H: 70, V: 90	H: 57, V: 86	[0.0, 5.00]	[0.3, 0.6]

Regarding the actuation noise, we find out that the use of the incremental pose estimator (employed in the occupancy anticipation model and described in our baseline architecture) is not optimal, especially when combined with the actuation noise typical of real-world scenarios. Luckily, we can count on more precise and reliable information coming from the LoCoBot actuation system. By checking the actual rotation of each wheel at every time step, the robot can update its position step by step. We adapt the odometry sensor of the LoCoBot platform to be compliant with our architecture. To that end, the pose returned by the sensor is converted by resetting it with respect to its state at the beginning of the episode. We name $\chi_0 = (x_0, y_0, \theta_0)$ the coordinate triplet given by the odometry sensor at $t = 0$. We then define:

$$\mathbf{A} = \begin{pmatrix} \mathbf{R}_0 & \mathbf{t}_0 \\ \mathbf{0} & 1 \end{pmatrix} = \begin{pmatrix} \cos\theta_0 & -\sin\theta_0 & x_0 \\ \sin\theta_0 & \cos\theta_0 & y_0 \\ 0 & 0 & 1 \end{pmatrix}. \tag{6}$$

Let us define $\mathbf{x}_t$ as the augmented position vector $(x_t, y_t, 1)$ containing the agent position at each step t. We compute the relative position of the robot as:

$$\tilde{\mathbf{x}}_t = \mathbf{A}^{-1}\mathbf{x}_t, \qquad \tilde{\theta}_t = \theta_t - \theta_0 \tag{7}$$

where $\tilde{\mathbf{x}}_t = (\tilde{x}_t, \tilde{y}_t, 1)$ contains the agent position after the conversion to episode coordinates. The relative position and heading is given by $\tilde{\chi}_t = (\tilde{x}_t, \tilde{y}_t, \tilde{\theta}_t)$. Note that, for $t = 0$, $\tilde{\chi}_0 = (\tilde{x}_0, \tilde{y}_0, \tilde{\theta}_0) = (0, 0, 0)$.

Hyperparameters. Finally, we noticed that typical hyperparameters employed in simulation do not match the real robot characteristics. For instance, the camera height is set to 1.25 m in previous works, but the RealSense camera on the LoCoBot is placed only 0.6 m from the floor. During the adaptation to the real-world robot, we change some hyperparameters to align the observation characteristics of the simulated and the real world and to match real robot constraints. These parameters are listed in Table 1.

4 Experiments

Testing Setup. We run multiple episodes in the real environment, in which the agent needs to navigate from a starting point A to a destination B. The goal is specified by using relative coordinates (in meters) with respect to the agent's starting position and heading. Although the agent knows the position of its destination, it has no prior knowledge of the surrounding environment.

Because of this, it cannot immediately plan a direct route to the goal and must check for obstacles and walls before stepping ahead. After each run, we reset the agent memory so that it cannot retain any information from previous episodes. We design five different navigation episodes that take place in three different office rooms and the corridor connecting them (Fig. 2a). For each episode, we run different trials with different configurations: obstacles are added/moved, or people are sitting/standing in the room. In total, we run 50 different experiments, resulting in more than 10 h of real-world testing.

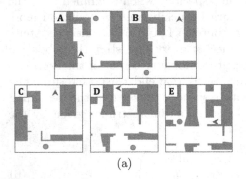

Path	Length(m)	Time(s)	# Step
A	3.80	124	23
B	6.75	239	45
C	5.95	223	43
D	6.55	217	42
E	4.20	227	33

(a)   (b)

Fig. 2. Layout of the navigation episodes (a). Path-specific information, as obtained with human supervision (b).

Evaluation Protocol. An episode is considered successful if the agent sends a specific *stop* signal within 0.2 m from the goal. This threshold corresponds to the radius of the robot base. For every navigation episode, we also track the number of steps and the time required to reach the goal. Since the absolute number of steps is not comparable among different episodes, we ask human users to control the LoCoBot and complete each navigation path via a remote interface (we report human performance in Fig. 2b). We then normalize these measures using this information so that results close to 1.00 indicate human-like performances. We provide absolute and normalized length and time for each episode, as well as the popular SPL metric (Success rate weighted by inverse Path Length). We employ a slightly modified version of the SPL, in which the normalization is made basing on the number of steps and not on the effective path length to penalize purposeless rotations. Additionally, we set a boolean flag for each episode that signals whether the robot has bumped into an obstacle, and we report the average Bump Rate (BR). We also report the Hard Failure Rate (HFR) as the fraction of episodes terminated if the agent gets stuck and cannot proceed, or if the episode length exceeds the limit of 300 steps.

Real-world Navigation. In this experiment, we test our robot on five different realistic navigation paths (Fig. 2a). We report the numerical results for these experiments in Table 2, and we plot the main metrics in Fig. 3 to allow for a better visualization of navigation results across different episodes. When a path

Table 2. Navigation results. Numbers after ± denote the standard error of the mean.

Path	SR ↑	SPL ↑	HFR ↓	BR ↓	Abs. steps	Norm. steps ↑	Abs. time	Norm. time ↑
A	1.0	0.718	0.0	0.30	32.70 ± 1.73	0.717 ± 0.033	176.11 ± 10.39	0.718 ± 0.031
B	0.8	0.711	0.10	0.22	51.67 ± 1.72	0.880 ± 0.027	273.70 ± 8.24	0.879 ± 0.030
C	0.5	0.205	0.10	0.78	123.44 ± 10.66	0.374 ± 0.034	631.15 ± 50.09	0.372 ± 0.036
D	0.5	0.318	0.10	0.89	65.67 ± 3.90	0.645 ± 0.037	344.00 ± 20.08	0.657 ± 0.038
E	0.2	0.060	0.40	1.00	135.17 ± 29.97	0.290 ± 0.049	722.76 ± 162.01	0.38 ± 0.066
Overall	0.6	0.402	0.14	0.60	–	0.608 ± 0.036	–	0.617 ± 0.034

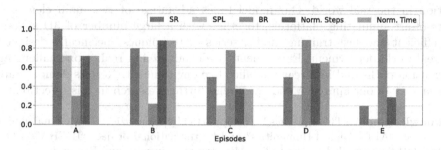

Fig. 3. Comparison of the main navigation metrics on different episodes.

is contained in a single room (A), the agent achieves optimal results, as it always stops within the success threshold from the goal. The number of steps is slightly higher than the minimum required by the episode (33 instead of 23), but this overhead is necessary as the agent must rotate and "look around" to build a decent map of the surrounding before planning a route to the goal. Paths that involve going outside the room and navigating different spaces (B, C, D, E) are fairly complicated, but the agent can generally terminate the episode without hard failures. When the shortest path to the goal leads to a wall or a dead-end, the agent needs to find an alternative way to circumvent this obstacle (e.g. a door). This leads to a higher episode length because the robot must dedicate some time to general exploration of the surroundings. Finally, we find out that the most challenging scenario for our LoCoNav is when reaching the goal requires to get out of a room and then enter a door immediately after, on the same side of the corridor (as in E). Since the robot sticks to the shortest path, the low parallax prevents it from identifying the second door correctly. Even in these cases, a bit of general exploration helps to solve the problem.

Discussion and Failure Cases. Overall, our experimental setup provides a challenging test-bed for real-world robots. We find out that failures are due to two main issues. First, when the agent must navigate to a different room, it has no access to a map representing the general layout of the environment. This prevents the robot from computing a general plan to reach the long-term goal and forces it to explore the environment before proceeding. If a map was given to the agent, this problem would have been greatly alleviated. A second problem arises when the goal is close in terms (x, y) coordinates but is physically placed in an adjacent room. To solve this problem, one could decompose the navigation between rooms in a multi-goal problem where neighboring nodes are closer. In this way, it is possible to reduce a complex navigation episode in simpler sub-episodes (like A or B), in which our agent has proved to be successful.

5   Conclusion

We have presented LoCoNav, an out-of-the-box architecture for embodied navigation in the real world. Our model takes advantage of two main elements: state-of-art simulating platforms, together with a large number of 3D spaces, for efficient and fast training, and a series of techniques specifically designed for real-world deployment. Experiments are conducted in reality on challenging navigation paths and in a realistic office-like environment. Results demonstrate the validity of our approach and encourage further research in this direction.

Acknowledgment. This work has been supported by "Fondazione di Modena" under the project "AI for Digital Humanities" and by the national project "IDEHA" (PON ARS01_00421), cofunded by the Italian Ministry of University and Research.

References

1. Bigazzi, R., Landi, F., Cornia, M., Cascianelli, S., Baraldi, L., Cucchiara, R.: Explore and explain: self-supervised navigation and recounting. In: ICPR (2020)
2. Cascianelli, S., Costante, G., Bellocchio, E., Valigi, P., Fravolini, M.L., Ciarfuglia, T.A.: A robust semi-semantic approach for visual localization in urban environment. In: ISC2 (2016)
3. Chaplot, D.S., Gandhi, D., Gupta, S., Gupta, A., Salakhutdinov, R.: Learning to explore using active neural SLAM. In: ICLR (2019)
4. Chaplot, D.S., Gandhi, D.P., Gupta, A., Salakhutdinov, R.R.: Object goal navigation using goal-oriented semantic exploration. In: NeurIPS (2020)
5. Chen, T., Gupta, S., Gupta, A.: Learning exploration policies for navigation. In: ICLR (2019)
6. Choi, S., Zhou, Q.Y., Koltun, V.: Robust reconstruction of indoor scenes. In: CVPR (2015)
7. Deitke, M., et al.: RoboTHOR: an open simulation-to-real embodied AI platform. In: CVPR (2020)
8. Kadian, A., et al.: Sim2Real predictivity: does evaluation in simulation predict real-world performance? IEEE Robot. Autom. Lett. **5**(4), 6670–6677 (2020)
9. Landi, F., Baraldi, L., Cornia, M., Corsini, M., Cucchiara, R.: Multimodal attention networks for low-level vision-and-language navigation. CVIU (2021)
10. LoCoBot: An Open Source Low Cost Robot. https://locobot-website.netlify.com
11. Murali, A., et al.: PyRobot: an open-source robotics framework for research and benchmarking. arXiv preprint arXiv:1906.08236 (2019)
12. Ramakrishnan, S.K., Al-Halah, Z., Grauman, K.: Occupancy anticipation for efficient exploration and navigation. In: Vedaldi, A., Bischof, H., Brox, T., Frahm, J.-M. (eds.) ECCV 2020. LNCS, vol. 12350, pp. 400–418. Springer, Cham (2020). https://doi.org/10.1007/978-3-030-58558-7_24

13. Ramakrishnan, S.K., Jayaraman, D., Grauman, K.: An exploration of embodied visual exploration. Int. J. Comput. Vis. **129**(5), 1616–1649 (2021). https://doi.org/10.1007/s11263-021-01437-z
14. Rosano, M., Furnari, A., Gulino, L., Farinella, G.M.: On embodied visual navigation in real environments through habitat. In: ICPR (2020)
15. Savva, M., et al.: Habitat: a platform for embodied AI research. In: ICCV (2019)
16. Schulman, J., Wolski, F., Dhariwal, P., Radford, A., Klimov, O.: Proximal policy optimization algorithms. arXiv preprint arXiv:1707.06347 (2017)
17. Telea, A.: An image inpainting technique based on the fast marching method. J. Graph. Tools **9**(1), 23–34 (2004)
18. Wijmans, E., et al.: DD-PPO: learning near-perfect PointGoal navigators from 2.5 billion frames. In: ICLR (2019)
19. Xia, F., Zamir, A.R., He, Z., Sax, A., Malik, J., Savarese, S.: Gibson env: real-world perception for embodied agents. In: CVPR (2018)

Toward a Novel LSB-based Collusion-Secure Fingerprinting Schema for 3D Video

Karama Abdelhedi[1]([✉]) [iD], Faten Chaabane[1] [iD], William Puech[2] [iD], and Chokri Ben Amar[1]

[1] Research Groups in Intelligent Machines ENIS Sfax, Sfax, Tunisia
[2] LIRMM, Université de Montpellier, CNRS, 860 rue de St Priest, 34095 Montpellier, France

Abstract. Securing multimedia content and preventing it from being maliciously manipulated has developed at a rapid pace, and researchers have been studying the traitor tracing as an appropriate solution. This approach consists in retrieving back the actors who contributed to the construction of an illegal release of a multimedia product. It includes two major steps which are the fingerprinting step and the tracing one. The fingerprinting step relies on the watermarking technique whereas the efficiency of the tracing scheme depends on several requirements: the robustness of the watermarking technique, the type of the media content, and even the computational complexity. In this paper, we propose a new collusion-secure fingerprinting scheme for 3D videos. It has essentially a twofold purpose: at a first step, we propose to embed the watermark in the video copy by applying a standard Least Significant Bit (LSB) substitution to all the frames of both the 2D video and the depth map components in order to ensure simultaneously and independently the protection of these two parts. In the second step, we apply the tracing process whose target is the identification of eventual colluders by extracting the hidden identifier from the suspicious video and analyse it. Experimental assessments show that the proposed scheme provides interesting results in terms of speed and tracing accuracy constraints.

Keywords: Collusion-secure · Fingerprinting · 3D video · LSB · Tardos · Traitors tracing

1 Introduction

The availability of the Internet and the evolution of the digital era facilitate the sharing of the content media. Unfortunately, the emerging digital technology is obliquely the origin of several unauthorized manipulations of digital content: multiple copies, illegal re-distributions and arbitrary modifications. Indeed, sharing, copying and redistributing illegally the content of a media may lead to a dangerous phenomenon, well-known as the digital piracy. Consequently, it

© Springer Nature Switzerland AG 2021
N. Tsapatsoulis et al. (Eds.): CAIP 2021, LNCS 13052, pp. 58–68, 2021.
https://doi.org/10.1007/978-3-030-89128-2_6

becomes crucial to protect any distributed media by preventing illegal use of shared release and detecting eventually malicious users. Henceforth, the digital rights management systems (DRM) have involved a set of measures to control the use of digital content [1]. One key approach is the digital fingerprinting approach which involves the presence of a watermarking technique to embed the identifier of the media copy and a tracing algorithm to retrieve malicious users. Indeed, the watermark embedding in the media content step has to respect some requirements: it should make the copyright infringement harder without altering its quality [2]. On the other hand, one media content that is attracting interest and is increasingly under threat is 3D video content. This type of media can be classified into two major categories according to the adapted on their archiving format: the side by side videos composed of right and left views taken by two cameras with the same characteristics, and the Depth-Image-Based Rendering videos (DIBR) based on respectively 2D video frames and their corresponding depth maps. Most of stored videos use the 3D-DIBR format because of its less storage and transmission bandwidth costs compared to the first category.

In this context, we focus our work on DIBR-based 3D videos, and our main contribution is to propose a robust watermarking scheme to protect both the two components of the 3D video and a suitable tracing technique to identify the illegal users in case of collusion attacks.

The paper is arranged as follows. Section 2 reviews the related work in multimedia tracing systems. Section 3 provides an overview of the general tracing system. In Sect. 4 we detail the different steps of the proposed tracing framework. In Sect. 5 we present the different experimental assessments we carry out to validate the performance of the proposed approach. Finally, we summarize with a conclusion and future work in Sect. 6.

2   Related Work

Handling a great number of shared videos and surviving different types of unauthorized manipulations present crucial challenges of the majority of fingerprinting schemes. In this context, several techniques were proposed in the literature. This section is divided in two sections. Section 2.1 focus on the watermarking techniques suitable for 3D videos while Sect. 2.2 is reserved to the tracing techniques.

2.1   Overview on the Existing 3D Video Watermarking Techniques

Several watermarking schemes were suggested for DIBR-based videos [3,4]. Mainly, these schemes are divided into three classes according to the watermark embedding positions: 2D video frame-based watermarking, the depth map-based watermarking and the third one is an hybrid scheme. Among the 2D video frame-based watermarking proposed schemes in the literature, a scheme proposed by [5] consists in constructing the Depth Perceptual Region of Interest (DP-ROI) by extracting some relevant characteristics as gray contour regions, the foreground, and the depth-edge, to improve the embedding strength. According to [6], the main weakness of 2D frames-based watermarking schemes to address is tied to the generated distortions. To cope with that issue, the depth-map-based watermarking schemes were proposed. The particularity of this kind of approach is that watermarks are embedded into the depth maps, which guarantees that no distortions can be seen on the synthesized 3D videos [7]. A great deal of research has been carried out on depth-map schemes, ranging from the Unseen Visible (UVW) schemes [8]; where watermarks are embedded after estimating computations and are perceptible only in few views, to the Unseen Extractable (UEW) schemes; where watermarks are hidden once DC quantization is performed. Similarly, it has been noticed [7] that depth-map based watermarking schemes do not respond to the requirements of DRM for of DIBR-based 3D videos. Consequently, the third family of watermarking schemes based on embedding watermarks in both 2D frames and depth maps were proposed. Among these techniques, the zero-watermarking schemes were suggested where the watermark is not embedded in the signal host. The main steps of this type of watermarking schemes are the copyright registration step and its identification step [9]. In [7], the main contribution is to improve the traditional zero-watermarking schemes to be suitable for DIBR-based 3D videos. Although, it proposed to protect both 2D frames and depth map to ensure efficient robustness and good imperceptibility, its performance is reduced noticeably for high watermark bandwidth. In [10], a new SVM-based zero-watermarking technique for DIBR-based 3D videos is proposed, it has proven good results of robustness and transparency but its does not make any trace in the video copy which makes its tracing process harder.

2.2 The Tracing Traitor: A Brief Review

In the literature, several fingerprinting techniques have been proposed to improve codes to ameliorate their detection rates with fair lengths, for a large number of users and pirates [11]. A massive research was investigated on the Tardos tracing process [12] which has proposed a good trade-off between the code length and the tracing rates [13]. In this context, several researchers focus on optimizing Tardos accusation's functions to ameliorate its robustness against the collusion attacks [14–21]. In other fingerprinting schemes, the target was to find a good trade-off between the tracing code and the watermarking technique to provide a tracing scheme able to resist to different types of collusion attacks [10,22]. But the robustness of these schemes was checked only against the averaging collusion attacks for a small number of users. In this paper, we propose a good trade-off between the adapted watermarking scheme and the Tardos-based tracing process that provides good tracing results for several collusion attacks.

3   The General Tracing System

The tracing system is the whole system that ensures the protection of the delivered media in a multimedia distribution platform. The tracing system consists of three essential steps: the copyright registration step, collusion attacks and the copyright identification step [23].

3.1 The Copyright Registration Step

This step is closely tied to the users of platform. It includes the generation, by the media holder, of a unique fingerprint codeword, $I \in \{1, \cdots, m\}$ to each media release buyer. Moreover, the fingerprint, or identifier should be unique and should identify the media owner in order to protect the digital content from any unauthorized treatment [24]. The second part in the copyright registration step consists in embedding the fingerprint using a watermarking technique. In advance of each sale of the media release, a fingerprint is affected to the customer and embedded in the media by using a robust watermark embedding process. Henceforth, to have an efficient tracing process, it is crucial to have a robust fingerprint generation scheme and subsequently a robust watermark technique.

3.2 The Collusion Attacks

These attacks [25] consist in the eventual attacks that can be operated by dishonest users to generate a suspicious copy in order to redistribute it. The efficiency of the tracing scheme and its accusation accuracy depends on its robustness against these types of attacks.

3.3 The Copyright Identification Step

Once the suspicious copy is detected, the copyright identification process aims to detect the embedded fingerprint by using the appropriate watermark extraction process. The extracted fingerprint is then used in the last step to retrieve back at least one of colluders who have cooperate in constructing the suspicious copy.

4 The Proposed Traitor Tracing Framework

In this section, we present the proposed traitor tracing framework which is using an LSB-based watermarking technique for 3D DIBR-based videos. We detail each step of the system separately.

4.1 The Proposed Copyright Registration Step

As mentioned previously, the main target of the proposed tracing scheme is to provide an efficient tracing scheme for 3D-DIBR-based videos. To cope with that issue, we propose to use a traditional Least Significant Bits watermarking approach with a high imperceptibility. As depicted in Fig. 1, a preprocessing operation is made. We propose to split respectively the 2D frames and the depth maps to k components. Then, to protect the 2D video or the depth, we must protect all their corresponding frames. Hence, we must integrate the identifier

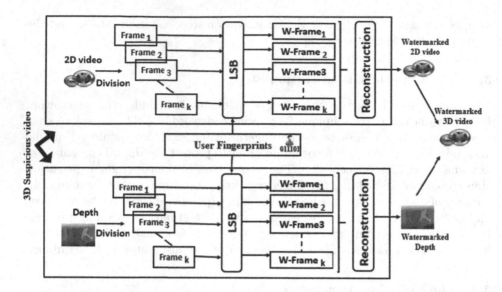

Fig. 1. The copyright registration scheme.

code in each frame of both of them by applying a LSB embedding approach. We remind that a LSB substitution [26] is a simple approach to embed a secret key in an image. A LSB technique embeds each bit of the identifier code (IC) in one pixel of the image (Im). As an example, in case of image size equal to $N \times M$ and an identifier code of L size, we must choose one pixel from P pixels:

$$P = \frac{N \times M}{L}. \tag{1}$$

For each P pixel of the frames, we choose randomly one pixel to embed the corresponding bit of the identifier code and this position is saved to be used in the identification phase. The embedding position of the bit is different from P pixels to others and from frames to others. After embedding the identifier in each frame, the video is reconstructed.

4.2   The Proposed Copyright Identification Scheme

The fingerprint extraction step is the reverse process of its fingerprint embedding step as shown in Fig. 2. The input of this step can be a fingerprinted 3D video, or an intentional modified fingerprinted one. Hence, the LSB technique is applied on each component of both 2D video frames and depth maps to extract respectively the corresponding embedded identifier. As a result of the LSB technique, we have k codewords extracted from the k 2D video frames and the k other codewords extracted from the depth maps frames. Then, by using a correlation coefficient, we conserve one codeword among the k extracted ones from the 2D video frames and another one among the codeword extracted from the depth maps.

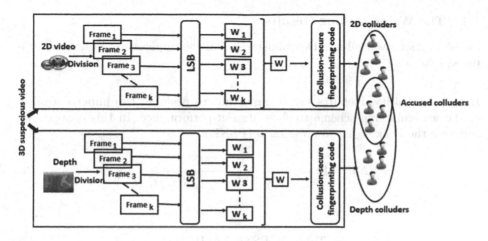

Fig. 2. The copyright identification scheme.

The coefficient correlation used to compare two codes X, Y which have the same length m:

$$r = \frac{\sum_{i=1}^{m} x_i y_i - 2\bar{X}\bar{Y}}{\sqrt{\sum_{i=1}^{m}(x_i - \bar{X})^2 - \sum_{i=1}^{m}(y_i - \bar{Y})^2}}. \tag{2}$$

In the last step of the copyright identification scheme, we apply the collusion-secure fingerprinting code to both the two detected identifiers. We propose in our approach to use the well-known Tardos [12] as a tracing code.

Step 1: Select the suspicious video (V_s)
Step 2: Split the video in 2D video (V_{2d}) and Depth map (V_D)
Step 3: Split V_{2d} to 25 frames and similarly to V_D
Step 4: For 1 to 25 frames of V_{2d}: Extract the identifier code (IC) from the corresponding bits
Step 5: Keep one (IC) from the 25 codes extracted from V_{2d} by using the coefficient correlation
Steps 6 and 7: Same processing of Step 4 and 5 with V_D
Step 8: Apply the Tardos code to (IC) extracted from V_{2d} and V_D.

5   Experimental Results

In this section, we evaluate firstly the robustness of the proposed LSB-based watermarking scheme. Then, we evaluate the efficiency of its tracing process against different collusion attacks. The tested database contains *150* different 3D video clips with a total amount of *3750* different video frames collected from the database [27]. Others 2D video clips are selected from different movies, with their corresponding depth maps calibrated using the technique in [28]. Each video of the database contains respectively *100* frames of size $320 \times 180 \times 25$.

5.1   The Watermarking Results

In this section, we evaluate the robustness and the imperceptibility of the proposed scheme.

Imperceptibility Results. It is important to evaluate the imperceptibility of the watermarking scheme to show its out-performance. In this context, we compute the Peak signal to noise ratio (PSNR):

$$PSNR = 10 \log \frac{255^2}{\frac{1}{M \times N} \sum_{i=1}^{M} \sum_{j=1}^{N} I(i,j)I'(i,j)}. \tag{3}$$

Table 1. PSNR VALUES.

Length m=	9817	7000	5000	4096
PSNR (dB)	71.56	74.79	78.34	80,25

Table 1 presents the different PSNR values depending on the watermark length embedded in the 3D video. The high rates of PSNR show that the proposed scheme guarantees an efficient imperceptibility of the embedded watermarks.

Robustness to Signal Processing Attacks. To evaluate the robustness of the proposed scheme against collusion attacks, we calculate the Normalized Correlations (NC) criterion an the Bit Correction Rate (BCR) between the original and the recovered watermarks with the same length L. Higher NC and higher BCR indicate stronger watermarking rebustness.

$$NC_{2d} = \frac{\sum W_{2d}(i)W'_{2d}(i)}{\sqrt{\sum W_{2d}(i)^2}\sqrt{\sum W'_{2d}(i)^2}}, \tag{4}$$

$$BCR(W, W') = 1 - \frac{1}{2}\sum_{k=1}^{n}\left|W_k - W'_k\right|, \tag{5}$$

where $1 \leq i \leq L$ and $1 \leq k \leq L$.

5.2   The Tracing Results

In this section, we present the tracing results against different collusion attacks. In order to prove the performance of the proposed system in terms of tracing rates, we compare the number of recovered colluders in three cases: using only 2D videos in a first case, only the depth maps in the second case and both 2D

Table 2. NC AND BCR VALUES.

	NC		BCR	
	2D frames	Depth map	2D frames	Depth map
Resize 1/4	0.6	0.48	0.75	0.75
Resize 1/25	0.63	0.45	0.75	0.75
Gaussien 0.05	0.96	0.87	0.98	0.93
Salt 0.01	0.98	0.99	0.99	0.99
Salt 0.05	0.96	0.97	0.98	0.98
Average 9	0.6	0.47	0.75	0.75
Average 15	0.6	0.47	0.74	0.75
Brightness +30%	0.5	0.5	0.74	0.75
Rotation3°	0.56	0.47	0.75	0.76
Translate [5, 10]	0.6	0.48	0.75	0.74

video frames and theirs depth maps in a third case. Then, we take respectively the number of users to $n = 50$, the collusion size to $c = 6$, the false positive probability to $\varepsilon_1 = 10^{-6}$ and the fingerprinting length to m. Table 3, present very good detection rates of the colluders respectively with $m = 9817$, $m = 7000$, $m = 5000$ and $m = 4096$ Table 2.

Table 3. Majority, average, All-one and All-zero attack.

Majority Attack					
Length m=		9817	7000	5000	4096
2D Video	Nb accused users	10	6	8	8
	Nb pirates	6	6	6	6
Depth	Nb accused users	7	19	6	9
	Nb pirates	6	6	6	6
Our approach	Nb accused users	6	7	6	6
	Nb pirates	6	6	6	6

Average Attack					
Length m=		9817	7000	5000	4096
2D Video	Nb accused users	6	6	13	11
	Nb pirates	6	6	6	6
Depth	Nb accused users	11	13	6	6
	Nb pirates	6	6	6	6
Our approach	Nb accused users	6	6	6	6
	Nb pirates	6	6	6	6

All-one Attack					
Length m=		9817	7000	5000	4096
2D Video	Nb accused users	11	15	24	10
	Nb pirates	2	3	4	3
Depth	Nb accused users	9	2	6	8
	Nb pirates	5	2	2	4
Our approach	Nb accused users	2	2	5	2
	Nb pirates	2	2	2	2

All-zero Attack					
Length m=		9817	7000	5000	4096
2D Video	Nb accused users	3	15	6	10
	Nb pirates	2	1	2	4
Depth	Nb accused users	15	7	1	12
	Nb pirates	4	1	1	3
Our approach	Nb accused users	3	2	1	2
	Nb pirates	2	1	1	2

The experimental results prove the good and the accuracy tracing performance of our system against Majority vote, Average, All-one and All-zero attacks. With Majority vote and Average attacks, we can detect all the colluders but with All-one and All-zero attacks, we detect only nearly 50% of the colluders. As seen in Table 3 the proposed technique allows to reduce efficiently the number of false accused users.

6   Conclusions and Future Work

3D videos which are shared on the Internet or P2P networks are susceptible to be manipulated illegally. Henceforth, it becomes interesting for researchers to provide suitable measures to protect this type of videos and to try to detect any infringement actor. In this context, some previous works focus on proposing watermarking techniques that embed fingerprints in the 2D video or in the depth map components, and has proven limitations for the watermarking efficiency. In our work, we propose a tracing system which is based firstly on a LSB-based watermarking technique that embeds the identifier of each user in his appropriate copy, then we used this code to achieve a Tardos-based process to retrieve back eventual colluders responsible of the suspicious copies. Regarding the experimental assessments, the proposed scheme has provided very good watermarking and tracing results for different types of collusion attacks.

As a future work, we propose to test this tracing system for 3D video games that are subject to countless hacking attempts.

References

1. de Rosnay, M.D.: Digital rights management systems and European law: between copyright protection and access control. In: Second International Conference on Web Delivering of Music, 2002. WEDELMUSIC 2002. Proceedings, pp. 117–124 (2002)
2. Thilagavathi, N., Saravanan, D., Kumarakrishnan, S., Punniakodi, S., Amudhavel, J., Prabu, U.: A survey of reversible watermarking techniques, application and attacks. In: Proceedings of the 2015 International Conference on Advanced Research in Computer Science Engineering & Technology (ICARCSET 2015), ICARCSET 2015. Association for Computing Machinery, New York (2015) https://doi.org/10.1145/2743065.2743102
3. Lee, M., Lee, J., Lee, H.: Perceptual watermarking for 3d stereoscopic video using depth information. In: 2011 Seventh International Conference on Intelligent Information Hiding and Multimedia Signal Processing, pp. 81–84 (2011)
4. Kim, H., Lee, J., Oh, T., Lee, H.: Robust DT-CWT watermarking for DIBR 3D images. IEEE Trans. Broadcast. 58(4), 533–543 (2012)
5. Sheng-Li, F., Mei, Y., Gang-Yi, J., Feng, S., Zong-Ju, P., Sheng-li, F.: A digital watermarking algorithm based on region of interest for 3D image. In: Eighth International Conference on Computational Intelligence and Security 2012, pp. 549–552 (2012)
6. Lin, Y., Wu, J.: Unseen visible watermarking for color plus depth map 3D images. In: 2012 IEEE International Conference on Acoustics, Speech and Signal Processing (ICASSP), pp. 1801–1804 (2012)
7. Liu, X., Zhao, R., Li, F., Liao, S., Ding, Y., Zou, B.: Novel robust zero-watermarking scheme for digital rights management of 3D videos. Signal Process. Image Commun. 54, 140–151 (2017). http://www.sciencedirect.com/science/article/pii/S0923596517300371
8. Pei, S.-C., Wang, Y.-Y.: A new 3D unseen visible watermarking and its applications to multimedia. In: 2014 IEEE 3rd Global Conference on Consumer Electronics, GCCE 2014, pp. 140–143, February 2015

9. Gao, G., Jiang, G.: Bessel-fourier moment-based robust image zero-watermarking. Multimedia Tools Appl. **74**(3), 841–858 (2015). https://doi.org/10.1007/s11042-013-1701-8

10. Abdelhedi, K., Chaabane, F., Ben Amar, C.: A SVM-based zero-watermarking technique for 3D videos traitor tracing. In: Blanc-Talon, J., Delmas, P., Philips, W., Popescu, D., Scheunders, P. (eds.) ACIVS 2020. LNCS, vol. 12002, pp. 373–383. Springer, Cham (2020). https://doi.org/10.1007/978-3-030-40605-9_32

11. He, S., Wu, M.: Collusion-resistant video fingerprinting for large user group. IEEE Trans. Inf. Forensics Secur. **2**(4), 697–709 (2007)

12. Tardos, G.: Optimal probabilistic fingerprint codes. J. ACM **55**(2) (2008). https://doi.org/10.1145/1346330.1346335

13. Peikert, C., Shelat, A., Smith, A.: Lower bounds for collusion-secure fingerprinting. In: SODA 2003, pp. 472–479, January 2003

14. Chaabane, F., Charfeddine, M., Puech, W., Ben Amar, C.: Towards a blind map-based traitor tracing scheme for hierarchical fingerprints. Neural Inf. Process. **11**, 505–512 (2015)

15. Chaabane, F., Charfeddine, M., Puech, W., Ben Amaf, C.: A QR-code based audio watermarking technique for tracing traitors. In: 2015 23rd European Signal Processing Conference (EUSIPCO), pp. 51–55 (2015)

16. Craver, S., Memon, N., Yeo, B., Yeung, M.M.: Resolving rightful ownerships with invisible watermarking techniques: limitations, attacks, and implications. IEEE J. Sel. Areas Commun. **16**(4), 573–586 (1998)

17. Desoubeaux, M., Guelvouit, G.L., Puech, W.: Fast detection of Tardos codes withboneh-shaw types. In: SPIE 8303, Media Watermarking, Security, and Forensics (2012). http://www.rmit3dv.com

18. El'arbi, M., Amar, C.B., Nicolas, H.: Video watermarking based on neural networks. In: IEEE International Conference on Multimedia and Expo 2006, pp. 1577–1580 (2006)

19. Fernandez, M., Soriano, M., Cotrina, J.: Tracing illegal redistribution using errorsanderasures and side information decoding algorithms. Inf. Secur. IET **1**, 83–90 (2007)

20. Furon, T., Pérez-Freire, L.: Worst case attacks against binary probabilistic traitor tracing codes. In: First IEEE International Workshop on Information Forensics and Security (WIFS) 2009, pp. 56–60 (2009)

21. Chaabane, F., Charfeddine, M., Ben Amar, C.: An enhanced hierarchical traitor tracing scheme based on clustering algorithms, pp. 379–390, June 2017

22. Hayashi, N., Kuribayashi, M., Morii, M.: Collusion-resistant fingerprinting scheme based on the CDMA-technique. In: Miyaji, A., Kikuchi, H., Rannenberg, K. (eds.) IWSEC 2007. LNCS, vol. 4752, pp. 28–43. Springer, Heidelberg (2007). https://doi.org/10.1007/978-3-540-75651-4_3

23. Chaabane, F., Charfeddine, M., Puech, W., Amar, C.B.: A two-stage traitor tracing scheme for hierarchical fingerprints. Multimedia Tools Appl. **76**(12), 14 405–14435 (2017). https://doi.org/10.1007/s11042-016-3749-8

24. Wagner, N.R.: Fingerprinting. In: Proceedings of the 1983 IEEE Symposium on Security and Privacy, Oakland, California, USA, 25–27 April 1983. IEEE Computer Society, 1983, pp. 18–22. https://doi.org/10.1109/SP.1983.10018

25. Chaabane, F., Charfeddine, M., Ben Amar, C.: A survey on digital tracing traitors schemes. In: 2013 9th International Conference on Information Assurance and Security (IAS), pp. 85–90 (2013)

26. Sharma, V., Shrivastava, V.: A steganography algorithm for hiding image in image by improved LSB substitution by minimize detection. J. Theor. Appl. Inf. Technol. **36**, 1–8 (2012)
27. Cheng, E., Burton, P., Burton, J., Joseski, A., Burnett, I.: RMIT3DV: pre-announcement of a creative commons uncompressed HD 3D video database. In: 2012 Fourth International Workshop on Quality of Multimedia Experience, pp. 212–217 (2012)
28. Scharstein, D., Szeliski, R., Zabih, R.: A taxonomy and evaluation of dense two-frame stereo correspondence algorithms. In: Proceedings IEEE Workshop on Stereo and Multi-Baseline Vision (SMBV 2001), pp. 131–140 (2001)

A Combinatorial Coordinate System for the Vertices in the Octagonal $C_4C_8(R)$ Grid

Lidija Čomić[(✉)]

Faculty of Technical Sciences, University of Novi Sad, Novi Sad, Serbia
comic@uns.ac.rs

Abstract. The octagonal C_4C_8 grid is a tessellation of the plane into regular octagons and squares. It is one of the eight semiregular grids, which have been receiving an increasing amount of research attention as a viable alternative to the traditional square grid.

We present an integer-valued combinatorial coordinate system for the vertices in the $C_4C_8(R)$ grid. We review the existing coordinate systems for this grid proposed in the literature and we provide formulas for the conversion between this coordinate system and the existing ones, including the Cartesian coordinates. Adjacency relation between the vertices can be easily obtained from their coordinates through simple integer arithmetics.

Keywords: Discrete geometry · Combinatorial image analysis · Nontraditional grids · Octagonal $C_4C_8(R)$ grid · 4.8.8 semiregular grid · Truncated square (8, 8, 4) grid · Khalimsky grid · Combinatorial coordinate system

1 Introduction

Non-traditional 2D and 3D grids have been recognized in many application domains, ranging from topological image analysis [9–11,18,19] to discrete geometry [5,12,26], as a viable alternative to the traditional square and cubic grids. A convenient coordinate system for these grids eases addressing and navigating the grid elements.

The octagonal C_4C_8 grid is a tessellation of the plane into squares and regular octagons, usually considered in one of two orientations. The rhomboidal $C_4C_8(R)$ grid (called also the truncated square (8, 8, 4) grid [13]), which we consider here, has squares (rhombi) with sides parallel to the lines $y = \pm x$ (see Fig. 1). Its dual, where polygons correspond to vertices and vice versa, is the Khalimsky grid [20]. The square $C_4C_8(S)$ grid is equal to $C_4C_8(R)$ rotated by $\pi/4$, i.e., it has the squares with axes-parallel sides. The C_4C_8 grid is one of the eight semiregular grids in the plane (the 4.8.8 grid), composed of regular polygons, such that the circular order of polygons is the same around each vertex [6,7]. The C_4C_8 grid is one of the central and most commonly studied families of chemical

© Springer Nature Switzerland AG 2021
N. Tsapatsoulis et al. (Eds.): CAIP 2021, LNCS 13052, pp. 69–78, 2021.
https://doi.org/10.1007/978-3-030-89128-2_7

graphs, and many coordinate systems have been proposed both for $C_4C_8(S)$ [1,2,4,8,14,16,17,28] and for $C_4C_8(R)$ [3,15,23–25,27] grids.

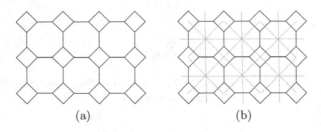

(a)   (b)

Fig. 1. (a) A rectangular part of the $C_4C_8(R)$ grid (the 4.8.8 semiregular grid) and (b) overlay with the dual Khalimsky grid (in dashed green). (Color figure online)

We present an integer-valued combinatorial coordinate system for the vertices in the $C_4C_8(R)$ grid, in which each vertex is uniquely and unambiguously represented through two integer coordinates. This coordinate system has recently been used to addresses the pixels (which are triangles) in the dual Khalimsky (tetrakis square) grid [24]. We show how the adjacency relation between the vertices can be obtained from their coordinate values. We review the relevant literature and we provide a conversion between these combinatorial coordinates for the vertices and their coordinates in the other known coordinate systems [3,15,25,27], as well as their Cartesian coordinates. Unlike this, most of the existing coordinate systems use more than two coordinates [15,23,25,27], or they use two coordinates which are not integer-valued [27] or have the axes which are asymmetric with respect to the grid [3]. This coordinate system enables easy computation of the graph distance between the grid vertices [24], defined as the length of the shortest paths connecting them. Distance is a basic ingredient of many graph descriptors, widely used for interconnection, social, chemical and other types of graphs. Our future aim will be to extend the distance formula [24] to tubes and tori (rectangular parts of the grid with identified pairs of parallel sides), and to use it for the computation of distance-based graph indices.

Throughout the paper, sgn(a), $a \in \mathbb{R}$, denotes the sign function (equal to 0 if $a = 0$, equal to 1 or -1 if $a >$ or $a < 0$, respectively), $[a]$, $a \in \mathbb{R}$, denotes the integer part of a, $(m)_{/n}$, $m \in \mathbb{Z}$, $n \in \mathbb{N}$, denotes the remainder of m modulo n and $(k, l)_{/n} = ((k)_{/n}, (l)_{/n})$, $k, l \in \mathbb{Z}$, $n \in \mathbb{N}$.

2   Related Work

The intense research of the $C_4C_8(R)$ grid resulted in several coordinate systems (indexing or labeling schemes) for its vertices, using two [3,27], three [15,23,25] or four [27] coordinates, which may not all be integer-valued.

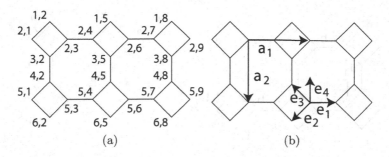

Fig. 2. (a) The 2-valued labelling by Ashrafi and Loghman [3]. (b) Four linearly dependent vectors e_1, e_2, e_3, e_4 and two linearly independent orthogonal vectors a_1, a_2 proposed by Taghizadeh et al. [27].

2.1   The 2-Valued Labelling by Ashrafi and Loghman [3]

The vertices lying on the same Euclidean horizontal line have the same coordinate $u \in \mathbb{Z}$, those on the same vertical line have the same $v \in \mathbb{Z}$ (see Fig. 2 (a)). The origin $(u, v) = (0, 0)$ is asymmetric with respect to the grid, and is located inside one of the octagons. Intuitively, the labels u and v are the indexes of the rows and columns containing the vertices in the $C_4C_8(R)$ grid.

2.2   The 4-Valued Coordinate System by Taghizadeh et al. [27]

Each vertex in the $C_4C_8(R)$ grid can be expressed through four linearly dependent vectors $e_1 = (1, 0)$, $e_2 = (-\sqrt{2}/2, -\sqrt{2}/2)$, $e_3 = (-\sqrt{2}/2, \sqrt{2}/2)$ and $e_4 = (0, 1)$ (see Fig. 2 (b)) as

$$(\alpha, \beta, \gamma, \delta) = \alpha e_0 + \beta e_1 + \gamma e_2 + \delta e_3, \text{ where } \alpha + \beta + \delta, \ \alpha + \gamma - \delta \in \{0, 1\}.$$

The graph distance between any two vertices $P_1(\alpha_1, \beta_1, \gamma_1, \delta_1)$ and $P_2(\alpha_2, \beta_2, \gamma_2, \delta_2)$ in the grid is equal to their d_1 (taxicab) distance

$$d(P_1, P_2) = |\alpha_2 - \alpha_1| + |\beta_2 - \beta_1| + |\gamma_2 - \gamma_1| + |\delta_2 - \delta_1|$$

(i.e., to the sum of the absolute values of the differences of their coordinates).

2.3   The 2-Valued Coordinate System by Taghizadeh et al. [27]

An alternative 2-valued coordinate system expresses each vertex in the $C_4C_8(R)$ grid through two orthogonal linearly independent vectors $a_1 = e_1 - e_2 - e_3 = (1 + \sqrt{2}, 0)$ and $a_2 = e_2 - e_3 - e_4 = (0, -1 - \sqrt{2})$ (see Fig. 2 (b)). Only one quarter of grid vertices have integer coordinates.

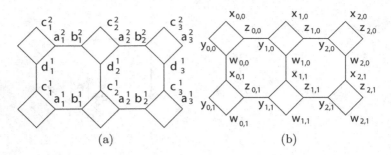

Fig. 3. The 3-valued coordinate system by (a) Siddiqui et al. [25] and Naeem et al. [23] and (b) by Heydari and Taeri [15].

2.4 The 3-Valued Coordinate System by Siddiqui et al. [25] and Naeem et al. [23]

Each vertex has a label of the form l_i^j, $l \in \{a, b, c, d\}$, $i, j \in \mathbb{Z}$. For the right, left, upper and lower vertex of each rhombus, l is equal to a, b, c and d, respectively. Two endpoints of each horizontal and each vertical line have the same index i. For each rhombus, its left vertex has the index i smaller by 1 than the remaining three vertices, which all have the same index i. For each octahedron, the two rightmost vertices have the index i greater by 1 than the remaining six vertices, which all have the same i (see Fig. 3 (a)). The index j behaves similarly.

2.5 The 3-Valued Coordinate System by Heydari and Taeri [15]

Each vertex has a 3-valued labelling $l_{i,j}$, $l \in \{x, y, z, w\}$, $i, j \in \mathbb{Z}$. The label l is equal to x, y, z and w for the upper, left, right and lower vertex of each rhombus, and i and j are the indices of the row and column containing the rhombus, respectively (see Fig. 3 (b)).

The graph distance between two vertices is obtained from

$$
d(a_{k,t}, b_{r,0}) = \begin{cases} 3(k-t) + t + \alpha, & 0 \le t < k - r < \left\lceil \frac{p+1}{2} \right\rceil \text{ or } a_{k,t} \in \{y_{t,t}, w_{t,t}\} \\ 3t + (k-t) + \alpha, & k - r < t \le p < \left\lceil \frac{p+1}{2} \right\rceil \text{ or } a_{k,t} \in \{z_{t,t}, x_{t,t}\} \end{cases},
$$

where $0 \le r \le k$ and $\alpha = d(a_{r,0}, b_{r,0})$.

3   The Combinatorial Coordinate System

We present the combinatorial coordinate system, we describe the connection between combinatorial and Cartesian coordinates for each vertex and we explain how the three neighbors of each vertex can be retrieved from its coordinates.

3.1   Definition

In the combinatorial coordinate system, each vertex has two integer coordinates (p, q). They are obtained from its Cartesian coordinates in the deformed $C_4C_8(R)$ grid, in which each oblique edge has been scaled by the factor $\sqrt{2}$, i.e., each oblique edge has length $\sqrt{2}$ (instead of 1), and its projections on coordinate axes have length 1 (instead of $\sqrt{2}/2$). The length of the horizontal and vertical edges remains 1 (see Fig. 4 (a)). All vertices on the same (Euclidean) vertical line have the same first coordinate p, and those on the same horizontal line have the same second coordinate q (see Fig. 4 (b)). The origin $(0,0)$ is at the center of one rhombus.

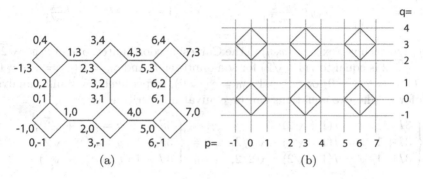

Fig. 4. (a) The combinatorial coordinates for the vertices in the $C_4C_8(R)$ grid. (b) Vertical lines $p = i$ and horizontal lines $q = j$, $i, j \in \mathbb{Z}$.

The left and right vertex of each rhombus have the first coordinate p equal to 2 and 1 (mod 3), respectively, and their second coordinate q is equal to 0 (mod 3). The upper and lower vertex have the q coordinate equal to 1 and 2 (mod 3), respectively, and their p coordinate is equal to 0 (mod 3). For each vertex, one of the two coordinates is equal to 0 (mod 3), and the other is equal to 1 or 2 (mod 3). Thus, the possible coordinate values (mod 3) are $(0,1)$, $(0,2)$, $(1,0)$, $(2,0)$.

3.2   Connection with the Cartesian Coordinates

Proposition 1. *The Cartesian coordinates (x, y) of a vertex T with combinatorial coordinates $T = (p, q)$ are given by*

$$
x = \frac{\sqrt{2}}{2}\left(p - \left[\frac{p+1}{3}\right]\right) + \left[\frac{p+1}{3}\right], \quad y = \frac{\sqrt{2}}{2}\left(q - \left[\frac{q+1}{3}\right]\right) + \left[\frac{q+1}{3}\right].
$$

Proof. Each interval $[k, k + 1]$, $k \in \mathbb{Z}$ on the p axis is covered by orthogonal projections of either horizontal or oblique edges, depending on k: for $(k)_{/3} = 1$, the interval is covered by projections of horizontal edges, otherwise (for $(k)_{/3} = 0$

or 2) of oblique ones (see Fig. 4 (b)). The number of intervals in $[0, p]$, $p \geq 0$, covered by horizontal edges is equal to $[(p + 1)/3]$. For $p \leq 0$, this number is $-[(p + 1)/3]$. The remaining intervals are covered by oblique edges. Thus,

$$x = \frac{\sqrt{2}}{2}\left(p - \left[\frac{p+1}{3}\right]\right) + \left[\frac{p+1}{3}\right].$$

A similar reasoning applies to q and y.

Proposition 2. *The combinatorial coordinates (p, q) of a vertex T with Cartesian coordinates $T = (x, y)$ are given by*

$$p = \sum_{k \in \{-1,0,1\}} \gamma_k(x)\left(\frac{3x+k}{1+\sqrt{2}} + k\right), \quad \gamma_k(x) = (1 - \mathrm{sgn}(\frac{x+k}{1+\sqrt{2}} - [\frac{x+k}{1+\sqrt{2}}]))$$

$$q = \sum_{k \in \{-1,0,1\}} \gamma_k(y)\left(\frac{3y+k}{1+\sqrt{2}} + k\right), \quad \gamma_k(y) = (1 - \mathrm{sgn}(\frac{y+k}{1+\sqrt{2}} - [\frac{y+k}{1+\sqrt{2}}]))$$

Proof. For each vertex $T = (x, y)$ in the Cartesian coordinates, one of x, $x - \sqrt{2}/2$ or $x + \sqrt{2}/2$ is equal to $I(1 + \sqrt{2})$ for some integer I, and $\gamma_0 = 1$ ($\gamma_1 = \gamma_{-1} = 0$), $\gamma_1 = 1$ ($\gamma_0 = \gamma_{-1} = 0$) or $\gamma_{-1} = 1$ ($\gamma_0 = \gamma_1 = 0$), respectively. A similar analysis holds for y. The two equations can be equivalently written as

$$p = \begin{cases} 3I & x = I(1 + \sqrt{2}), \\ 3I + 1 & x = I(1 + \sqrt{2}) + \sqrt{2}/2, \\ 3I - 1 & x = I(1 + \sqrt{2}) - \sqrt{2}/2, \end{cases} \quad q = \begin{cases} 3J & y = J(1 + \sqrt{2}), \\ 3J + 1 & y = J(1 + \sqrt{2}) + \sqrt{2}/2, \\ 3J - 1 & y = J(1 + \sqrt{2}) - \sqrt{2}/2, \end{cases}$$

thus proving the claim.

3.3 Neighbors

Proposition 3. *The neighbors N_1, N_2, N_3 of a vertex $T = (p, q)$ are*

$$\{N_1, N_2, N_3\} = \begin{cases} \{(p+1, q), (p-1, q-1), (p-1, q+1)\} & (p, q)/3 = (1, 0) \\ \{(p-1, q), (p+1, q+1), (p+1, q-1)\} & (p, q)/3 = (2, 0) \\ \{(p, q+1), (p+1, q-1), (p-1, q-1)\} & (p, q)/3 = (0, 1) \\ \{(p, q-1), (p-1, q+1), (p+1, q+1)\} & (p, q)/3 = (0, 2) \end{cases}$$

Proof. Each vertex $T = (p, q)$ has two diagonal and one horizontal or vertical neighbor. If $(p)/3 = 1$, the horizontal neighbor is to the right of T, and the diagonal neighbors are to the left (they have the smaller first coordinate than T). If $(p)/3 = 2$, the horizontal neighbor is to the left, and the diagonal neighbors are to the right. If $(q)/3 = 1$, the vertical neighbor is above T, and the diagonal ones are below. If $(q)/3 = 2$, the vertical neighbor is below and the diagonal ones are above. This analysis proves the claim.

4   Conversion to/from Existing Coordinate Systems

We provide conversion formulas between the presented and existing coordinate systems, and we illustrate them in Fig. 5 with a point T (with $p = 3$, $q = 5$) shown with respect to these systems.

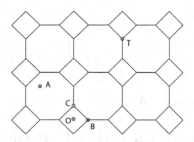

Fig. 5. The origins O (green) of the presented coordinate system, A (red) of the coordinate system by Ashrafi and Loghman [3], B (blue) of the system by Taghizadeh et al. [27], C (orange) of the systems by Siddiqui et al. [25] and Naeem et al. [23] and Heydari and Taeri [15] and the point T (violet). (Color figure online)

4.1 The 2-Valued Labelling by Ashrafi and Loghman [3]

If we take the origin A of this coordinate system to be at the point $(p, q) = (-2, 2)$, then

$$
\begin{aligned}
u &= -q + 2 & p &= v - 2 \\
v &= p + 2 & q &= -u + 2
\end{aligned}
$$

For example, for the point T with $p = 3$, $q = 5$, we have that $u = -3$, $v = 5$.

4.2 The 4-Valued Coordinate System by Taghizadeh et al. [27]

Taking the origin B of the 4-coordinate system to be at the point $(p, q) = (1, 0)$, we have that

$$
p = \alpha - \beta - \gamma + 1, \quad q = -\beta + \gamma + \delta
$$

and

$$
\alpha = \left\lceil \frac{p+1}{3} \right\rceil \quad \beta = \frac{1}{2}\left(\left\lceil \frac{p+1}{3} \right\rceil + \left\lceil \frac{q+1}{3} \right\rceil - p - q + 1 \right)
$$

$$
\gamma = \frac{1}{2}\left(\left\lceil \frac{p+1}{3} \right\rceil - \left\lceil \frac{q+1}{3} \right\rceil - p + q + 1 \right) \quad \delta = \left\lceil \frac{q+1}{3} \right\rceil.
$$

For the point T with $p = 3$, $q = 5$, we have that $\alpha = 1$, $\beta = -2$, $\gamma = 1$, $\delta = 2$.

4.3 The 3-Valued Coordinate System by Siddiqui et al. [25] and Naeem et al. [23]

If we take the point B with $(p, q) = (1, 0)$ to be at the point $l_i^j = a_1^1$, then

$$
l = \begin{cases} a, & (p)/3 = 1 \\ b, & (p)/3 = 2 \\ c, & (q)/3 = 1 \\ d, & (q)/3 = 2 \end{cases}, \quad i = \left\lceil \frac{p}{3} \right\rceil + 1, \quad j = \left\lceil \frac{q}{3} \right\rceil + 1.
$$

and

$$p = \begin{cases} 3i - 2, & l = a \\ 3i - 1, & l = b \\ 3i - 3, & l = c, d \end{cases} \qquad q = \begin{cases} 3j - 3, & l = a, b \\ 3j - 2, & l = c \\ 3j - 1, & l = d. \end{cases}$$

For the point T with $p = 3$, $q = 5$, we have that $l = d$, $i = 2$, $j = 2$.

4.4   The 3-Valued Coordinate System by Heydari and Taeri [15]

If we take the point C with $(p, q) = (0, 1)$ to be at the point $z_{0,0}$, then

$$p = \begin{cases} 3i, & l = x, w \\ 3i - 1, & l = y \\ 3i + 1, & l = z \end{cases} \qquad q = \begin{cases} -3j, & l = y, z \\ -3j + 1, & l = x \\ -3j - 1, & l = w \end{cases}$$

and

$$l = \begin{cases} x, & (q)/3 = 1 \\ y, & (p)/3 = 2 \\ z, & (p)/3 = 1 \\ w, & (q)/3 = 2 \end{cases} \qquad i = \left[\frac{p+1}{3}\right] \qquad j = -\left[\frac{q+1}{3}\right].$$

For $p = 3$, $q = 5$, we have that $l = w$, $i = 1$, $j = -2$.

5   Discussion

We presented a combinatorial coordinate system for the vertices in the $C_4C_8(R)$ grid. This work parallels or recent work on the $C_4C_8(S)$ grid [8]. An equivalent coordinate system (with the q-axis pointing down instead of up) has been recently proposed for the pixels (triangles) in the dual Khalimsky grid [24] and used for computing distance directly from coordinate values. The advantage of this coordinate system is that it is intuitively pleasing, it uses only two coordinates instead of three [27] or four [15], and it is symmetric with respect to the grid. Our future aim will be to extend this coordinate system and the distance formula [24] to $C_4C_8(R)$ tubes and tori.

Computing distances is of great interest in mathematical chemistry, as many numerical graph descriptors, reflecting physical and chemical properties of the molecule modeled by the graph, are distance-based. Such indices have also found applications in other types of networks, like interconnection or social networks. Distance is ubiquitous in discrete geometry as well. The 2-valued Cartesian coordinate system for the vertices in the dual of the $C_4C_8(R)$ grid (for the Khalimsky grid), i.e., for the pixels (octagons and squares) in the $C_4C_8(R)$ grid, has been used in the study of weighted distances and digital discs in this grid [22]. Recently, a 6-values coordinate system and distance formula have been developed for another non-traditional grid, the Cairo pattern [21], which is the dual of the 3.3.4.3.4 semiregular grid. We plan to define combinatorial coordinate systems and derive distance formulas also on other types of grids.

Acknowledgments. This work has been partially supported by the Ministry of Education, Science and Technological Development of the Republic of Serbia through the project no. 451-03-68/2020-14/200156.

References

1. Arezoomand, M.: Energy and Laplacian spectrum of $C_4C_8(S)$ nanotori and nanotube. Dig. J. Nanomater. Nanostruct. **4**(6), 899–905 (2009)
2. Arezoomand, M.: Random walks on infinite $C_4C_8(S)$ net, nanotube and nanotori. MATCH Commun. Math. Comput. Chem. **65**, 231–240 (2011)
3. Ashrafi, A.R., Loghman, A.: Computing Padmakar-Ivan index of a $TC_4C_8(R)$ nanotorus. J. Comput. Theor. Nanosci. **5**, 1–4 (2008)
4. Ashrafi, A.R., Rezaei, F., Loghman, A.: PI index of the $C_4C_8(S)$-nanotorus. Revue Roumaine de Chimie **54**(10), 823–826 (2009)
5. Biswas, R., Largeteau-Skapin, G., Zrour, R., Andres, E.: Digital objects in rhombic dodecahedron grid. Math. Morphol. Theory Appl. **4**(1), 143–158 (2020)
6. Borgefors, G.: A semiregular image grid. J. Vis. Commun. Image Represent. **1**(2), 127–136 (1990)
7. Brlek, S., Labelle, G., Lacasse, A.: Properties of the contour path of discrete sets. Int. J. Found. Comput. Sci. **17**(3), 543–556 (2006)
8. Čomić, L.: A combinatorial coordinate system for the vertices in the octagonal $C_4C_8(S)$ grid. In: Petković, T., Petrinović, D., Lončarić, S. (eds.) 12th International Symposium on Image and Signal Processing and Analysis, ISPA, pp. 235–240. IEEE (2021). https://doi.org/10.1109/ISPA52656.2021.9552147
9. Čomić, L., Magillo, P.: Repairing 3D binary images using the BCC grid with a 4-valued combinatorial coordinate system. Inf. Sci. **499**, 47–61 (2019)
10. Čomić, L., Magillo, P.: Repairing 3D binary images using the FCC grid. J. Math. Imaging Vis. **61**(9), 1301–1321 (2019). https://doi.org/10.1007/s10851-019-00904-0
11. Čomić, L., Magillo, P.: Repairing binary images through the 2D diamond grid. In: Lukić, T., Barneva, R.P., Brimkov, V.E., Čomić, L., Sladoje, N. (eds.) IWCIA 2020. LNCS, vol. 12148, pp. 183–198. Springer, Cham (2020). https://doi.org/10.1007/978-3-030-51002-2_13
12. Čomić, L., Zrour, R., Largeteau-Skapin, G., Biswas, R., Andres, E.: Body centered cubic grid - coordinate system and discrete analytical plane definition. In: Lindblad, J., Malmberg, F., Sladoje, N. (eds.) DGMM 2021. LNCS, vol. 12708, pp. 152–163. Springer, Cham (2021). https://doi.org/10.1007/978-3-030-76657-3_10
13. Conway, J.H., Sloane, N.J.A.: Sphere Packings, Lattices and Groups. Grundlehren der Mathematischen Wissenschaften, vol. 290. Springer New York (1999). https://doi.org/10.1007/978-1-4757-6568-7
14. Deng, H.: Wiener index on tori $T_{p,q}[C_4C_8]$ covered by C_4 and C_8. MATCH Commun. Math. Comput. Chem. **56**, 357–374 (2006)
15. Heydari, A., Taeri, B.: Szeged index of $TUC_4C_8(R)$ nanotubes. MATCH Commun. Math. Comput. Chem. **57**, 463–477 (2007)
16. Heydari, A., Taeri, B.: Wiener and Schultz indices of $TUC_4C_8(S)$ nanotubes. MATCH Commun. Math. Comput. Chem. **57**, 665–676 (2007)
17. Heydari, A., Taeri, B.: Szeged index of $TUC_4C_8(S)$ nanotubes. Eur. J. Comb. **30**(5), 1134–1141 (2009)

18. Kardos, P.: Characterizations of simple points on the body-centered cubic grid. In: Lukić, T., Barneva, R.P., Brimkov, V.E., Čomić, L., Sladoje, N. (eds.) IWCIA 2020. LNCS, vol. 12148, pp. 62–72. Springer, Cham (2020). https://doi.org/10. 1007/978-3-030-51002-2_5

19. Kardos, P., Palágyi, K.: On topology preservation of mixed operators in triangular, square, and hexagonal grids. Discrete Appl. Math. **216**, 441–448 (2017)

20. Khalimsky, E.D., Kopperman, R., Meyer, P.R.: Computer graphics and connected topologies on finite ordered sets. Topol. Appl. **36** (1990)

21. Kovács, G., Nagy, B., Turgay, N.D.: Distance on the Cairo pattern. Pattern Recognit. Lett. **145**, 141–146 (2021)

22. Kovács, G., Nagy, B., Vizvári, B.: Weighted distances and digital disks on the Khalimsky grid - disks with holes and islands. J. Math. Imaging Vis. **59**(1), 2–22 (2017)

23. Naeem, M., Siddiqui, M.K., Guirao, J.L.G., Gao, W.: New and modified eccentric indices of octagonal grid O_n^m. Appl. Math. Nonlinear Sci. **3**(1), 209–228 (2018)

24. Saadat, M., Nagy, B.: Digital geometry on the dual of some semi-regular tessellations. In: Lindblad, J., Malmberg, F., Sladoje, N. (eds.) DGMM 2021. LNCS, vol. 12708, pp. 283–295. Springer, Cham (2021). https://doi.org/10.1007/978-3-030-76657-3_20

25. Siddiqui, M.K., Naeem, M., Rahman, N.A., Imran, M.: Computing topological indices of certain networks. J. Optoelectron. Adv. Mater. **18**(9–10), 884–892 (2016)

26. Strand, R., Nagy, B., Borgefors, G.: Digital distance functions on three-dimensional grids. Theor. Comput. Sci. **412**, 1350–1363 (2011)

27. Taghizadeh, M., Arezoomand, M., Gazor, H.: Computation of full symmetry group of $C_4C_8(R)$ nanotubes using a mathematical model. Optoelectron. Adv. Mater. Rapid Commun. **6**, 1046–1048 (2012)

28. Xu, S., Zhang, H.: Hosoya polynomials of $TUC_4C_8(S)$ nanotubes. J. Math. Chem. **45**, 488–502 (2009)

Bilingual Speech Recognition by Estimating Speaker Geometry from Video Data

Luis Sanchez Tapia[1]([✉]), Antonio Gomez[1], Mario Esparza[1], Venkatesh Jatla[1], Marios Pattichis[1], Sylvia Celedón-Pattichis[2], and Carlos LópezLeiva[2]

[1] Department of Electrical and Computer Engineering,
The University of New Mexico, Albuquerque, NM, USA
{luis2sancheztapia,agsuper,javesparza,venkatesh369,pattichi}@unm.edu
[2] Department of Language, Literacy, and Sociocultural Studies,
The University of New Mexico, Albuquerque, NM, USA
{sceledon,callopez}@unm.edu

Abstract. Speech recognition is very challenging in student learning environments that are characterized by significant cross-talk and background noise. To address this problem, we present a bilingual speech recognition system that uses an interactive video analysis system to estimate the 3D speaker geometry for realistic audio simulations. We demonstrate the use of our system in generating a complex audio dataset that contains significant cross-talk and background noise that approximate real-life classroom recordings. We then test our proposed system with real-life recordings.

In terms of the distance of the speakers from the microphone, our interactive video analysis system obtained a better average error rate of 10.83% compared to 33.12% for a baseline approach. Our proposed system gave an accuracy of 27.92% that is 1.5% better than Google Speech-to-text on the same dataset. In terms of 9 important keywords, our approach gave an average sensitivity of 38% compared to 24% for Google Speech-to-text, while both methods maintained high average specificity of 90% and 92%.

On average, sensitivity improved from 24% to 38% for our proposed approach. On the other hand, specificity remained high for both methods (90% to 92%).

Keywords: Speech recognition · Projection geometry · Bilingual · Video processing.

1 Introduction

Human activity recognition can strongly benefit from the combined use of audio and video data. More recently, audio processing has been used to identify visual

This material is based upon work supported by the National Science Foundation under Grant No.1613637, No.1842220, and No.1949230.

N. Tsapatsoulis et al. (Eds.): CAIP 2021, LNCS 13052, pp. 79–89, 2021.
https://doi.org/10.1007/978-3-030-89128-2_8

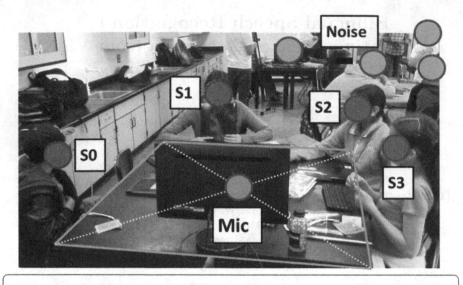

S1: - "Okay, so, binary numbers only go to zero to one."

S2: - "What? Zero to one?"

S1: - "Yeah, that's all the computer knows, zero and one."

S3: - "Three... where is three? Three, where is the zero?"

Fig. 1. Example setup of a typical AOLME group interaction. Blue dots mark the speaker position and the Yellow dot is assumed to be at the center of the table (marked by red). Cross-talk is expected among speakers S0 to S3, background noise is also captured by the microphone (green dots in the back). Under the picture, we depict a sample of a transcript from the current session. Keywords can be identified like "zero", "one", "computer" and "three" . (Color figure online)

events [4,9]. For our paper, we want to investigate the use of video data to reconstruct the speaker geometry in 3D and then use this information to develop a speaker recognition system. Our approach addresses the strong need to develop a speech recognition system that can help transcribe student conversations from video recordings of collaborative learning environments.

We present an example in Fig. 1. In this example, a small group of students is sitting around the table, using the keyboard to program the Raspberry Pi. The video has been recorded as part of the Advancing Out-of-School Learning in Mathematics and Engineering (AOLME) after-school program [3]. The speech recognition problem requires that we recognize what each of the students is saying as shown in the transcription of Fig. 1. More specifically, the speaker geometry requires that we identify the 3D locations of the speakers (S_0, S_1, S_2, S_3) with respect to the omnidirectional microphone placed on the center of the table.

In Fig. 1, we also see several other speakers talking in the background (refer to green dots). The students speak in both Spanish and English.

Student speech recognition in this environment is very challenging due to cross-talk, background noise, and the use of multiple languages. Current deep learning systems are hence ineffective in such environments. To address the issue, we use the estimated 3D speaker geometry and video audio transcriptions to generate a large, acoustic model based audio dataset that can be used to train a bilingual speech recognition system for this collaborative learning environment. As we demonstrate in this paper, although we train on synthetic datasets, we are still able to match and slightly exceed state-of-the-art systems.

The current paper significantly extends our previous research on analyzing such videos. More specifically, prior research has been focused on face and back of the head detection in [12–16], face recognition was also targeted in [18]. Furthermore, authors in [6] provided an early approach to context-based activity detection using deep learning. The research on video activity detection was significantly extended in [7]. The object detection system developed by [17] will be the baseline system for estimating 3D speaker geometry from the AOLME videos. For completeness, we will also explain the approach in [17] in our methodology.

The paper uses video object detection and projective geometry to locate the 3D speaker geometry from still video frames. The 3D speaker geometry is input to Pyroomacoustics [10] to simulate how the speakers will be recorded by the omnidirectional microphone located on the center of the table. We use the audio transcriptions with the AWS text-to-speech system to generate the ground truth audio datasets for training our speech recognition system. The proposed approach obtained a 27.92% recognition rate on Spanish words that was slightly better than Google Speech-to-text [1] at 26.12%. In addition, the Bilingual Keyword Classifier obtained an average of 38% sensitivity on Spanish Keywords.

The rest of the paper is organized as follows. We define the 3D speaker geometry problem in Sect. 2. We then describe the underlying methods in Sect. 3. Results are given in Sect. 4. We then provide concluding remarks in Sect. 5.

2   3D Speaker Geometry Estimation

We use projective geometry to estimate 3D coordinates from still image frames. Our basic assumption is to use cross-ratios along the projections of 3D lines to estimate 3D distances. We begin by assuming the basic concept and showing how to apply cross-ratios to define the problem for our videos.

We illustrate the concept of cross-ratios in Fig. 2 [2]. The basic assumption is that we know the actual physical distances between three consecutive, co-linear points A, B, C. In our example, let these distances be AB and BC. Then, to estimate the distance to another point D along the same line, we use the cross-ratio R defined by [2]:

$$R = \frac{AC}{CB} \bigg/ \frac{AD}{DB} = \frac{AC \cdot BD}{BC \cdot AD} = \frac{(AB + BC) \cdot (BC + CD)}{BC \cdot (AB + BC + CD)} \tag{1}$$

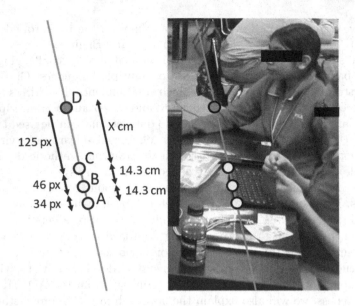

Fig. 2. Physical distance estimation using cross-ratios.

where CD is the physical distance to be estimated. To estimate CD from equation (1), we first estimate the ratio R using pixel ratios of AD/DB. Then, we substitute the value for R and solve for CD.

To estimate the 3D locations of the speakers using cross-ratios, we will first need to estimate distances along 3D planes where our colinear points lie. In our example of Fig. 2, we assume that we know the physical dimensions of the keyboard (given as distance AC). Then, we estimate the midpoint B of the keyboard. We then assume that the keyboard is parallel to the sides of the table (1 to 2 or 3 to 4), and estimate the distance CD to the edge of the table using cross-ratios. Unfortunately, we cannot use the side of the keyboard to estimate the width of table that is depicted as a near-horizontal line in Fig. 1. This is because the keyboard side, compared against the table width is too small, and estimation can be very inaccurate.

We define all of the points that are needed to estimate the 3D speaker geometry in Fig. 3. Here, we estimate all physical distances along with the table defined by points $1, 2, 3, 4$ using cross-ratios. The basic idea is to define a 2D grid on the the table that is defined through the intersection of lines parallel to the keyboard (points $5, 6, 7, 8$) and the computer monitor ($9, 10, 11, 12, 13$). Here, we assume monitor points $8, 9, 10$ lie on the table to eliminate the need to map these points to the table surface. These lines are also assumed to be parallel to the corresponding sides of the table.

Fig. 3. Speaker geometry estimation setup.

Since the table is not always fully visible, we also extend the estimated depth of the table (points 1 to 2) by 5% to account for mild occlusion. Here, we note that the size of the table is needed because we assume that the microphone is located in the center of the table.

Similarly, we define 3D planes associated with each speaker (assumed to be about 4 in. away from the table edge) and assume that the mouths and hands lie on the same 3D plane that is orthogonal to the table. In terms of object recognition, we require hand detection, head detection as depicted in Fig. 3(b).

We refer to [5] as a base for the assumptions at building the system of projections of parallel lines. We plan to test at the real scenario from AOLME videos (around 1000 h).

3   Methodology

We summarize our methodology in Fig. 4. Our 3D speaker geometry estimation requires detection of keyboard, hands and monitor. We based the detection on [17] with added post processing to detect necessary features to establish 3D geometry. We provide more details on our object detection methods in Sect. 3.1.

Through the use of an interactive system, the users select specific frames, select the table corners and corners from the detected keyboard and monitor. Then, out system uses cross-ratios to reconstruct the 3D speaker geometry as summarized in Sect. 2. As shown in the bottom branch of Fig. 4(a), the AOLME transcripts are pre-processed to serve as input to the speech synthesis module. We then use the reconstructed 3D geometry and the synthesized dialogues to provide an acoustic-based generation of the audio dataset. We input the 3D speaker and microphone geometry, and synthesized speech into our acoustic simulation framework based on Pyroomacoustics [10]. The result is the acoustics-based simulated dataset for training our bilingual speech recognition system.

The speech recognition system is shown in Fig. 4(b). The system is trained using the generated audio dataset. We provide more details of our speech recognition system in Sect. 3.2.

3.1   Object Detection

As shown in Fig. 3, we require detection of the keyboard and monitor in order to estimate the location of the speakers with respect to the table. Furthermore, to estimate the 3D locations of the speaker's mouths, we also assume that their hands and mouths are on the same 3D plane and further require hand and head detection. Here, we are only interested in hands that are located near the table as shown in Fig. 3.

We next summarize the methods that we will use to detect each object. For head detection, we use the latest version of YOLO [8] pre-trained on the crowd human data set for head detection [11]. To restrict head detection within the current student group, we use a minimum area threshold that successfully rejects smaller faces of people outside the group. For detecting hands, monitors, and keyboards, we use faster R-CNN pre-trained on the COCO dataset. The results of faster R-CNN are post-processed using clustering, time-projections (adding detections through time), and small area removal to remove distant hands (see [17] for details). Among the hand detections, we then manually select hands that lie on the table. Furthermore, we manually select the edges of the Table, the monitor, and the keyboard.

We assume that we can learn the scales, number of pixels per inch for each speaker using manual measurements during training. Later, we will look at estimating the scales for each image. Here, we note that our assumption is very restrictive. It does not account for strong scale variations when the speakers move to new positions not reflected in the training set.

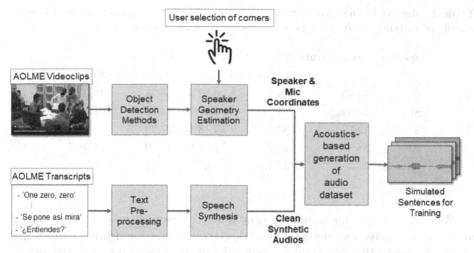

(a) Acoustics-based dataset generator based on 3D speaker geometry.

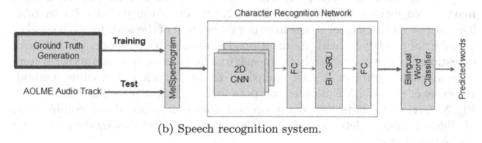

(b) Speech recognition system.

Fig. 4. Bilingual speech recognition system using 3D speaker geometry estimated from the video dataset.

3.2 Speech Recognition System

We summarize the speech recognition system in Fig. 4(b). The acoustic-based generated dataset is used to train a phoneme-based recognition network composed of a 2D CNN (a single layer of 8 filters of size 3×3 with stride=2) processing Mel-spectrograms, a two-layer bi-directional GRUs with 64 units per layer, and a fully connected layer with an output for each phoneme. The system generates a sequence of phonemes characters that are post-processed by a bilingual word classifier based on minimum distance.

4 Results

We first summarize results from 3D speaker geometry estimation using a baseline approach and our proposed methods. We then summarize our results for speech recognition system using the 3D speaker geometry.

Table 1. Results for 3D speaker geometry estimation. The error is given as a percentage of the distance to the microphone. All distances are given in inches.

Speaker	Ground truth	Our method		Baseline	
		Estimation	Error	Estimation	Error
S0	36.70	34.16	6.92%	19.74	46.21%
S1	35.59	41.27	15.96%	24.32	31.67%
S2	42.12	43.88	4.18%	27.79	34.02%
S3	34.99	29.29	16.29%	27.79	20.58%
Average	37.35	37.15	10.84%	24.91	33.12%

We define a baseline approach that does not require projective geometry or any object detection method. Assuming the keyboard and table corners are given, we assume that speakers sit around the table, equidistant from each other.

Our proposed approach performed significantly better. We present a summary of our estimates for Fig. 3 in Table 1. Our error ranges from 7% to 16%. The largest source of error comes from our estimation of the scale for each speaker (number of pixels per inch). As mentioned earlier, in future work, we will work on estimating the scale directly from each image. Overall, our interactive system gave a reduced error of 10.84% compared to 33.12% for the baseline method. In terms of the AOLME dataset, we present an example of object detection in Fig. 5. Overall we note that our proposed approach required the combination of different object detections from different video frames to establish the 3D speaker geometry.

Fig. 5. Object detection for 3D speaker geometry estimation. We use blue bounding boxes for head detection, orange bounding boxes for hand detection, and purple bounding boxes for keyboard detection.

Table 2. Keyword recognition results. Here, we note that our system does not recognize accents.

Keywords	Our system		Google Speech-to-Text	
	Sensitivity	Specificity	Sensitivity	Specificity
uno	0.50	0.95	0.13	1.00
dos	0.24	0.91	0.06	1.00
tres	0.63	0.92	0.00	1.00
cuatro	0.30	0.99	0.00	1.00
cinco	0.25	0.99	0.23	1.00
cero	0.36	0.93	0.00	1.00
computadora	0.25	0.99	0.25	1.00
numero	0.27	0.97	0.45	1.00
Others	0.65	0.67	1.00	0.13
Average	0.38	0.92	0.24	0.90

The output of 3D speaker geometry system is the complex simulated audio dataset, used to train the speech recognition system. The training dataset was generated using audio transcriptions of 720 min extracted from 54 video sessions, and a typical AOLME 3D speaker geometry. For testing, we selected 517 sentences from unseen video sessions. We then assessed the character error rate for recognizing the 517 sentences. For this test, our proposed approach gave an accuracy of 27.92% compared to 26.12% by Google speech-to-text.

We also present comparative results for the recognition of 9 Spanish keywords that were used in the number representations lessons. We summarize our results in terms of sensitivity and specificity as given in Table 2. From the results, it is clear that Google Speech-to-text fails to detect any instances of tres, cuatro, and cero. Overall, Google Speech-to-text is insensitive to the target keywords, it is prone to discard noisy samples as 'Others'. By comparison, our proposed method is much better at detecting our targeted keywords because it will try to classify even the noisy samples. On average, sensitivity improved from 24% to 38% for our proposed approach. On the other hand, specificity remained high for both methods (90% to 92%).

Our proposed approach produces more false positives and fewer false negatives than Google Speech-to-text. Hence, in terms of using our method, we note that the users would have to reject our false positive detections. On the other hand, Google Speech-to-text requires noise-free examples and fails to detect important AOLME type keywords (e.g., tres, cuatro, and cero).

5   Conclusions and Future Work

We presented an interactive system for estimating 3D speaker geometries from a single-camera video recording. We then used a typical 3D speaker geometry based on AOLME videos to generate a complex, acoustics-based, simulated dataset based on 11.66 h of audio dataset. Then, when tested on actual audio datasets, the proposed system slightly outperformed Google Speech-to-text. Ultimately, the detection of meaningful keywords can be used by educational researchers to identify moments of interest for further analysis.

For future work, we are currently developing multi-objective optimization methods for improving our sensitivity while maintaining high specificity.

References

1. Google cloud speech-to-text API. https://cloud.google.com/speech-to-text
2. Brannan, D.A., Esplen, M.F., Gray, J.J.: Geometry, 2nd edn. Cambridge University Press, Cambridge (2011). https://doi.org/10.1017/CBO9781139003001
3. Celedón-Pattichis, S., LópezLeiva, C.A., Pattichis, M.S., Llamocca, D.: An interdisciplinary collaboration between computer engineering and mathematics/bilingual education to develop a curriculum for underrepresented middle school students. Cultural Stud. Sci. Educ. **8**(4), 873–887 (2013). https://doi.org/10.1007/s11422-013-9516-5
4. Ephrat, A., et al.: Looking to listen at the cocktail party. ACM Trans. Graph. (2018)
5. Hartley, R.I., Zisserman, A.: Multiple View Geometry in Computer Vision. Cambridge University Press, Cambridge, 2nd edn (2004). ISBN: 0521540518
6. Jacoby, A.R., Pattichis, M.S., Celedón-Pattichis, S., LópezLeiva, C.: Context-sensitive human activity classification in collaborative learning environments. In: 2018 IEEE Southwest Symposium on Image Analysis and Interpretation (SSIAI), pp. 1–4, April 2018. https://doi.org/10.1109/SSIAI.2018.8470331
7. Jatla, V., LópezLeiva, C.: Long-term human video activity quantification of student participation. Asilomar Conference on Signals, Systems, and Computers, Invited (2021)
8. Jocher, G., et al.: ultralytics/yolov5: v5.0 - YOLOv5-P6 1280 models, AWS, Supervisely and YouTube integrations, April 2021. https://doi.org/10.5281/zenodo.4679653
9. Owens, A., Efros, A.A.: Audio-visual scene analysis with self-supervised multisensory features. CoRR (2018)
10. Scheibler, R., Bezzam, E., Dokmanic, I.: Pyroomacoustics: a python package for audio room simulations and array processing algorithms. CoRR abs/1710.04196 (2017). http://arxiv.org/abs/1710.04196
11. Shao, S., et al.: Crowdhuman: a benchmark for detecting human in a crowd. arXiv preprint arXiv:1805.00123 (2018)
12. Shi, W., P.M.C.P.S., LópezLeiva, C.: Person detection in collaborative group learning environments using multiple representations. Asilomar Conference on Signals, Systems, and Computers, Accepted (2021)
13. Shi, W., LópezLeiva, C.: Talking detection in collaborative learning environments. In: The 19th International Conference on Computer Analysis of Images and Patterns (CAIP), accepted (2021)
14. Shi, W., Pattichis, M.S., Celedón-Pattichis, S., LópezLeiva, C.: Dynamic group interactions in collaborative learning videos. In: 2018 52nd Asilomar Conference on Signals, Systems, and Computers, pp. 1528–1531, October 2018

15. Shi, W., Pattichis, M.S., Celedón-Pattichis, S., LópezLeiva, C.: Robust head detection in collaborative learning environments using AM-FM representations. In: 2018 IEEE Southwest Symposium on Image Analysis and Interpretation (SSIAI), pp. 1–4, April 2018. https://doi.org/10.1109/SSIAI.2018.8470355
16. Shi, W.: Human Attention Detection Using AM-FM Representations. Master's thesis, University of New Mexico (2016)
17. Teeparthi, S., LópezLeiva, C.: Fast hand detection in collaborative learning environments. In: The 19th International Conference on Computer Analysis of Images and Patterns (CAIP), accepted (2021)
18. Tran, P., LópezLeiva, C.: Facial recognition in collaborative learning videos. In: The 19th International Conference on Computer Analysis of Images and Patterns (CAIP), accepted (2021)

Cost-Efficient Color Correction Approach on Uncontrolled Lighting Conditions

Pedro H. Carvalho[1]([✉]), Inês Rocha[1,2], Fábio Azevedo[1,3],
Patrícia S. Peixoto[4]([iD]), Marcela A. Segundo[4]([iD]), and Hélder P. Oliveira[1,2]([iD])

[1] INESC TEC - Institute for Systems and Computer Engineering,
Technology and Science, Porto, Portugal
`pedro.h.carvalho@inesctec.pt`
[2] Faculty of Sciences, University of Porto, Porto, Portugal
[3] Faculty of Engineering, University of Porto, Porto, Portugal
[4] LAQV - REQUIMTE, Faculty of Pharmacy, University of Porto, Porto, Portugal

Abstract. The misuse and overuse of antibiotics lead to antibiotic resistance becoming a serious problem and a threat to world health. Bacteria developing resistance results in more dangerous infections and a more difficult treatment. To monitor the antibiotic pollution of environmental waters, different detection methods have been developed, however these are normally complex, costly and time-consuming. In a previous work, we developed a method based on digital colorimetry, using smartphone cameras to acquire sample images and color correction to ensure color constancy between images. A reference chart with 24 colors, with known ground truth values, is included in the photographs in order to color correct the images using least squares minimization. Then, the color of the sample is detected and correlated to antibiotic concentration. Although achieving promising results, the method was too sensitive to contrasting illumination conditions, with high standard deviations in these cases. Here, we test different methods for improving the stability and precision of the previous algorithm. By using only the 13 patches closest to the color of the targets and more parameters for the least squares minimization, better results were achieved, with an improvement of up to 83.33% relative to the baseline. By improving the color constancy, a more precise, less influenced by extreme conditions, estimation of sulfonamides is possible, using a practical and cost-efficient method.

Keywords: Digital colorimetry · Color correction · Image processing

1 Introduction

Bacteria with antibiotic resistance is an increasingly serious problem for world health. Antibiotics misuse is resulting in environmental waters being polluted

This work is financed by the ERDF - European Regional Development Fund through the Operational Programme for Competitiveness and Internationalisation - COMPETE 2020 Programme and by National Funds through the Portuguese funding agency, FCT - Fundação para a Ciência e Tecnologia within project POCI-01-0145-FEDER-031756.

N. Tsapatsoulis et al. (Eds.): CAIP 2021, LNCS 13052, pp. 90–99, 2021.
https://doi.org/10.1007/978-3-030-89128-2_9

with them, causing bacteria that are exposed to this pollution to develop resistance to antibiotics and spread antibiotic resistance genes [1]. Therefore, infections become much more dangerous, with treatments not being as effective due to an increase in treatment time, risk of the spread of the infection and, ultimately, risk of death[1]. This pollution needs to be monitored and various detection methods have been developed. High performance liquid chromatography (HPLC) tandem mass spectrometry or other detectors, electrophoresis with different detectors, immunoassays, and colorimetry are the most commonly used techniques [2]. These are costly and challenging to deploy on a large scale, making them difficult to use in monitoring programs. However, analytical chemistry based on computer vision is starting to be explored for this purpose. Using smartphones [3], which are readily available, these monitoring methods are cheaper and more practical.

Colorimetry is a technique for detecting a compound in a solution and determining its concentration by analyzing its color. A spectrophotometer (device for measuring absorbance of ultra violet or visible light in a given wavelength) is normally used to quantify the color of a sample. Recently, ways to digitally analyze the color of a sample from a photograph have been researched, which is called digital colorimetry. Although these methods are faster, they are less precise due to different cameras capturing photographs of the same target with different color values, and different illuminations also affecting these values. A color correction step is needed to minimize these differences, which is a challenging problem, due to approximations and assumptions color correction methods use, which can fail if certain conditions are not met [4].

Smartphone use in environmental applications has been proposed for pollution monitoring. For example, air pollution exposure, taking into account temporal and spatial variation of population in an area, was monitored using wireless and mobile devices. By detecting when and where people are present, patterns of activity were established utiizing connections to the cellular network. Thus, estimating population-weighted exposure to fine particulate matter is made possible. This was implemented in New York City, USA [5]. Furthermore, in the state of Oregon, USA, an application was proposed to calculate health risks at a smartphones location, considering levels of fine and coarse particulate matter and ozone concentrations [6].

Combining the use of smartphones and chemistry is an under exploited area for environmental monitoring, although some methods have been proposed. For example, an application for determining pH and nitrite concentration using a low-cost paper-based microfluidic device and a photograph was proposed. The application detects seven sensing areas with reagents that exhibit color when a sample is placed. The acquisition of the photograph is done under controlled illumination conditions and using the flash of the smartphone as lighting. Then, an image processing algorithm lowers the influence of the light source and the seven colored sensing areas are detected using a multi-detection algorithm. There is no mention of application on the field [3]. Another example is a portable

---

[1] https://www.who.int/drugresistance/documents/surveillancereport/en/.

system for detecting chromium (III) ion based on ELISA protocol and using a smartphone readout, although there was no field analysis [7].

It is, therefore, important to keep innovating and develop cheaper, faster and more practical methods for monitoring antibiotic pollution, that can be used on the field, as it is a growing and serious problem worldwide.

2 Previous Work

In a previous work [8], we proposed a new approach for antibiotic pollution monitoring, using digital colorimetry in conjunction with smartphones for an accessible and mobile application. The study focused on the estimation of sulfonamides (a family of antibiotics) concentration in contaminated water samples. The color of the sample was corrected using a reference target, so that there is color constancy between images with different illuminations and from different devices. To estimate the concentration of sulfonamides, a calibration curve is used, correlating concentrations to a color value. We found that the $a*$ color value from the *CIELab* color space and the hue (H) color value from the *HSV* color space provided the best correlation between color and sulfonamide concentrations [8].

For [8], a dataset with photographs of the *x-rite ColorChecker Passport* next to a sample of contaminated water was prepared in a laboratory setting, with the color chart used as the reference for the color correction and the target being the sample. Figure 1 is an example of a photograph from this dataset and Fig. 2 shows the colors the samples assume depending on the concentration of sulfonamides.

Fig. 1. Example photograph from dataset of [8].

Although we achieved good results, we still detected a high standard deviation between color corrected images, especially when more contrasting illumination conditions were present. Therefore, we wanted to explore other variations of our color correction method to minimize the influence of different illuminations, providing a more stable color correction even in extreme conditions. Besides,

Fig. 2. Examples of samples of varying sulfonamide concentrations, from 0 μg/L (left) to 150 μg/L (right).

we wanted to validate this approach by testing independently of the samples. Using water samples might contribute to more variance in the results, since it's preparation in a laboratory is more prone to human error. To do this, we used different patches from the *x-rite ColorChecker Passport* as the targets, instead of the samples.

3   Methodology

The objective of this work is to improve upon the color correction algorithm of [8]. Color correction is needed to ensure color is consistent between photographs under different illuminations or from different devices. In [8], color correction is done using the classic color chart of an *x-rite ColorChecker Passport*, which is a reference chart with 24 patches of different colors arranged in a 6 by 4 grid, with known ground truth values. With a least squares method, we minimize the difference between the detected color of the 24 patches and the ground truth. Other works have demonstrated different approaches such as:

- **Using only 13 of the 24 color chart patches:** Alsam and Finlayson showed that color correction with only 13 patches is comparable to using the 24 patches [9];
- **Using more parameters in the least squares minimization:** In [10], it is shown that more parameters in the least squares minimization for the color correction leads to better results.

Inspired by these two works, we hypothesize that choosing the 13 patches closest to the color of the target and using more parameters in the least squares minimization will improve upon our previous work.

3.1   Data

For this work, a database was built with 24 photographs (see Fig. 3 for examples) of the *x-rite ColorChecker Passport* in various illumination conditions:

- inside with different white lights,
- inside with different yellow lights,
- inside by the window,
- outside in the sun,
- outside in the shade.

These new images provide more contrasting conditions of illumination when compared to the first dataset that was captured in a laboratory setting, making the color correction more challenging, but more prepared for real world use. To further simulate practical use, the photographs were captured free hand, with the only restriction being encompassing the reference and target color patches from above, as close to 90° possible.

Fig. 3. Example photographs from the new dataset. Taken inside with a white light (left), inside with a yellow light (middle) and outside in the sun (right). (Color figure online)

In [8], the target of the color correction was the sample of contaminated water. Here, the targets are the 8 colored patches (red, orange, yellow, green, cyan, blue, violet and magenta) in the page opposite of the classic color chart that is used for the color correction, see Fig. 4.

Fig. 4. Example of the 8 targets of the color correction (top) and the 24 patches used as reference for the color correction (bottom). (Color figure online)

3.2 Color Correction

A color correction matrix M_{cc} is used to transform the original image I, resulting in a color corrected image I_{cc}, as shown in Eq. (1).

$$I_{cc} = M_{cc}I \tag{1}$$

M_{cc} is found by solving the following minimization problem:

$$M_{cc} = \arg\min_{T}\|C_{XYZ} - M_{cc}C_{RGB}\|^2 \tag{2}$$

M_{cc} is calculated using a least-squares method to find the difference between the measured RGB values of the color chart patches, C_{RGB}, in each image and the corresponding ground truth XYZ values, C_{XYZ}, see Eq. (3). Transforming the image using M_{cc} will result in a color corrected image in the XYZ color space, which is a device independent color space.

$$M_{cc} = (C_{RGB}^T C_{RGB})^{-1} C_{RGB}^T C_{XYZ} \tag{3}$$

For a 3×3 color correction matrix and using all color chart patches, C_{RGB} and C_{XYZ} are both the same size, 24×3 matrices. Each line represents a patch RGB or XYZ value. In [8], all 24 patches were used to find M_{cc}, and for this work, we compare it with using only 13 patches, making C_{RGB} and C_{XYZ} 13×3 matrices.

13 Patches. To choose the 13 patches, the RGB values of each patch are assumed as a 3-dimensional coordinate and the euclidean distance between each of the 24 patches and the RGB value of each of the 8 target colors is calculated, and the 13 closest patches are chosen. Eq. (4) shows how each distance D is calculated, with P_R, P_G and P_B being the RGB values of each patch and T_R, T_G and T_B the RGB values of the targets. Figure 5 shows the 13 patches chosen for each target.

$$D = \sqrt{(P_R - T_R)^2 + (P_G - T_G)^2 + (P_B - T_B)^2} \tag{4}$$

Polynomial Extensions. More complex least squares minimization can be done by adding polynomial terms to C_{RGB}, making it a $24 \times N$ or $13 \times N$ matrix, with N depending on the number of parameters added. In this work, the more commonly used $\{R, G, B\}$ ($P1$) is tested along with the polynomial extension ($P2$) and the root-polynomial extension ($P3$). The terms for these extensions are shown in Eq. (6) and (7), respectively.

$$P1 = \{R, G, B\} \tag{5}$$

$$P2 = \{R, G, B, R^2, G^2, B^2, RG, GB, RB\} \tag{6}$$

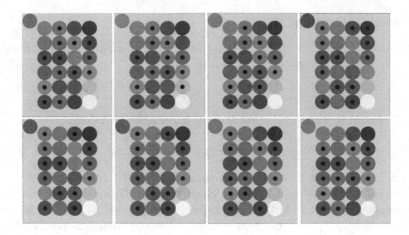

Fig. 5. Selected patches for each of the 8 targets. Each target indicated on the top left, and the black dots indicating the 13 chosen patches from the 24 color chart (Color figure online)

$$P3 = \{R, G, B, \sqrt{RG}, \sqrt{GB}, \sqrt{RB}\} \tag{7}$$

Other polynomial terms were tested, however the results were consistently worse than these three and, therefore, discarded from this study.

The baseline method employing all 24 patches of the color chart and with $C_{RGB} = \{R, G, B\}$ is compared with the use of only the 13 closest patches to the target and the polynomial and root-polynomial extensions.

Metrics. The metric used to compare the results is the standard deviation (Std) of the target values between images. Lower values meaning that the colors of the targets in different color corrected images are closer to each other.

4 Results

In [8], we found that the color components that can best differentiate sulfonamides concentrations are the a^* component from the *CIELab* color space and the hue (H) from the *HSV* color space, therefore the results are focused on these two values.

Tables 1 and 2 showcase the Std of the target values when using 24 or 13 patches and different polynomial terms, P1 being $\{R, G, B\}$, P2 the polynomial extension $\{R, G, B, R^2, G^2, B^2, RG, GB, RB\}$ and P3 the root-polynomial extension $\{R, G, B, \sqrt{RG}, \sqrt{GB}, \sqrt{RB}\}$.

Table 1 shows that, for a^*, 13 patches usually provide better results, with only the orange, yellow and violet patches having the best result with 24 patches. The polynomial extension $P2$ with 13 patches has the lowest average, with a 2.56% improvement over the baseline ($P1$ with 24 patches). The Std for this method shows a 32.01% improvement relative to the baseline. Table 2 shows that for H, the results are similar, with 13 patches providing lower standard deviations, with the exception of the Red target. On average the best case for H is the simplest, $P1$ with 13 patches. Compared to the baseline, the average improves by 18.58%, while the Std improves by 59.85%. These results prove that the new methods are more stable, as the Std is significantly lower than in the baseline.

Table 1. Standard deviations of the a* color value for each of the target patches

a^*	Patches	Red	Orange	Yellow	Green	Cyan	Blue	Violet	Magenta	Average (Std)
P1	24	2.632	**0.613**	**0.915**	0.830	1.100	1.205	0.825	1.242	1.170(0.628)
	13	2.554	1.035	0.987	**0.558**	**0.675**	**0.947**	1.329	1.721	1.226(0.648)
P2	24	1.477	1.309	1.275	1.196	1.417	2.046	**0.651**	0.919	1.286(**0.410**)
	13	1.111	0.803	1.382	1.606	0.884	1.848	0.696	**0.791**	**1.140**(0.427)
P3	24	3.552	1.135	1.031	0.826	0.916	1.271	1.007	1.640	1.422(0.896)
	13	**0.905**	1.614	1.046	1.751	0.785	1.912	0.936	0.863	1.227(0.454)

Table 2. Standard deviations of the H color value for each of the target patches

H	Patches	Red	Orange	Yellow	Green	Cyan	Blue	Violet	Magenta	Average (Std)
P1	24	0.635	0.947	0.552	1.628	0.439	0.758	1.259	0.759	0.872(0.396)
	13	0.880	0.796	0.547	**0.716**	0.436	**0.640**	0.788	0.880	**0.710(0.159)**
P2	24	**0.550**	0.530	0.562	0.869	0.741	1.233	0.708	0.818	0.751(0.232)
	13	0.627	**0.427**	0.651	1.812	**0.435**	0.900	**0.544**	0.534	0.741(0.458)
P3	24	1.265	0.639	0.548	1.021	0.507	0.934	1.518	0.800	0.904(0.356)
	13	0.561	0.569	**0.513**	1.532	0.436	1.105	0.611	**0.494**	0.728(0.386)

With these results, it is clear that using the 13 patches closest to the target's RGB values produces a more precise color correction, with less variation between the targets in different photographs. However, the best polynomial extension varies for each color target, making it a case by case selection.

In our case, the real world samples can vary in color from yellow to a dark magenta, almost red, in the highest sulfonamides concentrations (Fig. 2). We decided to study the results with only the closest targets to our samples. Using the euclidean distance shown in Eq. (4), we calculated the distance of the 8 targets to random samples (with concentrations 0 µg/L, 20 µg/L, 50 µg/L and 100 µg/L) and found the closest targets to our real world application to be the red, orange, yellow, violet and magenta patches. Tables 3 and 4 show the results focused on these 5 targets.

Table 3. Standard deviations of the a* color value for the red, yellow, violet and magenta target patches

a*	Patches	Red	Orange	Yellow	Violet	Magenta	Average (Std)
P1	24	2.632	**0.613**	**0.915**	0.825	1.242	1.245(0.807)
	13	2.554	1.035	0.987	1.329	1.721	1.525(0.645)
P2	24	1.477	1.309	1.275	**0.651**	0.919	1.126(0.335)
	13	1.111	0.803	1.382	0.696	**0.791**	**0.956(0.284)**
P3	24	3.552	1.135	1.031	1.007	1.640	1.673(1.081)
	13	**0.905**	1.614	1.046	0.936	0.863	1.073(0.310)

Table 4. Standard deviations of the H color value for the red, yellow, violet and magenta target patches

H	Patches	Red	Orange	Yellow	Violet	Magenta	Average (Std)
P1	24	0.635	0.947	0.552	1.259	0.759	0.830(0.282)
	13	0.880	0.796	0.547	0.788	0.880	0.778(0.136)
P2	24	**0.550**	0.530	0.562	0.708	0.818	0.634(0.125)
	13	0.627	**0.427**	0.651	**0.544**	0.534	0.557(0.088)
P3	24	1.265	0.639	0.548	1.518	0.800	0.954(0.419)
	13	0.561	0.569	**0.513**	0.611	**0.494**	**0.550(0.047)**

For the a^* color value, the polynomial extension $P2$ with 13 patches gives the lower average, while, for the H color value, the root-polynomial extension $P3$ with 13 patches is the lowest. In this case, the improvement for a^* is 23.21% on average and 64.81% on the Std. As for H, we can see a decrease of 33.73% on average and 83.33% for the Std. These improvements are proof that these methods are more stable, meaning less influenced by contrasting illumination conditions.

Thus, for our application, the best setup for color correction depends on which color value provides the best correlation between color and sulfonamides concentration.

5 Conclusions

Promising results were achieved in this work, with a lower standard deviation between different color corrected images compared to previous work. Confirming that, with fewer but more relevant patches, the color correction has more resolution in the targets color range. This improvement means that smaller differences in sulfonamides concentration can be detected, making digital colorimetry a more reliable method for monitoring environmental water pollution. As for the polynomial expansions, the results are not as straight forward, with each choice ($P1$, $P2$ and $P3$) being the best for different targets. This makes it a case by case decision on which to use, depending on the real world application.

In the future, tests with contaminated water will be conducted to confirm that this new method can better differentiate sulfonamides concentrations.

References

1. Huerta, B., et al.: Exploring the links between antibiotic occurrence, antibiotic resistance, and bacterial communities in water supply reservoirs. Sci. Total Environ. **456**, 161–170 (2013)
2. Dmitrienko, S.G., Kochuk, E.V., Apyari, V.V., Tolmacheva, V.V., Zolotov, Y.A.: Recent advances in sample preparation techniques and methods of sulfonamides detection - a review. Anal. Chim. Acta **850**, 6–25 (2014)
3. Lopez-Ruiz, N., et al.: Smartphone-based simultaneous pH and nitrite colorimetric determination for paper microfluidic devices. Anal. Chem. **86**(19), 9554–9562 (2014)
4. Kaur, H., Sharma, S.: A comparative review of various illumination estimation based color constancy techniques. In: 2016 International Conference on Communication and Signal Processing (ICCSP), pp. 0486–0490. IEEE (2016)
5. Nyhan, M., et al.: The impact of mobile-device-based mobility patterns on quantifying population exposure to air pollution. Environ. Sci. Technol. **50**(17), 9671–9681 (2016)
6. Larkin, A., Williams, D.E., Kile, M.L., Baird, W.M.: Developing a smartphone software package for predicting atmospheric pollutant concentrations at mobile locations. Comput. J. **58**(6), 1431–1442 (2014)
7. Yu, S., et al.: A portable chromium ion detection system based on a smartphone readout device. Anal. Methods 8(38), 6877–6882 (2016)
8. Carvalho, P.H., Bessa, S., Silva, A.R.M., Peixoto, P.S., Segundo, M.A., Oliveira, H.P.: Estimation of sulfonamides concentration in water based on digital colourimetry. In: Morales, A., Fierrez, J., Sánchez, J.S., Ribeiro, B. (eds.) IbPRIA 2019. LNCS, vol. 11867, pp. 355–366. Springer, Cham (2019). https://doi.org/10.1007/978-3-030-31332-6_31
9. Alsam, A., Finlayson, G.: Integer programming for optimal reduction of calibration targets. Color. Res. Appl. **33**(3), 212–220 (2008)
10. Finlayson, G.D., Mackiewicz, M., Hurlbert, A.: Color correction using root-polynomial regression. IEEE Trans. Image Process. **24**(5), 1460–1470 (2015)

HPA-Net: Hierarchical and Parallel Aggregation Network for Context Learning in Stereo Matching

Wei Chen, Jun Peng, Ziyu Zhu, and Yong Zhao[✉]

School of Electronic and Computer Engineering, Shenzhen Graduate School of Peking
University, Shenzhen, China
{llshenwei,jun_peng,vaccyzhu}@pku.edu.cn, yongzhao@pkusz.edu.cn

Abstract. Accurate disparity estimation with regard to rectified stereo image pairs is essential for many computer vision tasks. Current deep learning-based stereo networks generally construct single-scale cost volume to regularize and regress the disparity. However, these methods do not take advantage of multi-scale context information, leading to the limited performance of disparity prediction in ill-posed regions. In this paper, we propose a novel stereo network named HPA-Net, which provides an efficient representation of context information and lower error rates in ill-posed regions. First, we propose a hierarchical aggregation module to fuse context information from multi-scale cost volumes into an integrated cost volume. Then, we apply the integrated cost volume to the proposed parallel aggregation module, which utilizes several 3D dilated convolutions simultaneously to capture global and local clues of context information for disparity regressions. Experimental results show that our proposed HPA-Net achieves state-of-the-art stereo matching performances on KITTI datasets.

Keywords: Stereo macthing · Context information · Aggregation

1 Introduction

Stereo matching has been extensively studied because it is one of the core technologies in computer vision. Stereo matching is indispensable for many applications such as autonomous driving, robotic navigation, and 3D reconstruction. Given two rectified images, the task of stereo matching is to compute the disparity of each pixel. Recently, many end-to-end neural networks have been developed for stereo matching and have achieved impressive accuracy on several datasets.

According to the seminal work [1], a traditional stereo matching algorithm typically consists of four steps: matching cost computation, cost aggregation, disparity regression, and disparity refinement. Matching cost computation leads to the cost volume construction. Cost volume plays a significant role in convolution neural networks based stereo matching methods. Cost volume contains numerous context information of the left and right images. Multi-scale context

© Springer Nature Switzerland AG 2021
N. Tsapatsoulis et al. (Eds.): CAIP 2021, LNCS 13052, pp. 100–109, 2021.
https://doi.org/10.1007/978-3-030-89128-2_10

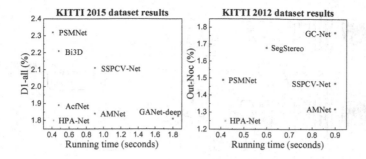

Fig. 1. KITTI datasets results comparison. HPA-Net outperforms published methods on KITTI 2015 (Left) and KITTI 2012 dataset (Right) in terms of **error rates** (D1-all in KITTI 2015 and Out-Noc in KITTI 2012) and running time.

information has been proven to be essential for stereo matching tasks, especially in some ill-posed regions. Leveraging useful context information from multi-scale architecture is a challenging problem. To deal with the problem, many different methods have been proposed. DispNet [2] utilizes dot products to compute a correlation volume from the left and right image features. GCNet [3] concatenates the left and shifted right features, resulting in a 4D concatenation volume. Gwc-Net [4] splits the features into groups and computes the correlation maps group by group and then all the correlation maps are packed into a 4D matching cost volume. However, these methods mentioned above only consist of a single-scale cost volume and lose much context information especially in ill-posed regions. Moreover, underutilization of context information may result in wrong predictions in ill-posed regions or losing global consistency. Thus, how to construct and process cost volume properly for the improvement of stereo matching is a challenging task.

In this paper, we propose a novel hierarchical and parallel aggregation network (HPA-Net) to effectively exploit and integrate multi-scale context information in stereo matching. We introduce a hierarchical method, extracting the cross-scale features from multi-scale cost volumes, to improve the network's ability to understand multi-level contexts. Moreover, we design a parallel aggregation module with dilated convolutions for cost volume filtering, which improves the utilization of global context information. Therefore, the local and global clues can propagate and integrate effectively in HPA-Net, making a highly reliable prediction.

In summary, our main contributions are threefold:

- We propose a novel stereo matching network named HPA-Net. The proposed hierarchical aggregation (HA) module exploits multi-scale cost volumes to propagate context across different scales and obtains an efficient representation of context information.
- We propose a 3D parallel aggregation (PA) module in cost filtering to learn global and local context information for effective disparity regression.

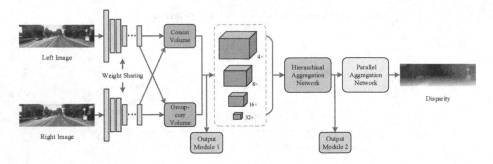

Fig. 2. The pipeline of the proposed HPA-Net. The whole network consists of four parts: feature extraction, hierarchical aggregation (HA) network, parallel aggregation (PA) network and disparity regression.

- Extensive experiments show that our model significantly improves the state-of-the-art performance of stereo matching on the KITTI datasets (as shown in Fig. 1).

2   Related Work

2.1   Deep Neural Networks for Stereo Matching

In recent years, deep neural networks have seen great success in stereo matching tasks [5,6]. Zbontar and LeCun [7] introduced a Siamese network to compare two image patches. Pang *et al.* [8] proposed a cascaded CNN architecture by first obtaining an initial disparity map, and then employing residual learning for refinement purposes. Mayer *et al.* [2] demonstrated an end-to-end disparity estimation neural network with a correlation layer for stereo cost volume construction. Kendall *et al.* [3] introduced a method of filtering 4D costs using a 3D cost filtering approach and using the soft argmax process to regress depth. Chang *et al.* [9] improved upon [3] by utilizing a SPP module to consider global context information by aggregating context at different scales and locations to form a cost volume, and it stacked multiple hourglass networks with 3D convolutions to regress the disparity. Guo *et al.* [4] adopted two complementary cost volumes, a concatenation volume and a group-wise correlation volume, to provide efficient representations for measuring feature similarities. One network related to our work is [10], which utilized pyramid cost volumes for disparity estimation.

2.2   Multi-scale Information

Multi-scale information has been used in many learning-based computer vision applications. Zhao *et al.* [11] embed the multi-scale features of scenes to improve semantic segmentation. Sun *et al.* [12] used multi-scale features to compute optical flow with a single branch. Gu *et al.* [13] cascaded cost volumes in a

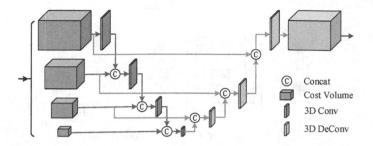

Fig. 3. Architecture of the Hierarchical Aggregation (HA) Network. Multi-scale cost volumes are aggregated hierarchically by 3D convolutions, 3D deconvolutions and concatenations.

coarse-to-fine manner for multi-view stereos and stereo matching. Yang *et al.* [14] demonstrated that building a cost volume pyramid instead of constructing a cost volume at a fixed resolution leads to better reconstruction results. [15] was presented as a robust, unified descriptor network with a large context region and receptive field. We share a similar idea with the abovementioned methods in that we exploit multi-scale cost volumes and aggregate context information by using a compact and efficient network.

3   Proposed Method

3.1   Network Architecture

The architecture of our proposed HPA-Net is shown in Fig. 2. Our network can be mainly divided into four modules: feature extraction, Hierarchical Aggregation (HA) Network, Parallel Aggregation (PA) Network and disparity regression.

The feature extraction process starts with a shared-weight Siamese network that takes a pair of images as inputs. Following the ResNet-like [16] network proposed in GwcNet [4], we first use three convolution layers with 3×3 kernels and 4 residual blocks with 2 dilated blocks. Two convolution layers with a stride of 2 are included to obtain a 1/4 scale feature map. Additionally, three other downsampling blocks with strides of 2 are successively employed to obtain lower resolution feature maps with 1/8, 1/16, and 1/32 scales. To construct the concatenated cost volume, we use two more convolution layers to regularize the four scaled feature maps. Specifically, each of our convolutions is followed by a batch normalization layer and a ReLU activation layer except for the last convolution. The rest modules are described in later sections.

3.2   Hierarchical Aggregation (HA) Network

Recent works [4] and [9] have shown that using a group-wise correlation volume combined with a concatenation volume can improve the network performance. Given the concatenation volume V_{concat} and group-wise correlation volume V_{gwc}, the combination volume is computed as:

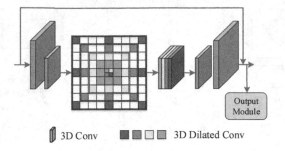

Fig. 4. Architecture of the Parallel Aggregation (PA) module. Four parallel convolutions with different dilation rates are exploited to aggregate context information.

$$V_{i,combine} = <V_{i,concat}, V_{i,gwc}> \tag{1}$$

where $<,>$ is the concatenation operator, i denotes the i-th scale, and $i = 1$ corresponds to the largest scale (1/4 of the original image size). The multi-scale cost volumes have sizes of $C_i \times \alpha(D \times H \times W)$ with $\alpha = 1/4, 1/8, 1/16$, and $1/32$ respectively. The channels C_i from the highest to the lowest scale are 32, 64, 128, and 128. A short summary of the HA module is as follows:

$$V_i' = \begin{cases} \downarrow< V_i, V_{i-1}' >, i = 2,3,4 \\ \uparrow< V_{i+1}', V_i >, i = 1,2,3 \end{cases} \tag{2}$$

where $\downarrow$ and $\uparrow$ represent the downsampling and upsampling convolution layers, respectively. V' represents the downsampled or upsampled cost volume.

As shown in Fig. 3, the 1/4 combination volume (V_1) is first downsampled to 1/8 scale (V_1') by a 3D convolution with a stride of 2. The 1/8 combination volume (V_1') is concatenated with the original 1/8 volume (V_2) to form a new 1/8 scale volume (V_2'). Then a $1 \times 1 \times 1$ convolution is employed to halve the channels of the new volume to the corresponding channels of that scale. During the downsampling procedure, the four volumes are hierarchically aggregated until the lowest scale (1/32) is obtained. Vice versa is a breeze. Cooperating with the other three larger-scale volumes and the concatenation operation, the new lowest-scale volume (1/32) is hierarchically aggregated to the highest scale (1/4). 3D deconvolutions with strides of 2 are employed to upsample the corresponding cost volume.

After the hierarchical aggregation module, multi-scale cost volumes with four different scales are aggregated into a 1/4 size volume for disparity regression in a later step.

3.3   Parallel Aggregation (PA) Network

We propose a Parallel Aggregation Network to aggregate the cost volume from the former network. The Parallel Aggregation Network consists of **three** cascaded parallel aggregation (PA) modules. Details about the PA module are shown in Fig. 4.

Table 1. Ablation study for network architecture settings on Scene Flow dataset. The effectiveness of HA and PA modules are evaluated.

Model	Scene Flow (EPE)	3px (%)
base	0.79015	3.368
base + HA	0.75391	3.257
base + PA	0.75399	3.238
base + HA + PA	**0.74262**	**3.131**

To learn additional context information, we first use a 3D convolution with a stride of 2 followed by another 3D convolution to reduce the feature size to 1/8. Then, four parallel dilated convolutions with increasing dilation rates, output four feature maps of the same size. After the concatenation, the four feature maps are combined together and then fed into two 3D convolutions that later one is a deconvolution with a stride of 2. The final 1/4 size feature map is processed by an output module to predict the disparity map.

3.4 Output Module and Loss Function

For each output module, we use two stacked 3D convolutions and an upsampling operator to generate a 1-channel 4D volume. Then the 1-channle 4D volume is converted into a probability volume with softmax along the disparity dimension. For each pixel, the regressed disparity $\tilde{\mathbf{d}}$ is defined as a weighted softmax function:

$$\tilde{\mathbf{d}} = \sum_{d=0}^{D_{\max}} d \times \sigma\left(-c_d\right) \tag{3}$$

where c_d is the predicted cost and $D_{\max}$ is the maximum dispatiry. The predicted disparity maps from five output modules are denoted as $\tilde{\mathbf{d}}_1, \tilde{\mathbf{d}}_2, \tilde{\mathbf{d}}_3, \tilde{\mathbf{d}}_4, \tilde{\mathbf{d}}_5$, and the loss function is defined as:

$$L = \sum_{j=1}^{j=5} \lambda_j \cdot \text{Smooth}_{L_1}\left(\tilde{\mathbf{d}}_j - \mathbf{d}^*\right) \tag{4}$$

in which

$$\text{Smooth}_{L_1}(x) = \begin{cases} 0.5x^2, & \text{if } |x| < 1 \\ |x| - 0.5, & \text{otherwise} \end{cases} \tag{5}$$

where the λ_j denotes the coefficient of the j-th disparity map and $\mathbf{d}^*$ represents the ground truth disparity map.

4 Experimental Results

4.1 Datasets and Evaluation Metrics

The proposed method is trained and tested on several benchmark datasets, including Scene Flow, KITTI 2012, and KITTI 2015.

Table 2. Ablation study for the dilation rates in PA module.

Model	$d1$	$d2$	$d3$	$d4$	Scene Flow (EPE)
base	1	1	1	1	0.75391
HPA-Net*	1	4	8	–	0.74825
HPA-Net	1	2	3	4	0.76211
HPA-Net	1	2	4	8	0.74579
HPA-Net	1	4	8	16	0.74305
HPA-Net	1	4	16	32	**0.74262**

Scene Flow is a large synthetic dataset that consists of dense ground truth disparity maps for 35454 training and 4370 testing stereo pairs of size H×W=540×960. We use the 'finalpass' version of the Scene Flow datasets.

KITTI 2015 and KITTI 2012 are two real-world datasets with street views captured from a driving car. KITTI 2015 provides 200 training and 200 testing image pairs. KITTI 2012 provides 194 training and 195 testing image pairs. All image sizes are H×W=376×1240. All the training image pairs are provided with sparse ground-truth disparity maps. For both datasets, we use 180 training image pairs as a training set and the rest as a validation set.

The performance of our method is measured using two standard metrics:(1) EPE: End-Point-Error, the average difference between the predicted disparity and the true disparity. (2) 3px: The percentage of pixels for which the estimation error is 3 pixels larger than the ground-truth disparity.

4.2   Implementation Details

The proposed HPA-Net is implemented using the PyTorch framework and trained using the Adam optimizer ($\beta_1 = 0.9$, $\beta_2 = 0.999$). We train the networks with 2 Nvidia GTX 2080Ti GPUs, and the batch size is fixed to 4. The coefficients of the five outputs are set as $\lambda_1 = 0.5$, $\lambda_2 = 0.5$, $\lambda_3 = 0.5$, $\lambda_4 = 0.7$, and $\lambda_5 = 1.0$. For the Scene Flow dataset, we train our network from scratch over 16 epochs. The initial learning rate is 0.001 and is downscaled by 2 after epochs 10, 12, and 14. Then, we fine-tune our pretrained model on KITTI 2015 and KITTI 2012 for another 300 epochs. The learning rate begins at 0.001 and is downscaled by 10 after 200 epochs. During training, we randomly crop the images to size $H \times W = 256 \times 512$. The maximum disparity ($D_{\max}$) is set to 192. Furthermore, we extend the training process to 20 epochs for Scene Flow, and we use another 400 epochs for KITTI to obtain the best test results for the KITTI submission.

4.3   Ablation Studies

In this section, we conduct several ablation studies on the Scene Flow dataset to evaluate the key components in the HPA-Net.

Table 3. KITTI 2015 test set results. "All" denotes all regions and "Noc" denotes non occluded regions. D1-bg, D1-fg, and D1-all denote that the pixels in the background, foreground, and all areas, respectively.

Methods	All (%)			Noc (%)			Runtime (s)
	D1-bg	D1-fg	D1-all	D1-bg	D1-fg	D1-all	
GC-Net [3]	2.21	6.16	2.87	2.02	5.58	2.61	0.9
PSMNet [9]	1.86	4.62	2.32	1.71	4.31	2.14	0.41
GwcNet-g [4]	1.74	3.93	2.11	1.61	3.49	1.92	**0.32**
SSPCVNet [10]	1.75	3.89	2.11	1.61	3.4	1.91	0.9
CSN [13]	1.59	4.03	2.00	1.43	3.55	1.78	0.6
AcfNet [17]	1.51	3.80	1.89	**1.36**	3.49	1.72	0.48
HPA-Net (Ours)	**1.50**	**3.31**	**1.80**	1.37	**2.99**	**1.63**	0.42

Ablation Study for the HA and PA Modules. To evaluate HPA-Net, we conduct experiments with several settings. We first propose a base model as our baseline. In the "base" model, the hierarchical aggregation (HA) module is removed, and the dilation rates in the parallel aggregation (PA) module are all set as 1. It needs to be emphasized that we use four PA modules in "base + PA" to achieve parameters that are comparable to those in other settings. As listed in Table 1, the models with either the HA or PA module work better than the base model and reduce the end-point-error by 0.03624 and 0.03616, respectively. This indicates that our proposed hierarchical and parallel aggregation module can significantly improve estimation accuracy. By combining the HA and PA modules together, our model yields the best results with the lowest EPE and 3PE, at 0.74262 and 3.131%, respectively. These remarkable improvements are mainly due to the fact that our proposed HA and PA modules aggregate global and local context information effectively, thus obtaining an optimal understanding ability for our network.

Ablation Study for the Dilation Rates in PA Module. The proposed PA module helps optimize the learning accuracy with four parallel dilated convolution layers. We experiment by setting the four dilation rates between 1 and 32, and the results are shown in Table 2. We first use four normal convolution layers without dilation as a baseline for comparison purposes, in which $d1 \sim d4 = 1$. The parallel aggregation module with four dilated convolutions can achieve better results than the module with only three (denoted as HPA-Net*). The results show that a setting with $d1 \sim d4 = 1, 4, 16, 32$ yields the best performance, which is a 0.74262 EPE on the Scene Flow dataset. This demonstrates that a PA module with relatively large dilation rates can capture more context information than a module with small dilation rates.

4.4 KITTI Datasets Results

We evaluate our method on both the KITTI 2015 and KITTI 2012 datasets. For the KITTI 2015 dataset, the three columns D1-bg, D1-fg, and D1-all indicate

Table 4. KITTI 2012 test set results. "Noc" denotes non-occluded regions and "All" denotes all regions, in which the 2, 3, 4, and 5 pixel errors are evaluated.

Methods	>2px (%)		>3px (%)		>4px(%)		>5px (%)		Mean Error		Runtime (s)
	Noc	All	Noc	All	Noc	All	Noc	All	Noc	All	
GC-Net [3]	2.71	3.46	1.77	2.30	1.36	1.77	1.12	1.46	0.6	0.7	0.9
PSMNet [9]	2.44	3.01	1.49	1.89	1.12	1.42	0.90	1.15	0.5	0.6	0.41
GwcNet-gc [4]	2.16	2.71	1.32	1.70	0.99	1.27	0.80	1.03	0.5	0.5	**0.32**
SSPCVNet [10]	2.47	3.09	1.47	1.90	1.08	1.41	0.87	1.14	0.5	0.6	0.9
AMNet [18]	2.12	2.71	1.32	1.73	0.99	1.31	0.80	1.06	0.5	0.5	0.9
HPA-Net (ours)	**2.03**	**2.57**	**1.25**	**1.64**	**0.94**	**1.23**	**0.76**	**0.99**	**0.4**	**0.5**	0.42

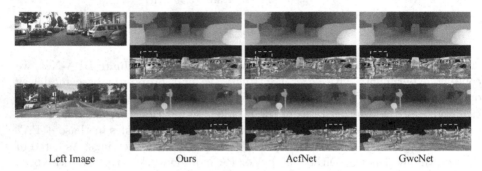

| Left Image | Ours | AcfNet | GwcNet |

Fig. 5. Two testing results from KITTI 2015 test set. For each left input image in the first column, the predicted disparity (above) and corresponding error map (below), obtained by HPA-Net (ours), AcfNet [17], and GwcNet [4], are presented.

the pixels in the background, foreground, and all areas, respectively. For the KITTI 2012 dataset, "Noc" indicates non-occluded regions and "All" indicates all regions, in which the 2, 3, 4, and 5 pixel errors are evaluated.

5   Conclusion

Recent studies attempt to exploit multi-scale cost volumes to improve stereo matching performance. Nevertheless, it is challenging to learn much context information for accurate disparity estimation. In this paper, we have proposed HPA-Net, a novel stereo matching network that contains two main modules: Hierarchical Aggregation (HA) Network and Parallel Aggregation (PA) Network. The HA module fuses cross-scale context information from multi-scale cost volumes to build up an integrated cost volume. The PA module utilizes four parallel dilated convolutions with increasing dilation rates to capture global and local clues of context information. The ablation study demonstrates the effectiveness of the two proposed modules. Comprehensive experiments on Scene Flow, KITTI 2015, and KITTI 2012 datasets show that our proposed HPA-Net significantly improves the disparity prediction accuracy. The visualization results demonstrate that the HPA-Net can reduce estimation errors in the ill-posed regions.

References

1. Scharstein, D., Szeliski, R.: A taxonomy and evaluation of dense two-frame stereo correspondence algorithms. IJCV **47**(1–3), 7–42 (2002)
2. Nikolaus, M., et al.: A large dataset to train convolutional networks for disparity, optical flow, and scene flow estimation. In: CVPR, pp. 4040–4048 (2016)
3. Kendall, A., et al.: End-to-end learning of geometry and context for deep stereo regression. In: ICCV, pp. 66–75 (2017)
4. Guo, X., Yang, K., Yang, W., Wang, X., Li, H.: Group-wise correlation stereo network. In: CVPR, pp. 3273–3282 (2019)
5. Nie, G.Y., et al.: Multi-level context ultra-aggregation for stereo matching. In: CVPR, pp. 3283–3291 (2019)
6. Liang, Z., et al.: Stereo matching using multi-level cost volume and multi-scale feature constancy. PAMI (2019)
7. Zbontar, J., LeCun, Y.: Stereo matching by training a convolutional neural network to compare image patches. JMLR **17**(1), 2287–2318 (2016)
8. Pang, J., Sun, W., Ren, J.S.J., Yang, C., Yan, Q.: Cascade residual learning: a two-stage convolutional neural network for stereo matching. In: CVPRW, pp. 887–895 (2017)
9. Chang, J.R., Chen, Y.S.: Pyramid stereo matching network. In: CVPR, pp. 5410–5418 (2018)
10. Wu, Z., Wu, X., Zhang, X., Wang, S., Ju, L.: Semantic stereo matching with pyramid cost volumes. In: ICCV, pp. 7484–7493 (2019)
11. Zhao, H., Shi, J., Qi, X., Wang, X., Jia, J.: Pyramid scene parsing network. In: CVPR, pp. 2881–2890 (2017)
12. Sun, D., Yang, X., Liu, M., Kautz, J.: PWC-Net: CNNs for optical flow using pyramid, warping, and cost volume. In: CVPR, pp. 8934–8943 (2018)
13. Gu, X., Fan, Z., Zhu, S., Dai, Z., Tan, F., Tan, P.: Cascade cost volume for high-resolution multi-view stereo and stereo matching. In: CVPR, pp. 2495–2504 (2020)
14. Yang, J., Mao, W., Alvarez, J.M., Liu, M.: Cost volume pyramid based depth inference for multi-view stereo. In: CVPR, pp. 4877–4886 (2020)
15. Schuster, R., Wasenmuller, O., Unger, C., Stricker, D.: SDC-stacked dilated convolution: a unified descriptor network for dense matching tasks. In: CVPR, pp. 2556–2565 (2019)
16. He, K., Zhang, X., Ren, S., Sun, J.: Deep residual learning for image recognition. In: CVPR, pp. 770–778 (2016)
17. Zhang, Y., et al.: Adaptive unimodal cost volume filtering for deep stereo matching. In: AAAI, pp. 12926–12934 (2020)
18. Du, X., El-Khamy, M., Lee, J.: AmNet: deep atrous multiscale stereo disparity estimation networks, arXiv preprint arXiv:1904.09099 (2019)

MTStereo 2.0: Accurate Stereo Depth Estimation via Max-Tree Matching

Rafaël Brandt[1(✉)], Nicola Strisciuglio[2], and Nicolai Petkov[1]

[1] Bernoulli Institute, University of Groningen, Groningen, The Netherlands
r.brandt@rug.nl
[2] Faculty of Electrical Engineering, Mathematics and Computer Science
at University of Twente, Enschede, The Netherlands

Abstract. Efficient yet accurate extraction of depth from stereo image pairs is required by systems with low power resources, such as robotics and embedded systems. State-of-the-art stereo matching methods based on convolutional neural networks require intensive computations on GPUs and are difficult to deploy on embedded systems. In this paper, we propose MTStereo2.0, an improved version of the MTStereo stereo matching method, which includes a more robust context-driven cost function, better detection of incorrect matches and the computation of disparity at pixel level. MTStereo provides accurate sparse and semi-dense depth estimation and does not require intensive GPU computations. We tested it on several benchmark data sets, namely KITTI 2015, Driving, FlyingThings3D, Middlebury 2014, Monkaa and the TrimBot2020 garden data sets, and achieved competitive accuracy. The code is available at https://github.com/rbrandt1/MaxTreeS.

Keywords: Stereo matching · Max-Tree

1 Introduction

Estimation of scene depth from stereo image pairs is deployed as a building block in various high level computer vision applications [8] in which the three-dimensional structure of a scene is recovered by matching corresponding pixels.

The similarity of two pixels is quantitatively computed by a matching cost function, e.g. absolute image gradient or gray-level difference [27]. Substantial matching ambiguity can be caused by repetitive patterns and uniformly colored regions. Hence, costs are aggregated over neighbor pixels to strengthen the robustness of the matching. For instance, color similarity and proximity were used for weighted cost aggregation in [37] and [39], while the strength of image boundaries in [3]. Early methods performed exhaustive matching search [27], which requires many

This work was partially funded by the EU H2020 program, under the project Trim-Bot2020 (grant No. 688007).

N. Tsapatsoulis et al. (Eds.): CAIP 2021, LNCS 13052, pp. 110–119, 2021.
https://doi.org/10.1007/978-3-030-89128-2_11

computations. Later approaches reduced the disparity search-range by computing a coarse disparity map first and then refining it iteratively [10]. Image pyramids were also used to reduce disparity search range in [17,29]. A coarse disparity map is estimated considering the full disparity range. Then, increasingly higher-resolution disparity maps are constructed whereby the disparity search range is dictated by the coarser map. To increase efficiency and reduce ambiguity, matching (hierarchically structured) image regions instead of individual pixels was proposed [1,6,30].

Recent approaches use convolutional neural networks (CNNs) to compute aggregated matching costs. One of the first CNN-based methods deployed a siamese network [38], which works with small patch inputs. Increasing the receptive field while maintaining details in estimated disparity maps was investigated [4]. Pairs of siamese networks, each receiving as input a pair of patches at different scales were used and the matching cost was computed as the inner product between their responses. Although CNN-based methods are very accurate, they need power-consuming dedicated hardware (GPUs) to compute the many convolutions. This limits their usability on embedded or power-constrained systems. This applies, for instance, to battery-powered robots or drones, for which depth estimation has to trade-off between accuracy and computational efficiency. Furthermore, for robot navigation and obstacle avoidance, the very high accuracy and density of estimation achieved by CNNs are not strictly necessary.

We present a stereo matching method, named MTStereo 2.0, that balances efficiency with effectiveness, making it appropriate for devices where limited computational and energy resources are available. The MTStereo 1.0 algorithm, that we proposed in [2], exploits contrast information of objects in a hierarchical fashion to efficiently and effectively perform stereo matching. It constructs a hierarchical representation of image scan-lines using Max-Trees [24], and performs disparity estimation via tree matching cost computation that takes into account contextual image structural information. The MTStereo 2.0 algorithm that we propose improves on the 1.0 version as it *a)* deploys a more robust cost function, *b)* performs more thorough incorrect match detection, *c)* computes disparity maps with pixel-level rather than node-level precision. We carried out an extensive experimental benchmark on several data sets, namely the KITTI, Driving, FlyingThings3D, Middlebury, Monkaa and TrimBot2020 data sets.

2 MTStereo 2.0

The matching procedure performed by our method is shown in Fig. 1. Matching the finest structures directly results in non-efficient yet precise stereo matching, while matching coarse structures results in disparity maps that lack precision but can be efficiently obtained. To tackle this trade-off and efficiently perform precise stereo matching, our method matches increasingly finer regions in an iterative manner, and only compares regions contained in earlier matched coarser ones.

Fig. 1. A (left column) pair of stereo images and their (middle column) version resulted by edge-detection (darker regions contain more contrast). The Max-Tree representations of the horizontal scan-lines highlighted in the middle column are illustrated in the right column: MTstereo 2.0 computes the disparity for coarse structure first (nodes represented by darker bars) and increasingly matches only finer structures (with lighter color) contained in earlier matched coarser ones.

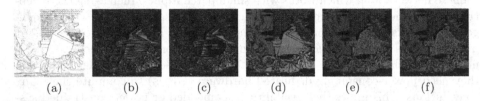

(a)	(b)	(c)	(d)	(e)	(f)

Fig. 2. Outputs after intermediate algorithm stages: (a) pre-processed image of Middlebury training data set, (b) coarse-to-fine matching (c) outlier removal, (d) reliable node extrapolation, (e) guided pixel matching, (f) outlier removal.

2.1   Steps Performed by MTStereo 2.0

MTStereo 2.0 is composed of the following steps: pre-processing, Max-Tree construction, cost volume computation, cost aggregation, consistency check, disparity map computation, confidence check and map refinement.

Pre-processing. We process the input rectified image pair images with a median filter for noise removal. Subsequently, we detect edges by a horizontal and a vertical Sobel operator, and average their absolute response images pixel-wise. We perform contrast-stretch and color quantization on the inverted image, see an example in Fig. 2a. We compute a 1D Max-Tree for each scan-line of the pre-processed images. A parameter q controls the number of colors for color quantization and influences the size of the constructed trees. Shallower trees are less expensive to match, but represent less precisely the image structures.

Max-Tree Construction. The coarse-to-fine matching is facilitated by a hierarchical representation of both images in a rectified stereo pair: we compute 1D Max-Trees on the scan-lines of the images using the algorithm in [33]. We store regions with less contrast being contained in regions with more contrast [30]. The Max-Tree, proposed by [24], allows storing the hierarchy of connected components resulting from different thresholds and is efficiently constructed. A 1D connected component set contains the 1-valued pixels for which no 0-valued pixel exists in between any of the pixels in the binary image resulting from applying a threshold t to a 1D gray-scale image (i.e. a scan-line).

Cost Volume. We construct a cost volume by computing the cost of matching each pixel in the left image with those in the right image at all possible disparity levels. We compute the weighted-average of the absolute difference of the pixel gray-level, horizontal Sobel, and vertical Sobel values. We process each slide of the cost volume with a Gaussian blur operator, which smooths the estimation. A parameter ω_{cv} controls the size of the Gaussian blur kernel ($\omega_{cv} \times \omega_{cv}$). MTStereo 1.0 does not make use of a smoothed cost volume and does not consider gray-level difference in the cost computation.

Matching Cost Aggregation. We then use the cost volume and tree structures for matching of regions in image pairs: nodes that correspond to coarse image structures are matched first, followed by finer structures. Only the finest nodes and their descendants are matched, the width of which is greater than ω_α and less than ω_β (see [2] for details).

We define the matching cost as a context cost and an intensity cost. The context cost is the average relative difference between the area of corresponding ancestors of a node pair. Given a node pair (n_1, n_2), we define that its ancestors (n_3, n_4), where n_3 is an ancestor of n_1 and n_4 is an ancestor of n_2, can be matched if in the Max-Trees the number of nodes between n_3 and n_1, and n_4 and n_2 is equal. We define the intensity cost as the average of the cost volume matching costs at aligned pixels part of the two nodes. Given a node pair, we compute the matching cost at the location of the left node's left (right) endpoint and disparity between the left (right) endpoints of both nodes, as well as at linearly interpolated disparity and pixel values in between the two endpoints. This definition of intensity cost is more robust than that of MTStereo 1.0. Parameter α controls the relative weight of the intensity cost (α) and context cost ($1 - \alpha$).

Let $P_y = (n_{L,y}, n_{R,y})$ denote a pair of nodes, respectively in the left and right image, both at row y. Matching cost is aggregated over their node neighborhood. We define the neighborhood of P_y as the nodes with the same coarseness level as P_y that have similar x-coordinates and are in scan-lines next to each other in the original image. Two nodes have the same coarseness level when the distance (i.e. the difference in tree level) between the leaf nodes with the greatest level out of the leaf nodes which are descendants of the nodes, and the nodes themselves are equal. Recursively, node pair P_{y+1} is part of the neighborhood of P_y, if both $n_{L,y+1}$ crosses the x-coordinate of the center of node $n_{L,y}$, $n_{R,y+1}$ crosses the x-coordinate of the center of node $n_{R,y}$, and both $n_{L,y+1}$, and $n_{R,y+1}$ have a y-coordinate which is one lower or higher than that of ($n_{L,y}$ and $n_{R,y}$). At most, the ω_γ nodes above and below a node pair are included in the neighborhood of P_y. In the coarse-to-fine matching procedure, we compute a disparity search range for each node in each iteration. Given a node pair that has likely been correctly matched in a previous iteration, only descendants of this node pair are matched in subsequent iterations. Nodes are considered likely correctly matched when they pass a left-right consistency check [9] and a confidence check (peak-ratio used in [36]). The confidence check, not used in MTStereo 1.0, provides more accurate disparity maps as it filters out ambiguous/incorrect matches.

Disparity Map. The coarse to fine matching procedure results in a list of matched node pairs. For each of the finest nodes, we compute the disparity at the left and right endpoints of matched nodes by linear interpolation, see Fig. 2b for an example disparity map. We detect and remove outliers, when in the $(2 \cdot 21) \times (2 \cdot 21)$ local neighborhood the number of pixels with a disparity difference that exceeds their x-offset surpasses the number of pixels with a disparity difference less than or equal to their x-offset (see Fig. 2c for a post-processed map).

We densify sparse disparity maps by computing the median disparity values of neighbor nodes across scan-lines. This procedure fills-in some missing disparity values and smooths out eventual remaining outliers. Furthermore, as in the tree disparity values are computed only for the left- and rightmost nodes, our densification process also includes the linear interpolation of the disparity values of the pixels within the node, thus obtain the disparity of an entire 1D region. Inside-node interpolation is only performed by the semi-dense variant. A disparity map after reliable node extrapolation is illustrated in Fig. 2d. Different from MTStereo 1.0, we perform pixel-level matching to recover surface shape. Disparity search range for pixels is set such that only disparities which differ at most a fraction ω_ω from the previously computed disparity value are considered. The matching cost of pixel pairs is derived from the constructed cost volume.

Confidence Check and Map Refinement. We perform a confidence check on the estimated pixel disparities, which evaluates the relative difference between the matching cost of the best match and that of the second-best match. If this difference is more than ω_Π the check is passed. Areas which contain more texture are more likely to pass the confidence check. We obtain the final disparity map by removing outliers. A disparity map after noise removal is shown in Fig. 2f.

3 Experiments

We evaluated MTStereo 2.0 on the following benchmarks: Middlebury 2014 [26], Kitti 2015 [19], Trimbot2020 Synthetic Garden [23] and Real Garden [25] (test sets used by [23]), Driving (cleanpass, fast, 35mm_focallength subset), Monkaa (cleanpass subset), and Flying Things 3D [18] (cleanpass, test set - stereo pairs excluded by [18] were excluded in our experiments as well). Marked results were taken from the online Kitti[1] and Middlebury[2] leaderboard. For Middlebury, accuracy results were weighted by the official weights, while average density results were not weighted. We used full-size images for our experiments. We ran our algorithm on an Intel® Core™ i7-2600K CPU @3.40GHz. We used 4 cores.

3.1 Evaluation

We computed standard metrics for the concerned benchmarks, allowing direct comparison with existing methods:

---

[1] http://www.cvlibs.net/datasets/kitti/eval_scene_flow.php?benchmark=stereo.
[2] https://vision.middlebury.edu/stereo/.

Table 1. Results on the Middlebury training and test set in *avgerr(Density)*, compared with those of other methods, taken from the online leaderboard. Results of MTStereo 2.0 are bold. * Result was computed on downsized input image pairs.

Middlebury 2014 (train set)							
MotionStereo [31]	SGM [11]	**Sparse**	SNCC [7]	LPSM [34]	SDR [35]	Cens5 [13]	LS-ELAS [15]
1.72*(46%)	2.06(58%)	2.35(3%)	3.25*(62%)	4.22*(100%)	4.32*(100%)	4.32*(67%)	4.35(61%)
ELAS [10]	SLCCF	SED [22]	**Semi-Dense**	MANE [32]	SRM [21]	SGBM1 [12]	REAF [5]
4.94(73%)	5.27*(100%)	5.38(1%)	6.51(24%)	6.59*(67%)	6.92*(100%)	7.83*(78%)	8.49*(100%)
Middlebury 2014 (test set)							
SGM [11]	MotionStereo [31]	SNCC [7]	Cens5 [13]	SDR [35]	SLCCF	MANE [32]	LS-ELAS [15]
2.50(50%)	3.30*(38%)	3.96*(55%)	5.31*(61%)	6.16*(100%)	6.36*(100%)	8.70*(61%)	9.10(49%)
LPSM [34]	**Sparse**	SRM [21]	ELAS [10]	TSGO [20]	SED [22]	ELAS_RVC [10]	SGBM1 [12]
9.29*(100%)	9.36(3%)	10.40*(100%)	10.6(66%)	11.00(100%)	12.3(2%)	13.40*(100%)	14.20*(73%)

- **avgerr**: average absolute disparity error (in pixels) among all pixels of which both a disparity value was in the ground truth and estimated disparity map.
- **D-all-est**: the percentage of stereo disparity outliers (the disparity error is $\geq$3px and $\geq$5% of the true disparity) among all pixels of which both a disparity value was in the ground truth and estimated disparity map.
- **Density**: the rounded percentage of pixels with a disparity prediction with respect to the total number of pixels in the reference image.

We defined a single set of parameters that contributes to achieving robust performance across the different benchmark data sets. The number of color quantization levels q was set to 16 (8) for sparse (semi-dense) disparity maps. The weight α of the context cost relative to that of the gradient cost was set to 0.8. The minimum (or maximum) width of nodes to be matched ω_α (or ω_β) was set to 0 (or 1/2 of the image width). Matched node levels S was set to $\{1, 0\}$. The maximum neighborhood size ω_γ was set to 10. The size of the Gaussian kernel used to aggregate the cost volume ω_{cv} was 21. The minimum confidence ω_Π was set to 12 during coarse-to-fine matching. In guided pixel refinement, ω_Π was set to 12% (4%) for sparse (semi-dense) disparity maps. ω_ω was set to 15%.

3.2 Results and Comparison

In Table 1, we report the error achieved by our method on the Middlebury training and test data, as well as those of other methods that have low average execution times and do not run on a GPU. The lower (higher) the avgerr (Density) values, the better the performance. In Table 2, we report the results achieved by our method on the other considered benchmark data sets, as well as those of other methods that do not formulate the stereo matching as a learning problem (upper part) and those that deploy neural networks to tackle disparity estimation (lower part). Both the sparse and semi-dense versions of our method achieved better or comparable accuracy than those of many existing methods. The results on the Middlebury training data set show, for example, the avgerr in sparse disparity maps produced by MTStereo 2.0 is lower than that of all other listed methods except MotionStereo and SGM. Also, the avgerr on the

Table 2. Results in *avgerr*, except Kitti Test which is expressed in *D-all-est*. Average prediction *Density* is reported in brackets. Entries without density specification are 100% dense. "-SD" denotes Semi-Dense. $^{+}$: Result was taken from benchmark website. ^{d}: Result was computed on downsized input images. ^{g}: Generalization study result. *: Result was (possibly) computed on a different set of images than we used for evaluation. ^{t}: Result was (possibly) computed using a different metric computation approach.

	Middlebury		Kitti 2015		Real garden	Synth garden	Driving	Monkaa	Flying things
	Train	Test	Train	Test					
MTS2-Sparse	**2.35**$^{+}$(3)	**9.36**$^{+}$(3)	1.35(4)	7.83^{+}(4)	3.08(5)	5.22(6)	**5.36**(3)	**3.83**(4)	**1.7**(6)
MTS2-SD	6.51(24)	–	2.24(25)	–	3.72(20)	7.04(19)	7.38(15)	6.54(25)	2.61(30)
MTS1-Sparse	5.47^{+}(2)	15.5^{+}(2)	1.58(2)	8.92^{+}(3)	2.48(2)	7.32(2)	8.8(1)	7.22(3)	3.03(4)
MTS1-SD	17.47(57)	–	4.47(44)	–	3.78(18)	12.8(14)	16.2(38)	15.41(40)	6.93(58)
SGBM1 [12]	7.83^{+}(67)	16.3^{+}(63)	1.45(84)	–	2.61(89)	4.77(92)	16.06(70)	11.37(81)	5.14(86)
SGBM2 [12]	8.92^{+d}(83)	15.9^{+d}(77)	**1.27**(82)	**5.86**$^{+}$(90)	2.16(90)	**4.67**(90)	15.72(64)	10.5(78)	4.58(85)
ELAS_ROB [10]	10.5^{+d}	13.4^{+d}	1.49(99)	9.67^{+}	**2.06**	7.02	11.71	17.75	7.46
FPGA Stereo [14]	–	–	–	–	2.94 [23]	11.41 [23]	–	–	–
DispNet [18]	–	–	**0.68** [18]	4.34^{+}	**1.35** [23]	**6.28** [23]	**15.62*** [18]	**5.78** [18]	**1.68** [18]
EdgeStereo [28]	**2.00**$^{+}$	**3.72**$^{+}$	2.07^{g} [28]	**2.08**$^{+}$	–	–	–	–	**0.74**^{*t} [28]
iResNet [16]	–	–	1.21^{g} [28]	2.44 [28]	–	–	–	–	**0.95**^{*t} [28]

Kitti training data set of MTStereo 2.0 sparse is lower than that of all other listed methods except DispNet, iResNet, and SGBM2. Furthermore, the avgerr of disparity maps produced by MTStereo 2.0 sparse is lower than all other non-learning based methods when evaluated on Driving, Monkaa, and Flying Things 3D. MTStereo 2.0 sparse produced more accurate disparity maps than some of the listed CNN based methods on the Kitti2015 training, Synthetic garden, Driving, and Monkaa data sets. The density of the disparity maps of MTStereo2.0 is lower than CNN-based methods. However, it can be regulated by manipulating the parameters that trade-off accuracy and density. These results indicate that MTStereo2.0 estimates very precisely the depth of regions for which a robust match in the max-trees is found, and can be used as an unsupervised matching strategy to serve as prior for more sophisticated methods.

The results of MTStereo 2.0 sparse and semi-dense are generally more accurate than those of their counterpart based on MTStereo 1.0. The cases where MTStereo 2.0 produces more accurate and dense disparity maps than MTStereo 1.0 (e.g. for the sparse variants when evaluated on the Middlebury, Kitti2015, Synth garden, and Driving data sets) suggest that MTStereo 2.0 is a more robust method than MTStereo 1.0.

In Fig. 3, we show example images from the Middlebury, Trimbot2020 Synthetic Garden, Monkaa, and Flying Things 3D data sets, with corresponding ground truth depth images, our semi-dense and sparse depth reconstruction. One can notice that our method can robustly extract depth information also in image regions with little texture, whose depth ambiguity can be successfully handled by MTStereo2.0. For example, the depth of the gray cubes in the Flying Things example is reasonably well estimated although the cube surfaces are rendered with little texture. When the parameter ω_Π is set to a low value, an assumption is made that regions with little texture are flat. The dis-ambiguous

Fig. 3. Example images from the Middlebury, Trimbot2020 Synthetic Garden, Monkaa, and Flying Things 3D data sets, with corresponding ground truth images, our semi-dense and sparse reconstruction. Black disparity map pixels have no assignment. Morphological dilatation was applied to sparse outputs for visualization purposes.

disparity values at the left and right side edges of regions with little or no texture are then linearly interpolated within the region. When the assumption is not correct, such as between the tree branches in the MTStereo semi-dense disparity map of the Synth Garden example, the parameter ω_Π can be increased such that no disparity values are estimated in ambiguous regions. Edge blurring artifacts are frequently filtered out, as can be seen for the edges of the pipes in the first row or those of the tree in the second row of Fig. 3. The disparity maps produced by MTStereo2.0 tend to have very few outliers.

The methodological improvements of MTStereo2.0 with respect to version 1.0 increase the computation intensity of the method, which requires a longer time to process the images. The processing times of MTStereo 2.0 are generally longer than those of other methods listed in this paper. However, the efficiency of our implementation (publicly available) can likely be increased through code-level optimization, e.g. of our cost volume construction and coarse-to-fine matching procedures that take up to about 70% of the processing time.

4   Conclusions

We proposed a stereo matching method, MTStereo 2.0, for systems with limited computational and energy resources that require efficient and accurate depth estimation. It improves on its predecessor MTStereo 1.0 as it: *a)* deploys a more robust cost function, *b)* performs more thorough detection of incorrect matches, *c)* computes disparity maps with pixel-level rather than node-level precision.

MTStereo 2.0 produces disparity maps which are generally more accurate than those produced by MTStereo 1.0, and does not require intensive GPU computations like methods based on CNNs. It can thus run on devices with low-power resources. Our method achieves competitive results on several benchmark data sets: Middlebury 2014, KITTI 2015, Driving, FlyingThings3D, Monkaa, and the TrimBot2020 garden data sets.

References

1. Arnold, R.D.: Automated stereo perception. Stanford Univ CA Dept of Computer Science, Technical report (1983)
2. Brandt, R., Strisciuglio, N., Petkov, N., Wilkinson, M.H.: Efficient binocular stereo correspondence matching with 1-D max-trees. Pattern Recogn. Lett. **135**, 402–408 (2020)
3. Chen, D., Ardabilian, M., Wang, X., Chen, L.: An improved non-local cost aggregation method for stereo matching based on color and boundary cue. In: IEEE ICME, pp. 1–6 (2013)
4. Chen, Z., Sun, X., Wang, L., Yu, Y., Huang, C.: A deep visual correspondence embedding model for stereo matching costs. In: IEEE ICCV, pp. 972–980 (2015)
5. Cigla, C.: Recursive edge-aware filters for stereo matching. In: CVPRW, pp. 27–34 (2015)
6. Cohen, L., Vinet, L., Sander, P.T., Gagalowicz, A.: Hierarchical region based stereo matching. In: IEEE CVPR, pp. 416–421 (1989)
7. Einecke, N., Eggert, J.: A two-stage correlation method for stereoscopic depth estimation. In: DICTA, pp. 227–234. IEEE (2010)
8. Engel, J., Stückler, J., Cremers, D.: Large-scale direct slam with stereo cameras. In: IEEE/RSJ IROS, pp. 1935–1942. IEEE (2015)
9. Fua, P.: A parallel stereo algorithm that produces dense depth maps and preserves image features. Mach. Vis. Appl. **6**(1), 35–49 (1993)
10. Geiger, A., Roser, M., Urtasun, R.: Efficient large-scale stereo matching. In: Kimmel, R., Klette, R., Sugimoto, A. (eds.) ACCV 2010. LNCS, vol. 6492, pp. 25–38. Springer, Heidelberg (2011). https://doi.org/10.1007/978-3-642-19315-6_3
11. Hirschmuller, H.: Accurate and efficient stereo processing by semi-global matching and mutual information. In: CVPR, vol. 2, pp. 807–814 (2005)
12. Hirschmuller, H.: Stereo processing by semiglobal matching and mutual information. IEEE Trans. Pattern Anal. Mach. Intell. **30**(2), 328–341 (2008)
13. Hirschmüller, H., Innocent, P.R., Garibaldi, J.: Real-time correlation-based stereo vision with reduced border errors. IJCV **47**(1–3), 229–246 (2002)
14. Honegger, D., Sattler, T., Pollefeys, M.: Embedded real-time multi-baseline stereo. In: ICRA 2017, pp. 5245–5250. IEEE (2017)
15. Jellal, R.A., Lange, M., Wassermann, B., Schilling, A., Zell, A.: LS-ELAS: line segment based efficient large scale stereo matching. In: ICRA, pp. 146–152 (2017)
16. Liang, Z., et al.: Learning deep correspondence through prior and posterior feature constancy. arXiv preprint arXiv:1712.01039 (2017)
17. Luo, X., Bai, X., Li, S., Lu, H., Kamata, S.I.: Fast non-local stereo matching based on hierarchical disparity prediction. arXiv preprint arXiv:1509.08197 (2015)

18. Mayer, N., et al.: A large dataset to train convolutional networks for disparity, optical flow, and scene flow estimation. In: CVPR 2016, pp. 4040–4048 (2016)
19. Menze, M., Heipke, C., Geiger, A.: Joint 3D estimation of vehicles and scene flow. In: ISPRS Workshop on Image Sequence Analysis (ISA) (2015)
20. Mozerov, M.G., van de Weijer, J.: Accurate stereo matching by two-step energy minimization. IEEE Trans. Image Process. 24(3), 1153–1163 (2015)
21. Okae, J., Du, J., Hu, Y.: Robust statistical approach to stereo disparity maps denoising and refinement. Control Theory Technol. 18(4), 348–361 (2020). https://doi.org/10.1007/s11768-020-00014-y
22. Peña, D., Sutherland, A.: Disparity estimation by simultaneous edge drawing. In: Chen, C.-S., Lu, J., Ma, K.-K. (eds.) ACCV 2016. LNCS, vol. 10117, pp. 124–135. Springer, Cham (2017). https://doi.org/10.1007/978-3-319-54427-4_10
23. Pu, C., Song, R., Tylecek, R., Li, N., Fisher, R.: SDF-MAN: semi-supervised disparity fusion with multi-scale adversarial networks. Remote Sens. 11(5), 487 (2019)
24. Salembier, P., Oliveras, A., Garrido, L.: Antiextensive connected operators for image and sequence processing. IEEE Trans. Image Process. 7(4), 555–570 (1998)
25. Sattler, T., Tylecek, R., Brox, T., Pollefeys, M., Fisher, R.B.: 3D reconstruction meets semantics-reconstruction challenge 2017. In: ICCV Workshop (2017)
26. Scharstein, D., et al.: High-resolution stereo datasets with subpixel-accurate ground truth. In: Jiang, X., Hornegger, J., Koch, R. (eds.) GCPR 2014. LNCS, vol. 8753, pp. 31–42. Springer, Cham (2014). https://doi.org/10.1007/978-3-319-11752-2_3
27. Scharstein, D., Szeliski, R.: A taxonomy and evaluation of dense two-frame stereo correspondence algorithms. Int. J. Comput. Vis. 47(1–3), 7–42 (2002)
28. Song, X., Zhao, X., Fang, L., Hu, H.: Edgestereo: an effective multi-task learning network for stereo matching and edge detection. arXiv:1903.01700 (2019)
29. Sun, C.: A fast stereo matching method. In: DICTA, pp. 95–100. Citeseer (1997)
30. Todorovic, S., Ahuja, N.: Region-based hierarchical image matching. Int. J. Comput. Vision 78(1), 47–66 (2008)
31. Valentin, J., et al.: Depth from motion for smartphone AR. In: SIGGRAPH Asia, p. 193. ACM (2018)
32. Vázquez-Delgado, H.D., et al.: Real-time multi-window stereo matching algorithm with fuzzy logic. IET Comput. Vision 15(3), 208–223 (2021)
33. Wilkinson, M.H.: A fast component-tree algorithm for high dynamic-range images and second generation connectivity. In: IEEE ICIP, pp. 1021–1024 (2011)
34. Xu, C., Wu, C., Qu, D., Xu, F., Sun, H., Song, J.: Accurate and efficient stereo matching by log-angle and pyramid-tree. IEEE Trans. Circuits Syst. Video Technol. (2020). https://doi.org/10.1109/TCSVT.2020.3044891
35. Yan, T., Gan, Y., Xia, Z., Zhao, Q.: Segment-based disparity refinement with occlusion handling for stereo matching. IEEE TIP 28(8), 3885–3897 (2019)
36. Yang, Q., Ji, P., Li, D., Yao, S., Zhang, M.: Fast stereo matching using adaptive guided filtering. Image Vis. Comput. 32(3), 202–211 (2014)
37. Yoon, K.J., Kweon, I.S.: Adaptive support-weight approach for correspondence search. IEEE Trans. Pattern Anal. Mach. Intell. 28(4), 650–656 (2006)
38. Zbontar, J., LeCun, Y., et al.: Stereo matching by training a convolutional neural network to compare image patches. J. Mach. Learn. Res. 17(1–32), 2 (2016)
39. Zhang, K., Lu, J., Lafruit, G.: Cross-based local stereo matching using orthogonal integral images. IEEE Trans. Circuits Syst. Video Technol. 19(7), 1073–1079 (2009)

Biomedical Image and Pattern Analysis

H-OCS: A Hybrid Optic Cup Segmentation of Retinal Images

Abdullah Sarhan[1]([✉]), Jone Rokne[1], and Reda Alhajj[1,2,3]

[1] Department of Computer Science, University of Calgary, Calgary, AB, Canada
asarhan@ucalgary.ca
[2] Department of Computer Engineering, Istanbul Medipol University,
Istanbul, Turkey
[3] Department of Health Informatics, University of Southern Denmark,
Odense, Denmark

Abstract. Glaucoma is the second leading cause of irreversible vision loss. Early diagnosis and treatment can, however, slow the progression of the disease. Specialists making this diagnosis rely on several tests and examinations such as visual field tests and examinations of retinal images and optical coherence tomography images. One of the regions examined by specialists when checking for retinal conditions is the optic nerve head region, which is the brightest region in retinal images. Within this region, the ratio between the cup and the disc can be used when diagnosing for glaucoma. Calculating the cup–disc ratio requires the segmentation of both the disc and the cup from retinal images. In a previous paper, a method for segmenting the disc was proposed. Here another deep learning model, H-OCS, is proposed for segmenting the cup from retinal images. A customized InceptionV3 model with transfer learning and image augmentation is used. Additionally, the output of H-OCS is refined and enhanced using a series of post-processing steps. H-OCS is tested on six publicly available datasets: RimOneV3, Drishti, Messidor, Refuge, Riga, and Magrebia and several ablation studies are conducted to evaluate the effectiveness of the proposed approach. Additionally, the performance of H-OCS is compare with other studies. An overall average accuracy of 97.86%, DC of 88.37%, Sensitivity of 89.09% and IoU of 79.66% was achieved.

Keywords: Cup · Deep learning · Inception · Segmentation · Retinal image · Transfer learning · Loss function

1 Introduction

Glaucoma is the world's second leading cause of irreversible vision loss after cataracts [4]. The optic nerve head (ONH) is one of the most important anatomical features where glaucomatous conditions may be diagnosed. The ONH is most often damaged due to elevated intraocular pressure. To assess the damage an specialists need to locate the disc and cup and estimate the ratio between them

© Springer Nature Switzerland AG 2021
N. Tsapatsoulis et al. (Eds.): CAIP 2021, LNCS 13052, pp. 123–134, 2021.
https://doi.org/10.1007/978-3-030-89128-2_12

as one of the indicators to assess the damage to the ONH. An abnormal cup–disc ratio (CDR), calculated by dividing the cup diameter over the disc diameter, indicates the possibility of having glaucoma.

Several deep learning approaches have been developed for segmenting the optic cup which is further then used to calculate the CDR as one of the measure for glaucoma detection. Examples are ensemble learning CNNs [20], a two staged model [14], a fully connected deep learning model [15], patch based network [7], and multi scale input/output models [3].

Most studies have tended to test their approaches on non-diverse datasets, and hence they face the issue of domain shift: that is, their models may not perform well on images from a variety of sources. Another limitation is related to the imbalance between the cup and the background. Such a limitation can be handled by using a proper loss function that penalizes an incorrect classification more severely for a pixel related to the object to be segmented rather than a pixel from the background.

In this work, we propose an approach for cup segmentation from retinal images while taking into consideration the challenges listed above. A deep learning model for cup segmentation from retinal images using the InceptionV3 model is proposed. This model is similar to the one used for segmenting the vessels [13]. As there are insufficient annotated cup datasets for deep learning, the idea of using transfer learning (TL) and image augmentation (IA) is adopted. Instead of using random weights to initialize the model, weights trained on millions of images for semantic segmentation from the ImageNet dataset [10] are used. These weights are then fine-tuned to match the optic cup. Additionally, a series of post-processing steps to improve the output of the proposed model are adopted. H-OCS was tested on six publicly available datasets.

2   Proposed Method

The goal is to segment the optic cup given the disc region from the retinal image. To achieve this, the disc region is first extracted and then feed to the model. H-OCS is a deep learning model with an architecture similar t the one developed by Ronneberger et al. [9]. A customized InceptionV3 model is used as the encoder. In this section, architecture of the model, pre-processing steps, data augmentation, model initialization, and post-processing steps are discussed.

2.1   Region of Interest Extraction

To segment the cup, the optic disc is first located and then the disc region is cropped using the approach developed by Sarhan et al. [11], which has been shown to be effective in segmenting discs from diverse datasets. Once the disc is located, the disc region, including a surrounding a region, is extracted as shown in Fig. 1. The segmented optic disc is mapped back to the original image and a sub-image based on the center of the predicted optic disc is cropped. The images are then resized to 224 × 224 pixels and normalized so that all pixel values are within the range (0,1). Finally, the images undergo binary thresholding. Hence, the region to be analyzed is reduced to be only the disc region as shown in Fig. 1. The cup labels are adjusted using the same process as used for the original retinal image so that the ground truth of the cup is aligned with the ROI.

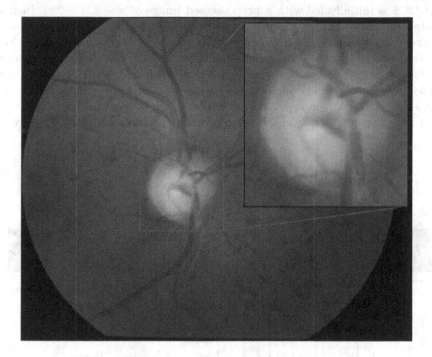

Fig. 1. Region of interest extraction from a retinal image.

2.2   Network Architecture

The challenges in segmenting the cup as opposed to segmenting the disc are now discussed. First, the model used for the segmenting disc region was tested while retraining it on images with the cup ground truth. The model did not perform well, however, mainly because of the variations in the cup and how fuzzy it is when blood vessels are present. Hence, it was decided to use a deeper network, namely the InceptionV3 network, for H-OCS. The network utilized is similar to the one utilized by Sarhan et al. [13].

The InceptionV3 model was adopted as the encoder while customized the normalization, and reducing the convolutional layers. The architecture of H-OCS is presented in Fig. 2. The original architecture of the InceptionV3 model requires four instances of block C to be presented in block A (marked as 3 in Block A in Fig. 2) but only three was used here. Another change was that, while in the original architecture, block B is required to have four instances of block C, but here five were used instead (marked as 3 in Block B in Fig. 2). The final concatenation layer and other layers as shown in Fig. 2 were used to construct our encoder. A drop-out was added at the end with a factor of 0.3, and a sigmoid activation was used. The rectified linear used is used as the activation method for all convolutional layers [13]. Max pooling layers are also utilized to reduce the spatial of the feature maps.

H-OCS is initially fed with a prepossessed image of size 224 × 224. Instead of using the whole image, the disc region referred to as the region of interest is extract, and fed to H-OCS after applying a set of pre-processing steps. Five Skip connections are then utilized to transfer the features from some layers in the encoder to the decoder as shown in Fig. 2. After the skip layers have been selected, upsampling and concatenation is done so that the output image will have a 224 × 224 dimensions. Short connections from early to later layers are useful in preserving high-level information about the positioning of the cup.

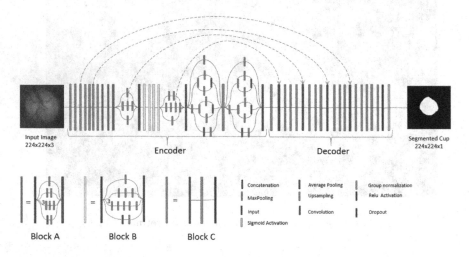

Fig. 2. Architecture of the model adopted for cup segmentation.

2.3   Loss Function

The binary cross entropy (BCE) loss function is one of the common loss functions for semantic segmentation. However, this loss function does not perform well when the data are imbalanced, as they are in this case. The output will be biased toward the background, resulting in improper cup segmentation. That is, the output may suggest a misleading result of 90% accuracy. Accurate cup segmentation therefore requires additional processing. This problem was avoided using the Jaccard distance, which measures the dissimilarity between two datasets. In order to check if the model had improved or not on the value returned from the loss function during the training of the network it was run using the validation data. The Jaccard loss function defined as:

$$L_j = 1 - \frac{|Y_d \cap \hat{Y}_d|}{|Y_d \cup \hat{Y}_d|} = 1 - \frac{\sum_{d \in Y_d}(1 \wedge \hat{y}_d)}{|Y_d| + \sum_{b \in Y_b}(0 \vee \hat{y}_b)} \tag{1}$$

is used where Y_d represents the ground truth of the cup, Y_b the ground truth of the background, $\hat{Y}_d$ represents the predicted pixels of the cup and $\hat{Y}_b$ represents the background pixels. The cardinalities of the cup Y_d are $|Y_d|$ and the cardinality background $|Y_b|$ is $|\hat{Y}_b|$ and $\hat{y}_d \in \hat{Y}_d$ and $\hat{y}_b \in \hat{Y}_b$. The values of $\hat{y}_d$ and $\hat{y}_b$ will always be between 0 and 1 since they are both probabilities. This loss function can now be approximated as shown in Eq. 2. The result is then updated by Eq. 3 where j represents the jth pixel in the input image and $\hat{y}_j$ represents the predicted value for that pixel.

$$\tilde{L}_j = 1 - \frac{\sum_{d \in Y_d} min(1, \hat{y}_d)}{|Y_d| + \sum_{b \in Y_b} max(0, \hat{y}_b)} = 1 - \frac{\sum_{d \in Y_d} \hat{y}_d}{|Y_d| + \sum_{b \in Y_b} \hat{y}_b} \tag{2}$$

$$L_j y_i \{ -\frac{1}{|Y_d| + \sum_{b \in Y_b} \hat{y}_b} \quad for \quad i \in Y_d - \frac{\sum_{d \in Y_d} \hat{y}_d}{|Y_d| + \sum_{b \in Y_b} \hat{y}_b} \quad for \quad i \in Y_b \tag{3}$$

2.4 Transfer Learning

To overcome this problem related to insufficient images for training, an approach referred to as transfer learning is used. This approach can alleviate the issues caused by insufficient training data by using weights generated by training on millions of images. In this study, the weights generated when training the InceptionV3 model on the ImageNet dataset are adopted which contains around 14 million labeled images.

Using TL reduces the problem of over-fitting caused by training on a limited number of images and improves the overall performance of the model. The weights of the encoder network component are initialized using the ImageNet weights, and other layers are randomly initialized using a Gaussian distribution. H-OCS is then trained using a mini-batch gradient to tune the weights of the whole network.

In addition to transfer learning, we applied random augmentation to each image. We randomly applied one or combination of the following: shifting vertically and/or horizontally, rotation within a range of 360 °C, flipping vertically and/or horizontally. We tested the evaluation effectiveness of data augmentation with and without transfer learning.

2.5 Postprocessing

To handle the issue of false positives and smooth the boundaries of the predicted cup, some image postprocessing steps are performed. The first step is to remove small regions in the image that were clearly not related to the cup. This step is implemented by applying contour edge detection, in which all contours are first detected and then sorted from largest to smallest. The cup is the largest contour, and hence all other regions are removed.

Fig. 3. Predicted cup and the resulting ellipse fitted image.

The next step involves smoothing the boundaries of the predicted cup using least square ellipse fitting [5]. The least square fitting approach identifies the smallest circle that can traverse all given points. Given Eq. 4, where A, B, C, D, E, and F are the coefficients of the ellipse and $B^2 - 4AC < 0$, generate an ellipse by minimizing the square of the euclidean distance of the points to the ellipsoid plane. As the cup may have an ellipsoid shape, this approach helped in smoothing the boundaries of the cup and improve the results [12]. Figure 3 shows the cup predicted by H-OCS and how the prediction would be adjusted after applying ellipse fitting. It can also be seen how the DC has increased from the original predicted image to the one in which ellipse fitting has been applied.

$$Ax^2 + Bxy + Cy^2 + Dx + Ey + F = 0 \qquad (4)$$

3   Experimental Analysis

The H-OCS model was evaluated on six datasets. Cups of different sizes, orientations, and resolutions were fed to the model. In this section, the datasets and the experiments are discussed and the performance of the model is compared with those of other approaches.

3.1   Datasets

To verify the robustness of H-OCS, it was tested on six publicly available datasets. Table 1 provides an overview of these datasets along with the machines used to capture these images. These datasets contain information regarding multiple retinal conditions: namely, glaucoma and diabetic retinopathy. Moreover, retinal images that belong to these datasets were acquired at different angles and resolutions, as can be seen in Fig. 4. For datasets that contained multiple annotations the average of the tracings were used when training and evaluating H-OCS, which is the common technique used in such scenarios [8,16].

Table 1. Dataset properties and machines used to capture their images.

Dataset	Images		Dimensions	Machine
	Train	Test		
Drishti-GS [16]	50	51	2049*1757	–
Refuge [6]	400	400	2124*2056	Zeiss Visucam 500
RimOneV3 [8]	128	31	1072*712	Nidek AFC-210
BinRushed [2]	147	35	2376*1584	Canon CR2 non-mydriatic
Magrebia [2]	52	11	2743*1936	Topcon TRC 50DX mydriatic
Messidor [2]	365	92	2240*1488	Topcon TRC NW6 non-mydriatic

In total 1,142 images were used for training and 620 were used for testing. To test H-OCS, the data had to be split into training and testing portions. The model could only see the training images, and the performance was checked by evaluating the model's predictions on the test images and comparing it to labels. Doing so makes it fair to compare H-OCS with other approaches, as they would test their approach on the same test images used here. However, not all datasets are split in this manner. For the Drishti and Refugee dataset, [6,16], they were already split into training and testing datasets. However, this was not the case with the RimOneV3, Magrabia, Riga, and Messidor datasets [2,8]. Hence, the same split approach as proposed by [11] was used. Following the same split approach allows other researchers to compare their approaches by standardizing the set of test images.

3.2   Implementation Details

H-OCS was implemented on a linux machine with a NVIDIA GeForce RTX 2080 with 8 GB dedicated memory. When training, we have utilized the NAdam optimizer with learning rate set to 0.0001 and 8 images per batch. During training, the H-OCS checkpoints would save the model whenever a decrease in the loss value has occurred on the validation dataset. While training, the training and validation sets were split by 80% and 20% respectively. These sets were selected randomly at each epoch. Four evaluation methods were then utilized to evaluate and compare the approach: namely, accuracy (Acc), dice coefficient (DC), sensitivity (Sen), intersection over union (IOU).

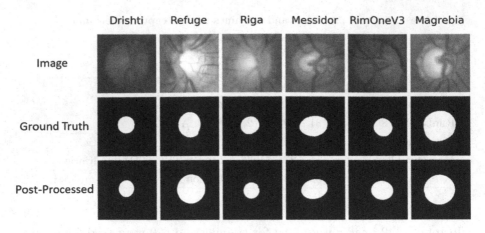

Fig. 4. A sample image from each dataset used in our study. The first row shows the actual retinal image. The second shows the related cup ground truth and the third row shows the post processed segmented cup using the mode. The dataset name of which each image belongs to is written on the top.

3.3   Effectiveness of TL and IA

To test the impact of using TL and IA when training H-OCS, a series of experiments were conducted and then evaluated using the test images from all the datasets. In this section, the overall performance, is discussed, not the performance on each dataset.

Four experiments were conducted to test the effectiveness of TL and IA. In the first experiment, H-OCS was trained on the data without using TL and IA and initialized the model with random weights. The second experiment did not use TL but did use IA. The third and fourth experiments both used TL, and H-OCS was initialized with weights from imagenet trained on image segmentation using millions of images. One of these final two experiments was run with IA and one without. The results from each of these experiments are presented in Table 2.

Table 2. Performance of the proposed approach with and without using TL and/or IA.

Experiment	Acc	DC	Sen	IoU
No TL and No IA	96.36	82.00	86.91	71.57
No TL and IA	97.40	87.11	89.58	78.11
TL with No IA	97.80	87.15	89.09	78.30
TL with IA	**97.86**	**88.37**	**89.09**	**79.66**

The least accurate results were obtained when TL and IA were not used. H-OCS only required 474 epochs to complete the training, however, with each epoch requiring 66 s. With IA and no TL, 1,014 epochs were required to complete the training. When using TL with and without IA, the models required 760 and 638 epochs, respectively, to finish training. The most accurate results were achieved using TL and IA together, especially for DC and IoU values, which reflect how precisely the cup is segmented. Moreover, using TL and IA together was slightly faster than when not using TL with IA when training the model. Although adding IA to the training results increased the number of epochs until convergence, it yielded more accurate results than if it were not used.

3.4   Effectiveness of Loss Functions

Three experiments were conducted to test which loss function would achieve the most accurate results. First, H-OCS was trained using the BCE loss function alone, which is a built-in loss function in the Keras library. It was then trained using only the Jaccard loss function. Finally, H-OCS was trained using a combination of both loss functions. The results are shown in Table 3.

Table 3. Performance of the proposed approach when using the BCE and/or Jaccard loss functions.

Loss Function	Acc	DC	Sen	IoU
BCE	95.19	79.06	97.26	66.46
Jaccard+BCE	97.04	87.54	86.82	78.56
Jaccard	**97.86**	**88.37**	**89.09**	**79.66**

Using the Jaccard loss function alone yielded more accurate results than using BCE and Jaccard combined. This was due to the morphological appearance of the cup, which is not the same as the disc [11]. Additionally, using BCE alone, the model always overestimated the cup region, which was not the case using Jaccard alone. Combining both loss functions yielded a very similar performance to using Jaccard alone. However, combining both functions helped our model to converge faster, taking only 530 epochs compared to 759 using Jaccard alone. Using BCE alone required only 318 epochs. Throughout the rest of the experiments only the Jaccard loss function was used.

3.5   Comparison with Other Approaches

To evaluate H-OCS, it was compared with approaches which were tested on some of the same datasets as used here, as shown in Table 4. Unfortunately, these approaches did not evaluate using all available datasets and hence when comparing results were split per dataset to be able to do a fair comparison. An

Table 4. Performance of the proposed approach when compared with other approaches when using similar datasets.

Method	Dataset	Performance				Time(s)
		Acc	DC	Sen	IoU	
StackUNet [14]	Drishti	-	89.00	-	80.00	2*–
	RimOneV3	-	84.00	-	73.00	
Depth Estimation Network [15]	Drishti	-	84.80	-	-	2*–
	RimOneV3	-	87.60	-	-	
pOsal [17]	Drishti	-	85.80	-	-	3*0.08
	RimOneV3	-	78.70	-	-	
	Refuge	-	**87.50**	-	-	
ResNet34 [18]	Drishti	-	88.77	-	80.42	2*–
	RimOneV3	-	84.45	-	74.29	
Attention UNET [19]	Drishti	**99.53**	87.93	87.65	78.46	2*0.33
	RimOneV3	**99.69**	83.97	81.33	72.37	
M-Net [3]	Refuge	-	84.39	-	73.00	-
GlaucoNet [7]	Drishti	-	89.99	-	82.29	2*–
	RimOneV3	-	85.07	-	74.01	
FBLS [1]	Drishti	-	88.00	-	-	-
Proposed Approach	Drishti	96.70	**90.05**	87.34	**82.29**	0.08
	RimOneV3	97.78	**88.13**	89.25	79.24	0.17
	Refuge	**97.11**	86.57	92.09	76.75	0.03
	Riga	**98.11**	**86.56**	87.07	76.62	0.11
	Magrabia	**97.32**	**87.27**	89.65	78.04	0.35
	Messidor	**97.95**	**88.67**	87.21	80.03	0.05

overall average accuracy of 97.86%, DC of 88.37%, Sensitivity of 89.09% and IoU of 79.66% was achieved. This approach outperformed other approaches tested on some of the online publicly available datasets as shown in Table 4 except two approaches for some of the dataset they used. Further, a prediction time was achieved that is the best among the current state of the art approaches with average segmentation time is 0.05 s.

Using the Refuge dataset, we achieved higher results than those reported by Fu et al. [3] although the results were slightly lower than those reported by Wang et al. [17] who reported 87.5% for DC in comparison to the 88.57% attained here. For the Drishti-GS and RimOneV3 datasets, this model outperformed other approaches in precisely segmenting the cup. However, Zhao et al. [19] the authors achieved slightly higher accuracy, but H-OCS achieved better in DC, sensitivity, and IoU, indicating that H-OCS performs better in precisely segmenting the cup. Wang et al. [17] achieved an average segmentation time of 0.08 s when using the Drishti, Refuge, and RimOneV3 datasets. On these three datasets, an average segmentation time of 0.04 s was achieved.

The dataset of images provided by Almazroa et al. [2] are so new that to our knowledge no studies have been published using these images. Thus, the performance of H-OCS cannot be compared to those datasets with that of any other model. To ensure the continuity of this research and allow researchers to perform fair comparisons to H-OCS, all the test images use will be published in the supporting material. In general, H-OCS demonstrated high performance in segmenting the cup from images obtained from different sources with different cup angles and resolutions.

4   Conclusion

In this study, we proposed H-OCS, a deep learning-based approach for cup segmentation. We showed the effectiveness of H-OCS when combined with TL, IA, and the Jaccard loss function. Additionally, H-OCS involved a series of post-processing steps that improved the precision of the segmented cup. H-OCS outperforms other approaches to cup segmentation and it achieve state of the art performance on cup segmentation when compared to other approaches. An overall average accuracy of 97.86%, DC of 88.37%, Sensitivity of 89.09% and IoU of 79.66% with an average segmentation time of 0.05 s was achieved with H-OCS.

In the future, we would like to expand our approach to include measurement of the CDR, which can be used as an indicator when diagnosing glaucoma. The CDR alone, however, is not sufficient to diagnose glaucoma. In some cases, the patient may have a low CDR but still have glaucoma. Thus, we would like to combine measurement of the CDR with analysis of the disc region.

References

1. Ali, R., et al.: Optic disc and cup segmentation through fuzzy broad learning system for glaucoma screening. IEEE Trans. Industr. Inform. **17**(4), 2476–2487 (2020)
2. Almazroa, A., et al.: Retinal fundus images for glaucoma analysis: the RIGA dataset. In: Medical Imaging 2018: Imaging Informatics for Healthcare, Research, and Applications, vol. 10579, p. 105790B. International Society for Optics and Photonics (2018)
3. Huazhu, F., Cheng, J., Yanwu, X., Wong, D.W.K., Liu, J., Cao, X.: Joint optic disc and cup segmentation based on multi-label deep network and polar transformation. IEEE Trans. Med. Imaging **37**(7), 1597–1605 (2018)
4. Fu, H., et al.: Segmentation and quantification for angle-closure glaucoma assessment in anterior segment OCT. IEEE Trans. Med. Imaging **36**(9), 1930–1938 (2017)
5. Halır, R., Flusser, J.: Numerically stable direct least squares fitting of ellipses. In: Proceedings of 6th International Conference in Central Europe on Computer Graphics and Visualization. WSCG, vol. 98, pp. 125–132. Citeseer (1998)
6. Orlando, J.I., et al.: Refuge challenge: a unified framework for evaluating automated methods for glaucoma assessment from fundus photographs. Med. Image Anal. **59**, 101570 (2020)

7. Panda, R., Puhan, N.B., Mandal, B., Panda, G.: Glauconet: patch-based residual deep learning network for optic disc and cup segmentation towards glaucoma assessment. SN Comput. Sci. **2**(2), 1–17 (2021)
8. Pena-Betancor, C., et al.: Estimation of the relative amount of hemoglobin in the cup and neuroretinal rim using stereoscopic color fundus images. Investig. Ophthalmol. Vis. Sci. **56**(3), 1562–1568 (2015)
9. Ronneberger, O., Fischer, P., Brox, T.: U-Net: convolutional networks for biomedical image segmentation. In: Navab, N., Hornegger, J., Wells, W.M., Frangi, A.F. (eds.) MICCAI 2015. LNCS, vol. 9351, pp. 234–241. Springer, Cham (2015). https://doi.org/10.1007/978-3-319-24574-4_28
10. Russakovsky, O., et al.: Imagenet large scale visual recognition challenge. Int. J. Comput. Vision **115**(3), 211–252 (2015)
11. Sarhan, A., et al.: Utilizing transfer learning and a customized loss function for optic disc segmentation from retinal images. arXiv preprint arXiv:2010.00583 (2020)
12. Sarhan, A., Rokne, J., Alhajj, R.: Glaucoma detection using image processing techniques: a literature review. Comput. Med. Imaging Graph. **78**, 101657 (2019)
13. Sarhan, A., Rokne, J., Alhajj, R., Crichton, A.: Transfer learning through weighted loss function and group normalization for vessel segmentation from retinal images. arXiv preprint arXiv:2012.09250 (2020)
14. Sevastopolsky, A., Drapak, S., Kiselev, K., Snyder, B.M., Keenan, J.D., Georgievskaya, A.: Stack-u-net: refinement network for image segmentation on the example of optic disc and cup. arXiv preprint arXiv:1804.11294 (2018)
15. Shankaranarayana, S.M., Ram, K., Mitra, K., Sivaprakasam, M.: Fully convolutional networks for monocular retinal depth estimation and optic disc-cup segmentation. IEEE J. Biomed. Health Inform. **23**(4), 1417–1426 (2019)
16. Sivaswamy, J., Krishnadas, S.R., Joshi, G.D., Jain, M., Tabish, A.U.S.: Drishti-GS: retinal image dataset for optic nerve head (ONH) segmentation. In: International Symposium on Biomedical Imaging (ISBI), pp. 53–56. IEEE (2014)
17. Wang, S., Lequan, Yu., Yang, X., Chi-Wing, F., Heng, P.-A.: Patch-based output space adversarial learning for joint optic disc and cup segmentation. IEEE Trans. Med. Imaging **38**(11), 2485–2495 (2019)
18. Shuang, Yu., Xiao, D., Frost, S., Kanagasingam, Y.: Robust optic disc and cup segmentation with deep learning for glaucoma detection. Comput. Med. Imaging Graph. **74**, 61–71 (2019)
19. Zhao, X., Wang, S., Zhao, J., Wei, H., Xiao, M., Ta, N.: Application of an attention U-Net incorporating transfer learning for optic disc and cup segmentation. Signal Image Video Process. **15**(5), 913–921 (2021)
20. Zilly, J., Buhmann, J.M., Mahapatra, D.: Glaucoma detection using entropy sampling and ensemble learning for automatic optic cup and disc segmentation. Comput. Med. Imaging Graph. **55**, 28–41 (2017)

Retinal Vessel Segmentation Using Blending-Based Conditional Generative Adversarial Networks

Suraj Saxena[1], Kanhaiya Lal[1], and Sharad Joshi[2]([✉])

[1] Department of Computer Science and Engineering, Birla Institute of Technology, Mesra, Ranchi, India
{mtcs15008.19,klal}@bitmesra.ac.in
[2] Multimedia Analysis and Security (MANAS) Lab, Electrical Engineering, Indian Institute of Technology Gandhinagar (IIT-GN), Gandhinagar, India
sharad.joshi@iitgn.ac.in

Abstract. With a critical need for faster and more accurate diagnosis in medical image analysis, artificial intelligence plays a critical role. Precise artery segmentation and faster diagnosis in retinal blood vessel segmentation can be beneficial for the early detection of acute diseases such as diabetic retinopathy and glaucoma. Recent advancements in deep learning have led to some exciting improvements in the field of medical image segmentation. However, one common problem faced by such methods is the limited availability of labelled data to train a suitable deep learning model. The publicly available dataset for retinal vessel segmentation contains less than 50 images. On the other hand, deep learning is a data-hungry process. We propose a method to generate synthetic images to augment the training needs of the deep learning model. Specifically, we propose a blending and enhancement-based strategy to learn a conditional generative adversarial model. The network synthesizes high-quality fundus images used along with the real images to learn a convolutional neural network-based segmentation model. Experimental evaluation shows that the proposed synthetic generation method improves segmentation performance on the real test images of the vascular extraction (DRIVE) dataset achieving 97.01% segmentation accuracy.

Keywords: Retinal vessel segmentation · Image synthesis · Medical image segmentation · Generative adversarial networks · Convolutional neural network

1 Introduction

Diabetic retinopathy, glaucoma, and other retinal diseases rely heavily on retinal blood vessels for diagnosis and treatment [1]. The automated segmentation of medical images is a significant step for collecting valuable information that can help doctors diagnose. For instance, it can be used to segment retinal vessels so

N. Tsapatsoulis et al. (Eds.): CAIP 2021, LNCS 13052, pp. 135–144, 2021.
https://doi.org/10.1007/978-3-030-89128-2_13

that their structure can be interpreted and their width calculated, which can help diagnose retinal diseases [2]. The Manual inspection by medical practitioners is a tedious and painful task, and until complex tests like MRI (magnetic resonance imaging) scans are not performed, which can be proven costly [3]. Despite this, the mechanism is nevertheless hampered by a scarcity of datasets. Fluorescein angiography [4] is the standard method for photographing blood vessels. A dye is injected into the bloodstream, which then travels through the blood vessels, exposing them and being photographed later. This procedure appears to be painless and straightforward, but it has several negative consequences, including breathing difficulties, laryngeal swelling, nausea, fainting and cardiac arrest [5]. Vessel segmentation is an important medical imaging problem. However, since CNN needs a large number of images to produce an appropriate model, dataset scarcity is a significant challenge. Traditionally, Chaudhuri et al. [6] accomplished this aim by using the Gaussian curve to estimate the grey-level characterizations of the retinal vessel and the Otsu-thresholding method to segment the vessels. Zana and Klein [7] came up with a vessel segmentation approach based on mathematical structure. Li et al. [8] proposed a cross-modality learning approach in deep learning techniques that learns from a copy of a retinal image and its corresponding vessel map. Several exciting CNN-based approaches have been proposed recently [9–13]. Our proposed approach addresses the issue of restricted dataset availability by employing a blending and enhancement-based synthetic image generation technique to increase the size of the original dataset. Specifically, a retinal fundus image is synthetically generated from an original image by combining ground truth and mask from publicly accessible datasets to develop an improved blended image. Further, we train a CNN-based segmentation model to retrieve segmented vessel images. The major highlights of the proposed approach are: (1) it employs a blending, and enhancement-based strategy for producing synthetic images resulting in high-quality images, (2) experiments show that training the CNN model with a mixture of blended-enhanced synthetic images and original images improves performance on the real test images of the vascular extraction (DRIVE) dataset achieving 97.01% segmentation accuracy.

2   Proposed Method

The overview of the proposed method is in Fig. 1. The proposed method addresses the issue of the very low amount of training data by producing high-quality synthetic fundus images. To achieve this, we propose a blending and enhancement based technique (step 1) to generate a synthetic image using a conditional generative adversarial network (step 2). The synthetic images, along with the original images, are further used to learn a CNN-based segmentation model (step 3). Our synthetic image generation approach uses image-to-image translation GAN [14]. A detailed explanation of the process is provided in the following subsections.

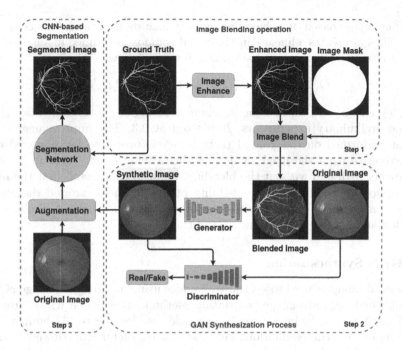

Fig. 1. Oveview of our proposed method

2.1 Datasets

Our method is put to the test on two benchmark datasets covering a wide range of cylindrical organized images, including retinal blood vessels and the optic disc. They are DRIVE [15] and CHASE-DB1 [16], respectively. Both DRIVE and CHASE-DB1 contain 20 training examples and 20 testing images of size 584 × 565 and 999 × 965, respectively. Official FoV masks for the test images are included in the DRIVE dataset. On the other hand, the CHASE-DB1 dataset does not provide such masks with the fundus images. For adequate evaluation, we generate the FoV masks for CHASE-DB1 using simple binary thresholding of the red channel of the colour fundus image. Images manually segmented by trained human observers are included in the datasets and serve as ground truth during testing. We employ three alternative rotation angles: 90, 180, and 270 °C. As a result, there are three actual image variations and three synthetic image variations for a total of 160 images to train the segmentation network.

2.2 Blending and Enhancement-Based Strategy

Before synthesizing a high-quality image using Conditional GAN (cGAN), we use a unique blending and enhanced step as shown in Fig. 1 (step 1). First, we apply contrast enhancement to the segmented ground truth of retinal fundus images. Then, we perform a blending operation on a pair of contrast-enhanced ground truth images and masks of the retinal fundus images. This generates an output image based on a blending ratio, which defines the influence of each image. Thus, it creates a linear combinatorial composite blended image as shown in Fig. 1 (step 2). Blending operation is defined as follows [17],

$$Q(i,j) = X \times P_1(i,j) + (1-X) \times P_2(i,j) \qquad (1)$$

where, P_1 and P_2 are input images, whereas X denotes blending ratio. Based on some preliminary experiments, R is fixed at 0.3. The improvement of the contrast performed on the ground truth image allows it to be more reflective and distinctive from the mask, which acts as a foreground. Some preliminary experiments with and without the blending operation suggested that the image blending operation increased the visibility of the perimeter around the generator input image by severing the black pixel values, which helped eliminate the distortion and conceal the colours around the edges.

2.3   GAN Synthesization

The blended image is used to synthesize images using an image-to-image cGAN, an established image-to-image translation method. It has been widely used for several applications, including texture transfer of clothing to 3D human models [18] and generating animations for a cartoon character [19]. However, unlike traditional GAN, which uses a random noise vector, the parameters of a cGAN are learned from an observed image. Similar to a traditional GAN architecture, the generator and discriminator act as adversaries. The work of the discriminator remains unchanged, which is to work in opposition to the generator and maximize the loss. The loss function of discriminator is defined as [14],

$$\mathcal{L}_{cGAN}(G,D) = \mathbb{E}_{x,y}[\log D(x,y)] + \mathbb{E}_{x,z}[\log(1 - D(x, G(x,z)))] \qquad (2)$$

On the other hand, to minimize the loss and blurring generator, it uses L1 loss multiplied by lambda value which acts as a regularizer [20]. Equation 3 shows the generator function.

$$\mathcal{L}_{L1}(G) = \mathbb{E}_{x,y,z}\left[\|y - G(x,z)\|_1\right] \qquad (3)$$

Equation 4 can summarize the overall process of the cGAN, which learns the mapping from the contrast-enhanced input image (z) and ground truth fundus image (x) to synthesized image (y)

$$G^* = \arg \min_G \max_D \mathcal{L}_{cGAN}(G,D) + \lambda \mathcal{L}_{L1}(G) \qquad (4)$$

As shown in Fig. 1, step 2, the process begins with a combined blended image that includes a segmented vessel and a masked image. The ground truth is the fundus Image, which corresponds to the same vessel and mask. cGAN generates a synthesized image from the blended image and the original fundus image, respectively.

Image-to-Image cGAN maps a high-resolution input to a high-resolution output, with the underlying structure of both being similar. Based on these inputs

and output aligned parameters, an encoder-decoder [21] network in the shape
of a UNet [22] network was formed, consisting of a sequential layer comprised
of convolution layers (conv2D & conv2D Transpose), batch normalization layer
(BN) [23], and activation layer (ReLU & leaky ReLU) [24] and skip-connections
to bypass data across the bottleneck layer in the case of a generator. A simple
decreasing order sequential layer-based model was used to evaluate real and fake
in the discriminator, followed by binary cross-entropy loss [25]. The network was
trained using the Adam optimizer [26] with a learning rate of $2e-4$ and $\beta_1 = 0.5$
and a batch size of 1. The synthesized images (1024 × 1024) are resized to the
original image size for adaptability during CNN training.

2.4 CNN-Based Segmentation

During this stage, the augmented images are fed into the convolutional neural network. This study employed a network with a spatial attention module-
based [27] encoder-decoder network with a specific layer like DropBlock [28] to
minimize overfitting and a spatial attention module [29] that leverages an inter-
spatial relationship between features to build a spatial attention map. Binary
cross-entropy loss functions are used to reduce the distance between the ground
truth and expected outputs in the semantic segmentation task. We trained the
model for 100 epochs using the Adam optimizer with a learning rate of $1e-3$
and a kernel size of 20.

3 Experimental Evaluation

The proposed method's efficacy was assessed using two main criteria: qualitative
and quantitative performance using two publicly available datasets. Qualitative
results visually compare various methods and the images produced by their performance on test datasets. The findings are discussed quantitatively in terms
of parametric values, which are measured pixel-wise test accuracy between predicted and ground truth image, sensitivity (SN), specificity (SP), and AUC Score.
We compared our improved approach with several state-of-the-art methods. The
sample segmentation results are shown in Figs. 2 and 3. The current state-of-
the-art approach has limitations when it comes to capturing light-colored blood
vessels. The proposed approach is marginally better at detecting some portions
of retinal vessels, as seen in Figs. 3 (a) and (b). The quantitative results confirm
this as discussed in the following subsections.

3.1 Qualitative Result

The qualitative findings demonstrate how our synthetically generated images
influence segmentation as compared to training the segmentation network, i.e.,
RSAN [27] using only original images. The comparison is shown using colour-
coded vessel segments that are illuminated, with the colour codes illustrated in
Table 1. Figure 2 (a–e) shows images from the DRIVE [15] dataset, from left to

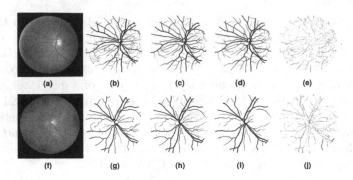

Fig. 2. The qualitative segmentation results of a sample test image from the DRIVE [15] and CHASE-DB1 [16] dataset. (a) and (f) original input image; (b) and (g) inverted ground truth image; (c) and (h) RSAN-segmented image; (d) and (i) and proposed method-segmented image; and (e) and (j) color coded difference between segmented images obtained using RSAN [27] and proposed method as colour-coded in Table 1.

right: original image, human perception-based ground reality, RSAN [27] segmented image, proposed system segmented image, and the difference between highlighted colour-coded image vessels (the zoomed version shown in Fig. 3 (a)). As seen in the zoomed version for CHASE-DB1 [16] dataset (Fig. 3 (b)), our methods trained model, based on a mixture of synthesised and original images, detects certain parts of vessels better RSAN-based [27] approach.

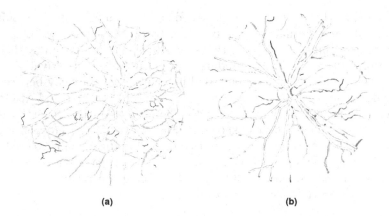

Fig. 3. Zoomed version of color coded difference between segmented images obtained using RSAN [27] and proposed method as colour-coded in Table 1 for DRIVE [15] and CHASE-DB1 [16] dataset.

Table 1. Comparison of ground truth, RSAN [27], and proposed system for vessel segmentation using a confusion matrix

Type	Ground truth	RSAN	Proposed	Color
FN → TP	Vessel	Background	Vessel	Red
FP → TN	Background	Vessel	Background	Green
TP → FN	Vessel	Vessel	Background	Blue
TN → FP	Background	Background	Vessel	Yellow

Table 2. Comparison of proposed method with the state-of-the-art Methods for DRIVE [15] dataset

Method	Test Acc (%)	SP (%)	SN(%)	AUC (%)
DRIU [10]	91.35%	90.88%	**94.54%**	97.84%
VGN [13]	92.71%	93.82%	92.55%	98.02%
Yan et al. (2019)	95.38%	98.20%	76.31%	97.50%
DU-Net [31]	95.66%	98.00%	79.63%	98.02%
Iter-Net [32]	95.73%	98.38%	77.35%	98.16%
RSAN [27]	96.91%	98.39%	81.49%	98.55%
AGN-Net [33]	96.92%	98.48%	81.00%	98.56%
SA-UNet [30]	96.98%	98.40%	82.12%	**98.64%**
Proposed Method	**97.01%**	**98.60%**	80.44%	98.62%

3.2 Quantitative Result

The proposed approach is compared to existing methods for related problems in the DRIVE and CHASE [16] datasets in quantitative results in Tables 2 and 3. Table 2 shows that the proposed approach outperforms all other methods of test accuracy and specificity. DRIU [10] achieves the highest sensitivity, and SA-UNET [30] is achieving the highest AUC score for the DRIVE [15] Dataset. Although the proposed approach provides lower sensitivity, it provides second-best AUC Score. Similarly, Table 3 shows that the proposed method achieves the highest specificity when applied to the CHASE dataset. SA-Unet [30], on the other hand, has the highest test accuracy and AUC ranking. In terms of sensitivity, VGN [13] is at the top of the list.

Table 3. Comparison of proposed method with the state-of-the-art Methods for CHASE-DB1 [16] dataset

Method	Test Acc (%)	SP (%)	SN(%)	AUC (%)
DRIU [10]	93.39%	93.32%	94.11%	98.10%
VGN [13]	93.73%	93.64%	**94.63%**	98.30%
Yan et al. (2019)	96.07%	98.06%	76.41%	97.76%
DU-Net [31]	96.10%	97.52%	81.55%	98.04%
Iter-Net [32]	96.50%	98.23%	79.70%	98.51%
RSAN [27]	97.22%	97.83%	88.07%	98.92%
AGN-Net [33]	97.43%	98.48%	81.86%	98.63%
SA-UNet [30]	**97.55%**	98.35%	85.73%	**99.05%**
Proposed Method	97.28%	**98.55%**	78.34%	98.21%

4   Conclusion

This paper defines a method for improving the capability of any CNN model for vessel segmentation from fundus images considering the practical scenario of the very low amount of labelled training data (20 original images). When compared to the state-of-the-art process, the findings are encouraging. First, we devised a novel method for creating an input image for GAN-based synthesis of the original image by combining image enhancement and blending. This result helped to constrain the perimeter of the produced synthetic image and displayed more highlighted light vessels more dynamically. When using the DRIVE dataset [15], our cGAN-based synthesis takes just 90 min to generate 20 synthetic images. We used an Nvidia GTX 1060 with 6GB memory and a Ryzen 5 3600 processor with 32GB RAM, indicating that synthesizing requires significantly less computing power. Furthermore, the CNN model's result demonstrated state-of-the-art accuracy on the DRIVE dataset [15] when compared against several recently proposed methods. Future research will focus on devising methods for synthesising high-quality, high-definition images using similar methods like pix2pixHD [34], which uses an image translational cGAN approach.

References

1. Diabetic Eye Disease — NIDDK. https://www.niddk.nih.gov/health-information/diabetes/overview/preventing-problems/diabetic-eye-disease
2. Mansar, Y.: Vessel segmentation with python and keras (2020). https://towardsdatascience.com/vessel-segmentation-with-python-and-keras-722f9fb71b21
3. MRI Scan - Cost & Safety measures before undergoing the Test. https://www.tesladiagnostics.com/blog/mri-scan-cost-safety-measures-before-undergoing-the-test

4. Marmor, M.F., Ravin, J.G.: Fluorescein angiography: insight and serendipity a half century ago. Arch. Ophthalmol. **129**(7), 943–948 (2011). https://doi.org/10.1001/archophthalmol.2011.160
5. Musa, F., Muen, W.J., Hancock, R., Clark, D.: Adverse effects of fluorescein angiography in hypertensive and elderly patients. Acta Ophthalmol. Scand. **84**(6), 740–742 (2006)
6. Chaudhuri, S., Chatterjee, S., Katz, N., Nelson, M., Goldbaum, M.: Detection of blood vessels in retinal images using two-dimensional matched filters. IEEE Trans. Med. Imaging **8**(3), 263–269 (1989)
7. Zana, F., Klein, J.C.: Segmentation of vessel-like patterns using mathematical morphology and curvature evaluation. IEEE Trans. Image Process. **10**(7), 1010–1019 (2001)
8. Li, Q., Feng, B., Xie, L., Liang, P., Zhang, H., Wang, T.: A cross-modality learning approach for vessel segmentation in retinal images. IEEE Trans. Med. Imaging **35**(1), 109–118 (2016)
9. Ganin, Y., Lempitsky, V.: n^4-fields: neural network nearest neighbor fields for image transforms (2014)
10. Maninis, K.-K., Pont-Tuset, J., Arbeláez, P., Van Gool, L.: Deep retinal image understanding. In: Ourselin, S., Joskowicz, L., Sabuncu, M.R., Unal, G., Wells, W. (eds.) MICCAI 2016. LNCS, vol. 9901, pp. 140–148. Springer, Cham (2016). https://doi.org/10.1007/978-3-319-46723-8_17
11. Liskowski, P., Krawiec, K.: Segmenting retinal blood vessels with deep neural networks. IEEE Trans. Med. Imaging **35**(11), 2369–2380 (2016)
12. Oliveira, A., Pereira, S., Silva, C.A.: Retinal vessel segmentation based on fully convolutional neural networks. Expert Syst. Appl. **112**, 229–242 (2018). https://doi.org/10.1016/j.eswa.2018.06.034
13. Shin, S.Y., Lee, S., Yun, I.D., Lee, K.M.: Deep vessel segmentation by learning graphical connectivity. Medical Image Anal. **58**, 101556 (2019). http://www.sciencedirect.com/science/article/pii/S1361841519300982
14. Isola, P., Zhu, J., Zhou, T., Efros, A.A.: Image-to-image translation with conditional adversarial networks. In: 2017 IEEE Conference on Computer Vision and Pattern Recognition (CVPR), pp. 5967–5976 (2017)
15. Staal, J., Abramoff, M.D., Niemeijer, M., Viergever, M.A., van Ginneken, B.: Ridge-based vessel segmentation in color images of the retina. IEEE Trans. Med. Imaging **23**(4), 501–509 (2004)
16. Fraz, M.M., Remagnino, P., Hoppe, A., Uyyanonvara, B., Rudnicka, A.R., Owen, C.G., Barman, S.A.: An ensemble classification-based approach applied to retinal blood vessel segmentation. IEEE Trans. Biomed. Eng. **59**(9), 2538–2548 (2012)
17. Poornima, N., Annapurna, V.K.: Image blending and its importance in image processing. Int. J. Sci. Res. Comput. Sci. Eng. Inf. Technol. IJSRCSEIT **6**(4), 888–891 (2018). http://www.ijsrcseit.com/
18. Mir, A., Alldieck, T., Pons-Moll, G.: Learning to transfer texture from clothing images to 3D humans. In: Proceedings of the IEEE/CVF Conference on Computer Vision and Pattern Recognition, pp. 7023–7034 (2020)
19. Poursaeed, O., Kim, V., Shechtman, E., Saito, J., Belongie, S.: Neural puppet: generative layered cartoon characters. In: Proceedings of the IEEE/CVF Winter Conference on Applications of Computer Vision, pp. 3346–3356 (2020)
20. Ng, A.Y.: Feature selection, regularization, and rotational invariance. In: Proceedings of the Twenty-First International Conference on Machine Learning, ICML 2004, p. 78. Association for Computing Machinery, New York (2004). https://doi.org/10.1145/1015330.1015435

21. Hinton, G.E., Salakhutdinov, R.R.: Reducing the dimensionality of data with neural networks. Science **313**(5786), 504–507 (2006)
22. Ronneberger, O., Fischer, P., Brox, T.: U-Net: convolutional networks for biomedical image segmentation. In: Navab, N., Hornegger, J., Wells, W.M., Frangi, A.F. (eds.) MICCAI 2015. LNCS, vol. 9351, pp. 234–241. Springer, Cham (2015). https://doi.org/10.1007/978-3-319-24574-4_28
23. Ioffe, S., Szegedy, C.: Batch normalization: Accelerating deep network training by reducing internal covariate shift (2015)
24. Xu, B., Wang, N., Chen, T., Li, M.: Empirical evaluation of rectified activations in convolutional network (2015)
25. Zhang, Z., Sabuncu, M.R.: Generalized cross entropy loss for training deep neural networks with noisy labels (2018)
26. Kingma, D.P., Ba, J.: Adam: a method for stochastic optimization (2017)
27. Guo, C., Szemenyei, M., Yi, Y., Zhou, W., Bian, H.: Residual spatial attention network for retinal vessel segmentation (2020)
28. Ghiasi, G., Lin, T.Y., Le, Q.V.: Dropblock: a regularization method for convolutional networks. arXiv preprint arXiv:1810.12890 (2018)
29. Woo, S., Park, J., Lee, J.-Y., Kweon, I.S.: CBAM: convolutional block attention module. In: Ferrari, V., Hebert, M., Sminchisescu, C., Weiss, Y. (eds.) ECCV 2018. LNCS, vol. 11211, pp. 3–19. Springer, Cham (2018). https://doi.org/10.1007/978-3-030-01234-2_1
30. Guo, C., Szemenyei, M., Yi, Y., Wang, W., Chen, B., Fan, C.: SA-UNet: spatial attention u-net for retinal vessel segmentation (2020)
31. Jin, Q., Meng, Z., Pham, T.D., Chen, Q., Wei, L., Su, R.: DUNet: a deformable network for retinal vessel segmentation. Knowl. Based Syst. **178**, 149–162 (2019). https://www.sciencedirect.com/science/article/pii/S0950705119301984
32. Li, L., Verma, M., Nakashima, Y., Nagahara, H., Kawasaki, R.: Iternet: retinal image segmentation utilizing structural redundancy in vessel networks. In: Proceedings of the IEEE/CVF Winter Conference on Applications of Computer Vision, pp. 3656–3665 (2020)
33. Zhang, S., et al.: Attention guided network for retinal image segmentation. In: Shen, D. (ed.) MICCAI 2019. LNCS, vol. 11764, pp. 797–805. Springer, Cham (2019). https://doi.org/10.1007/978-3-030-32239-7_88
34. Wang, T.C., Liu, M.Y., Zhu, J.Y., Tao, A., Kautz, J., Catanzaro, B.: High-resolution image synthesis and semantic manipulation with conditional GANs. In: Proceedings of the IEEE Conference on Computer Vision and Pattern Recognition (2018)

U-Shaped Densely Connected Convolutions for Left Ventricle Segmentation from CMR Images

Khouloud Boukhris[1,2](✉) ⓘ, Ramzi Mahmoudi[2,3] ⓘ, Asma Ben Abdallah[2] ⓘ,
Mabrouk AbdelAli[4] ⓘ, Badii Hmida[4] ⓘ, and Mohamed Hédi Bedoui[2] ⓘ

[1] Faculty of Sciences Monastir, University of Monastir, Monastir, Tunisia
[2] Faculty of Medicine Monastir, Medical Imaging Technology Lab – LTIM-LR12ES06,
University of Monastir, Monastir, Tunisia
[3] Gaspard-Monge Computer-Science Laboratory, Paris-Est University, Mixed Unit
CNRS-UMLV-ESIEE UMR8049, BP99, ESIEE Paris City Descartes,
93162 Noisy Le Grand, France
[4] Radiology Service- UR12SP40 CHU Fattouma Bourguiba, 5019 Monastir, Tunisia

Abstract. Segmentation of cardiac magnetic resonance images (cMRI) remains
a challenging task in the field of scientific research due to its significance in the
medical assessment of cardiovascular diseases. Ensuring accurate segmentation
of the heart structures, mainly the left ventricle cavity, serves to extract important
information and has a major impact on the quantitative analysis of the heart function which helps to conduct the proper diagnosis of doctors. The present paper
introduces a simple and efficient U-shaped convolutional neural network aiming
to accurately segment the LV from cMR images. We applied our architecture for
Left Ventricle (LV) segmentation on cardiac MR images (cMRI), from the Automated Cardiac Diagnosis Challenge (ACDC). Obtained results are promising. This
simple model based on CNN has significantly fewer parameters rendering it less
demanding in terms of computation. Nevertheless, it has provided accurate segmentation. The tested method achieved LV Dice scores of 0.958 at end-systolic
time (ES) and 0.979 at end-diastolic time (ED), which yields a mean Dice score
of 0.968 on the ACDC dataset.

Keywords: cMRI Segmentation · CNN · U-Net · Dense net · Convolutions

1 Introduction

Cardiovascular diseases represent the leading cause of death according to the World
Health Organization. Therefore, they have become a major healthcare issue over past
years worldwide. There are different cardiac imaging techniques for viewing the heart
structures that help in making the right diagnosis of these diseases. One of them is
Cardiovascular magnetic resonance imaging (cMRI) which represents the current gold
standard reference for assessing cardiac function [1]. Indeed, the accurate segmentation
of the left ventricle (LV) from these cardiac images is required to retrieve information

© Springer Nature Switzerland AG 2021
N. Tsapatsoulis et al. (Eds.): CAIP 2021, LNCS 13052, pp. 145–153, 2021.
https://doi.org/10.1007/978-3-030-89128-2_14

on ventricular function, such as left ventricular end-systolic volume (LVESV), the left ventricle end-diastolic volume (LVEDV) and the left ventricle ejection fraction (LVEF) [2]. Consequently, major advances have been made in the field of cardiac image segmentation aiming to evaluate the heart function and establish the right diagnosis and treatment of cardiac diseases.

Before the advent of deep learning, a wealth of techniques had been developed to segment and evaluate the heart function from cardiovascular images including level sets, dynamic programming, active contour, graph cuts, and atlas registration [1, 3, 4]. These early approaches required significant manual intervention by the expert in order to achieve their goals. These first techniques may show promising results on limited datasets, but they generally tend to underperform on large variable datasets. In contrast, deep learning based approaches have proven to be able to overcome these limitations by automatically discovering intricate features from data for object detection and segmentation.

Convolutional neural networks (CNN), which were first introduce by Yann LeCun et al. in 1998 [5], are currently the most widely used techniques in the field of biomedical image classification and segmentation. U-Net [6], which is one of the most remarkable extensions of FCN [7] and therefore of CNN, has proven to be a gold-standard in the field of biomedical segmentation while achieving the highest accuracy [8]. U-Net has received much attention with the field of cardiovascular analysis in the last two years and therefore, several U-shaped architectures have been proposed in the literature for fully automated segmentation of the LV from cine MRI [9–14].

In this paper, we propose a fully automatic deep learning approach for left ventricle LV segmentation in cine MRI. Our proposed method is a U-Net-based architecture using Dense connections [15] in order to achieve fewer parameters while ensuring higher accuracy. This paper is organized as follows. A brief overview of related works is introduced in the next section. Then, the proposed method is presented in Sect. 3. Next, experimental results are provided in Sect. 4. And finally the conclusion and future work are drawn.

2 Related Works

U-Net [6], such as SegNet [16] and PspNet [17], is an encoder-decoder-based architecture that uses skip connections between encoder and decoder blocks. This skip connection consists of concatenating the high-level feature maps from the decoder and the low-level feature maps from the corresponding encoder which have the same spatial resolution (see Fig. 1). In the original U-Net, the encoder is down-sampled in total of 4 times, symmetrically to the decoder which is also up-sampled 4 times. This symmetry enables the model to restore the same size as the input image.

WenjunYan et al. [12] proposed a U-net-based method (OF-net) that integrates temporal information from cine MRI into LV segmentation. They incorporated an optical flow (OF) field to capture the cardiac motion towards adding temporal dimension. For this to happen, they used Res-Blocks [18] incrementing, thereby, the number of parameters and so the execution time.

Isensee et al. [11] used a 3D-U-Net inspired architectures for the segmentation of the left and the right ventricles at the end-systolic and the end diastolic time. Zhang

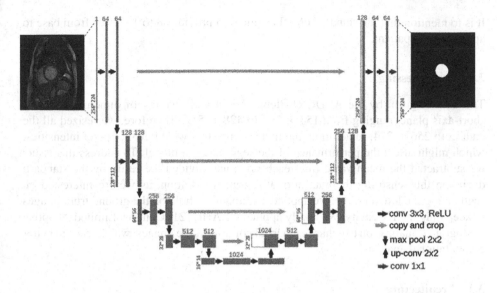

Fig. 1. An illustration of the original U-Net architecture towards LVC segmentation

et al. [19] also combined U-net with SE-Net model in order to reweight the channels of the feature map by giving higher weight to the relevant information and lower weight to the disabled one. Many approaches regarding U-Net have led to good results in LV segmentation from cMR images.

3 Proposed Method

3.1 Dataset

The dataset we adopted in this work is that of The Automated Cardiac Diagnosis Challenge (ACDC). It contains short-axis cMR images along with their corresponding ground truth images of Left Ventricle LV, LV myocardium, and Right Ventricle RV for 100 patients. The ACDC dataset results from clinical examinations acquired at the University Hospital of Dijon France [20].

The 100 patients of the ACDC dataset constitute a total number of 1902 labeled images at both end-systole (ES) and end-diastole (ED) time. In order to enable the evaluation of our method, we divided the labeled data into 80% and 20% which makes 1700 images for the training and 202 for the test. The giving dataset was divided into five subgroups according to the patient's pathology: 20 normal patients, 20 patients with previous myocardial infarction, 20 patients with dilated cardiomyopathy, 20 patients with hypertrophic cardiomyopathy and 20 patients with abnormal right ventricle. The training-test split we have just proposed maintains this subdivision, which means that the 202 test images are composed of four patients from each of these five subgroups.

It is to mention that the standard cMRI acquisition provides 8 to 12 slices from base to apex for each patient.

3.2 Preprocessing

The dataset given by the ACDC challenge has a wide variety of dimensions in the short-axis plane, ranging from 154×224 to 428×512. Therefore, we resized all the dataset to 256×224. In addition, the images present a wide range of pixel intensities, which might affect the performance of the segmentation model. To address this issue, we subtracted the mean value from each pixel and divided the result by the standard deviation thus ensuring the data normalization. In addition, as we are interested on segmenting the left ventricle, we applied a simple threshold on the ground truth images to keep only the LV cavity. We finally applied CLAHE [21] Contrast Limited Adaptive Histogram Equalization to enhance the local contrast of the images, which leads to better computational analysis.

3.3 Architecture

In this study we aim to achieve higher accuracy while considerably reducing the number of trainable parameters. For this to happen, we propose a U-shaped model using Dense Blocks for LV segmentation from cMR images. Our architecture is shown in the figure below (Fig. 2).

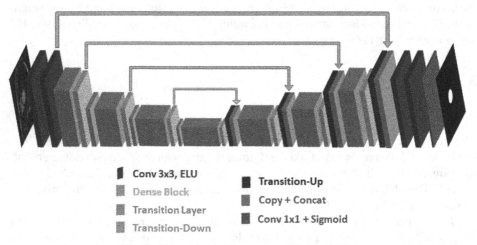

	Conv 3x3, ELU		Transition-Up
	Dense Block		Copy + Concat
	Transition Layer		Conv 1x1 + Sigmoid
	Transition-Down		

Fig. 2. Illustration of the proposed Dense U-Net

As with U-Net, our architecture is down-sampled then up-sampled symmetrically 4 times. In the first level, the input images are fed into two successive 3×3 unpadded convolutions using Exponential Linear Unit (ELU) and followed by a 2×2 max pooling operation with stride 2 for down-sampling.

The next levels are composed of Dense Blocks followed by Transition layers (same depth) that are down-sampled in the contraction path and up-sampled in the symmetric expanding path. Each Dense Block consists of four consecutive convolution layers having the same resolution, each followed by batch normalization (BN), Exponential Linear Unit (ELU) and a dropout layer of 0.2. The output of each convolution in the dense block is concatenated with the input of the following convolutions. The structure of a Dense Block followed by a Transition-Down is illustrated in the figure below (Fig. 3).

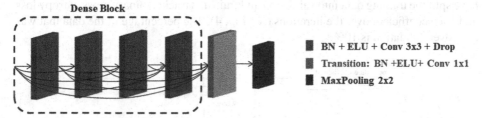

Fig. 3. Illustration of the Dense Block and the Transition-Down

In the contracting path, the filter size of the first dense block starts with 16 and is been duplicated after each down-sampling operation, whilst ensuring symmetry with the expanding path.

Eventually, to obtain the final binary segmentation, the resulting feature maps from the last 3 × 3 convolution layer of the proposed architecture, are agglomerated and averaged by employing a 1 × 1 convolution with a sigmoid activation to predict the probability of each output class. In our case, the number of classes is 1, indicating the LV (left Ventricle).

3.4 Post-processing

The resulting masks are resized to their initial dimensions. And no further post-processing is applied to the resulting segmented images.

3.5 Evaluation Metrics

Several metrics were used in order to evaluate the performance of our method, including accuracy, sensitivity, specificity and dice coefficient. To obtain these metrics, we first need to go through the computation of true Positive (TP), True Negative (TN), False Positive (FP) and False Negative (FN).

$$Accuracy = (TN + TP)/(TN + TP + FN + FP) \tag{1}$$

$$Sensitivity = TP/(TP + FN) \tag{2}$$

$$Specificity = TN/(TN + FP) \tag{3}$$

$$Dice\ coefficient = 2TP/(2TP + FP + FN) \tag{4}$$

4   Experiments and Results

The model was trained using binary cross-entropy as loss function and Adam [22] optimizer with its default parameters, starting with its default learning rate which is set to 0.001. We adopted the "reduce learning rate on the plateau" strategy with the aim of automatically reducing the learning rate. The learning rate was reduced by a constant factor of 0.1 when the loss metric has reached a plateau on the validation set, which varied the learning rate from 0.001 to $1e - 6$ over 32 epochs. For model evaluation, we have split the training data into validation and train and tracked binary cross entropy loss and Dice coefficient over the iterations (see Fig. 4). The percentage of the data that was held over validation is 10%.

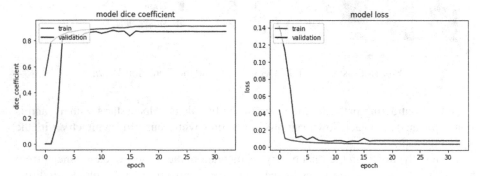

Fig. 4. Visualization of the proposed model history with training and validation

Table 1 presents the evaluation results of the two models (U-Net and the proposed U-shaped densely connected Convolutions) on the previously described test data (202 test images). Both Models were trained using the same preprocessing, the same post-processing and the same hyper parameters including loss function, batch size, learning rate and number of epochs. The U-Net architecture used in this comparison is detailed in the first figure (see Fig. 1).

Table 1. Comparison of LV segmentation performance in terms of Accuracy, Sensitivity, Specificity and Dice coefficient at the end-systolic (ES) and the end-diastolic (ED) time

Method	Accuracy		Sensitivity		Specificity		Dice		Parameters
	ES	ED	ES	ED	ES	ED	ES	ED	
U-Net	0.98	0.98	0.73	0.81	0.97	0.98	0.79	0.81	**31M**
Proposed method	0.99	0.99	0.97	0.98	0.99	0.99	0.95	0.97	**3M**

The large margin of difference between the proposed networks and U-Net could be explained by the use of dense blocks in the lower levels of U-Net which enables extracting abundant local features via densely connected convolutional layers. This has played a crucial role in improving the quality of the segmentation especially when dealing with basal and apical slices (see Fig. 5), in cMRI images, that are found to perform poorly with U-Net and other existing methods in the literature. Basal and apical slices have always been challenging in the literature when it comes to left ventricular segmentation. It is worth mentioning that this improvement is achieved despite a reduced number of trainable parameters that is divided by 10 when compared with with U-Net.

As it may be observed, the number of trainable parameters has decreased from 31 million parameters with U-Net to only 3 million parameters with the proposed method. Our model is less computationally intensive and therefore helps to gain in terms of time.

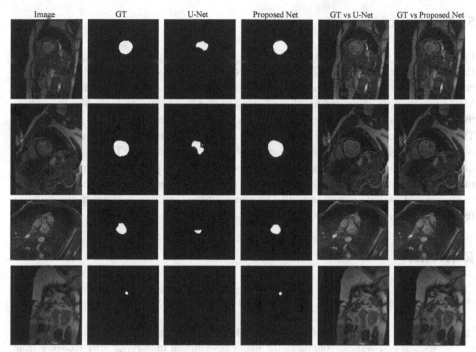

Fig. 5. Qualitative segmentation results of U-Net and the proposed model on the ACDC dataset. The experimental results show that the proposed Dense-U-shaped-Net yields better segmentation masks than the original U-Net especially when dealing with basal and apical slices. The ground truth (GT) is delined in green color with both comparisons.

Even though we established our test on 20% of the ACDC training data, we conducted a comparison with existing state-of-the-art methods set for the left ventricle (LV). Table 2 shows that our approach outperforms other existing methods.

Table 2. Comparison of LV segmentation performance of the proposed method with the state of the art in terms of Dice coefficient at the end-systolic (ES) and the end-diastolic (ED) time

Method	Dice	
	ES	ED
Isensee et al. [11]	0.93	0.96
Simantiris et al. [14]	0.92	0.96
Proposed Method	0.95	0.97

5 Conclusion

In this paper, a simple efficient method for segmenting LV cMR images is proposed. Experimental results on the ACDC dataset show that our U-shaped method with densely connected Convolutions has proven its ability to enhance the performance of cardiac MRI segmentation compared to other existing methods. The use of dense blocks enables the model extracting abundant features, which led to achieve impressive performance. This improvement is provided with reduced number of trainable parameters compared with other existing approaches that make it less time consuming. The obtained results demonstrated the effectiveness of our proposed method in performing precise LV segmentation, which may help establishing an early diagnosis of heart diseases. Further studies could include combining dilated convolutions and dense connections to learn features at different scales.

References

1. Petitjean, C., Dacher, J.-N.: A review of segmentation methods in short axis cardiac MR images. Med. Image Anal. **15**(2), 169–184 (2011). https://doi.org/10.1016/j.media.2010.12.004
2. White, H.D., Norris, R.M., Brown, M.A., Brandt, P.W., Whitlock, R.M., Wild, C.J.: Left ventricular end-systolic volume as the major determinant of survival after recovery from myocardial infarction. Circulation **76**(1), 44–51 (1987). https://doi.org/10.1161/01.CIR.76.1.44
3. Pluempitiwiriyawej, C., Moura, J.M.F., Lin Wu, Y.-J., Ho, C.: STACS: new active contour scheme for cardiac MR image segmentation. IEEE Trans. Med. Imaging **24**(5), 593–603 (2005). https://doi.org/10.1109/TMI.2005.843740
4. Feng, C., Zhang, S., Zhao, D., Li, C.: Simultaneous extraction of endocardial and epicardial contours of the left ventricle by distance regularized level sets. Med. Phys. **43**(6(Part 1)), 2741–2755 (2016)
5. Lecun, Y., Bottou, L., Bengio, Y., Haffner, P.: Gradient-based learning applied to document recognition. Proc. IEEE **86**(11), 2278–2324 (1998). https://doi.org/10.1109/5.726791
6. Ronneberger, O., Fischer, P., Brox, T.: U-Net: convolutional networks for biomedical image segmentation. In: Navab, N., Hornegger, J., Wells, W.M., Frangi, A.F. (eds.) MICCAI 2015. LNCS, vol. 9351, pp. 234–241. Springer, Cham (2015). https://doi.org/10.1007/978-3-319-24574-4_28

7. Long, J., Shelhamer, E., Darrell, T.: Fully convolutional networks for semantic segmentation, pp. 3431–3440 (2015). Accessed 28 Oct 2020
8. Rizwan, I., Haque, I., Neubert, J.: Deep learning approaches to biomedical image segmentation. Inform. Med. Unlocked **18**, 100297 (2020)
9. Jang, Y., Hong, Y., Ha, S., Kim, S., Chang, H.-J.: Automatic segmentation of LV and RV in cardiac MRI. In: Pop, M., et al. (eds.) STACOM 2017. LNCS, vol. 10663, pp. 161–169. Springer, Cham (2018). https://doi.org/10.1007/978-3-319-75541-0_17
10. Khened, M., Alex, V., Krishnamurthi, G.: Densely connected fully convolutional network for short-axis cardiac cine MR image segmentation and heart diagnosis using random forest. In: Pop, M., et al. (eds.) STACOM 2017. LNCS, vol. 10663, pp. 140–151. Springer, Cham (2018). https://doi.org/10.1007/978-3-319-75541-0_15
11. Isensee, F., Jaeger, P.F., Full, P.M., Wolf, I., Engelhardt, S., Maier-Hein, K.H.: Automatic cardiac disease assessment on cine-MRI via time-series segmentation and domain specific features. In: Pop, M., et al. (eds.) STACOM 2017. LNCS, vol. 10663, pp. 120–129. Springer, Cham (2018). https://doi.org/10.1007/978-3-319-75541-0_13
12. Yan, W., Wang, Y., van der Geest, R.J., Tao, Q.: Cine MRI analysis by deep learning of optical flow: adding the temporal dimension. Comput. Biol. Med. **111**, 103356 (2019). https://doi.org/10.1016/j.compbiomed.2019.103356
13. He, Y., et al.: Automatic left ventricle segmentation from cardiac magnetic resonance images using a capsule network. J. X-Ray Sci. Technol. **28**(3), 541–553 (2020)
14. Simantiris, G., Tziritas, G.: Cardiac MRI segmentation with a dilated CNN incorporating domain-specific constraints. IEEE J. Sel. Top. Signal Process. **14**(6), 1235–1243 (2020). https://doi.org/10.1109/JSTSP.2020.3013351
15. Huang, G., Liu, Z., van der Maaten, L., Weinberger, K.Q.: Densely connected convolutional networks, pp. 4700–4708 (2017). Accessed 01 May 2021
16. Badrinarayanan, V., Kendall, A., Cipolla, R.: SegNet: a deep convolutional encoder-decoder architecture for image segmentation. IEEE Trans. Pattern Anal. Mach. Intell. **39**(12), 2481–2495 (2017). https://doi.org/10.1109/TPAMI.2016.2644615
17. Zhao, H., Shi, J., Qi, X., Wang, X., Jia, J.: Pyramid scene parsing network, pp. 2881–2890 (2017). Accessed 01 May 2021
18. He, K., Zhang, X., Ren, S., Sun, J.: Deep Residual Learning for Image Recognition, pp. 770–778 (2016). Accessed 02 May 2021
19. Zhang, J., Du, J., Liu, H., Hou, X., Zhao, Y., Ding, M.: LU-NET: an improved U-Net for ventricular segmentation. IEEE Access **7**, 92539–92546 (2019). https://doi.org/10.1109/ACCESS.2019.2925060
20. Bernard, O., et al.: Deep learning techniques for automatic MRI cardiac multi-structures segmentation and diagnosis: is the problem solved? IEEE Trans. Med. Imaging **37**(11), 2514–2525 (2018). https://doi.org/10.1109/TMI.2018.2837502
21. Zuiderveld, K.: Contrast limited adaptive histogram equalization. Graph. Gems **4**, 474–485 (1994)
22. Kingma, D.P., Ba, J.: Adam: A Method for Stochastic Optimization (2017). https://arxiv.org/abs/1412.6980. Accessed 03 May 2021

Deep Learning Approaches for Head and Operculum Segmentation in Zebrafish Microscopy Images

Navdeep Kumar[1]([✉]), Alessio Carletti[2], Paulo J. Gavaia[2], Marc Muller[3],
M. Leonor Cancela[4], Pierre Geurts[1], and Raphaël Marée[1]

[1] Montefiore Institute, Department of EE & CS, University of Liège, Liège, Belgium
nkumar@uliege.be
[2] Centre of Marine Sciences (CCMAR), University of Algarve, Faro, Portugal
[3] Laboratory of Organogenesis and Regeneration (GIGA Research),
University of Liège, Liège, Belgium
[4] Centre of Marine Sciences (CCMAR), Faculty of Medicine and Biomedical Sciences
and Algarve Biomedical Centre, University of Algarve, Faro, Portugal

Abstract. In this paper, we propose variants of deep learning methods
to segment head and operculum of the zebrafish larvae in microscopy
images. In the first approach, we used a three-class model to jointly
segment head and operculum area of zebrafish larvae from background.
In the second, two-step, approach, we first trained binary segmentation
model to segment head area from the background followed by another
binary model to segment the operculum area within cropped head area
thereby minimizing the class imbalance problem. Both of our approaches
use a modified, simpler, U-Net architecture, and we also evaluate different
loss functions to tackle the class imbalance problem. We systematically
compare all these variants using various performance metrics. Data and
open-source code are available at https://uliege.cytomine.org.

Keywords: Image segmentation · Deep learning · Zebrafish images

1 Introduction

Biomedical research heavily uses Zebrafish (*Danio rerio*) as a model to study
developmental processes. In the earlier stages of their lifecycle, zebrafish embryos
and larvae are translucent, which greatly facilitates monitoring of their developmental organs such as operculum and vertebral column using microscopy techniques [3,5,10]. Biomedical researchers also rely on microscopy to study the
effects of various chemical compounds on the developing parts of the fish model
in toxicological studies [1]. Such analyses often involve segmenting different categories of regions of interest (ROI) within images in order to quantify their morphological changes. For example, the analysis of *Head* and *Operculum* (a series
of bone) regions of Zebrafish larvae and the quantification of the operculum-
to-head ratio is considered as a good marker of increased bone formation and

© Springer Nature Switzerland AG 2021
N. Tsapatsoulis et al. (Eds.): CAIP 2021, LNCS 13052, pp. 154–164, 2021.
https://doi.org/10.1007/978-3-030-89128-2_15

mineralization and it is a validated method to screen for bioactive compounds which have effects on bones [7,12]. It also gives an additional information on the possible toxicity of a compound at the organism level. However, the visual examination and area quantification are a bottleneck and prevent applying such a workflow at high throughput.

In this paper, supervised deep learning strategies are proposed and evaluated to segment head and operculum regions, as evaluation of such approaches has not been proposed previously. We describe image acquisition settings, our dataset, and our methods in Sect. 2, and we present our results in Sect. 3.

2 Methodology

In the section, we describe image acquisition procedure and dataset description followed by two deep learning strategies and provide more details about convolutional neural network (CNN) architectures that have been used to segment head and operculum areas.

2.1 Image Acquisition and Dataset Description

Zebrafish larvae stained with alizarin red S were imaged using a MZ 7.5 fluorescence stereomicroscope (Leica, Wetzlar, Germany) equipped with a green light filter (λex = 530–560 nm and λem = 580 nm) and a black-and-white F-View II camera (Olympus, Hamburg, Germany). Images were acquired using the following parameters: exposure time 1 s, gamma 1.00, image format 1376×1032 pixels, binning 1×1. For morphometric analysis, color channels of the RGB images were split. Red channel (8-bit) images were used for further analyses.

We follow a supervised deep learning approach that requires original images and corresponding head and operculum ground-truth masks to design and validate the approach. Our dataset consists of 8-bit single channel (red channel) fluorescence images of zebrafish larvea at 6 days post fertilization (dpf). Red channel fluorescence images were first transformed into greyscale images (with contrast and brightness enhancement) to ease the manual annotations by experts of head and operculum areas. Manual annotations (illustrated in Fig. 1) consist of contours of head area and operculum area as the main objective is to compute the operculum-to-head ratios for different experimental conditions. A total of 2293 zebrafish images of 1376×1032 resolution have been collected and manually annotated over a period of one year. The dataset consists of 28 different sets of experiments using 5 different compounds, to analyse their effect on the operculum of the zebrafish larvea. Each set has been acquired with the same acquisition settings. Manual annotations were imported into Cytomine open-source software [9] to centralize data and ease binary masks creation to be further used as inputs of deep learning algorithms.

2.2 Two Deep Learning Strategies

One-Step Segmentation with a Three-Class Model. Following this strategy, original size images without cropping are used. Since typical CNN networks require input images of small size (see below), original sized images are first downsized to the size required by the network, keeping their original aspect ratio to avoid any kind of distortions in the predictions, while upsizing the predicted masks. Since our images are rectangular but network require square images, we padded the rectangular images with zeros to make them square. A three-class output segmentation model is then trained to segment both head and operculum from background areas as illustrated in Fig. 1 (top).

Two-Step Segmentation with Two Binary Models. In this approach, a first binary segmentation model is trained to segment the head from the background in original full images downsized appropriately (as in the three-class approach). A second binary segmentation model is trained to segment the operculum area using resized cropped images (rectangular box around the head). At prediction phase, the first model is applied to segment the head, then a rectangular bounding box is automatically extracted. Using these box coordinates, we apply the second model to the resized cropped images (around the head) to segment the operculum area. The two-step approach is illustrated in Fig. 1 (bottom).

2.3   U-Net Implementation

For both approaches, the U-Net architecture [11] has been adapted to segment areas of the zebrafish larvea. The main idea of U-Net is its two parts: the convolution (encoder) or contracting operations, and deconvolutional (decoder) or expanding operations. In the first part, convolutional operations are applied in successive layers with the max pooling operations at the end of each layer, thereby contracting the input resolution. In the second part, an expanding resolution path is adopted using upsampling or deconvolutional layers. The first part is considered as a traditional stack of convolutional and max pooling layers to capture context information within the image. In the second part, deconvolutional operations are applied along a symmetric expanding path to capture the precise localized information. One more important thing about this architecture is its symmetric concatenation of the previous activations from the first part to the activations of the second part.

As preliminary results with the original U-Net architecture on the training set were unsatisfactory (including a tendency to predict only the majority class, i.e. the background), we implemented some modifications in U-Net architecture including the input size and output size of the network and number of layers and filters. Figure 2 shows our "modified U-Net" network architecture.

In our experiments, we used two versions of modified U-Net, one that accepts 512×512 images as input and another that accepts 256×256 images. In both the cases, the output size of the masks is same as the input size whereas in [11], authors used 572×572 inputs and 388×388 outputs. The reason behind using two variants of the network is to assess whether using less parameters will negatively impact recognition performance. Using smaller networks indeed reduces execution times which can be useful in real-time applications. With

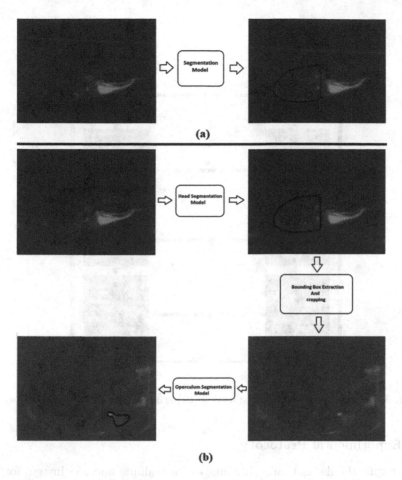

Fig. 1. (a) One-step segmentation approach with three classes: head (yellow contour), operculum (pink contour), background. (b) Two-step binary segmentation approach with a first binary model (head vs background) followed by a second binary model (operculum vs other). (Color figure online)

the small size variant of the U-Net architecture (with 256 × 256 input size), we used fewer filters in each convolutional block as compared to the larger network thereby reducing the network size and the number of parameters by 5 folds. For better optimization, we used "Adam" [6] optimizer and batch normalization in each convolutional block before max pooling. Adam uses *gradient descent with momentum* combined with an adaptive learning rate using exponential moving averages, which makes it more computationally and memory efficient than "Stochastic Gradient Descent" used in the original *U-Net* paper. During training, we also used data augmentation (random flips and rotations, brightness, and contrast changes). We implemented these networks in Python using Tensorflow and Keras [2].

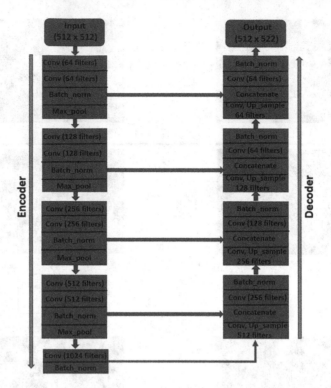

Fig. 2. Modified U-Net architecture used in experiments for 512×512 sized images.

2.4   Experimental Protocol

We first split the dataset into 2105 images for training and 188 images for final evaluation. To assess variability, the set of 2105 images is split into five equally sized folds. Each fold is used in turn as the validation set and the remaining folds as the training set. Five models are trained independently on each training set for 1000 epochs and the five best models on their corresponding validation set across the epochs are finally retained as the final models. In addition, we used early stopping, which forces the training to stop when there is no improvement in the training loss for 100 consecutive epochs. We report below the average performance and standard deviation of these five models estimated on the test set.

2.5   Model Training with Different Loss Functions

In both approaches, we used deep learning based semantic segmentation approach in which a model predicts the class of every pixel in the image (dense predictions). In such a setting, we are faced with a problem of class imbalance as less than 2% of the image area is occupied by operculum region while around 90% is background region. In such situation, the contribution of the majority class (in our case, the background) in the loss during training is more important than that of the minority class, which biases the model in favor of the majority class while ignoring minority class. While the two-step approach tends to reduce this phenomenon (by cropping then predicting operculum only within the head region), a certain class imbalance still persists. Therefore, for both approaches we propose to evaluate different loss functions during training to handle class imbalance. Namely, we evaluated the Cross Entropy Loss, Dice Loss, Tversky Loss, Focal Loss and Jaccard Loss [8].

3   Results and Discussion

Tables 1 and 3 show the results of the first (*three-class* segmentation) app-roach whereas Tables 2 and 4 of the second (*two-step binary class segmentation*) approach using 256×256 input size and 512×512 input size networks, respec-tively. In both the variants, we report several performance metrics that take into account class imbalance, namely *Precision, Recall, F1 Score* and *Dice score*, computed at the pixel level and averaged over the 5 models. To get a single score with which to compare the models, the Dice score is further averaged over head and operculum. Its standard deviation over the 5 models is also provided to assess variability.

Table 1. Segmentation results with the one-step, three-class approach using different loss functions for input size 256.

Avg. scores with three-class output based segmentation					
Loss function	Class	Precision	Recall	F1 Score	Dice Score ± Std
Cross Entropy	Head	0.9806	0.9796	0.9801	0.9412 ± 0.0043
	Operculum	0.8780	0.9263	0.9014	
Tversky loss	Head	0.9779	0.9806	0.9792	0.9470 ± 0.0017
	Operculum	0.9086	0.9190	0.9136	
Dice loss	Head	0.9819	0.9806	0.9813	0.9462 ± 0.0024
	Operculum	0.9120	0.9092	0.9106	
Jaccard Loss	Head	0.9678	0.9789	0.9733	0.49 ± 0.0002
	Operculum	0.0	0.0	0.0	
Focal loss	Head	0.9820	0.9798	0.9809	0.9442 ± 0.0046
	Operculum	0.9060	0.9076	0.9066	

In the three-class approach, the Tversky Loss seems to better cope with the strong class imbalance in both 512×512 and 256×256 settings. The worst performer in the three-class approach is Jaccard loss as it only predicts the majority class (90% background) and no operculum area. This loss leads however to good predictions with the two-step binary approach in both input size settings.

Table 2. Segmentation results with the two-step, binary approach using different loss functions for input size 256.

Avg. scores with two binary-class output based segmentation					
Loss function	Class	Precision	Recall	F1 Score	Dice Score ± Std
Cross Entropy	Head	0.9832	0.9805	0.9819	0.9540 ± 0.0015
	Operculum	0.9196	0.9340	0.9267	
Tversky loss	Head	0.9824	0.9806	0.9815	0.9524 ± 0.0024
	Operculum	0.9104	0.9374	0.9237	
Dice loss	Head	0.9828	0.9826	0.9827	0.9511 ± 0.0046
	Operculum	0.9175	0.9276	0.9225	
Jaccard Loss	Head	0.9782	0.9826	0.9804	0.9513 ± 0.0012
	Operculum	0.9124	0.9355	0.9238	
Focal loss	Head	0.9835	0.9815	0.9825	0.9516 ± 0.0018
	Operculum	0.9213	0.9261	0.9236	

Table 3. Segmentation results with the one-step, three-class approach using different loss functions for input size 512.

Avg. scores with three-class output based segmentation					
Loss function	Class	Precision	Recall	F1 Score	Dice Score ± Std
Cross Entropy	Head	0.9815	0.9747	0.9781	0.9358 ± 0.0064
	Operculum	0.8992	0.8934	0.8953	
Tversky loss	Head	0.9812	0.9789	0.9800	0.95 ± 0.0011
	Operculum	0.9090	0.9308	0.9196	
Dice loss	Head	0.9822	0.9744	0.9783	0.9428 ± 0.0043
	Operculum	0.9085	0.9066	0.9074	
Jaccard Loss	Head	0.9678	0.9789	0.9733	0.49 ± 0.0002
	Operculum	0.0	0.0	0.0	
Focal loss	Head	0.9817	0.9768	0.9792	0.9364 ± 0.007
	Operculum	0.9078	0.8846	0.8946	

In the two-step binary segmentation approach, all losses are very close except cross entropy in 512 × 512 setting. Overall, the two-step approach for 512 × 512 inputs with Jaccard loss has a slight edge over other losses. We believe that the improved performance of the two-step approach is due to the fact that the second segmentation model works with a cropped, head-focused, dataset. Because of the cropping, the class imbalance is not as severe and the operculum image is not downscaled as much as with the three-class approach. Predictions are thus more precise and less influenced by the class imbalance. Regarding the two input sizes,

Table 4. Segmentation results with the two-step, binary approach, using different loss functions for input size 512.

Avg. scores with two binary-class output based segmentation					
Loss function	Class	Precision	Recall	F1 Score	Dice Score ± Std
Cross Entropy	Head	0.9840	0.9780	0.9810	0.9189 ± 0.0159
	Operculum	0.9114	0.8428	0.8747	
Tversky loss	Head	0.9832	0.9785	0.9808	0.9505 ± 0.0024
	Operculum	0.9223	0.9245	0.9234	
Dice loss	Head	0.9828	0.9797	0.9812	0.9424 ± 0.0057
	Operculum	0.9256	0.8947	0.9097	
Jaccard Loss	Head	0.9818	0.99796	0.9807	0.9516 ± 0.002
	Operculum	0.9178	0.9311	0.9244	
Focal loss	Head	0.9841	0.9732	0.9786	0.9490 ± 0.0031
	Operculum	0.9207	0.9227	0.9217	

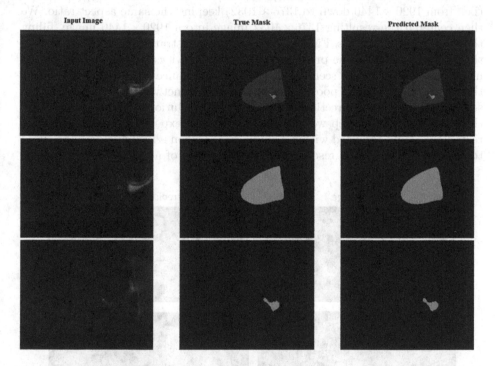

Fig. 3. Sample predictions with best performer on test images with three class (top row) and two-step binary class (last two rows). From the first to third column: input Image, true mask, predicted mask.

we see that they lead to almost identical performance in terms of Dice Score. Sample predictions from the best performing models are shown in Fig. 3.

3.1   Robustness to Image Acquisition with Another Microscope

In practice, microscopes with different acquisition settings might be used over-time by biomedical researchers which raises the issue of robustness of segmentation models to such variabilities, an issue known as domain shift [4]. As a first step towards robustness evaluation, we applied our best two-step binary approach on additional, unlabeled, images acquired with another microscope namely Leica MZ10F fluorescence stereomicroscope equipped with a green fluorescence filter (λex = 546/10 nm), a barrier filter (λem = 590 nm) and a DFC7000T camera (Leica, Wetzlar, Germany) with a different output image size (1920 × 1440). When run on these unprocessed new images, we observe that the performance of our model declines, as illustrated by Fig. 4 (first row). We hypothesized that this is due to the fact that, in the new microscope setting, ROIs (fish head and operculum) are larger in proportion to the size of the full image as compared to ROIs in the original training images. To address this issue, we applied a very simple *pre-processing* step to reduce the scale proportion of ROIs in the image. First, we downscaled the new images to the resolution of the original images (i.e., from 1920 × 1440 down to 1376 × 1032) keeping the same aspect ratio. We then centered the resulting 1376 × 1032 image into a 1920 × 1440 image, filling the new pixels with zeros. Figure 4 (second row) illustrates the positive effect of this pre-processing on the prediction. Note that downscaling further the image in the first step does not seem to affect the performance. We hypothesized that this is due to the use of pooling layers that makes network features somewhat scale invariant (in the direction of a decrease of resolution at least). In practice, this scale calibration step would require a human expert to manually draw a rectangle around the head within a single image when a new microscope is used to initiate the automatic rescaling for the whole set of new images (so that the

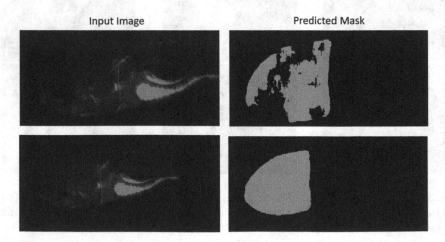

Fig. 4. Robustness evaluation: Predictions from best model using two-step binary class approach on a new image acquired with another microscope before pre-processing (first row), and after pre-processing (second row).

bounding box is rescaled down to the average size of the head in the learning set images). We consider this manual intervention to be acceptable.

4 Conclusions

We have evaluated deep learning based semantic segmentation variants on a new dataset of more than two thousands fluorescent microscopy images of Zebrafish larvae where the goal is to quantify operculum-to-head ratio. The dataset and prediction code compatible with Cytomine open-source web platform [9] is available to foster further research and to enable biomedical experts to routinely use our developments and proofread predictions. We plan to use such developments as the basis of large-scale morphological studies where the effects of different concentrations of many compounds on bone formation and mineralization will be evaluated thoroughly using various statistics (such as operculum-to-head ratio) derived from predicted masks. In the future, it may be necessary to investigate more advanced approaches for other image variations due to change of acquisition setting but ours was sufficient on the new microscope used by our collaborators.

Acknowledgments. This work, as well as N.K. and A. C. are supported by the EU MSCA-ITN project BioMedAqu (766347). R.M. was partially supported by ADRIC Wallonia Grant and EU IMI BIGPICTURE grant. M.M. is a "Maître de Recherche" at the Fund for Scientific Research (F.R.S.–FNRS).

References

1. Cassar, S., et al.: Use of zebrafish in drug discovery toxicology. Chem. Res. Toxicol. **33**(1), 95–118 (2019)
2. Chollet, F., et al.: Keras (2015). https://github.com/fchollet/keras
3. Evans, J.G., Matsudaira, P.: Linking microscopy and high content screening in large-scale biomedical research. High Content Screen. 33–38 (2007)
4. Guan, H., Liu, M.: Domain adaptation for medical image analysis: a survey. arXiv:2102.09508 (2021)
5. Hill, A.J., Teraoka, H., Heideman, W., Peterson, R.E.: Zebrafish as a model vertebrate for investigating chemical toxicity. Toxicol. Sci. **86**(1), 6–19 (2005)
6. Kingma, D.P., Ba, J.: Adam: a method for stochastic optimization. arXiv:1412.6980 (2014)
7. Lessman, C.A.: The developing zebrafish (danio rerio): a vertebrate model for high-throughput screening of chemical libraries. Birth Defects Res. C Embryo Today **93**(3), 268–280 (2011)
8. Ma, J., et al.: Loss odyssey in medical image segmentation. Med. Image Anal. **71**, 102035 (2021)
9. Marée, R., et al.: Collaborative analysis of multi-gigapixel imaging data using cytomine. Bioinformatics **32**(9), 1395–1401 (2016)
10. Mikut, R., et al.: Automated processing of zebrafish imaging data: a survey. Zebrafish **10**(3), 401–421 (2013)

11. Ronneberger, O., Fischer, P., Brox, T.: U-Net: convolutional networks for biomedical image segmentation. In: Navab, N., Hornegger, J., Wells, W.M., Frangi, A.F. (eds.) MICCAI 2015. LNCS, vol. 9351, pp. 234–241. Springer, Cham (2015). https://doi.org/10.1007/978-3-319-24574-4_28
12. Tarasco, M., Laizé, V., Cardeira, J., Cancela, M.L., Gavaia, P.J.: The zebrafish operculum: a powerful system to assess osteogenic bioactivities of molecules with pharmacological and toxicological relevance. Comp. Biochem. Physiol. Part C: Toxicol. Pharmacol. **197**, 45–52 (2017)

Shape Analysis Approach Towards Assessment of Cleft Lip Repair Outcome

Paul Bakaki[1,4](✉) , Bruce Richard[2] , Ella Pereira[1] , Aristides Tagalakis[1] ,
Andy Ness[3] , and Yonghuai Liu[1](✉)

[1] Faculty of Arts and Science, Edge Hill University, Lancashire L39 4QP, UK
{bakakip,pereirae,Aristides.Tagalakis,yonghuai.liu}@edgehill.ac.uk
[2] Birmingham Children's Hospital, Steelhouse Lane, Birmingham B4 6NH, UK
brucerichard@blueyonder.co.uk
[3] British Dental School, University of Bristol, Bristol BS1 2LY, UK
Andy.Ness@bristol.ac.uk
[4] Department of Computer Science, Makerere University,
P.O. Box 7062, Kampala, Uganda

Abstract. Current methods of assessing the quality of a surgically repaired cleft lip rely on humans scoring photographs. This is only practical for research purposes due to the resources necessary and is not used in routine audit. It has poor validity due to human subjectivity and thus low inter-rater reliability. An automatic method for aesthetic outcome assessment of cleft lip repair is required. The appearance and shape of the lips constitute the region of interest for analysis. The mouth borderline and corner points are detected using a bilateral semantic network for real-time segmentation. The bisector of the line linking the mouth corners is estimated as the vertical symmetric axis. By splitting the mouth blob into two parts, they are analyzed for similarity and a numeric score ranging from 1 to 5 is then generated. Pearson correlation coefficient between automatically generated scores and human-assigned ones serves as a validation metric. A correlation of about 40% indicates a good agreement between human and computer-based assessments. However, better automatic scoring correlation of 95.9% exists between the automatically detected mouth regions and those manually drawn by human experts, the third ground truth set in scenario two. Our method has the potential to automate an outcome estimation of the aesthetics of cleft lip repair with human bias reduced, easy implementation and computational efficiency.

Keywords: Cleft lip · Aesthetic assessment · Segmentation ·
Symmetry · Structural similarity · Correlation coefficient

1 Introduction

Cleft lip (CL) is one of the most common maxillofacial congenital deformities affecting about 1 in 500 Asians, 1 in 1000 Caucasians and 1 in 2500 Africans [1].

Supported by Graduate Teaching Assistantship, Edge Hill University.

N. Tsapatsoulis et al. (Eds.): CAIP 2021, LNCS 13052, pp. 165–174, 2021.
https://doi.org/10.1007/978-3-030-89128-2_16

Children with this condition face socio-economic challenges, including high costs for treatment and care (specialized feeding bottles and multiple surgeries), social integration with speech, hearing, and dental problems and potential rejection due to poor facial appearance [2].

Treatment of cleft lip and nose deformity is by surgical repair, usually when the child is 3 to 6 months old and again in early childhood and adolescence to revise or improve the facial appearance [3,4]. An attractive, symmetric and normal appearance of the lip repair is a primary purpose, since people with symmetric faces are more socially acceptable, more confident and have better educational and employment opportunities in life [5]. Current audit practice is to take standard 2D colour photographs to allow an evaluation of the aesthetic appearance of the lip when the child is five years old. But in practice, they are rarely evaluated unless the child is in a research project. Whilst other outcome measures such as mid facial growth, speech, hearing, dental and psychological well-being, which all have internationally accepted and validated outcome measures which are used for audit and research. If there was a validated, efficient outcome measure for the appearance of the lip, it would allow an effective evaluation of the surgical result, be a tool for comparisons of the techniques and protocols of care, and patient/parent satisfaction.

Predominantly, outcome assessment following CL repair is done through qualitative analysis of facial images of the patient. Whilst lip closure is necessary for normal eating, drinking and speaking, the facial beauty aspect is also a primary outcome of the procedure, and is referred to as facial aesthetics (facial appearance). Aesthetic outcome assessment is a research field, that has attracted attention because it has few objective measures. The different approaches for aesthetic outcome assessment are largely indirect in nature, although direct clinical assessment through physical expert observation of the patients is also possible.

Experts create a score of the facial aesthetics based on visualization of images presented to them, either as hard copies, projected on a screen or increasingly through a digital platform. This results in a descriptive qualitative assessment. The Asher-McDade method uses a five-point Likert scale [6] and has been widely used internationally. The image is described as either "Excellent", "Very Good" "Good", "Fair" or "Poor" as each individual expert or lay person may decide. A semi-automatic method, SymNose, was developed to improve objective scoring in [7]. Analyse It Doc (A.I.D.) [8] is an analysis software with modules for subjective and objective assessment/evaluation of aesthetic outcomes. These approaches are still subjective and rely on an emotive interpretation of what is "good" by different human subjects [9].

Given the advancement of computer vision and deep learning technology, this study advances the notion that minimizes human involvement in aesthetic outcome assessment, it will increase the objectivity and validity of any score derived [10]. This study leverages on the fact that digital aesthetic images contain a lot of useful information that can be used in aesthetic assessment research. Such information can be extracted and analyzed to support automatic aesthetic outcome assessment. This study proposes an automatic approach for aesthetic outcome assessment following CL surgery, based on low level features of the lips and mouth region. Our approach uses lip aesthetic assessment method based on the mouth boundary following successful lips segmentation, proven by ground truth.

2   Method

The method has the following main components/steps in the pipeline: mouth detection, symmetrical axis determination, similarity measurement, and numerical score estimation. Mouth detection is vital for clear determination of the visual features of lips, vermilion lines and mouth corners within a given image.

2.1   Dataset and Tools

The data set has 4 classes of 25 facial images, which have been anonymized for ethical reasons to reveal only the nose and mouth/lips. In addition, it was also intended that human assessors are not biased by any other facial features. The first class of 25 images constitute the raw data for aesthetic assessment. The other 3 classes (dubbed as GT1, GT2, and GT3) are ground truth (GT) whose mouth/lip region boundary was already manually drawn by three different human experts respectively using the open source ImageJ software. The 3 ground truths serve as validation for the segmentation approach and the assessment prediction mechanism discussed in this paper. Human numeric scores (HNS) were generated through a subjective aesthetic assessment process aided by statistical coding of assessor's description of the individual images in the raw dataset.

Using our method, all the images of the 4 classes are automatically assessed and a numeric score is then generated. These scores are named AENS, short form for automatically estimated numeric score, with a name prefix of the respective data set, for example, GT1-AENS, and so forth.

The implementation programming language is Python 3.7. The supporting libraries are OpenCV, Matplotlib, PyTorch and Keras.

2.2   Feature Description and Detection

All the images have been anonymized for ethical and other reasons stated previously, implying that some facial features are not available for detection, and thus only limited features can be identified. Our focus is on the features of the mouth region, starting with segmentation. The anatomy of the human mouth region consists of the following key parts: the vermillion border (upper and lower), oral commissures (left and right) and the philtra ridges (left and right, separated by philtrum) [11].

Ideals of facial beauty indicate that the mouth region should be in the lower third of a given facial image [4]. Because the skin color and the lips may be indistinguishable, contrast enhancement and selection of suitable color transform is inevitable. To mitigate this, the segmentation method we used considers the

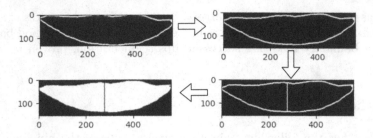

Fig. 1. An example for boundary extraction, rotation, and symmetry axis detection of a cropped mouth lip image. Top row - left: mouth corners are at different elevations from the horizontal axis. Top row - right: After anticlockwise rotation mouth corners are at the same elevation. Bottom row shows the symmetric axis (black and white).

semantics of individual pixels, first discussed in 1987 [12]. While traditional techniques which perform segmentation as a binarization task usually underperform at medical imagery analysis tasks [13], the deep learning based semantic segmentation method [14] has been employed in this study. Even so, residues such as scars, open mouth and runny nose still influence the segmentation outcome.

Semantic segmentation enhances edge detection by creating a sharper contrast between the surrounding skin and the mouth region, hence facilitating shape identification and feature extraction. The ideal mouth region mainly consists of soft tissue features defined below.

- PR_L and PR_R are philtra ridges identified as one of the upper most extreme pixels on the left-hand and right-hand sides of the philtrum, found along the mouth boundary.
- OC_L and OC_R are the left-hand and right-hand side mouth corners identified as the most extreme pixels on the left-hand and right-hand sides, located along the mouth boundary.
- VB_U and VB_B are a list of pixels constituting the upper and lower mouth region boundaries, stretching between OC_L and OC_R.

$$VB_U = \{u_1, u_2, ..., u_n\} \tag{1}$$
$$VB_B = \{b_1, b_2, ..., b_m\} \tag{2}$$

where u_i and b_j are pixels in a given $2D$ grayscale image I.
- The mouth boundary B is a combined list of VB_B and VB_U. Collectively, it is also known as the longest non-nested detected contour in the face, represented in Eq. 3 as

$$B = VB_U \cup VB_B \tag{3}$$

where $VB_U \cap VB_B = \{OC_L, OC_R\}$, $PR_L, PR_R \subset VB_U$ and $OC_L, OC_R \in B$.
- The linking line of OC_L and OC_R is not always parallel to the horizontal axis of the image. Its orientation angle θ to the horizontal axis dictates the magnitude of the rotational transformation (Fig. 1). if $\theta < 0$, rotate *anticlockwise*;

otherwise, rotate *clockwise*. Such orientation may influence how human subjects visualise and assign the numerical score to a given facial image, to be investigated in Sect. 3 below.

2.3 Symmetrical Axis Detection and Measurement

Several approaches have been previously used in general detection of symmetry. Related methods are discussed in [10]. However, those techniques utilized many more local and invariant object features with higher contrasts. This study utilizes basic lip and mouth features instead, similar to the perception of human assessors. The midpoint D given in Eq. 4 is a position where the vertical symmetric axis is plotted through in the image plane.

$$D = (OC_L + OC_R)/2. \tag{4}$$

A vertical straight line plotted through D and crossing the lower and upper mouth boundaries ensures slicing the mouth region into two shapes, left-side shape, sh_l and right-side shape, sh_r. The evenness or variance is computed and categorized using the structural similarity measure, S [15]. S is an aggregated rational number ranging between -1 and 1 for color images or 0 and 1 for binary images. We consider sh_l and sh_r as independent and unique shapes over which to compute S. S is an aggregate of luminance l, contrast c, and structure s, adapted from [15] and expressed in Eq. 5 below as:

$$S(sh_l, sh_r) = [l(sh_l, sh_r)^\alpha \cdot c(sh_l, sh_r)^\beta \cdot s(sh_l, sh_r)^\gamma] \tag{5}$$

where $\alpha = 1$, $\beta = 1$ and $\gamma = 1$ for easy implementation. Since the dimensions of sh_l and sh_r should be similar, sh_r is vertically flipped along the vertically plotted symmetric axis. Setting the different statistical parameters of l, c and s as described in [15] gives the usable form of the parameter S in Eq. 6 below as:

$$S(sh_l, sh_r) = \frac{(2\mu_{sh_l}\mu_{sh_r} + C_1)(2\sigma_{sh_l sh_r} + C_2)}{(\mu_{sh_l}^2 + \mu_{sh_r}^2 + C_1)(\sigma_{sh_l}^2 + \sigma_{sh_r}^2 + C_2)} \tag{6}$$

where μ_{sh_l}, σ_{sh_l}, μ_{sh_r}, and σ_{sh_r} are the mean and standard deviation of pixels in shapes sh_l and sh_r respectively, $\sigma_{sh_l sh_r}$ is the standard deviation of the pixels in sh_l and sh_r, $C_1 = (k_1 L)^2$, $C_2 = (k_2 L)^2$, $k_1 = 0.01$, $k_2 = 0.03$, $L = 2^p - 1$ and p is the number of bits per pixel.

2.4 Conversion of Similarity Measure to a Numeric Score

The structural similarity S is computed and converted to a numeric score in the range of 1 and 5, where 1 = "excellent", 2 = "Very good", 3 = "Good", 4 = "Fair" and 5 "Poor". The transformation $f(S)$ should fulfill the following boundary and monotonicity conditions: $f(0) = 5$, $f(1) = 1$, and $f(S)$ is monotonically decreasing. Therefore, $f(S)$ is thus the finally $AENS$. The following three models

(Eqs. 7, 8 and 9) are designed and selected for a comparative study about what relationship is between S and $AENS$.

$$f(S) = 5 - 4S \tag{7}$$
$$f(S) = 5 - 4S^3 \tag{8}$$
$$f(S) = 1/(0.2 + 0.8S^2) \tag{9}$$

Three scenarios are also considered in Fig. 2 for the generation of sh_l and sh_r for further comparative studies how the two shapes should be defined:

- Scenario 1: Parameters calculated over the entire mouth blob.
- Scenario 2: Parameters calculated over the entire mouth boundary only.
- Scenario 3: Parameters calculated over the upper lip blob only.

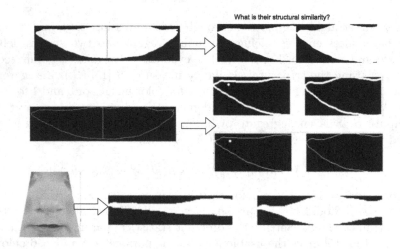

Fig. 2. Different scenarios for parameter calculation. Top: entire mouth region blob (upper and lower lips) has been split into right and left blobs, sh_l and sh_r. sh_r has been flipped. Middle: Scenario 2 with the boundaries defined with different thicknesses of 1 and 3 pixels respectively; Bottom: Scenario 3.

3   Experimental Results

In this section, we present both the qualitative and quantitative experimental results of the proposed automated programmed rating (PR) method compared with the others when applicable.

3.1   Image Segmentation

Facial images were segmented using the bilateral real-time semantic network (SN) [14]. Traditional approaches such as threshold-based segmentation (such as Otsu and moment preservation) and clustering method (K-means (KM) and mean shift (MS)) usually produce unsatisfactory results. A comparative study between the MS (spatial bandwidth = 20, color bandwidth = 7), KM (k = 3) and SN is presented in Fig. 3 where the performance is measured in F1-score in percentage: the higher the better. Clustering-based approaches yielded worse outputs with discontinuous areas and boundaries. Gaussian blurring, morphing and dilation were usually used to mitigate such issues. The segmented mouth region (our RoI) is found in the bottom third of the facial image. Standardization with a bounding box was also used to reduce the background from the image as seen in Fig. 1.

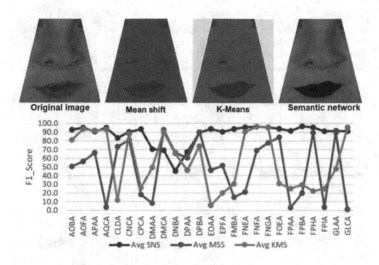

Fig. 3. Segmentation results of images using different techniques.

3.2 Evaluation of Aesthetic Assessment

After computing S based on Scenarios 1, 2, and 3 and converting it to an aesthetic numeric score, our method was evaluated using Pearson correlation coefficient against HNS: the higher the better. Table 1 shows that the orientation standardization is helpful to improve the $AENS$ using the bounding box. This shows that the mouth orientation may affect how the human subjects perceive and thus assign numerical scores to given images.

The performance metrics of our method are presented in Fig. 4 over 3 scenarios, 3 models, and 2 options of the symmetric axis crossing position, D and D_2:

$$D_2 = d(OC_L + OC_R)/2 \tag{10}$$

where shift factor d by 5% inward being most effective, considering that the mouth corners may not be accurately detected due to imaging noise and shadows.

It can be seen that the correlation between HNS and $PR - AENS$ is considered most significant because it is a test directly made between human and

Table 1. AENS before and after standardization of mouth orientation in Scenario 1.

Category	Range		Average	
	Before	After	Before	After
PR	$0.24 < S < 0.82$	$0.55 < S < 0.89$	0.60	0.79
GT1	$0.30 < S < 0.84$	$0.49 < S < 0.83$	0.60	0.72
GT2	$0.28 < S < 0.84$	$0.39 < S < 0.86$	0.60	0.69
GT3	$0.35 < S < 0.82$	$0.51 < S < 0.88$	0.64	0.72

Top two rows: Scenario 1

A (Equation 7) — AENS

	HNS	GT1	GT2	GT3
PR	0.199	0.819	0.862	0.896
GT1	0.180			
GT2	0.252	0.913		
GT3	0.252	0.943	0.926	

C (Equation 8) — AENS

	HNS	GT1	GT2	GT3
PR	0.339	0.769	0.826	0.857
GT1	0.246			
GT2	0.348	0.892		
GT3	0.331	0.935	0.901	

E (Equation 9) — AENS

	HNS	GT1	GT2	GT3
PR	0.123	0.830	0.881	0.914
GT1	0.138			
GT2	0.186	0.923		
GT3	0.196	0.942	0.939	

B (Equation 7) — AENS

	HNS	GT1	GT2	GT3
PR	0.209	0.868	0.817	0.936
GT1	0.201			
GT2	0.147	0.835		
GT3	0.219	0.918	0.885	

D (Equation 8) — AENS

	HNS	GT1	GT2	GT3
PR	0.399	0.844	0.725	0.924
GT1	0.257			
GT2	0.391	0.831		
GT3	0.336	0.924	0.825	

F (Equation 9) — AENS

	HNS	GT1	GT2	GT3
PR	0.154	0.845	0.766	0.944
GT1	0.129			
GT2	0.231	0.763		
GT3	0.188	0.927	0.801	

Middle two rows: Scenario 2

A (Equation 7) — AENS

	HNS	GT1	GT2	GT3
PR	0.299	0.919	0.772	0.953
GT1	0.162			
GT2	0.198	0.855		
GT3	0.286	0.950	0.882	

C (Equation 8) — AENS

	HNS	GT1	GT2	GT3
PR	0.296	0.933	0.793	0.959
GT1	0.176			
GT2	0.198	0.845		
GT3	0.274	0.965	0.868	

E (Equation 9) — AENS

	HNS	GT1	GT2	GT3
PR	0.302	0.891	0.796	0.935
GT1	0.184			
GT2	0.143	0.890		
GT3	0.240	0.912	0.909	

B (Equation 7) — AENS

	HNS	GT1	GT2	GT3
PR	0.306	0.915	0.915	0.953
GT1	0.168			
GT2	0.168	1.000		
GT3	0.276	0.948	0.948	

D (Equation 8) — AENS

	HNS	GT1	GT2	GT3
PR	0.296	0.938	0.938	0.957
GT1	0.182			
GT2	0.182	1.000		
GT3	0.272	0.963	0.963	

F (Equation 9) — AENS

	HNS	GT1	GT2	GT3
PR	0.298	0.903	0.903	0.956
GT1	0.177			
GT2	0.177	1.000		
GT3	0.248	0.905	0.905	

Bottom two rows: Scenario 3

(Equation 7) — AENS

	HNS	GT1	GT2	GT3
PR	0.276	0.936	0.654	0.799
GT1	0.226			
GT2	0.094	0.730		
GT3	0.183	0.890	0.648	

C (Equation 8) — AENS

	HNS	GT1	GT2	GT3
PR	0.298	0.897	0.481	0.712
GT1	0.275			
GT2	0.043	0.597		
GT3	0.215	0.878	0.528	

E (Equation 9) — AENS

	HNS	GT1	GT2	GT3
PR	0.252	0.947	0.740	0.832
GT1	0.219			
GT2	0.059	0.773		
GT3	0.194	0.908	0.683	

B (Equation 7) — AENS

	HNS	GT1	GT2	GT3
PR	0.279	0.945	0.611	0.820
GT1	0.236			
GT2	0.061	0.684		
GT3	0.205	0.898	0.579	

D (Equation 8) — AENS

	HNS	GT1	GT2	GT3
PR	0.298	0.895	0.375	0.718
GT1	0.290			
GT2	0.044	0.555		
GT3	0.219	0.870	0.466	

F (Equation 9) — AENS

	HNS	GT1	GT2	GT3
PR	0.262	0.952	0.693	0.841
GT1	0.214			
GT2	0.035	0.723		
GT3	0.190	0.907	0.640	

Equation 7   Equation 8   Equation 9

Fig. 4. Correlation coefficient results for different scenarios over different transformation models. Odd row: symmetric axis plotted at D; Even row: symmetric axis plotted at D_2. Top two rows: Scenario 1; Middle two rows: Scenario 2; Bottom two rows: Scenario 3.

fully automated computer-based assessment. The highest score is about 40% in Fig. 4 *Table D* in Scenario 1 and the lowest is about 15% in *Table F*, due to inconsistency for human subjects to assign scores from one image to another. Overall, shifting the mouth corners inward improves the most significant correlation across the three models. However, the model in Eq. 8 is the most robust, implying that the mapping from shape similarity measurements to their aesthetic scores is non-linear. In sharp contrast, correlation between $PR - AENS$ and either $GT_1 - AENS$, $GT_2 - AENS$ or $GT_3 - AENS$ is significantly higher, as high as 94% in *Table F* in Scenario 1. This implies that the automatic segmentation of the mouth regions is accurate, compared to human manually drawn ones.

In Scenario 2, the most significant correlation is about 31%, *Table B*. There is little difference in the various correlations over different setups, indicating that the mouth boundaries may not be as predictive as expected. This is somewhat contradictory to the practice that focuses on the vermilion lines and thus requires further investigation. Scenario 3 has produced the lowest correlation value in the category of $PR-AENS$ and either GT_1-AENS, GT_2-AENS or GT_3-AENS on record of as low as 38% in *Table D*. This indicates that the determination of the RoI is still challenging.

However, determining the symmetric axis using fewer features is a potential limitation of the proposed method, future research studies utilizing deep learning techniques such as transfer learning will target improving results. Additionally, the benchmark for the validity of our approach is based on a single method, spearheaded by human experts.

4   Conclusion

This paper proposed an automatic assessment approach that utilizes lips and mouth features. These features are considered appealing to humans and can be distinguishable to support aesthetics judgement. These include oral commissures and the vermillion border. Once the mouth region has been detected using the bilateral network segmentation method and split through the midpoint of the horizontal line linking the mouth corners, the two ensued blobs are analyzed for evenness or difference. To this end, the widely used structural similarity measure [15] was employed. The measure is a rational number, that was then converted non-linearly to a numeric score in the range of 1 and 5, like the Asher-McDade five-point Likert Scale used by human experts. A numerical similarity computation following a symmetric axis computation is a better objective aesthetics assessment of the repaired lips compared to the qualitative measures proposed in [7,8] and [9]. The experimental results show that the automatically estimated scores have relatively low correlation coefficients with human assigned ones but have high correlation coefficients with those estimated from the human manually drawn mouth regions.

It is also noted that inward shift of the mouth corners by 5% improves the accuracy of the midpoint D_2 and offers an alternative for a symmetric axis position to combat the challenging nature in identifying the mouth corners with improved aesthetics assessment scores. Further research will investigate more accurate estimation of the symmetrical axis and difference measurement between the two sides of the mouth regions.

Acknowledgments. The facial images are the cropped and anonymised anteroposterior (A/P) photos of 5-year-old children from the Cleft Care UK (CCUK). This publication presents data derived from the Cleft Care UK Resource (an independent study funded by the National Institute for Health Research (NIHR) under its Programme Grants for Applied Research scheme RP-PG-0707-10034).

References

1. Zhang, Q., et al.: A bibliometric analysis of cleft lip and palate-related publication trends from 2000 to 2017. Cleft Palate-Craniofacial J. **56**, 658–669 (2019)
2. Shkoukani, M.A., Chen, M., Vong, A.: Cleft lip - a comprehensive review. Front. Pediatr. **1**, 1–10 (2013)
3. Mosmuller, D.G.M., et al.: Scoring systems of cleft-related facial deformities: a review of literature. Cleft Palate-Craniofacial J. **50**, 286–296 (2013)
4. Kar, M., Muluk, N.B., Bafaqeeh, S.A., Cingi, C.: È Possibile Definire Le Labbra Ideali? Acta Otorhinolaryngol. Ital. **38**, 67–72 (2018)
5. Little, A.C., Jones, B.C., Debruine, L.M.: Facial attractiveness: evolutionary based research. Philos. Trans. R. Soc. B Biol. Sci. **366**, 1638–1659 (2011)
6. Asher-McDade, C., Roberts, C., Shaw, W.C., Gallager, C.: Development of a method for rating nasolabial appearance in patients with clefts of the lip and palate. Cleft Palate Craniofac. J. **28**(4), 385–390 (1991)
7. Piggot, R.W., Piggot, B.B.: Quantitative measurement of symmetry from photographs following surgery for unilateral cleft lip and palate. Cleft Palate Craniofac. J. **47**(4), 363–367 (2010)
8. Pietrski, P., Majak, M., Antoszewski, B.: Clinically oriented software for facial symmetry, morphology, and aesthetic analysis. Aesthet. Surg. J. **38**(1), NP19–NP22 (2017)
9. Deall, C.E., et al.: Facial aesthetic outcomes of cleft surgery: assessment of discrete lip and nose images compared with digital symmetry analysis. Plast. Reconstr. Surg. **138**, 855–862 (2016)
10. Deng, Y., Loy, C.C., Tang, X.: Image aesthetic assessment: an experimental survey. IEEE Signal Process. Mag. **34**, 80–106 (2017)
11. Berlin, N.F., et al.: Quantification of facial asymmetry by 2D analysis - a comparison of recent approaches. J. Cranio-Maxillofacial Surg. **42**, 265–271 (2014)
12. Chen, S., Leung, H.: Chaotic spread spectrum watermarking for remote sensing images. J. Electron. Imaging **13**(1), 220–230 (2004)
13. Kuruvilla, J., et al.: A review on image processing and image segmentation. In: Proceedings of 2016 International Conference on Data Mining and Advanced Computing, SAPIENCE 2016, pp. 198–203 (2016)
14. Yu, C., Wang, J., Peng, C., Gao, C., Yu, G., Sang, N.: BiSeNet: bilateral segmentation network for real-time semantic segmentation. In: Ferrari, V., Hebert, M., Sminchisescu, C., Weiss, Y. (eds.) ECCV 2018. LNCS, vol. 11217, pp. 334–349. Springer, Cham (2018). https://doi.org/10.1007/978-3-030-01261-8_20
15. Wang, Z., et al.: Image quality assessment: from error visibility to structural similarity. IEEE Trans. Image Process. **13**, 600–612 (2004)

MMEC: Multi-Modal Ensemble Classifier for Protein Secondary Structure Prediction

Gabriel Bianchin de Oliveira[ID], Helio Pedrini[✉][ID], and Zanoni Dias[ID]

Institute of Computing, University of Campinas, Campinas, SP, Brazil
{gabriel.oliveira,helio,zanoni}@ic.unicamp.br

Abstract. The protein secondary structure prediction is an important task with many applications, such as local folding analysis, tertiary structure prediction, and function classification. Driven by the recent success of multi-modal classifiers, new studies have been conducted using this type of method in other domains, for instance, biology and health care. In this work, we investigate the ensemble of three different classifiers for protein secondary structure prediction. Each classifier of our method deals with a transformation of the original data into a specific domain, such as image classification, natural language processing, and time series tasks. As a result, each classifier achieved competitive results compared to the literature, and the ensemble of the three different classifiers obtained 77.9% and 73.3% of Q8 accuracy on the CB6133 and CB513 datasets, surpassing state-of-the-art approaches in both scenarios.

Keywords: Deep learning · Protein secondary structure prediction · Multi-modal classification

1 Introduction

Proteins are responsible for several activities in the cells of living beings, such as immune response and defense systems, being heavily studied on biology. They are formed by basic units called amino acids, which, due to the peptide bonds between them, form a sequence of amino acids, known as protein primary structure. With the iterations between the atoms of the amino acids that form the chain, the protein folds, creating three-dimensional structures.

The three-dimensional structures of proteins are divided into local structures, that is, three-dimensional structures that each of the amino acids forms, called secondary structures [19], and the global folding of the protein, called tertiary structure [8]. The analysis of three-dimensional structures of the protein, mainly tertiary structure, has a great impact in the discovery of possible applications, such as the functions that each protein has and in the development of drugs, such as in the advances in the fight against the pandemic of COVID-19 (SARS-CoV-2) [7]. Even with advances in the direct prediction of tertiary structure from the primary structure [21], this remains an open task, and the main purpose of

© Springer Nature Switzerland AG 2021
N. Tsapatsoulis et al. (Eds.): CAIP 2021, LNCS 13052, pp. 175–184, 2021.
https://doi.org/10.1007/978-3-030-89128-2_17

analyzing the global folding of the protein is to first understand the secondary structures.

The sequencing of amino acids that form proteins became faster and cheaper with recent advances in technology. However, determining three-dimensional structures, such as secondary structures, remains costly, mainly due to the laboratory methods employed. In this sense, computational strategies can be used to predict and reduce costs when analyzing this type of structure.

The prediction of secondary structures consists of determining which of the 8 possible three-dimensional structures a given amino acid belongs: "B" (residue in isolated beta bridge), "E" (extended strand), "G" (3-helix), "H" (alpha helix), "I" (5-helix), "L" (loop), "S" (bend), and "T" (hydrogen-bonded turn) [18]. This type of prediction is known as Q8 classification.

Early studies approached the protein secondary structure prediction problem by analyzing of local structures, that is, focusing on the prediction from nearby structures [14]. However, new approaches were observed after 2010s, where new methods that use deep learning techniques emerged, mainly with recurrent networks [22], convolutional networks [16,20], and both types of architectures [6,10]. With the recent advances in machine learning and deep learning, many efforts have been made using multi-modal classification, that is, classifiers that utilize and combine different modalities of data to make predictions. Moreover, some methods in the literature employ this approach in biological and health care tasks [9,12,24].

In this work, we investigate the ensemble of Convolutional Neural Networks (CNN), Bidirectional Encoder Representations from Transformers (BERT), and Inception Recurrent Networks (IRN) for protein secondary structure prediction. Each classifier examines this task as a single view, that is, CNN predicts the secondary structure using images, BERT classifies the structures as natural language processing task, and IRN predicts the structures as time series data. The main contributions of our work are (i) our method is, to the extend of our knowledge, the first classifier that considers the protein secondary structure prediction task as a multi-modal task, and (ii) we explore and present an ensemble of classifiers with different properties (image classification, natural language processing, and time series tasks) that can deal with complementary characteristics of the data, surpassing state-of-the-art approaches on CB6133 and CB513 datasets.

The remainder of the paper is organized as follows. In Sect. 2, we describe our method for protein secondary structure prediction. In Sect. 3, we present the dataset used and the evaluation metric. In Sect. 4, we report and discuss the experimental results of our method. In Sect. 5, we present our final remarks and some directions for future work.

2   Methodology

In this section, we explain our multi-modal classifier for protein secondary structure prediction, which we called Multi-Modal Ensemble Classifier (MMEC).[1]

---
[1] https://github.com/gabrielbianchin/MMEC.

Our method is divided into Convolutional Neural Networks (CNN), Bidirectional Encoder Representations from Transformers (BERT), and Inception Recurrent Network (IRN). Furthermore, we detail our ensemble method to combine the three different classifiers.

2.1 Convolutional Neural Networks

Position-Specific Score Matrix (PSSM) have been the standard features in the last 20 years, as well as the sequence of the amino acids, for protein secondary structure prediction task. Jones [14] proposed the first method in the literature that employed such characteristic for this task, showing that it may help the prediction and can achieve better results than only using the amino acid sequence.

PSSM features are calculated using PSI-BLAST [1]. It generates $X \times L$ features matrix per protein, that is, X corresponding to the number of possible amino acids, typically 21, depending on the dataset, and L is the length of the sequence. In the end, each amino acid of the sequence has X features, showing the score for each possible amino acid, in the same position of the sequence, based on similar proteins of a search dataset. The standard dataset used to generate this information is UniRef90 [25]. The resulting values for the matrix range from 0 to 1.

Considering the data representation similarities between protein secondary structure prediction solvers and image classifiers, we investigate the transformation of PSSM features into images. We create images using the sliding window technique. That is, we generate images from local parts of the sequence to predict the central information of the image, with each feature of each amino acid corresponding to a row in the image.

As PSSM features extract exactly 21 values for each amino acid, we use this information to create and evaluate different image configurations, ranging from images with dimensions 11×21 up to 189×21. The best results were obtained with images of dimensions 21×21, 63×63, 105×105, and 147×147. As the minimum image input in the networks is 32×32, the 21×21 image was resized to 63×63. In the images, the first dimension corresponds to the sliding window size, that is, in a **21** $\times$ 21 image, there are 10 amino acid features before and after the central amino acid features. For the second dimension, the PSSM information is replicated to fill the image. That is, in a $63 \times$ **63** image, each value of each amino acid is triplicated to generate the specific image. For the padding in the beginning and end of the protein, we use symmetric zero-padding to normalize the sizes of all proteins.

We apply standard Convolutional Neural Network (CNN) architectures from the literature in recent years. The best results for protein secondary structure prediction task were achieved by EfficientNetB7 architecture [27], so we use this network in our classifier. With the four configuration predictions, that is, where each EfficientNetB7 evaluates different dimensions of images, we build the ensemble between them using the method presented in Sect. 2.5. The final result is considered as the final prediction of the CNN-based classifier.

All EfficientNetB7 configurations were trained by 50 epochs, early stopping and reduced learning rate on plateau techniques, learning rate equal to 10^{-3}, and Adam optimizer [15].

2.2 BERT

With the recent success of BERT in Natural Language Processing (NLP) tasks [5], such as sentiment analysis and named-entity recognition, many researches have been done to use this method in other domains. Elnaggar *et al.* [7] trained BERT on the BFD dataset [23], with more than 2 billion proteins, and provided this trained BERT for the community.

In the protein secondary structure prediction task, each amino acid, or token, has a secondary structure, or entity, related, therefore, this task can be considered as named-entity recognition. Moreover, as the sequence can be broken into parts, the sentiment analysis of each part, in this case, the prediction of the secondary structure of the central amino acid, can be used as a sliding window classifier.

In this context, we investigate the ensemble of BERT for this task. We evaluate different configurations of BERT, that is, original BERT [5], and the BERT trained on the BFD dataset [7], called BERT-prot, for name-entity recognition and sentiment analysis scenarios. The best results were achieved using BERT-prot for named-entity recognition task along with five BERT-prot fine-tuned for sliding window prediction. The window sizes employed were 21 (that is, 10 amino acids before and after the analyzed amino acid), 41, 61, 81, and 101. For the padding in the sliding window classifiers, we evaluate different configurations and the best results were obtained by repeating the first and the last amino acid of the sequence. For name-entity recognition and sliding window approaches, each BERT-prot-based classifier has a softmax layer to present the predictions.

After the prediction of BERT-prot for name-entity recognition task and for sliding window classification task, we ensemble them using the method presented in Sect. 2.5. The result of the ensemble was considered as the final prediction of BERT-based classifier.

All the BERT classifiers used only the amino acid sequence and they were fine-tuned by 5 epochs using ktrain [17], learning rate equal to 10^{-5}, and Adam optimizer [15].

2.3 Inception Recurrent Networks

In the protein secondary structure prediction task, each amino acid depends on the information of predecessors and successors amino acids of the sequence. With that, time series classifiers, such as convolutional and recurrent networks, can achieve good results on this type of data.

When considering convolutional networks, shallow layers can extract local information and deeper layer extracts global information, losing close interactions between amino acids. Concerning the recurrent networks, the prediction accuracy is intrinsically associated with the capability of the model to access information from predecessors and successors in each chain step, so bidirectional layers can obtain better results.

We create a classifier using the time series properties for protein secondary structure prediction. Our classifier is composed of convolutional layers followed by bidirectional recurrent layers. For the convolutional part, we employ Inception-v4 blocks [26], changing the two-dimensional convolutions to one-dimensional convolutions. We choose this architecture since it can deal with local information, from the shallow layers, passing this information until the deep layers. For bidirectional recurrent layers, we apply GRU memory modules [3]. As our method has inception blocks and recurrent layers, we called it Inception Recurrent Networks (IRN).

We evaluate different configurations of this classifier, that is, the type of Inception-v4 block, the number of stacked blocks, the number of bidirectional recurrent layers, and the number of neurons per bidirectional layer. The top-5 configurations use type B blocks (3 to 7 stacked blocks), with 3 bidirectional recurrent layers, and 100 neurons per layer. The output layer of each configuration is a softmax activation function. Finally, we ensemble the five configurations through the method described in Sect. 2.5. The result of the ensemble was considered as the final prediction of the IRN-based classifier.

In this classifier, we use both amino acid sequence information and PSSM features. Similarly to CNN-based classifier, we employ symmetric zero-padding to normalize the sizes of all proteins. Moreover, we trained each one of IRNs per 50 epochs, early stopping and reduced learning rate on plateau techniques, learning rate equal to 10^{-3}, and Adam optimizer [15].

2.4   Multi-Modal Ensemble Classifier

With the predictions made by CNN, BERT, and IRN classifiers, we build the ensemble among them using the Genetic Algorithm (GA) described in Sect. 2.5. Figure 1 illustrates our multi-modal method.

2.5   Genetic Algorithm

We apply Genetic Algorithm (GA) [11] to ensemble the CNN-based, BERT-based, and IRN-based classifiers, as well as the ensemble between them to form the MMEC classifier. We evaluate combining the predictions obtained by the classifiers using multiple algorithms: Random Forests, Support Vector Machines, Multilayer Perceptrons, and GA. The best results were obtained with GA.

The optimization process uses the probabilistic prediction from the predictors and it finds weights for each class of each classifier. The weights were found through the predictions in the validation set.

We performed a global search that uses the standard implementation of this algorithm, followed by a local search. In the global search, the algorithm starts with 2,000 individuals with weights ranging from 0 to 1. Then, the top-100

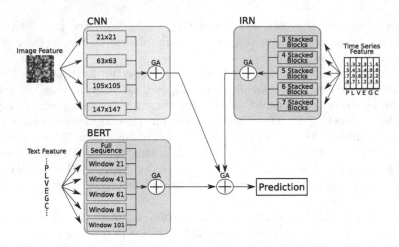

Fig. 1. Pipeline of the proposed Multi-Modal Ensemble Classifier (MMEC).

individuals are selected to be the parents of the next generation. With these parents, 900 individuals are generated through crossover and, with the parents plus the 900 new individuals, the algorithm generates more 1,000 individuals through mutation. This process is carried out for 1,000 iterations or until there are no improvements for 50 iterations.

After the global search, the algorithm starts the local search. The process begins by selecting the top-100 individuals and generates more 100 individuals through mutation. This process is carried out for 1,000 iterations or until there are no improvements for 50 iterations. In the end, the best individual is chosen as the weights for the ensemble.

3   Datasets and Evaluation Metric

In this section, we present the datasets and the evaluation metric.

3.1   Datasets

We use the CB6133 and CB513 datasets to train, validate and test our method. On both datasets, the PSSM features were generated via UniRef90 [25].

The CB6133 dataset [28] contains 6,133 proteins with less than 30% of similarity between them. In this dataset, all proteins have 21 features from the amino acid sequence in one-hot encoding, and 21 features from the Position-Specific-Score Matrix (PSSM), with values that range from 0 to 1. We split this dataset into 5,600 proteins for training, 256 proteins for validation, and 272 proteins for testing, as employed in many methods available in the literature. The classes of secondary structures are unbalanced.

A filtered version of the CB6133 database was applied for training and validating the proposed method when the CB513 dataset was applied for testing.

The filtered version has proteins with less than 25% of similarity with proteins from the CB513 dataset. We split the filtered CB6133 into 5,278 proteins for training and 256 proteins for validation.

The CB513 dataset [4] has 513 proteins and, for each protein, there are 21 features from the amino acid sequence form, and 21 features from the PSSM, as in the CB6133 dataset. This dataset was employed only for testing the proposed method. The classes of secondary structure are unbalanced.

3.2 Evaluation Metric

We apply the Q8 accuracy metric to evaluate our solution, as shown in Eq. 1. This is the standard metric employed by the methods in the literature to assess the correctness of the classifiers.

$$Q8 \text{ Accuracy} = \frac{\sum_{x \in \{B, E, G, H, I, L, S, T\}} \text{correct prediction in } x}{\sum_{x \in \{B, E, G, H, I, L, S, T\}} \text{residues in } x} \quad (1)$$

4   Experimental Evaluation

In this section, we present the experimental evaluation of our method on the CB6133 and CB513 datasets.

4.1   CB6133

In our first experiment on the CB6133, we evaluated the Q8 accuracy of each classifier, that is, CNN, BERT, and IRN, as well as the ensemble among them (MMEC). Table 1 presents the Q8 accuracy of each classifier and the ensemble. The results show that the ensemble achieved the best results compared to each classifier individually.

Table 1. Comparison of Q8 Accuracy of each classifier and the ensemble on the test set of CB6133 and CB513 datasets.

Method	Q8 Accuracy (%)	
	CB6133	CB513
MMEC	**77.9**	**73.3**
BERT	75.8	70.2
IRN	75.3	71.2
CNN	72.5	69.8

Then, we compared our method with the literature. Table 2 shows the Q8 accuracy of our classifier and the methods available in the literature. According to the results, MMEC surpassed the state-of-the-art approaches on the CB6133 dataset.

Table 2. Comparison of Q8 Accuracy on the CB6133 test set.

Method	Q8 Accuracy (%)
MMEC	**77.9**
PS8-Net [20]	76.9
Ensemble of Methods [6]	76.3
Ensemble of Classifiers [19]	75.8
2DConv-BLSTM [10]	75.7
biRNN-CRF [13]	74.8
Ensemble of RNN and RF [18]	73.4

4.2   CB513

Similar to what we did for CB6133, we first evaluated each classifier and the Multi-Modal Ensemble Classifier (MMEC) on the test set of CB513. Table 1 presents the results, demonstrating that MMEC achieved superior results compared to each classifier individually.

We then compared our method with the literature. Table 3 shows the results achieved by MMEC and the methods in the literature. Our method obtained 73.3% of Q8 accuracy, surpassing the state-of-the-art approaches.

Table 3. Comparison of Q8 Accuracy on the CB513 dataset.

Method	Q8 Accuracy (%)
MMEC	**73.3**
PS8-Net [20]	71.9
Conditioned CNN [2]	71.4
Ensemble of Classifiers [19]	71.2
biRNN-CRF [13]	70.9
Ensemble of Methods [6]	70.7
2DConv-BLSTM [10]	70.2
Ensemble of RNN and RF [18]	68.9

5   Conclusions and Future Work

Protein secondary structure prediction plays an important role in the analysis of protein characteristics and possible applications. As this type of task can be adapted to different domains, so multi-modal classifiers can be employed.

In this work, we present and discuss a method composed of convolutional neural networks-based, BERT-based, and inception recurrent networks-based classifiers to make the protein secondary structure prediction based on different views of the same task. Our method, Multi-Modal Ensemble Classifier (MMEC), surpasses state-of-the-art approaches on both CB6133 and CB513 datasets.

As directions for future work, we can highlight the use of techniques for data augmentation from image classification, natural language processing, and time series tasks. Finally, other ensemble methods, such as using deep features, have the potential to achieve good results.

Acknowledgments. The authors would like to thank FAPESP (grants #2015/11937-9, #2017/12646-3, #2017/16246-0, #2017/12646-3 and #2019/20875-8), CNPq (grants #304380/2018-0 and #309330/2018-1) and CAPES for their financial support.

References

1. Altschul, S.F., et al.: Gapped BLAST and PSI-BLAST: a new generation of protein database search programs. Nucleic Acids Res. **25**(17), 3389–3402 (1997)
2. Busia, A., Jaitly, N.: Next-step conditioned deep convolutional neural networks improve protein secondary structure prediction. arXiv:1702.03865 (2017)
3. Cho, K., et al.: Learning phrase representations using RNN encoder-decoder for statistical machine translation. arXiv:1406.1078 (2014)
4. Cuff, J.A., Barton, G.J.: Evaluation and improvement of multiple sequence methods for protein secondary structure prediction. Proteins: Struct. Funct. Bioinform. **34**(4), 508–519 (1999)
5. Devlin, J., Chang, M.W., Lee, K., Toutanova, K.: BERT: pre-training of deep bidirectional transformers for language understanding. arXiv:1810.04805 (2018)
6. Drori, I., et al.: High quality prediction of protein Q8 secondary structure by diverse neural network architectures. arXiv:1811.07143 (2018)
7. Elnaggar, A., et al.: ProtTrans: towards cracking the language of life's code through self-supervised deep learning and high performance computing. arXiv:2007.06225 (2020)
8. Fout, A., Byrd, J., Shariat, B., Ben-Hur, A.: Protein interface prediction using graph convolutional networks. In: Advances in Neural Information Processing Systems (NIPS), pp. 6530–6539. Curran Associates, Inc. (2017)
9. Gligorijević, V., Barot, M., Bonneau, R.: deepNF: deep network fusion for protein function prediction. Bioinformatics **34**(22), 3873–3881 (2018)
10. Guo, Y., Wang, B., Li, W., Yang, B.: Protein secondary structure prediction improved by recurrent neural networks integrated with two-dimensional convolutional neural networks. J. Bioinform. Comput. Biol. **16**(05), 1850021 (2018)
11. Holland, J.H.: Adaptation in Natural and Artificial Systems: An Introductory Analysis with Applications to Biology, Control, and Artificial Intelligence. MIT Press, London (1992)
12. Ieracitano, C., Mammone, N., Hussain, A., Morabito, F.C.: A novel multi-modal machine learning based approach for automatic classification of EEG recordings in dementia. Neural Netw. **123**, 176–190 (2020)
13. Johansen, A.R., Sønderby, C.K., Sønderby, S.K., Winther, O.: Deep recurrent conditional random field network for protein secondary prediction. In: 8th International Conference on Bioinformatics, Computational Biology, and Health Informatics (ACM BCB), pp. 73–78. ACM (2017)
14. Jones, D.T.: Protein secondary structure prediction based on position-specific scoring matrices. J. Mol. Biol. **292**(2), 195–202 (1999)
15. Kingma, D.P., Ba, J.: Adam: a method for stochastic optimization. arXiv:1412.6980 (2014)

16. Liu, Y., Cheng, J., Ma, Y., Chen, Y.: Protein secondary structure prediction based on two dimensional deep convolutional neural networks. In: 3rd International Conference on Computer and Communications (ICCC), pp. 1995–1999. IEEE (2017)
17. Maiya, A.S.: ktrain: a low-code library for augmented machine learning. arXiv preprint arXiv:2004.10703 (2020)
18. Oliveira, G.B., Pedrini, H., Dias, Z.: Ensemble of bidirectional recurrent networks and random forests for protein secondary structure prediction. In: 27th International Conference on Systems, Signals and Image Processing (IWSSIP), pp. 311–316. IEEE (2020)
19. Oliveira, G.B., Pedrini, H., Dias, Z.: Fusion of BLAST and ensemble of classifiers for protein secondary structure prediction. In: 33rd Conference on Graphics, Patterns and Images (SIBGRAPI), pp. 308–315. IEEE (2020)
20. Ratul, M.A.R., Elahi, M.T., Mozaffari, M.H., Lee, W.: PS8-Net: a deep convolutional neural network to predict the eight-state protein secondary structure. arXiv:2009.10380 (2020)
21. Senior, A.W., et al.: Improved protein structure prediction using potentials from deep learning. Nature **577**(7792), 706–710 (2020)
22. Sønderby, S.K., Winther, O.: Protein secondary structure prediction with long short term memory networks. arXiv:1412.7828 (2014)
23. Steinegger, M., Söding, J.: Clustering huge protein sequence sets in linear time. Nat. Commun. **9**(1), 1–8 (2018)
24. Sun, D., Wang, M., Li, A.: A multimodal deep neural network for human breast cancer prognosis prediction by integrating multi-dimensional data. IEEE/ACM Trans. Comput. Biol. Bioinform. **16**(3), 841–850 (2018)
25. Suzek, B.E., Huang, H., McGarvey, P., Mazumder, R., Wu, C.H.: UniRef: comprehensive and non-redundant UniProt reference clusters. Bioinformatics **23**(10), 1282–1288 (2007)
26. Szegedy, C., Ioffe, S., Vanhoucke, V., Alemi, A.: Inception-v4, Inception-ResNet and the impact of residual connections on learning. arXiv preprint arXiv:1602.07261 (2016)
27. Tan, M., Le, Q.V.: EfficientNet: rethinking model scaling for convolutional neural networks. arXiv preprint arXiv:1905.11946 (2019)
28. Zhou, J., Troyanskaya, O.: Deep supervised and convolutional generative stochastic network for protein secondary structure prediction. In: 31st International Conference on Machine Learning (ICML), pp. 745–753 (2014)

Patch-Level Nuclear Pleomorphism Scoring Using Convolutional Neural Networks

Leonardo O. Iheme[1]([✉])([iD]), Gizem Solmaz[2], Fatma Tokat[3], Sercan Çayir[1],
Engin Bozaba[1], Çisem Yazici[2], Gülşah Özsoy[2], Samet Ayalti[1,2],
Cavit Kerem Kayhan[3], and Ümit İnce[3]

[1] Artificial Intelligence Research Team, ViraSoft Inc., Istanbul, Turkey
`leonardo.iheme@virasoft.com.tr`
[2] Research and Development Team, ViraSoft Inc., Istanbul, Turkey
[3] Pathology Laboratory, Acibadem University Teaching Hospital, Istanbul, Turkey
`https://www.virasoft.com/`

Abstract. In an effort to ease the job of pathologists while examining Hematoxylin and Eosin stained breast tissue, this study presents a deep learning-based classifier of nuclear pleomorphism according to the Nottingham grading scale. We show that high classification accuracy is attainable without pre-segmenting the cell nuclei. The data used in the experiments was acquired from our partner teaching hospital. It consists of image patches that were extracted from whole slide images. Using the labeled data, we compared the performance of three state-of-the-art convolutional neural networks and tested the trained model on the unseen testing data. Our experiments revealed that the densely connected architecture (DenseNet) outperforms the residual network (ResNet) and the dual path networks (DPN) in terms of accuracy and F1 score. Specifically, we reached an overall validation accuracy and F1 score of over 0.96 and 0.94 respectively.

Keywords: Cancer · Nuclear pleomorhism · Deep learning · Classification · Histopathology

1 Introduction

Over the past decade, there has been tremendous progress in the development and application of deep learning algorithms to medical diagnosis, prognosis, and treatment. According to Jiang et al. [7], the data type that has been most considered is diagnostic imaging. This should not come as a surprise since most high-mortality-rate diseases such as cancer, are diagnosed by examining various imaging modalities. Naturally, the attention of researchers and practitioners alike is driven towards these diseases.

For neural networks training, where the amount of data significantly impacts the performance of the model, the number of training samples must be maximized. When Krizhevsky et al. [9] won the ImageNet LSVRC–2010 contest, the

© Springer Nature Switzerland AG 2021
N. Tsapatsoulis et al. (Eds.): CAIP 2021, LNCS 13052, pp. 185–194, 2021.
https://doi.org/10.1007/978-3-030-89128-2_18

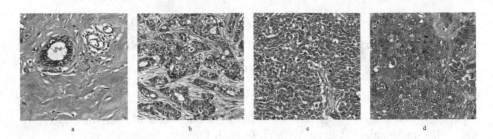

Fig. 1. H&E stained breast tissue image patches showing sample Nottingham scale nuclear pleomorphism scores. a: normal, b: Score 1, c: Score 2, d: Score 3.

potential of Convolutional Neural Networks (CNNs) for image classification was uncovered and has since been extended to the domain of digital pathology. The study by Cireşan et al. [2] is the earliest seminal work where the authors proposed using CNN for pixel-level classification of histopathology images. Since then, various CNN-based methods have been developed and/or adapted to histopathology image classification tasks [10, 15].

Tumors are scored based on some grading scale. For instance the Gleason score [4] and the Nottingham score [12] are widely accepted for prostate cancer and breast cancer respectively. Figure 1 shows a sample of non-tumoral breast tissue and one of each nuclear pleomorphism score (according to the Nottingham grading scale). The description of each score can be found in the CAP protocol [14]. Assigning a nuclear pleomorphism score to a region of interest denotes taking shape, chromatin distribution and size of nucleolus into account. This process is tedious, time-consuming, and may lead to visual fatigue. As a result, the likelihood of assigning erroneous or inconsistent scores is high. Therefore, the need for automating the scoring process cannot be over-emphasized.

In this work, we focus on the automatic assignment of nuclear pleomorphism scores on H&E stained histopathology breast tissue images using CNN. A novelty of our work is that it avoids hand-crafting features and pre-segmentation of nuclei. We rather propose an end-to-end solution which is independent of error-prone pre-processing steps. This study seeks to speed-up the job of pathologists while maintaining consistency and a high degree of accuracy.

1.1 Related Work

In an attempt to automate the way pathologists perform nuclear pleomorphism classification, Cosatto et al. [3] began by segmenting nuclei while ignoring those whose segmentation results were significantly floored. Then, using hand crafted features which expressed nuclei shape and texture, a Support Vector Machines (SVM) classifier was applied to assign the final pleomorphism score. The authors report that the segmentation of nuclei is likely to perform poorly or fail in certain conditions.

In a more recent study [11], the author employs a collection of neural networks to achieve the goal of automatic nuclear pleomorphism scoring according to the

Nottingham grading system. Similar to [3], nuclei were segmented prior to the final classification, albeit in a more rigorous approach. A cascade of CNNs were used in an attempt to overcome the aforementioned pitfalls. Nevertheless, the authors report that the results had shortcomings especially in detecting nuclei borders.

While the aforementioned studies have reported impressive results, they depend on the results of nuclear segmentation for the success. This could lead to inaccuracies accruing from loss of information. To the best of our knowledge, a method that is independent of nuclear segmentation as a pre-processing step is yet to be reported.

2 Data and Materials

The data used in our experiments consists of whole slide images (WSI) of H&E stained breast tissue which were collected from patients who had been diagnosed with breast cancer. The WSIs were scanned with a 3D HISTECH digital slide scanner at a ×20 magnification.

At the time this study was conducted, we had collected 86 WSI of which 24 have been classified by two experienced raters whose inter-rater Cohen Kappa score is 0.96. 1200 × 200 (pixels) sized patches were extracted from the slides. Each patch was then classified by the expert raters as normal or one of the three scores of the Nottingham grading scale of nuclear pleomorphism. The distribution of classes in our data-set is outlined in Table 1. To increase the number of samples and to train a more robust model, we performed a three-fold augmentation. Specifically, images from the training set were horizontally and vertically flipped then rotated 270° counter-clockwise. Finally, the data was split (in a stratified manner) 85% and 15% for training and validation respectively. After training, the model was tested with unseen images from the pool of WSIs.

3 Methods

At the core of our automatic nuclear pleomorphism classification system is a DenseNet [6] model which has been pre-trained on the 1000-class Imagenet data-set. The DenseNet architecture was introduced in 2017 and has out-performed

Table 1. The distribution of classes in the training and validation data-set

Grade	Quantity (before augmentation)
Normal	7768
Score 1	40
Score 2	7868
Score 3	7607
Total	23283

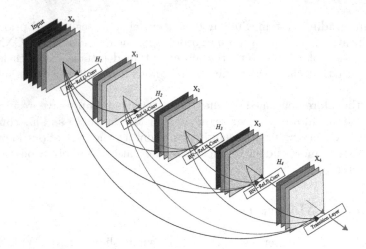

Fig. 2. A 5-layer dense block with a growth rate of k = 4. Each layer takes all preceding feature-maps as input [6].

other state-of-the-art CNN architectures on image classification tasks since then. Briefly, the DenseNet architecture connects each layer to every other layer in a feed-forward manner as shown in Fig. 2. As a result, unlike traditional convolutional networks with L connections and L layers, DenseNet architectures have $L(L+1)/2$ direct connections. It is an extension of the residual blocks concept introduced by ResNet [5], a detailed comparison of the two architectures can be found in [1]. We compared the performance of the DenseNet with the well-known ResNet and the Dual Path Network (DPN).

3.1   Model Training

In this work, we perform transfer learning. Specifically, we modified the final layer of the network, such that the number of neurons corresponds to the number of classes in our data-set, i.e. four neurons. In addition to fine-tuning the pre-trained weights of the model, a number of strategies and parameters were adopted during the training process. The networks were trained with images of size 244×244 and data augmentation was not applied.

Optimizer: We employ the Nesterov-type momentum-accelerated stochastic gradient descent (SGD) algorithm [13]. An initial learning rate of 10×10^{-4} and a momentum of 0.9 were used at the parameter-update step. An adaptive learning rate scheduler known as "reduce learning rate on plateau" was used to avoid sub-optimal performance.

Loss Function: We use the multi-class cross-entropy loss function to compute the loss of the model since we are dealing with a four-class problem. This loss is expressed in Eq. 1.

$$-\sum_{c=1}^{M} y_{o,c} \log(p_{o,c}) \tag{1}$$

where M is the number of classes, y is the binary indicator (0 or 1) if class label c is the correct classification for observation o, and p is the predicted probability observation o is of class c.

Dealing with Class Imbalance: As seen in Table 1, samples from the *Score 1* class are under-represented. To mitigate the effect of this imbalance, we performed an inverse class count wherein the minority class is assigned a higher weight. This *forces* the model to appropriately classify the minority samples. Accordingly, Eq. 1 may be re-formulated as:

$$w_o(-\sum_{c=1}^{M} y_{o,c} \log(p_{o,c})) \tag{2}$$

where w is the class weight of observation o.

This technique, as inspired by King et al. [8], heuristically balances the classes at the loss calculation stage.

For each architecture, the batch size and the initial learning rate were optimized with the validation loss as a criterion. A summary of the training hyperparameters is presented in Table 2.

4   Experimental Set-Up

The experiments performed in this work were carried out on a Dell PC equipped with an NVIDEA GEFORCE RTX 2080 Ti GPU and 32 GB memory. The following metrics were monitored: the mean loss, mean accuracy, per class mean true positive rate (TPR),

$$TPR = \frac{TP}{TP + FN} \tag{3}$$

where TP = number of true positives and FN = number of false negatives.

Per class mean positive predictive value (PPV),

$$PPV = \frac{TP}{TP + FP} \tag{4}$$

where FP = the number of false positives.

Table 2. Summary of hyperparameters used in training

Hyperparameter	Value
Batch size	[4, 8, 10, 12, 16]
Optimizer	SGD
Momentum	0.9
Initial learning rate	1×10^{-5}, 1.25×10^{-4}, 1×10^{-3}
Learning rate scheduler	Reduce on plateau

Per class F1 score which is the harmonic mean of PPV and TPR

$$F1 = 2\frac{TPR \times PPV}{TPR + PPV} \tag{5}$$

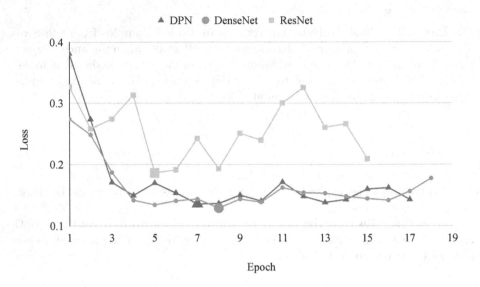

Fig. 3. Validation loss curves for DenseNet, DPN and ResNet. The best obtained losses are highlighted with larger data point sizes

5   Results

We present results of the classification capability of the DenseNet architecture and compare it with the performance of a ResNet and DPN. The results are expressed in terms of the learning curves as well as the statistical measures presented in Sect. 4. A 5-fold cross validation was performed to obtain the mean F1 score and mean accuracy of each network.

5.1   Validation Results

The graphs shown in Figs. 3 and 4 depict the learning progression of the respective architectures. Note that early stopping was triggered if the validation loss did not improve for 10 consecutive epochs. A high and low accuracy and loss, respectively, is desirable. The mean validation accuracy curves for the three network architectures are shown in Fig. 4. In Table 3 we present the mean accuracy and F1 scores obtained after cross validating each model.

5.2   Test Results

The confusion matrix obtained after running inference on the test data-set is depicted in Fig. 5. From the confusion matrix, we obtained test accuracy and F1 score of 0.81 and 0.65 respectively. In Fig. 6 we present a more detailed test result showing the per class metrics obtained by the trained DenseNet model.

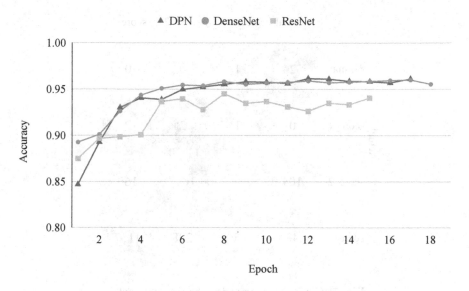

Fig. 4. Validation accuracy curves for DenseNet, DPN and ResNet.

6   Discussion

The DenseNet ($loss \approx 0.13$) architecture yielded the lowest validation loss, followed closely by the DPN architecture ($loss \approx 0.14$). Clearly, the feature reuse property of the DenseNet architecture is advantageous for accurately classifying H&E image patches into nuclear pleomorphism classes. As the DPN is a hybrid network, it owes its performance to the superiority of DenseNet but is degraded by the ResNet.

Table 3. The mean validation F1 scores and accuracy scores of the respective models. The best values are bold-faced

Trained model	Mean F1 score	Mean accuracy
ResNet	0.89	0.94
DPN	0.90	0.95
DenseNet	**0.94**	**0.96**

Overall, the DPN achieves a slightly higher validation accuracy than the DenseNet, albeit at a later epoch. Nevertheless, in an imbalanced data scenario, accuracy could be a misleading measure of performance. With a higher F1 score (Table 3), the DenseNet clearly outperforms the ResNet and the DPN. It would

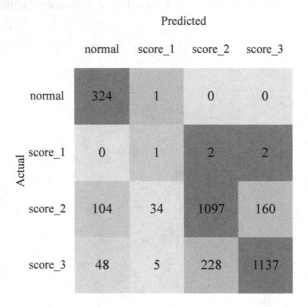

Fig. 5. Confusion matrix for the test data-set

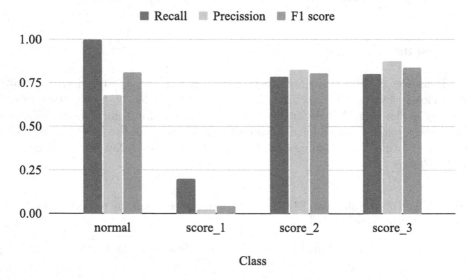

Fig. 6. The precision, recall, and F1 score after performing inference on the test data.

seem that for this classification problem, the DenseNet is better at handling class imbalance.

Although the overall test accuracy of the DenseNet was lower than the validation accuracy, a value over 0.75 is considered to be *good*. The relatively low F1 score of ≈ 0.65 can be attributed to the confusions especially between Score 2 and Score 3 and the relatively few samples present in the Score 1 class as seen in the confusion matrix of Fig. 5. While we acknowledge this seemingly low performance of the model on the minority class, we assert the Score 1 class is known to be extremely rare and is seldom encountered in practice.

7   Conclusion

This work has focused on the automatic classification of H&E stained breast tissue into nuclear pleomorphism scores based on the Nottingham grading scale. We validated three state-of-the-art CNNs and showed that the DenseNet outperforms its counterparts by reaching a validation accuracy of over 0.95 and a test accuracy of over 0.80. The performance of the model on each class gave insight into the lower F1 score obtained for the test set.

We acknowledge the severe class imbalance inherent in the dataset but assert that this reflects the scenario in practice. Therefore, we re-iterate that the aim of the study is to develop a system that will assist pathologists with decision making in terms of nuclear pleomorphism scoring.

As there exist numerous hyperparameters that can be fine-tuned while training modern neural networks, our future outlook is in that direction. This could be a daunting process that requires time and heavy compute. To that end, we continue to fine-tune our model while acquiring more data to make it more robust.

References

1. Chen, Y., Li, J., Xiao, H., Jin, X., Yan, S., Feng, J.: Dual path networks. In: Advances in Neural Information Processing Systems, vol. 2017-Dec, pp. 4468–4476 (2017)
2. Cireşan, D.C., Giusti, A., Gambardella, L.M., Schmidhuber, J.: Mitosis detection in breast cancer histology images with deep neural networks. In: Mori, K., Sakuma, I., Sato, Y., Barillot, C., Navab, N. (eds.) MICCAI 2013. LNCS, vol. 8150, pp. 411–418. Springer, Heidelberg (2013). https://doi.org/10.1007/978-3-642-40763-5_51
3. Cosatto, E., Miller, M., Graf, H.P., Meyer, J.S.: Grading nuclear pleomorphism on histological micrographs. In: Proceedings - International Conference on Pattern Recognition (2008). https://doi.org/10.1109/icpr.2008.4761112
4. Epstein, J.I., Allsbrook, W.C., Amin, M.B., Egevad, L.L.: The 2005 International Society of Urological Pathology (ISUP) consensus conference on gleason grading of prostatic carcinoma. Am. J. Surg. Pathol. **29**(9), 1228–1242 (2005). https://doi.org/10.1097/01.pas.0000173646.99337.b1, http://journals.lww.com/00000478-200509000-00015

5. He, K., Zhang, X., Ren, S., Sun, J.: Deep residual learning for image recognition. In: Proceedings of the IEEE Conference on Computer Vision and Pattern Recognition (CVPR), June 2016
6. Huang, G., Liu, Z., Weinberger, K.Q.: Densely connected convolutional networks. In: 2017 IEEE Conference on Computer Vision and Pattern Recognition (CVPR), pp. 2261–2269 (2017)
7. Jiang, F., et al.: Artificial intelligence in healthcare: past, present and future, December 2017. https://doi.org/10.1136/svn-2017-000101, https://www.ncbi.nlm.nih.gov/pmc/articles/PMC5829945/
8. King, G., Zeng, L.: Logistic regression in rare events data. Technical Report, vol. 2 (2001). https://doi.org/10.1093/oxfordjournals.pan.a004868, http://gking.harvard.edu
9. Krizhevsky, A., Sutskever, I., Hinton, G.E.: ImageNet classification with deep convolutional neural networks. Commun. ACM **60**(6), 84–90 (2017)
10. Qaiser, T., Rajpoot, N.M.: Learning where to see: a novel attention model for automated immunohistochemical scoring. IEEE Trans. Med. Imaging **38**(11), 2620–2631 (2019)
11. Qu, G.: Automatic pleomorphism grading for breast cancer image. Master's thesis, University of Florida (2018)
12. Rakha, E.A., et al.: Prognostic significance of Nottingham histologic grade in invasive breast carcinoma. J. Clin. Oncol. **26**(19), 3153–3158 (2008). https://doi.org/10.1200/JCO.2007.15.5986
13. Sutskever, I., Martens, J., Dahl, G., Hinton, G.: On the importance of initialization and momentum in deep learning. In: Dasgupta, S., McAllester, D. (eds.) Proceedings of Machine Learning Research, PMLR, Atlanta, Georgia, USA, 17–19 June 2013, vol. 28–3, pp. 1139–1147 (2013)
14. Wedemeyer, G.: Protocol for the examination of specimens from patients with invasive carcinoma of the breast. Technical Report, vol. 4 (2010). https://doi.org/10.1043/1543-2165-133.10.1515, www.cap.org/cancerprotocols
15. Zhang, Z., et al.: Pathologist-level interpretable whole-slide cancer diagnosis with deep learning. Nat. Mach. Intell. **1**, 236–245 (2019). https://doi.org/10.1038/s42256-019-0052-1

Automatic Myelofibrosis Grading from Silver-Stained Images

Lorenzo Putzu[1](✉), Maxim Untesco[2], and Giorgio Fumera[1]

[1] Department of Electrical and Electronic Engineering, University of Cagliari,
Piazza d'Armi, 09123 Cagliari, Italy
{lorenzo.putzu,fumera}@unica.it
[2] SmartPath S.R.L, via Borgo Palazzo, 24125 Bergamo, Italy
maxim.untesco@smart-path.it
https://pralab.diee.unica.it/en, https://www.smart-path.it/

Abstract. Histopathology studies the tissues to provide evidence of a disease, type and grade. Usually, the interpretation of these tissue specimens is performed under a microscope by human experts, but since the advent of digital pathology, the slides are digitised, shared and viewed remotely, facilitating diagnosis, prognosis and treatment planning. Furthermore, digital slides can be analysed automatically with computer vision methods to provide diagnostic support, reduce subjectivity and improve efficiency. This field has attracted many researchers in recent years who mainly focused on the analysis of cells morphology on Hematoxylin & Eosin stained samples. In this work, instead, we focused on the analysis of reticulin fibres from silver stained images. This task has been addressed rarely in the literature, mainly due to the total absence of public data sets, but it is beneficial to assess the presence of fibrotic degeneration. One of them is myelofibrosis, characterised by an excess of fibrous tissue. Here we propose an automated method to grade myelofibrosis from image patches. We evaluated different Convolutional Neural Networks for this purpose, and the obtained results demonstrate that myelofibrosis can be identified and graded automatically.

Keywords: Digital pathology · Silver staining · Myelofibrosis grading · Image classification · Convolutional neural networks

1 Introduction

Histopathology is the study of the microscopic structure of cells and tissues of organisms. It provides direct and reliable data that can be used as evidence of a disease or pathologic process along with its type and grade. Millions of tissue biopsies are performed annually in order to study the tissue structures; not only the cell morphology but also an extracellular matrix is assessed, as the distribution, shape and density of collagen and reticulin fibres are critical clues to healthcare providers in many diseases [8]. The biopsy samples first undergo tissue processing steps. Later, tissue sections are mounted on slides and directly

© Springer Nature Switzerland AG 2021
N. Tsapatsoulis et al. (Eds.): CAIP 2021, LNCS 13052, pp. 195–205, 2021.
https://doi.org/10.1007/978-3-030-89128-2_19

observed under a microscope for analysis or digitised through a digital scanner at various levels of magnification, such as 10×, 20×, 40×, 100×, etc. The analysis of tissue specimens is performed after a staining procedure, mainly Hematoxylin & Eosin (H&E) [30], that highlights the different tissue components with different colours [3]. The H&E is the most used stains since it allows the easy extraction of important features related to the cell population. On the contrary, the silver staining has been consolidated to emphasise and analyse reticulin fibres [10], like their number, density and thickness are very useful to assess the fibrotic degeneration dynamics [17].

Since the advent of histopathology and still today, the interpretation of these stained tissue specimens and the assessment of aberrant phenotypes has been performed under a microscope by human experts, [3,5]. The advent of digital pathology [1] has revolutionised modern histopathology: slides are digitised using scanners, shared and viewed remotely, facilitating diagnosis, prognosis and treatment planning [32]. Nevertheless, this task requires a substantial manual effort for pathologists, especially if the analysis consists of extracting statistics from whole slide images (WSIs), whose sizes can go up to 100k × 100k pixels while areas of interest are usually tiny (compared to the WSI size). In addition, about 80% of the biopsies are found to be benign, meaning that pathologists are spending much time searching for the diseased tissue that otherwise could be used for the treatment of patients [4]. Furthermore, the pathologist's analysis is subjective, and his level of expertise and tiredness can significantly influence the diagnosis. On the contrary, in this field, accuracy and efficiency are crucial.

The growing need for automatic techniques has attracted the attention of many researchers in this field, which who have applied many computer vision methods to develop Computer-Aided Diagnosis (CAD) systems, able to provide accurate and computationally efficient methods for the analysis of tissue samples [12,22,36]. CAD systems try to mimic the pathologist's work [14], but they can reduce subjectivity and the error rates and improve the efficiency for pathologists compared with manual examinations [15].

This paper focuses on analysing silver-stained images, which is a very challenging task due to chromatic stain variability, non-homogenous background, and differences in tissue morphology and stain density [13,35]. We focused on this challenging task during our work in the recent HistoDSSP project,[1] aimed at designing and implementing a prototype of Decision Support System for Pathologists (DSSP) to be integrated into a Laboratory Information System (LIS).

Despite the considerable effort spent so far by the research community and the high performances obtained in many CAD systems, especially those based on Convolutional Neural Networks (CNNs) [2,7,28,35], the task of automatic analysis of silver-stained images has been addressed rarely, and very few works are present in the literature [20,25,29]. Moreover, none of them addressed the analysis of reticulin fibres, but they focused only on other steps of the silver-stained image analysis process. This is, however, a necessary step since an increased

---

[1] Histopathology Decision Support Systems for Pathologists (HistoDSSP) based on whole slide imaging analysis.

number or density of reticulin fibre is regarded as an adverse prognostic factor in myeloproliferative neoplasms, such as polycythemia vera, essential thrombocythemia, and primary myelofibrosis [6].

In this work, we focused on myelofibrosis, which is characterised by the progressive, disordered replacement of bone marrow tissue with an excess of fibrous tissue produced by fibroblasts. In the initial phase, the marrow may be hypercellular and have few areas of fibrosis, but in the advanced phase, it will be hypo-cellular and diffusely fibrotic. The bone marrow fibrosis can be graded following a standard scheme included in the updated World Health Organisation classification system of tumours of the hematopoietic and lymphoid tissues for the purpose of grading primary myelofibrosis [31]. It consists of a qualitative (reticulin or collagen) and quantitative evaluation of bone marrow fibrosis. It distinguishes four increasing categories, ranging from MF-0, which corresponds to normal bone marrow, to MF-3, in which coarse bundles of collagen fibrosis are identifiable with significant overlaps.

With the aim of automatically analysing the quantity and distribution of reticulin fibres in each image, we empirically investigated different CNNs architectures, demonstrating that myelofibrosis can be identified and graded automatically from silver-stained images.

The remainder of our paper is organised as follows. In Sect. 2 we review some related work dealing with silver-stained image analysis and the challenges and the open issues in this field. The data set of silver-stained images and the CNNs used in this work are detailed in Sect. 3.1. In Sect. 4 we show in detail the performed experimental evaluation and the obtained results. Finally, conclusions and perspectives are drawn in Sect. 5.

2 Related Work and Open Issues

Very few works have been devoted so far to the automatic analysis of silver-stained images, mainly due to the total absence of public data sets for this task, which instead could be very attractive to the scientific community in this field. Furthermore, the presence of annotated training images would have favoured the comparisons, and the investigation of more sophisticated computer vision approaches, especially the ones based on supervised methods such as CNNs [9], that have recently demonstrated state-of-the-art performances on many histopathology support systems [2,7,28,35].

Among the existing works in the literature dealing with the analysis of reticulin fibres, none of them proposed a fully automatic procedure. Indeed, all the existing methods exploit image processing algorithms already present in commercial software, even embedded in the digital scanners used to acquire the slides [20,25,29]. One of these works focused on the selection of the Regions Of Interest (ROI) from the slides [25], by using a two-step approach. In the first step, the Aperio Image-Scope software[2] was used to draw a large ROI excluding

---

[2] https://www.leicabiosystems.com/digital-pathology/manage/aperio-imagescope/.

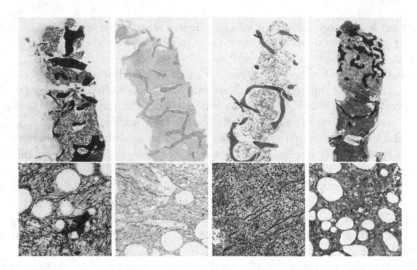

Fig. 1. Example of chromatic issues on WSI acquired during the HistoDSSP project: four slide portions (top) and four corresponding ROIs (bottom).

bone, large interstitial spaces or scanning artefact areas. In the second step, a different software was used to randomly select ten uniformly distributed smaller ROIs from the larger ROI, avoiding the areas of bone. The other methods in the literature focused on the analysis of reticulin fibres, instead [20,29]. Both methods are based on colour deconvolution, aimed either at assessing the total occupancy area of reticulin fibres regardless of their morphological features or distribution [29] or at analysing their shape and quantity after a segmentation process based on thresholding [20].

However, the use of colour deconvolution analysis is limited by the huge staining variability present in WSI, mainly caused by the silver impregnation that does not yield monochromatic reticulin fibres. For this reason, the automatic analysis of silver-stained images is a challenging task since the images can be acquired in different laboratories, using different staining protocols and different digital microscopes or scanners that add a further variability in optical image quality [20,21,24]. An example of such a chromatic issue is shown in Fig. 1, where four slide portions and four corresponding ROIs are compared. As it can be observed, the chromatic variability between the different WSI is very high, and sometimes this variability appears even in the same slide, as in the right-most one of Fig. 1. This work aims at proving that, even with the issues mentioned above, myelofibrosis can be identified and graded automatically from silver-stained images.

3   Materials and Methods

In this section, we firstly describe the used data set. Then, we describe the methods evaluated in this work to analyse and grade the silver-stained images.

3.1 Data Set

During the HistoDSSP project, we have collected almost 500 slides from the Laboratory of Pathological Anatomy of the Brotzu hospital in Cagliari, Sardinia, Italy, where they have been stained with standard techniques, mainly H&E, Silver and Trichrome staining [23], and different immunostaining techniques [18] (antibody-based methods to detect specific proteins). Slides were digitised using a Leica Aperio T2 WS Scanner at 20× magnification. From the digitised WSIs, we created our data set by extracting the ones stained with silver staining, 56 in total. The pathologists have labelled all the images with the associated degree of fibrosis. In detail, 16 slides with MF0, 26 with MF1, 2 with MF2 and 2 with MF3. We extracted a variable number of images from each slide that mainly depend on the quantity of tissue present in a slide and the ratio between compact bone and trabecular bone areas. We used a semi-automated sliding-windows procedure to extract the images, which select patches of size 512 × 512 from the tissue regions only. All the extracted images have been checked to discard those that present compact bone regions or cutting artefacts.

We have extracted a total number of 648 images; some samples are shown at the bottom of Fig. 1. In order to perform a fair comparison, we split the obtained data set into three parts, maintaining the proportions between the various classes. Indeed, in order to have a sufficient number of samples for the training process while preserving a sufficient number of samples for performance evaluation, we first split the data set into two equal parts, namely training and testing set. Then we further split the original training set into a training and a validation set, with about 80% and 20% of images, respectively. Even in this split, we have tried to keep intact the proportions between the various classes. Further details on the composition of the training, validation and testing sets can be found in Table 1. Unfortunately, we cannot share the data set yet, as the HistoDSSP project is still ongoing, but we plan to make publicly available all the sets of images used in this work, or even a more extensive set.

Table 1. Data set partitions and proportions.

Label	Total	Training set	Validation set	Testing set
All	648	261	63	324
MF-0	194	79	18	97
MF-1	394	159	38	197
MF-2	36	14	4	18
MF-3	24	9	3	12

3.2   Image Classification

For this task, we evaluated different well known CNN architectures that are: AlexNet, VGG-19, ResNet-50 and ResNet-152. They were all pre-trained on a well-known data set of natural images (ImageNet [16]) and adapted to medical image tasks, following an established procedure for transfer learning, and fine-tuning of CNN models [26]. AlexNet [16] and the VGG-19 [27] are very simple architectures, but, at the same time, are the most used for transfer learning and fine-tuning [26], since they gained popularity for their excellent performances in many classification tasks [16]. They are quite similar except for the number of layers 8 for AlexNet and 19 for VGG-19. Instead, the two ResNet architectures are much deeper: 50 and 152 layers for ResNet-50 ResNet-152, respectively. Nevertheless, being based on residual learning, they are easier to optimise and, with respect to other architectures, gain accuracy from their considerably increased depth while still having a lower complexity [11]. Indeed, ResNet architectures have been successfully used in different classification tasks involving medical images [19,33,34]. Each of the mentioned architecture has been used to build a new multi-class network using a common transfer learning procedure [26]. We preserved all the CNN layers of the original architectures except for the last fully-connected one, that we replaced with a new layer, which has been freshly initialised and set up in order to accommodate the new object categories that are 4, from MF-0 to MF-3. For each architecture, the fine-tuning process has been carried out for 100 epochs with a learning rate of 0.001, decreased by a factor of 0.1 every 30 epochs.

4   Experimental Evaluation

This section evaluates the effectiveness of the methods mentioned above by performing a series of experiments. In the following, we first describe the experimental set-up and finally the obtained results.

4.1   Experimental Set-Up

As it can be noted from Table 1 the four classes are considerably unbalanced. The class with the highest number of samples is MF1, which is the most common, while MF2 and MF3 present very few samples. For this reason, during CNN training, we used a weighted random sampling procedure. The weight values are obtained from the ratio between the total number of images and the number of images per class. Hence, the greater the number of images in a class, the lower its weight. This procedure guarantee that each batch contains a balanced number of images from each class, but at the same time, due to the data set imbalance, an image of the minority class could be drawn several times in a single epoch. Thus, to avoid over-fitting, we combined the mentioned sampling with an online data transformation procedure that exploits random padding, random cropping, as well as vertical and horizontal flipping with a probability of

50%. Finally, to fulfil the input layer requirements, the images have been resized to 224 × 224. We conducted all the experiments on a single machine with the following configuration: Intel(R) Core(TM) i9-8950HK @ 2.90 GHz CPU with 32 GB RAM and NVIDIA GTX1050 Ti 4 GB GPU. To evaluate the classification performance, we used four common metrics that are Accuracy (A), Precision (P), Recall (R) and F1-score (F1). They all range over the interval $[0, 1]$. For two-class problems they are computed according to the Eqs. (1), (2),(3) and (4):

$$A = \frac{TP + TN}{TP + FP + TN + FN} \quad (1) \qquad\qquad P = \frac{TP}{TP + FP} \quad (2)$$

$$R = \frac{TP}{TP + FN} \quad (3) \qquad\qquad F = \frac{2RP}{R + P} \quad (4)$$

where TP (True Positive) indicates the number of images correctly classified as positives, FP (False Positive) provides the number of negative images wrongly classified as positives, TN (True Negative) indicates the number of images correctly classified as negatives and FN (False Negative) gives the number of positive images wrongly classified as negatives. In our four-class problem, we used the above measures to compute the per-class performances, and then, to obtain a single performance measure, we compute the weighted average.

4.2 Results

Table 2 shows the results obtained by all the above mentioned CNN architecture on the testing set. As it can be observed, almost all the tested CNN architectures reached excellent performances. Just AlexNet does not benefit from the used balancing procedure since each testing sample is predicted as belonging to MF-1, which is the majority class. The other architectures instead present excellent results, in particular VGG-19, which curiously is the one that by design is most similar to AlexNet. Hence the reason behind these results is mainly related to the difference in depth between the two networks. This is also confirmed by the excellent results obtained by the two ResNet, even if, we cannot attribute all the credit to the greater depth, since these two networks differ from AlexNet also in terms of design. However, focusing on comparing the two ResNets, which are the same in terms of design, we can see that the deeper network performs better even in this case. In general, the obtained results are encouraging, demonstrating that myelofibrosis can be identified and graded automatically using CNNs from silver-stained images since that three models out of four obtained an F1 value greater than 93%. Obviously, these results must be consolidated on a larger set of images, but they are still an excellent starting point for creating a real diagnostic support system.

Table 2. A, P, R and F1 performances obtained by AlexNet, VGG-19, ResNet-50 and ResNet152.

CNN	MF	A	P	R	F1
AlexNet	MF-0	0.711	0.000	0.000	0.000
	MF-1	0.655	0.655	1.000	0.792
	MF-2	0.962	0.000	0.000	0.000
	MF-3	0.982	0.000	0.000	0.000
	Avg	0.655	0.430	0.655	0.519
VGG-19	MF-0	0.974	0.933	0.980	0.956
	MF-1	0.958	0.991	0.944	0.967
	MF-2	0.984	0.714	0.962	0.820
	MF-3	0.997	0.857	1.000	0.923
	Avg	0.956	0.961	0.956	0.957
ResNet-50	MF-0	0.934	0.820	0.990	0.897
	MF-1	0.928	0.995	0.895	0.943
	MF-2	1.000	1.000	1.000	1.000
	MF-3	0.994	0.750	1.000	0.857
	Avg	0.928	0.940	0.928	0.930
ResNet-152	MF-0	0.953	0.881	0.970	0.923
	MF-1	0.945	0.986	0.929	0.956
	MF-2	0.991	0.812	1.000	0.897
	MF-3	1.000	1.000	1.000	1.000
	Avg	0.945	0.949	0.945	0.945

5 Conclusions

In this work, we proposed an automated approach for myelofibrosis grading from silver-stained images. Myelofibrosis is characterised by an excess of fibrous tissue that can be detected through reticulin fibres analysis. This analysis has been addressed rarely in the literature and, mostly, none of the proposed methods presents a fully automatic procedure. Here instead, we proposed an automated method based on CNNs. We evaluated four different CNN architectures with different design and depth to assess their effectiveness in this task. The obtained results demonstrated that myelofibrosis could be identified and graded automatically using CNNs from silver-stained images. Indeed three models out of four obtained an F1 value greater than 93%, with the highest value obtained by the VGG-19 that obtained 95.7%. These results are really encouraging, but being obtained with a small number of images cannot represent real-world variability. For this reason, our ongoing work includes the acquisition of other WSI, possibly treated in different hospitals or clinics, since it is necessary to evaluate the performance of this method with a larger set of images. Further research could address the development of other steps for the analysis of WSI, such as automatic tissue segmentation and patch extraction.

Acknowledgement. This work was supported by the project HistoDSSP - Histopathology Decision Support System for Pathology based on whole slide imaging analysis, funded by Regione Autonoma della Sardegna (POR FESR Sardegna 2014–2020 Asse 1 Azione 1.1.3). We also thank the Laboratory of Pathological Anatomy of the Brotzu hospital for the material and the support in this research.

References

1. Al-Janabi, S., Huisman, A., van Diest, P.V.: Digital pathology: current status and future perspectives. Histopathology **61**, 1–9 (2012)
2. Alex, V., P., M.S.K., Chennamsetty, S.S., et al.: Generative adversarial networks for brain lesion detection. In: Styner, M.A., Angelini, E.D. (eds.) Medical Imaging: Image Processing, vol. 10133, pp. 113–121. SPIE (2017)
3. Alturkistani, H.A., Tashkandi, F.M., Mohammedsaleh, Z.M.: Histological stains: a literature review and case study. Glob. J. Health Sci. **8**, 72–79 (2015)
4. Bhattacharjee, S., Mukherjee, J., Nag, S., et al.: Review on histopathological slide analysis using digital microscopy. Int. J. Adv. Sci. Technol. **62**, 65–96 (2014)
5. Bándi, P., Geessink, O., Manson, Q., et al.: From detection of individual metastases to classification of lymph node status at the patient level: the camelyon17 challenge. IEEE Trans. Med. Imaging **38**(2), 550–560 (2019)
6. Vener, C., Fracchiolla, N.S., Gianelli, U., et al.: Prognostic implications of the European consensus for grading of bone marrow fibrosis in chronic idiopathic myelofibrosis. Blood **111**, 1862–1865 (2007)
7. Cireşan, D.C., Giusti, A., Gambardella, L.M., Schmidhuber, J.: Mitosis detection in breast cancer histology images with deep neural networks. In: Mori, K., Sakuma, I., Sato, Y., Barillot, C., Navab, N. (eds.) MICCAI 2013. LNCS, vol. 8150, pp. 411–418. Springer, Heidelberg (2013). https://doi.org/10.1007/978-3-642-40763-5_51
8. Dey, P.: Cancer nucleus: morphology and beyond. Diagn. Cytopathol. **38**(5), 382–390 (2010)
9. Glorot, X., Bengio, Y.: Understanding the difficulty of training deep feedforward neural networks. In: Proceedings of Machine Learning Research, vol. 9, pp. 249–256 (2010)
10. Géméri, G.: Silver impregnation of reticulum in paraffin sections. Am. J. Pathol. **13**, 993–1002 (1937)
11. He, K., Zhang, X., Ren, S., Sun, J.: Deep residual learning for image recognition. In: IEEE CVPR, pp. 770–778 (2016)
12. He, L., Long, L.R., Antani, S., et al.: Histology image analysis for carcinoma detection and grading. Comput. Meth. Prog. Biomed. **107**(3), 538–556 (2012)
13. Irshad, H., Veillard, A., Roux, L., et al.: Methods for nuclei detection, segmentation, and classification in digital histopathology: a review–current status and future potential. IEEE Rev. Biomed. Eng. **7**, 97–114 (2014)
14. Kothari, S., Phan, J.H., Young, A.N., et al.: Histological image classification using biologically interpretable shape-based features. BMC Med. Imaging **13**(1), 1–7 (2013)
15. Kowal, M., Filipczuk, P., Obuchowicz, A., Korbicz, J., Monczak, R.: Computer-aided diagnosis of breast cancer based on fine needle biopsy microscopic images. Comput. Biol. Med. **43**(10), 1563–1572 (2013)

16. Krizhevsky, A., Sutskever, I., Hinton, G.E.: ImageNet classification with deep convolutional neural networks. Commun. ACM **60**(6), 84–90 (2017)

17. Kuter, D., Bain, B., Mufti, G.: Bone marrow fibrosis: pathophysiology and clinical significance of increased bone marrow stromal fibres. Br. J. Haematol. **139**, 351–362 (2007)

18. Lawrence, S., Golubeva, Y.: Optimization of immunostaining for prospective image analysis. Meth. Mol. Biol. **1606**, 235–263 (2017)

19. Li, Y., Shen, L., Yu, S.: HEp-2 specimen image segmentation and classification using very deep fully convolutional network. IEEE Trans. Med. Imaging **36**(7), 1561–1572 (2017)

20. Lucero, H.A., Patterson, S., Matsuura, S., et al.: Quantitative histological image analyses of reticulin fibers in a myelofibrotic mouse. J. Biol. Meth. **3**(4), 60 (2016)

21. Naylor, P., Laé, M., Reyal, F., et al.: Nuclei segmentation in histopathology images using deep neural networks. In: IEEE ISBI, pp. 933–936 (2017)

22. Naylor, P., Laé, M., Reyal, F., et al.: Segmentation of nuclei in histopathology images by deep regression of the distance map. IEEE Trans. Med. Imaging **38**(2), 448–459 (2019)

23. Ozawa, A., Sakaue, M.: New decolorization method produces more information from tissue sections stained with hematoxylin and eosin stain and Masson-trichrome stain. Ann. Anat. **227**, 151431 (2020)

24. Qi, X., Xing, F., Foran, D.J., et al.: Robust segmentation of overlapping cells in histopathology specimens using parallel seed detection and repulsive level set. IEEE Trans. Biomed. Eng. **59**(3), 754–765 (2012)

25. Salama, M.E., et al.: Stereology and computer-based image analysis quantifies heterogeneity and improves reproducibility for grading reticulin in myeloproliferative neoplasms. In: Potts, S.J., Eberhard, D.A., Wharton, K.A. (eds.) Molecular Histopathology and Tissue Biomarkers in Drug and Diagnostic Development. MPT, pp. 117–126. Springer, New York (2014). https://doi.org/10.1007/7653_2014_36

26. Shin, H., Roth, H.R., Gao, M., et al.: Deep convolutional neural networks for computer-aided detection: CNN architectures, dataset characteristics and transfer learning. IEEE Trans. Med. Imaging **35**(5), 1285–1298 (2016)

27. Simonyan, K., Zisserman, A.: Very deep convolutional networks for large-scale image recognition. In: Bengio, Y., LeCun, Y. (eds.) International Conference on ICLR (2015)

28. Sirinukunwattana, K., Raza, S.E.A., Tsang, Y., et al.: Locality sensitive deep learning for detection and classification of nuclei in routine colon cancer histology images. IEEE Trans. Med. Imaging **35**(5), 1196–1206 (2016)

29. Teman, C.J., Wilson, A.R., Perkins, S.L., et al.: Quantification of fibrosis and osteosclerosis in myeloproliferative neoplasms: a computer-assisted image study. Leuk. Res. **34**(7), 871–876 (2010)

30. Titford, M.: The long history of hematoxylin. Biotech. Histochem. **80**(2), 73–78 (2005)

31. Vardiman, J.W., Thiele, J., Arber, D.A., et al.: The 2008 revision of the World Health Organization (WHO) classification of myeloid neoplasms and acute leukemia: rationale and important changes (2009)

32. Williams, S., Henricks, W., Becich, M., et al.: Telepathology for patient care: what am I getting myself into? Adv. Anat. Path. **17**, 130–149 (2010)

33. Wu, N., Phang, J., Park, J., et al.: Deep neural networks improve radiologists' performance in breast cancer screening. IEEE Trans. Med. Imaging **39**(4), 1184–1194 (2020)

34. Xie, Y., Xia, Y., Zhang, J., et al.: Knowledge-based collaborative deep learning for benign-malignant lung nodule classification on chest CT. IEEE Trans. Med. Imaging **38**(4), 991–1004 (2019)
35. Xing, F., Xie, Y., Yang, L.: An automatic learning-based framework for robust nucleus segmentation. IEEE Trans. Med. Imaging **35**(2), 550–566 (2016)
36. Zhou, Y., Dou, Q., Chen, H., et al.: SFCN-OPI: detection and fine-grained classification of nuclei using sibling FCN with objectness prior interaction. In: AAAI (2018)

A Deep Learning-Based Pipeline for Celiac Disease Diagnosis Using Histopathological Images

Farhad Maleki[1,2], Kevin Cote[3], Keyhan Najafian[1], Katie Ovens[1,2], Yan Miao[1], Rita Zakarian[1,2], Caroline Reinhold[1,2], Reza Forghani[1,2], Peter Savadjiev[2,3,4], and Zu-hua Gao[3(✉)]

[1] Augmented Intelligence and Precision Health Laboratory (AIPHL), The Research Institute of the McGill University Health Centre, McGill University, Montreal, QC, Canada
{farhad.maleki,katie.ovens,yan.miao}@mail.mcgill.ca, keyhan.najafian@usask.ca, rita.zakarian@affiliate.mcgill.ca, {caroline.reinhold,reza.forghani}@mcgill.ca
[2] Department of Radiology, McGill University, Montreal, QC, Canada
peter.savadjiev@mcgill.ca
[3] The Research Institute of McGill University Health Centre, Departments of Medicine and Pathology, McGill University, Montreal, QC, Canada
kevin.cote@mail.mcgill.ca, zu-hua.gao@mcgill.ca
[4] Departments of Pathology and School of Computer Science, McGill University, Montreal, QC, Canada

Abstract. With an increasing number of celiac disease diagnoses and the increasing number of misdiagnoses, automated approaches are valuable to aid pathologists in efficiently diagnosing this disease. Histopathological analysis of intestinal biopsy is considered the gold standard for diagnosis. Convolutional neural networks have achieved promising results for various image processing tasks. A common challenge in medical imaging analysis is obtaining a large number of samples, impeding the full potential of deep learning. In this paper, we propose a classification pipeline to train deep convolutional neural networks to accurately diagnosis celiac disease using models trained with a small number of samples. To show the utility of this approach, we compared it to a typical classification pipeline. The results indicate the superiority of our classification pipeline in distinguishing celiac disease from normal tissue with precision, recall, and accuracy of 0.941, 0.889, and 0.893, respectively. Although we showed the utility of the proposed pipeline for celiac diagnosis, it can also be used for other applications utilizing histopathological imaging.

Keywords: Small sample sizes · Histopathological classification · Convolutional neural networks · Deep learning

1 Introduction

Celiac disease (CD) is a chronic autoimmune disease that occurs in genetically predisposed individuals in whom the ingestion of gluten leads to damage of the

© Springer Nature Switzerland AG 2021
N. Tsapatsoulis et al. (Eds.): CAIP 2021, LNCS 13052, pp. 206–214, 2021.
https://doi.org/10.1007/978-3-030-89128-2_20

small bowel. CD is a clinicopathological diagnosis where the evaluation of histomorphology is considered the gold standard in confirming the diagnosis and evaluating its severity. Histologically, CD is characterized by villous blunting and intraepithelial lymphocytosis [1]. CD that goes undiagnosed and untreated can result in severe malnutrition and other health-related consequences such as the development of other autoimmune disorders [2]. Given the morbidity of CD and the number of conditions with high histopathological overlap, it is important to have a method of detection capable of distinguishing features specific to CD. Artificial intelligence techniques using deep learning methods and machine learning have emerged as a technology that can be used for computer-aided diagnosis with increased efficiency and reproducibility and could become a valuable tool at the point of clinical care [3, 4].

Gold-standard diagnostic confirmation of suspected CD is achieved by biopsying the small intestine. To confirm CD, pathologists use histopathological slides to review samples of the tissue taken from the intestinal biopsy [5]. Biopsied specimens can be viewed on a single or multiple tissue histopathology slides. Histopathology slides provide a more comprehensive view of disease and its effect on tissues since the preparation process preserves the underlying tissue architecture. As such, some disease characteristics may be deduced only from a histopathology image. After reviewing these histopathology slides, a pathologist makes a final decision about the condition under study. However, this process suffers from a high inter- and intra-observer variability due to the subjective nature of diagnosis made by pathologists [6–8]. Also, the recent increase in the number of incorrect detection of CD [9] highlights the need for a quantitative assessment of pathology slides.

Convolutional neural networks (CNNs) are machine learning models commonly used for analyzing images and have shown promising results in medical imaging applications [4]. A common approach used for image classification is to use a well-established architecture such as ResNet [10] with its classifier replaced with one that matches the number of classes in the underlying task. Figure 1 illustrates such a model. A common challenge in medical imaging studies using deep learning approaches is to acquire a sufficient number of samples for model development. In a recent work, Wei et al. using square patches of the whole slide images as inputs, trained a ResNet model to classify samples originated from normal, celiac disease, and nonspecific duodenitis [11]. This model was developed using a large number of samples (1230 slides from 1048 patients) in order to make an accurate distinction between each biopsy class. Often developing such a large medical dataset is impractical due to the limited access to and the high cost of medical expertise required for data labeling, the time-consuming nature of medical data labeling, or in some cases due to the rarity of the disease under study or even ethical concerns. This highlights the need for developing methods that could achieve comparable levels of performance with relatively smaller sample sizes.

In this study, we propose a deep learning-based pipeline to train and deploy a classification model for diagnosing CD from whole slide imaging of intestinal

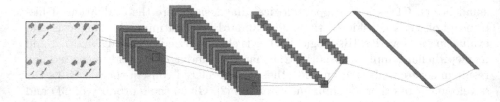

Fig. 1. In a typical convolutional neural network classifier, an image is used as input and the network aims to produce the accurate class label for that image. During model training, network weights are updated to decrease the error made by the model when applied to the images used for model training.

biopsies. We show this pipeline outperforms a typical supervised deep learning approach, which is commonly used for image classification tasks. Although, we use CD diagnosis to showcase the utility of the proposed pipeline, this pipeline is not limited to CD, and it can be used for other whole slide pathology image classification tasks as well. The pipeline could potentially be utilized as a useful approach for deployment in clinical settings in contexts where developing large-scale pathology datasets is impractical.

2   Materials and Methods

The primary aim of this study was to use histopathological images of biopsied specimens to develop diagnostic models for CD. In this section, we describe the data and methodology for developing deep learning-based classification models. The proposed pipeline synthesized images that are composed of randomly chosen subsets of tissue segments extracted from whole slide images. These tissue segments are then transformed and randomly placed on a blank canvas and used as inputs for model training. This approach allows us to increase both the number of available samples for training and the diversity of the data used for model development. The proposed pipeline synthesizes several images for each patient. Therefore, to be able to use the trained model for predicting unseen cases, an aggregation function is used to make patient-level predictions. We compared our pipeline with a typical classification pipeline in which each slide image is resized and used as a single input for the model. The source code for the proposed pipeline is available upon request.

2.1   Data

This study was approved by the Research Ethics Board of the Research Institute of the McGill University Health Center (Protocol Number: 2021-6924). We retrieved 426 histological slides from a single institute, the McGill University Health Center's Department of Pathology. The glass slides were digitalized using Aperio's ScanScope XT, for adult patients undergoing a duodenal biopsy to assess for the presence or absence of CD between 2016 and 2019, inclusively. The formalin-fixed, paraffin-embedded tissue samples were processed and stained with hematoxylin and eosin following a single, standardized tissue processing procedure. All cases were reviewed and considered for inclusion in our primary dataset. Excluded from this pool of subjects were those cases showing Helicobacter Pylori infection as demonstrated by a positive Giemsa stain on a concomitant gastric biopsy [10–13], features of peptic duodenitis, or any other significant pathological abnormality other than features associated with CD, i.e. intraepithelial lymphocytosis, villous blunting, or duodenal crypt hyperplasia. The pathological distribution of the samples according to the Marsh classification scheme can be summarized as follows:

- Histotype 1 (174 slides, representing 163 subjects): normal duodenal mucosa, including cases signed out as "no significant pathological abnormality," and/or "no evidence of celiac disease."
- Histotype 2 (104 slides, 91 subjects): intraepithelial lymphocytosis, no villous atrophy, including cases signed out as "intraepithelial lymphocytosis without villous atrophy," and/or "cannot rule out celiac disease."
- Histotype 3 (138 slides, 106 subjects): intraepithelial lymphocytosis, subtotal villous atrophy, including cases signed out as "intraepithelial lymphocytosis with subtotal/mild/moderate/focal/patchy/partial/some villous atrophy," and/or "suggestive of celiac disease in the right clinical context."
- Histotype 4 (109 slides, 66 subjects): intraepithelial lymphocytosis, complete villous atrophy, including cases signed out as "intraepithelial lymphocytosis with complete/total/severe/diffuse/near-complete/moderately to severe/marked/widespread and obvious villous atrophy," and/or "diagnosis of celiac disease in the right clinical context."

To design a CD diagnosis system, a binary classification model was developed. As the presence or absence of CD for the samples with histotype 2 is ambiguous, these samples were excluded from our analysis. Histotype 1 was used as a normal—"disease-free"—control group, and histotypes 3 and 4 were grouped together to represent those patients showing pathological evidence at least "suggestive of celiac disease in the right clinical context."

2.2 Methodology

In this section, we describe the proposed methodology for developing CD classification models.

Image Assembler. We develop Image Assembler (IA) as an approach to programmatically synthesize a large number of examples by assembling tissue segments from each slide while mitigating the lack of segment-level labels. Figure 2, illustrates a schematic view of IA that takes a slide from a patient as input and synthesizes a collection of images, which will later be used for model development. The label assigned to the synthesized images is set to be the same as their corresponding patient label.

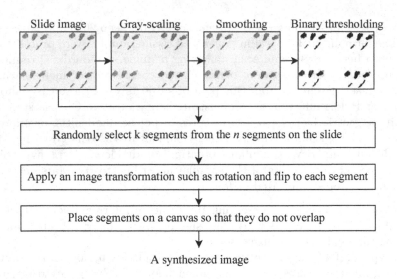

A synthesized image

Fig. 2. Image Assembler procedure. A foreground mask is created by converting the slide image to grayscale, followed by Gaussian and average smoothing, and applying OTSU thresholding. The detected foreground tissue segments are then used to synthesize several images for each slide/patient.

In order to create a mask for the foreground, first, we convert the slide image from RGB to grayscale. Then, to reduce noise, we apply a Gaussian smoothing followed by an average smoothing filter to each grayscale image. Next, an OTSU thresholding approach is used to provide a binary mask for segments. Disjoint foreground components, i.e. segments, are then detected and used as a mask for each segment. These masks will then be used to assemble a non-overlapping collage of k segments for synthesizing an image (see Fig. 3).

For a slide with n segments, both sampling with replacement and sampling without replacement can be used to choose k segments. The former results in n choose k distinct combinations of k segments with the condition that k is less than or equal to n. The latter results in n^k outcomes. For example, given $n = 12$ and $k = 4$, sampling without replacement results in 495 distinct sets of segments, and the sampling with replacement leads to $12^4 = 20736$ outcomes. Note that even when the sets of segments are not disjoint, the resulting synthesized images could be distinct due to the random placement of segments and the transformations applied to each segment randomly. This makes IA capable of synthesizing a large-scale and diverse set of images for model training. The transformations used by IA include random rotation, vertical flip, and horizontal flip.

Model Training and Evaluation. We randomly assigned patients to 3 disjoint sets: training (60%), testing (20%), and validation (20%), where the data for each patient was only assigned to one of these sets. For each slide (corresponding

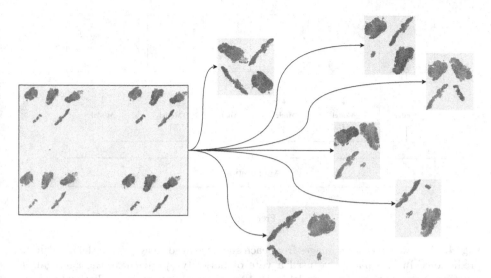

Fig. 3. Image Assembler synthesizes a set of examples from each slide by first selecting a random subset of segments on the slide, then applying some transformations such as rotation and flipping to each selected segment, and finally placing the transformed segments on a blank canvas randomly.

to a patient), IA was run to synthesize m images. A ResNet-50 architecture, pre-trained on ImageNet, with cross entropy loss was used for all experiments [10]. The classifier layer of the ResNet model was replaced to match the number of classes (2 classes here). It should be noted that the proposed approaches are independent of the deep learning model architecture. To deal with the imbalanced nature of the data (174 normal and 247 diseased samples), an oversampling approach using sampling with replacement was used [12]. To be consistent for the sake of comparison, the same set of hyperparameters was used for all models including a learning rate of 0.0001, momentum of 0.95, weight decay of 0.005, and a batch size of 16. All models were trained for 100 epochs and the model with the highest F1 score was selected as the best model. We used Albumentations package version 0.4.5 for image augmentation [13]. All experiments were conducted using Python 3.7 and PyTorch version 1.6 on a Titan RTX GPU machine.

As IA synthesizes m images from each slide, an aggregation function is required to achieve a consensus prediction for each patient, in the evaluation and test phases. Figure 4 depicts the aggregation step used to make a consensus prediction for each patient. As the aggregation function, we use a vote of majority among the predictions made for the m images generated from each slide. For all experiments, a value of $m = 101$ was used. The value of m is a hyperparameter for the model. We hypothesize that the proposed approach will lead to improving the classification performance. Also, we used values of $k = 1$ (single segment) and $k = 4$.

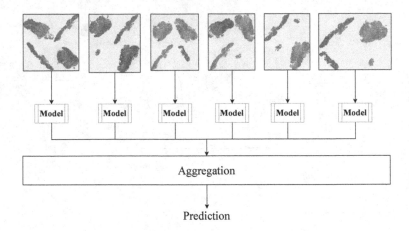

Prediction

Fig. 4. Since we generate n images from each slide, we need to aggregate the n resulting predictions. In this paper, we used a vote of majority approach as the aggregation function. Note that the same model is used for generating these n predictions.

3   Results

In this section, we present the results of three main experiments: (1) a typical CNN-based classification model where each slide image is resized and used as a single input for the model; (2) a binary classification model developed using the proposed pipeline with IA where $k = 1$; and (3) the binary classification developed using the proposed pipeline with IA where $k = 4$. Table 1 shows the model performance in these three scenarios. We use precision, recall (sensitivity), and accuracy as performance indicators.

Table 1. Model performance metrics using whole slide images (Approach 1) and images constructed using Image Assembler with $k = 1$ (Approach 2) and $k = 4$ (Approach 3).

Approach	Precision	Recall	Accuracy
1	0.833	0.833	0.786
2	0.917	0.815	0.831
3	0.941	0.889	0.893

Using IA improved all three performance metrics compared to using the original whole slide images for model construction. Using subsets of four segments ($k = 4$) to assemble additional samples resulted in the best precision, recall (sensitivity), and accuracy of 0.941, 0.889, and 0.893, respectively. Using single segments to construct additional samples also outperformed whole slide images. However, IA with $k = 4$ achieved better results in all performance metrics in comparison to the single segment approach.

4 Discussion

In this study, we proposed a deep learning-based pipeline for CD classification. Due to the small number of samples available when using whole slide images as input examples for model training, even after applying a comprehensive image augmentation pipeline, the model performance is lower than the segment-level approaches. Further, we showed that the number of segments used to generate the samples for model building substantially impacts classification performance. We expected that utilizing single segments to build models would result in lower model performance due to the increased risk of mislabelling samples compared to using a collection of larger number of segments (i.e. larger values of k) from a single patient. This also highlights the drawback of building models using histopathological biopsies using small patches of whole slide images as attempted in previous studies [11]. These results indicate that our approach improves upon the methodology typically used to classify histopathological images.

Although we performed a binary classification of CD versus healthy tissue, there are different histopathological classes of CD [14,15] as described in Sect. 2.1. Further works could be focused on exploring whether there are important morphological characteristics in the tissue segments representing a particular class of CD. This information could be used to automatically and efficiently identify different CD classes, which could impact treatment plans or make it possible to identify early or mild cases.

One limitation of our approach is that IA can be computationally intensive in comparison to using the whole slide images depending on how many images are assembled from different segments. However, the number of images assembled in this study (101) is a hyperparameter value, and it can be changed depending on the available data and compute power. Also, hyperparameter tuning can be done to find the optimal values for m and k in IA to achieve the highest model performance while meeting the computational requirements for a deployment setting.

This study also only utilized image resolution with width and height between 3000 to 4000 pixels; however, higher resolution images are available and can be leveraged to further improve the model performance. We suggest utilizing these high resolution images with the proposed pipeline for future research. In cases where such high resolution images are not available due to the technology used for imaging, this pipeline is still capable of generating high performing models.

5 Conclusion

In this work, we proposed a deep learning-based pipeline to diagnose CD from whole slide images of histopathological intestinal biopsies. Deep learning has shown promise in various medical applications; however, a small number of training examples can hinder developing deep learning models in the medical domain. The proposed approach in this paper makes it possible to synthesize a large number of samples for model training and leads to improving model performance in contexts where acquiring a large number of histopathological images is impractical.

The developed models for CD classification achieved high performance, as indicated by their high precision, recall, and accuracy in differentiating CD tissues from normal tissues. Although we showed the utility of the proposed approach in this paper using a CD classification, the proposed methodology has the potential to be applied to other pathological conditions.

References

1. Oxentenko , A.S., Rubio-Tapia, A.: Celiac disease. In: Mayo Clinic Proceedings, vol. 94, pp. 2556–2571. Elsevier (2019)
2. Rubio-Tapia, A., Murray, J.A.: Celiac disease beyond the gut. Clin. Gastroenterol. Hepatol. **6**(7), 722–723 (2008)
3. LeCun, Y., Bengio, Y., Hinton, G.: Deep learning. Nature **521**(7553), 436–444 (2015)
4. Le, W.T., Maleki, F., Romero, F.P., Forghani, R., Kadoury, S.: Overview of machine learning: part 2: deep learning for medical image analysis. Neuroimaging Clin. **30**(4), 417–431 (2020)
5. Robert, M.E., et al.: Statement on best practices in the use of pathology as a diagnostic tool for celiac disease. Am. J. Surg. Pathol. **42**(9), e44–e58 (2018)
6. Corazza, G.R., et al.: Comparison of the interobserver reproducibility with different histologic criteria used in celiac disease. Clin. Gastroenterol. Hepatol. 5(7), 838–843 (2007)
7. Arguelles-Grande, C., Tennyson, C.A., Lewis, S.K., Green, P.H.R., Bhagat, G.: Variability in small bowel histopathology reporting between different pathology practice settings: impact on the diagnosis of coeliac disease. J. Clin. Pathol. **65**(3), 242–247 (2012)
8. Montén, C., et al.: Validity of histology for the diagnosis of paediatric coeliac disease: a Swedish multicentre study. Scand. J. Gastroenterol. **51**(4), 427–433 (2016)
9. Caio, G., et al.: Celiac disease: a comprehensive current review. BMC Med. **17**(1), 1–20 (2019)
10. He, K., Zhang, X., Ren, S., Sun, J.: Deep residual learning for image recognition. In: Proceedings of the IEEE Conference on Computer Vision and Pattern Recognition, pp. 770–778 (2016)
11. Wei, J.W., Wei, J.W., Jackson, C.R., Ren, B., Suriawinata, A.A., Hassanpour, S.: Automated detection of celiac disease on duodenal biopsy slides: a deep learning approach. J. Pathol. Inform. **10**, 7 (2019)
12. He, H., Ma, Y.: Imbalanced Learning: Foundations, Algorithms, and Applications. Wiley, Hoboken (2013)
13. Buslaev, A., Iglovikov, V.I., Khvedchenya, E., Parinov, A., Druzhinin, M., Kalinin, A.A.: Albumentations: fast and flexible image augmentations. Information **11**(2), 125 (2020)
14. Oberhuber, G., Granditsch, G., Vogelsang, H.: The histopathology of coeliac disease: time for a standardized report scheme for pathologists. Eur. J. Gastroenterol. Hepatol. **11**(10), 1185 (1999)
15. Corazza, G.R., Villanacci, V.: Coeliac disease. J. Clin. Pathol. **58**(6), 573–574 (2005)

HEp-2 Cell Image Recognition with Transferable Cross-Dataset Synthetic Samples

Tomáš Majtner[✉]

Central European Institute of Technology (CEITEC), Masaryk University, Brno, Czech Republic
tomas.majtner@ceitec.muni.cz

Abstract. The paper examines the possibilities of using synthetic HEp-2 cell images as a means of data augmentation. The common problem of biomedical datasets is the shortage of annotated samples required for the training of deep learning techniques. Traditional approaches based on image rotation and mirroring have their limitations, and alternative techniques based on generative adversarial networks (GANs) are currently being explored. Instead of looking solely at a single dataset or the creation of a recognition model with applicability for multiple datasets, this study focuses on the transferability of synthetic HEp-2 samples among publicly available datasets. The paper offers a workflow where the quality of synthetic samples is confirmed via an independent fine-tuned neural network. The subsequent combination of synthetic samples with original images outperforms traditional augmentation approaches and leads to state-of-the-art performance on both publicly available HEp-2 cell image datasets employed in this study.

Keywords: Transfer learning · Generative adversarial network · Deep learning · DCGAN · HEp-2 cell images

1 Introduction

Clinicians commonly use the anti-nuclear antibody test to diagnose connective tissue diseases like systemic lupus erythematosus and rheumatoid arthritis [14, 22]. The gold standard for performing this test is the indirect immunofluorescence (IIF) protocol using human epithelial type 2 (HEp-2) cells [9,14]. The evaluation of the IIF test is done mainly by humans, and therefore, it is a subjective method too dependent on the physician's experience. Usually, multiple specialists need to analyse specimen images and make a decision about their staining patterns. This leads to low reproducibility and significant inter-laboratory differences. Thus, computer-aided systems aim to assist physicians with the diagnosis by automatic classification of HEp-2 images.

Nowadays, automated recognition is almost exclusively done via deep neural networks [6]. These networks are a powerful tool for image classification, but they require a large number of training samples [13]. In practice, it is difficult

© Springer Nature Switzerland AG 2021
N. Tsapatsoulis et al. (Eds.): CAIP 2021, LNCS 13052, pp. 215–225, 2021.
https://doi.org/10.1007/978-3-030-89128-2_21

to collect and annotate large biomedical image databases due to the lack of physicians' time and the cost of imaging devices. Therefore, it is a common practice to increase the number of training samples by various methods of so-called image augmentation. For biomedical samples like HEp-2 cell images, the mirroring operation and the rotation around the central image point are the most often used techniques.

However, these approaches do not produce new samples, only manipulate the ones that already exist. As an alternative, generative adversarial networks (GANs) [7] have proven to be an efficient tool for the unsupervised generation of new synthetic images with realistic visual appearance [13]. In this study, a deep convolutional GAN (DCGAN) [17] is employed to examine the application of synthetic samples between datasets. The question to be answered is if it is possible to take generated samples trained on one public HEp-2 dataset and use them as a form of image augmentation for another one. Could these synthetic images increase the classification accuracy and outperform the traditional forms of augmentation? There are multiple issues to be addressed, and this paper offers a clear workflow that leads to the state-of-the-art results on both employed datasets achieved by incorporating synthetic samples.

In the next section, the current HEp-2 image recognition methods are presented together with a brief introduction to GAN methods. Subsequently, two public datasets used in this study are introduced, followed by the proposed workflow in the methodology part of the paper. The last sections are dedicated to the evaluation and discussion of experiments and results, where the overall effectiveness of the suggested methods is presented.

2   Related Work

The automated methods for IIF image analysis have been a topic of a special thematic issue of Pattern Recognition Letters [8]. Multiple techniques, including those examining a multi-process system based on an ensemble of fifteen support vector machines [4] and the role of Gaussian Scale Space theory as a preprocessing approach [16], were introduced. Bayramoglu et al. [2] studied the influence of several preprocessing techniques on HEp-2 image classification, and more recently, Gao et al. [6] analysed the impact of hyper-parameter settings of the proposed fully-connected convolutional neural network on the classification accuracy. Deep residual networks were intensively studied by multiple authors [10,18,23], which led to the use of ResNet-50 architecture also in this study.

When looking specifically at the methods published for the *SNPHEp-2 dataset* that is also used in this study, William et al. [22] employed a system comprised of a dual-region codebook-based descriptor combined with the nearest convex hull classifier. Yang et al. [24] used learning filters from image statistics, where they trained a filter bank from unlabelled cell images by using independent component analysis. Faraki et al. [5] observed that HEp-2 cells can be efficiently described by symmetric positive definite matrices which lie on a Riemannian manifold and utilised them for classification.

Table 1. The number of samples in both public datasets used in this study before any form of augmentation.

			Ce	Sp		Ho	Nu	Nm	Go
ICPR2014	(i)ntermediate	Training	953	961		985	1,164	885	262
		Validation	136	137		140	166	126	37
		Testing	274	276		282	334	254	76
	(p)ositive	Training	965	1,020		761	653	659	245
		Validation	137	145		108	93	94	34
		Testing	276	292		218	188	190	70
			Ce	Co	Fi	Ho	Nu		
SNPHEp-2		Training	156	172	200	177	200		
		Testing	191	192	200	196	200		

More recently, Bajić et al. [1] incorporated texture information extracted by αLBP-maps, which enabled them to create a very efficient model that was tested on three datasets. The *SNPHEp-2 dataset* was also used by Vununu et al. [19]; however, some of their reported results were extracted for a different dataset.

Generative adversarial networks were introduced in 2014 [7], and since then, they have been successfully used for biomedical imaging tasks, including image synthesis and classification [25], and medical segmentation [3]. For HEp-2 images, the first application of GANs was demonstrated in 2019 [13]. This work uses a similar model; however, the focus is on the transferability of synthetic images between datasets. Multiple improvements are suggested, including the double-check process of generated samples, where a separate network confirmed the true category of images.

3 Datasets

Two publicly available datasets were employed in this study. The first one is known as the *ICPR2014 dataset* [9]. It utilises 419 unique positive sera extracted from 419 randomly selected patients to form 13,596 pre-segmented cell images in total. The dataset recognises two cell fluorescence intensity levels, namely positive (p) and intermediate (i). These intensity levels were rarely utilised before by automated methods, but in this study, they increase the intra-class compactness. All released cell samples form the original training set, while the original test set is not publicly available. Therefore, the public part of the dataset was randomly divided into 80 % for training (including 10 % for validation) and 20 % for independent testing. This approach was also chosen in other published studies [1,6]. The *ICPR2014 dataset* consists of six categories, namely Centromere (Ce), Speckled (Sp) (a combination of Coarse and Fine Speckled), Homogeneous (Ho), Nucleolar (Nu), Nuclear Membrane (Nm), and Golgi (Go).

The second dataset is known as *SNPHEp-2* [22] and it contains 1884 cell images from 40 specimens. The cell images were pre-segmented and divided into

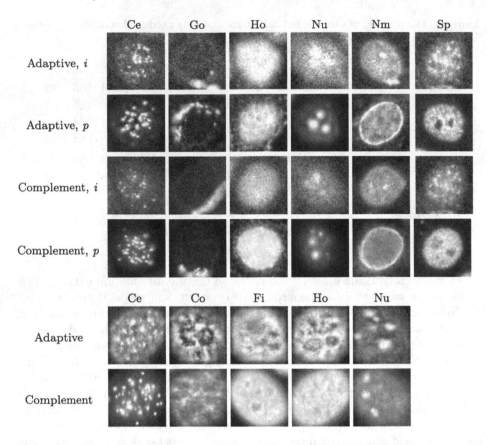

Fig. 1. The first four rows show example samples from the *ICPR2014 dataset* and the bottom rows show samples from *SNPHEp-2* after application of both enhancement methods.

five categories by their authors. These categories are Centromere (Ce), Coarse Speckled (Co), Fine Speckled (Fi), Homogeneous (Ho), and Nucleolar (Nu). SNPHEp-2 was partitioned by its authors into training (905 cell images) and testing (979 cell images) part. Table 1 summarizes both available datasets, and visual illustrations of samples are provided in Fig. 1.

4   Proposed Method

The simplified version of the employed workflow used in this study is in Fig. 2. In the first step, DCGANs needed to be trained in order to produce new artificial samples from each class. This network architecture was chosen because it was demonstrated by Radford *et al.* [17] to be stable for images of size 64×64 pixels, which is the approximate size of the HEp-2 samples used in this study. Since the *ICPR2014 dataset* has large intra-class variance, the distinction between p

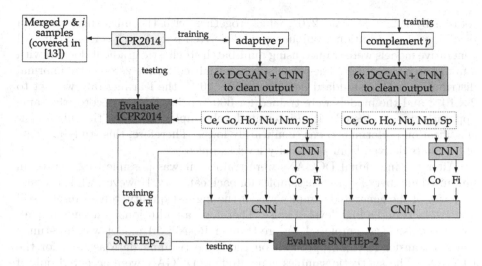

Fig. 2. Illustration of the simplified pipeline used in this study.

and i intensity level samples helps to keep the classes more compact. Previous studies showed that when both levels are mixed together, the network produces realistically looking samples, but the effect on the classification performance is only marginal [13].

In order to train the generative models, the training part of the positive intensity level samples from the *ICPR2014 dataset* was resampled using bicubic interpolation, and two contrast adjustment methods were applied to create two separate versions of the samples. The first one converts the RGB input image into the HSV colour space and applies the contrast-limited adaptive histogram equalization [26] on the value (V) channel. This method operates on small regions in the image and computes histograms corresponding to distinct image sections. Subsequently, it uses them to redistribute the lightness values of the image. The enhanced HSV image is then converted back into the RGB colour space. This method will be further referred to as the *adaptive* method.

The second method first derives a complement image of the original RGB input and then applies a dehazing algorithm [15]. The algorithm was originally designed to reduce the atmospheric haze, and it is based on the observation that unhazy images contain pixels that have low signal in colour channels, which is also the case for HEp-2 image samples. In the end, a complement image of the dehazed one was derived in order to convert the values back to the original range. This method will be further referred to as the *complement* method. The visual illustration of both methods applied on HEp-2 images is in Fig. 1.

For the application of DCGAN to augment HEp-2 images, an individual model was trained for each of the six p classes from *ICPR2014 dataset* in both the *adaptive* and *complement* version. However, as can be seen from Table 1, the training part of this dataset is relatively small, especially after excluding i intensity level samples. Therefore, in the training of DCGANs, each sample was rotated by 90°, 180°, and 270°, which together with the mirroring operation results in seven additional unique samples derived out of each original one. All generative models were trained using mini-batch stochastic gradient descent with a mini-batch size of 128. The weights were initialized from a zero-centred normal distribution with a standard deviation of 2×10^{-2}, the learning rate was set to 2×10^{-4} and the models were trained for 300 epochs. For the record, the same approach was also tested with i intensity level samples; however, the subsequent classification performance was significantly lower. Therefore, this study presents only results derived from p intensity level samples.

After the individual DCGANs were trained, it was possible to generate an unlimited number of synthetic samples for each category. However, while the vast majority of the generated samples was of the highest quality, some samples still had artefacts and a high level of noise. Therefore, an additional automated quality mechanism was employed. A pre-trained ResNet-50 model was fine-tuned using a transfer learning approach on the same training data as used for the DCGANs. The synthetic samples generated by DCGAN were accepted only if they were correctly categorised by this fine-tuned ResNet-50 model. In practice, it means that they went through a double-check process, where their true category was confirmed by a separate network.

The last question towards the transferability of synthetic samples between datasets is the class compatibility. Three classes, namely Ce, Ho, and Nu can be used directly, but the Sp class needs to be further divided into Co and Fi classes since *SNPHEp-2* distinguishes between them. For this purpose, another separate fine-tuned ResNet-50 model was trained exclusively on training the Co and Fi samples from *SNPHEp-2*. This model categorises synthetic input Sp samples into either the Co or Fi category. A similar approach was already employed to solve the class compatibility problem [1].

At the end, it was necessary to fuse the results from both the *adaptive* and *complement* method. Wetzer *et al.* [20] performed a comparison of fusion strategies and suggested to use a linear SVM on concatenated output probabilities. Later it was shown that multiplication of corresponding softmax probabilities can be more efficient [12]. Therefore, it is used here for each *adaptive* and *complement* version of the test image.

5   Evaluation and Discussion

In the evaluation part, two different models pre-trained in ImageNet were employed, namely ResNet-50 and GoogLeNet. Fine-tuning, also known as transfer learning, was applied to adjust them for HEp-2 image recognition. This implies that in both models, their last three layers were replaced with a fully-connected layer, a softmax layer, and a classification layer, which classifies images directly to categories of HEp-2 images. The stochastic gradient descent with a momentum optimiser, an initial learning rate of 0.001, and a mini-batch size of

Table 2. *ICPR2014 dataset* results including the comparison with other methods.

Method	OA			MCA		
Gao *et al.* [6]	97.2 %			96.8 %		
Bajić *et al.* [1]	98.1 %			96.0 %		
Shen *et al.* [18]	98.8 %			98.6 %		
Li *et al.* [11]	98.9 %			98.5 %		
Tested models	No augm		Traditional		Synthetic	
	OA	MCA	OA	MCA	OA	MCA
GoogLeNet$_6$	97.39 %	97.36 %	98.72 %	98.73 %	98.42 %	98.46 %
GoogLeNet$_{12strict}$	95.64 %	95.24 %	97.73 %	98.36 %	97.80 %	98.37 %
GoogLeNet$_{12relax}$	97.55 %	97.60 %	98.86 %	99.01 %	98.79 %	98.87 %
ResNet-50$_6$	97.06 %	97.06 %	98.60 %	98.71 %	98.57 %	98.69 %
ResNet-50$_{12strict}$	95.24 %	95.39 %	97.77 %	97.98 %	97.84 %	98.15 %
ResNet-50$_{12relax}$	97.18 %	97.18 %	98.83 %	98.92 %	**98.94 %**	**99.01 %**

8 images were utilised. Both architectures were trained for 30 epochs, and the convergence was controlled using the validation dataset. The final performance of each model reported in this study was derived using the independent test dataset. All tests were performed using MATLAB R2020a.

Before the synthetic samples were applied for the *SNPHEp-2 dataset*, they were first tested on their original *ICPR2014 dataset*. For comparison purposes and to demonstrate the effect of compact classes on the quality of generated synthetic samples, two different scenarios were employed. In the first one, p and i intensity level samples were merged together to form six classes, as it was done in previous studies. In the second scenario, p and i samples were considered separately, which led to the formation of twelve classes during the training. In this case, the evaluation was done under so-called *strict* and *relaxed* condition. In the strict condition, the intensity level is controlled strictly for all classes. For example, centromere p sample classified as centromere i and vice versa is considered a misclassification. In the relaxed condition, only the true class is important; therefore, the previous example will be considered as a correct class classification. This relaxed condition is compatible with previously published studies on six classes.

For the *ICPR2014 dataset*, the evaluation of the classification performance was done via the overall accuracy (OA) defined as the overall correct classification rate of all images and via the mean class accuracy (MCA) calculated as the mean of all individual class accuracies. The training was performed without any data augmentation (*no augm*), with traditional augmentation as it was described for the training of DCGANs (*traditional*), and with the inclusion of synthetic samples (*synthetic*), where their number matched the number of *traditional* samples. Results for the *ICPR2014 dataset* are summarized in Table 2, and it is observable

Table 3. *SNPHEp-2* results including the comparison with other methods.

Method	Overall accuracy				
Faraki *et al.* [5]	74.7%				
Yang *et al.* [24]	82.1%				
Wiliam *et al.* [21]	82.5%				
Bajić *et al.* [1]	83.0%				
Tested models	No augm	Augm 90°	Augm 30°	Synthetic	2 × synthetic
GoogLeNet	71.50%	74.67%	77.94%	77.63%	83.45%
ResNet-50	80.08%	80.18%	76.92%	**83.76%**	78.14%

that both models show high performance on this dataset even without any augmentation. This can be attributed to model architectures as well as to data pre-processing and post-processing steps, including the merging of *adaptive* and *complement* branches. However, both *traditional* and *synthetic* augmentation brought a slight increase in accuracy values. Especially for ResNet-50, the inclusion of synthetic samples outperformed all previously published methods.

The transferability of synthetic samples for classification purposes was tested on the *SNPHEp-2 dataset*. The results are summarised and compared to the state-of-the-art methods in Table 3. Here, the training of both models was first conducted without any data augmentation (*no augm*) and with a traditional augmentation using a 90° step in rotation (*augm 90°*) and a 30° step (*augm 30°*) that resulted in 3× more training samples in comparison to *augm 90°*. For GoogLeNet architecture, an increase in overall accuracy was measured when more samples were utilised. For ResNet-50, this traditional form of augmentation led to minimal increase or even a decrease in performance. It is a demonstration that the increase in dataset size does not automatically lead to higher accuracy.

The remaining two columns display the results with synthetic images from the *ICPR2014 dataset*. The number of images in the first one (*synthetic*) matches the number of the images in *augm 90°*, while the second one (*2 × synthetic*) uses double the amount of synthetic samples. A similar pattern in the results is observable, where more samples lead to an increase in performance for GoogLeNet. Moreover, the *2 × synthetic* variant significantly outperforms traditional augmentation. For ResNet-50, the *synthetic* variant outperforms all other results, including all published state-of-the-art methods. However, introducing more generated samples with *2 × synthetic* is not efficient, which confirms the observation with traditional augmentation. Figure 3 provides the confusion matrices for highlighted top-performing models on both datasets.

	Ce	Go	Ho	Nu	Nm	Sp
Ce	99.27	0.18	0	0.55	0	0
Go	0	99.32	0	0	0.68	0
Ho	0	0	99.60	0.20	0	0.20
Nu	0.19	0.77	0.19	98.08	00.19	0.57
Nm	0	0	00.23	0	99.55	0.23
Sp	0.18	0	0.70	0.53	0.35	98.24

	Ce	Co	Fi	Ho	Nu
Ce	95.81	3.66	0	0.52	0
Co	14.06	69.27	0	9.38	7.29
Fi	0	0.50	83.50	15.00	1.00
Ho	7.14	3.06	6.63	83.16	0
Nu	10.00	1.50	0	1.50	87.00

Fig. 3. Confusion matrices for the top performing ResNet-50 models highlighted in Table 2 for the *ICPR2014 dataset* (left) and in Table 3 for the *SNPHEp-2 dataset* (right). Presented values are in %.

6   Conclusion

In this paper, an efficient novel workflow for the generation of synthetic images with a focus on transferability between datasets is introduced. The paper demonstrates the limitations of traditional augmentation methods and offers an alternative approach based on synthetic samples that went through a double-check process. Two image texture enhancement methods are employed to highlight and better capture the inter-class variability. The evaluation was performed on two different public HEp-2 datasets, and the paper demonstrates that the inclusion of synthetic samples leads to increased classification performance on both of them. It was also shown that the presented workflow creates synthetic samples that are applicable also between datasets. There is a high potential in GANs as alternatives to traditional augmentation methods also for other medical and biomedical domains where large annotated datasets are missing. The future exploration of GAN-related approaches is therefore highly relevant in these domains.

Acknowledgement. The work was supported from European Regional Development Fund-Project "Postdoc2@MUNI" (No. CZ.02.2.69/0.0/0.0/18_053/ 0016952).

References

1. Bajić, B., Majtner, T., Lindblad, J., Sladoje, N.: Generalised deep learning framework for HEp-2 cell recognition using local binary pattern maps. IET Image Proc. **14**(6), 1201–1208 (2020)
2. Bayramoglu, N., Kannala, J., Heikkilä, J.: Human epithelial type 2 cell classification with convolutional neural networks. In: 15th International Conference on Bioinformatics and Bioengineering, pp. 1–6. IEEE (2015)
3. Bowles, C., et al.: GAN augmentation: augmenting training data using generative adversarial networks. arXiv preprint arXiv:1810.10863 (2018)
4. Cascio, D., Taormina, V., Cipolla, M., Bruno, S., Fauci, F., Raso, G.: A multi-process system for HEp-2 cells classification based on SVM. Pattern Recogn. Lett. **82**, 56–63 (2016)
5. Faraki, M., Harandi, M., Wiliem, A., Lovell, B.: Fisher tensors for classifying human epithelial cells. Pattern Recogn. **47**(7), 2348–2359 (2014)

6. Gao, Z., Wang, L., Zhou, L., Zhang, J.: HEp-2 cell image classification with deep convolutional neural networks. IEEE J. Biomed. Health Inform. **21**(2), 416–428 (2017)
7. Goodfellow, I.J., et al.: Generative adversarial networks. arXiv preprint arXiv:1406.2661 (2014)
8. Harandi, M., Lovell, B., Percannella, G., Saggese, A., Vento, M., Wiliem, A.: Executable thematic special issue on pattern recognition techniques for indirect immunofluorescence images analysis. Pattern Recogn. Lett. **82**, 1–2 (2016)
9. Hobson, P., Lovell, B., Percannella, G., Vento, M., Wiliem, A.: Benchmarking human epithelial type 2 interphase cells classification methods on a very large dataset. Artif. Intell. Med. **65**(3), 239–250 (2015)
10. Lei, H., et al.: A deeply supervised residual network for HEp-2 cell classification via cross-modal transfer learning. Pattern Recogn. **79**, 290–302 (2018)
11. Li, Y., Shen, L.: HEp-Net: a smaller and better deep-learning network for HEp-2 cell classification. Comput. Methods Biomech. Biomed. Eng. Imaging Vis. **7**(3), 266–272 (2019)
12. Majtner, T., Bajić, B., Herp, J.: Texture-based image transformations for improved deep learning classification. In: 25th Iberoamerican Congress on Pattern Recognition. Springer (2021)
13. Majtner, T., Bajić, B., Lindblad, J., Sladoje, N., Blanes-Vidal, V., Nadimi, E.S.: On the effectiveness of generative adversarial networks as HEp-2 image augmentation tool. In: Felsberg, M., Forssén, P.-E., Sintorn, I.-M., Unger, J. (eds.) SCIA 2019. LNCS, vol. 11482, pp. 439–451. Springer, Cham (2019). https://doi.org/10.1007/978-3-030-20205-7_36
14. Meroni, P.L., Schur, P.H.: ANA screening: an old test with new recommendations. Ann. Rheum. Dis. **69**(8), 1420–1422 (2010)
15. Park, D., Park, H., Han, D.K., Ko, H.: Single image dehazing with image entropy and information fidelity. In: International Conference on Image Processing, pp. 4037–4041. IEEE (2014)
16. Qi, X., Zhao, G., Chen, J., Pietikäinen, M.: HEp-2 cell classification: the role of Gaussian scale space theory as a pre-processing approach. Pattern Recogn. Lett. **82**, 36–43 (2016)
17. Radford, A., Metz, L., Chintala, S.: Unsupervised representation learning with deep convolutional generative adversarial networks. arXiv preprint arXiv:1511.06434 (2015)
18. Shen, L., Jia, X., Li, Y.: Deep cross residual network for HEp-2 cell staining pattern classification. Pattern Recogn. **82**, 68–78 (2018)
19. Vununu, C., Lee, S.H., Kwon, K.R.: A deep feature extraction method for HEp-2 cell image classification. Electronics **8**(1), 20 (2019)
20. Wetzer, E., Lindblad, J., Sintorn, I.-M., Hultenby, K., Sladoje, N.: Towards automated multiscale imaging and analysis in TEM: glomerulus detection by fusion of CNN and LBP maps. In: Leal-Taixé, L., Roth, S. (eds.) ECCV 2018. LNCS, vol. 11134, pp. 465–475. Springer, Cham (2019). https://doi.org/10.1007/978-3-030-11024-6_36
21. Wiliem, A., Sanderson, C., Wong, Y., Hobson, P., Minchin, R., Lovell, B.: Automatic classification of human epithelial type 2 cell indirect immunofluorescence images using cell pyramid matching. Pattern Recog. **47**(7), 2315–2324 (2014)
22. Wiliem, A., Wong, Y., Sanderson, C., Hobson, P., Chen, S., Lovell, B.: Classification of human epithelial type 2 cell indirect immunofluoresence images via codebook based descriptors. In: Workshop on Applications of Computer Vision, pp. 95–102. IEEE (2013)

23. Xie, H., He, Y., Lei, H., Han, T., Yu, Z., Lei, B.: Deeply supervised residual network for HEp-2 cell classification. In: 24th International Conference on Pattern Recognition (ICPR), pp. 699–703. IEEE (2018)
24. Yang, Y., Wiliem, A., Alavi, A., Hobson, P.: Classification of human epithelial type 2 cell images using independent component analysis. In: 20th International Conference on Image Processing, pp. 733–737. IEEE (2013)
25. Yi, X., Walia, E., Babyn, P.: Generative adversarial network in medical imaging: a review. Med. Image Anal. **58**, 101552 (2019)
26. Zuiderveld, K.: Contrast limited adaptive histogram equalization. In: Graphics Gems IV, pp. 474–485. Academic Press Professional (1994)

Clinically Guided Trainable Soft Attention for Early Detection of Oral Cancer

Roshan Alex Welikala[1]([✉]), Paolo Remagnino[1], Jian Han Lim[2],
Chee Seng Chan[2], Senthilmani Rajendran[3], Thomas George Kallarakkal[4],
Rosnah Binti Zain[4,5], Ruwan Duminda Jayasinghe[6], Jyotsna Rimal[7],
Alexander Ross Kerr[8], Rahmi Amtha[9], Karthikeya Patil[10],
Wanninayake Mudiyanselage Tilakaratne[4,6], Sok Ching Cheong[3,4],
and Sarah Ann Barman[1]

[1] Digital Information Research Centre, Kingston University, Surrey, UK
r.a.welikala@kingston.ac.uk
[2] Centre of Image and Signal Processing, University of Malaya,
Kuala Lumpur, Malaysia
[3] Head and Neck Cancer Research Team, Cancer Research Malaysia,
Subang Jaya, Malaysia
[4] Department of Oral and Maxillofacial Clinical Sciences, University of Malaya,
Kuala Lumpur, Malaysia
[5] Faculty of Dentistry, MAHSA University, Jenjarom, Malaysia
[6] Centre for Research in Oral Cancer, University of Peradeniya,
Peradeniya, Sri Lanka
[7] Department of Oral Medicine and Radiology, BP Koirala Institute of Health
Sciences, Dharan, Nepal
[8] Oral and Maxillofacial Pathology, New York University, New York, USA
[9] Faculty of Dentistry, Trisakti University, Jakarta, Indonesia
[10] Oral Medicine and Radiology, Jagadguru Sri Shivarathreeshwara University,
Mysuru, India

Abstract. Oral cancer disproportionately affects low- and middle-income countries, where a lack of access to appropriate medical care contributes towards late disease presentation. Using artificial intelligence to facilitate the automated identification of high-risk oral lesions can improve patient survival rates. With image classification using oral cavity images and other forms of medical images, the information to be classified can often be extremely localized. To address this problem, we propose the use of convolutional neural networks with trainable soft attention. Further to this, we incorporate the use of localization loss to penalize the difference between attention maps and clinically annotated mask. This effectively allows clinicians to help guide soft attention. Improvements to the baseline were made, with an accuracy of 0.8333 and a ROC AUC of 0.8632, which equates to increases of 0.0245 and 0.0394, respectively. This accuracy corresponds to a sensitivity of 0.8469 and a specificity of 0.8208. Perhaps of more importance, is a model that demonstrates better capability at paying attention to the lesions in its decision making. Furthermore, visualizing resulting attention maps can help to strengthen clinical confidence in AI decision making.

© Springer Nature Switzerland AG 2021
N. Tsapatsoulis et al. (Eds.): CAIP 2021, LNCS 13052, pp. 226–236, 2021.
https://doi.org/10.1007/978-3-030-89128-2_22

Keywords: Deep learning · Attention · Oral cancer · Oral potentially malignant disorders

1 Introduction

Oral cancer has a major impact on global health, with an estimated 177,384 deaths in 2018 [1]. It is most prevalent in low- and middle-income countries (LMICs), where a lack of access to appropriate medical care contributes towards late disease presentation, and as a result survival rates are low. However, oral cancer presents unique opportunities, with oral lesions called oral potentially malignant disorders (OPMDs) preceding oral cancer for many patients. These lesions are visible for early detection and close monitoring without the need for invasive procedures.

Telemedicine can aid early diagnosis. The use of mobile phones has been field tested in a rural community [2], enabling two-way communication between primary healthcare practitioners and specialists located off-site. Integration of artificial intelligence (AI) into such approaches to facilitate the automated identification of high-risk oral lesions, will reduce the pressure on the limited number of specialists.

Currently, deep convolutional neural networks (CNN) provide the state-of-the-art results for many computer vision tasks. For medical image classification, the information to be classified can often be extremely localized. Therefore, the trainable attention mechanism can play a big role in medical image analysis, highlighting areas of interest whilst suppressing irrelevant parts of the image. This replicates the ability of clinicians to know where to look when making decisions.

Object detection equates to a type of hard attention and has been used to classify lesions in mammograms [3]. Hard attention can be effective, but in some cases may lead to loss of useful information when outside of the cropped region. Trainable soft attention offers the suppression of irrelevant background information without the need for cropped regions. It was introduced in machine translation [4] and later in image captioning. Jetley [5] demonstrated an increase in CNN image classification performance with the use of multi-scale soft attention. Pesce [6] interestingly improved the detection of chest radiographs containing pulmonary lesions, by penalizing differences in the soft attention maps and a subset of clinically annotated masks.

We first propose to adapt the multi-scale soft attention model derived by Jetley [5] for the novel application of attention for the early detection of oral cancer. Further to this, we incorporate multiple task learning, with the use of localization loss and classification loss. This adaptation is inspired by Pesce [6], with the localization loss quantifying the difference between soft attention maps and clinically annotated data; thus, effectively allowing clinicians to help guide soft attention. Instead of the attention maps indirectly driving the attention mechanism [6], our attention maps directly weight the feature vectors [5]. Whist an increase in image classification performance is the primary target, providing a model that reliably pays attention to the lesions in its decision making is of importance. Model interpretability to support clinical confidence in AI decision making can be aided by visualizing attention maps, with clinical confidence strengthened when attention demonstrates clinical guidance.

2   Related Work

The following reported studies, related to oral cancer, were CNN based. Aubreville [7] used InceptionV3 [8] to classify laser endomicroscopy images as clinically normal and carcinogenic. Halicek [9] used InceptionV4 for cancer detection from histology slides of excised tissue from the head and neck (included the oral cavity). Uthoff [10] used pairs of autofluorescence and white light mobile phone captured images as inputs to a VGG model [11] to perform suspicious vs. not suspicious classification. These reported studies utilized images obtained using specialized clinical procedures and advanced imaging systems.

Of more relevance to screening in LMICs are methods applied to standard oral cavity images. Welikala [12] explored/compared image classification and object detection to automate the early detection of oral cancer. Image classification was shown to be the more viable approach, whilst object detection struggled due to the indistinct nature of lesion boundaries. A study [13] demonstrated that simpler CNN architectures are more suitable when fine-tuning on an oral lesion dataset of limited size, VGG-19 [11] performed the best for the classification of referral vs. non-referral. Shamim [14] used VGG-19 for benign vs. pre-cancerous classification and Jubair [15] used a lightweight CNN for benign vs. suspicious classifications; both restricted to tongue lesions only.

3   Materials

A library of well-annotated images of oral lesions is currently being built. Accompanied by metadata of age, gender, and smoking, alcohol, and betel quid chewing status. Lesions in the images have been annotated with bounding boxes and each box has been assigned multiple labels that include lesion type, morphology, site, referral decision etc. At this initial phase of construction, the dataset included 2155 images. Each image has been separately annotated by 3–7 clinicians.

For this study we were only concerned with the bounding boxes and their referral decision labels. The annotations from multiple clinicians were combined with a novel strategy proposed by [12]. The annotated lesion's referral decision label was used as a single image label and if an image contained multiple annotated lesions then that with the highest referral decision severity was used. There were five classes in total, although the data was simplified to 'non-referral' vs. 'referral' as this would be the first step towards translation into clinical practice. The dataset was split into training, validation, and test sets (see Table 1).

Table 1. Image numbers according to class and set type.

Class	Training	Validation	Testing	Total
Non-referral	949	125	106	1180
Referral	795	82	98	975
Total	1744	207	204	2155

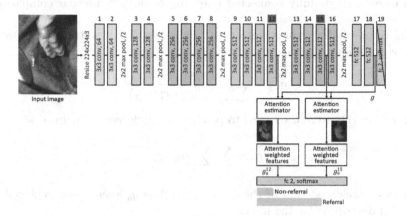

Fig. 1. Attention introduced at layer 12 and 15 of VGG-19.

4   Method

4.1   Attention Network

We adapted a soft trainable attention model proposed by Jetley [5], based on defining a compatibility score between local and global features. The intuition was that the compatibility score is intended to have a high value when the image patch described by the local features contained parts of the dominant image category. Therefore, a compatibility score assumed the role of attention values and was used to create a weighted combination of local features for performing image classification. This was a multi-scale approach achieved by leveraging local features from different intermediate stages of the CNN.

We implemented this approach using VGG-19 (reliable and outperforms newer architectures [13]), see Fig. 1. The softmax classification layer was reduced to 2 neurons ('non-referral' vs. 'referral') and the prior fully connected layers were reduced to 512 neurons. Transfer learning was used to address the limited amount of data, with the VGG-19 model pre-trained on the ImageNet dataset [16]. We only fine-tuned from layer 12 and up to help avoid overfitting [13]. The attention mechanism was introduced at convolutional layers 12 and 15 (most effective), limited to fine-tuned layers.

Consider the local feature vector $l_{i,j}$ which was the output activations at the spatial location (i, j) of $n \times n$ spatial locations at a specific convolutional layer. Consider the global feature vector g which was normally fed into the final classification layer of the original VGG architecture.

The alignment model from [4] was re-purposed to calculate a compatibility score, defined as

$$c_{i,j} = f(l_{i,j} + g) \tag{1}$$

where vectors $l_{i,j}$ and g were of equal dimensions (i.e. 1×512) and the function learnt using a single fully connected mapping to output a scalar compatibility score.

The compatibility scores were then normalized by a softmax operation to produce attention weights, defined as

$$a_{i,j} = \frac{exp(c_{i,j})}{\sum_{i=1}^{n} \sum_{j=1}^{n} exp(c_{i,j})} \tag{2}$$

Attention weights were then used to produce a single vector, defined as

$$g_a = \sum_{i=1}^{n} \sum_{j=1}^{n} a_{i,j} l_{i,j} \tag{3}$$

which was a weighted sum of the $l_{i,j}$ vectors. The g_a vector now replaced g as the global descriptor for the image.

This process was done separately for layers 12 and 15 (prior to max pooling), with 28×28 and 14×14 spatial locations, respectively. The resultant two vectors (g_a^{12} and g_a^{15}) were then concatenated into a single vector and passed through a softmax classification layer to produce class predictions.

4.2 Guided Attention

As standard, we minimized the classification loss (with cross entropy) of

$$L_{cls} = -\frac{1}{M} \sum_{m=1}^{M} \sum_{s=1}^{S} y_{s,m} log(\hat{y}_{s,m}) \tag{4}$$

where $y_{s,m}$ is the label from the one-hot class vector, $\hat{y}_{s,m}$ is the class probability from the prediction, S is the number of classes, and M is the number of images.

In addition to this, we also minimized attention based localization error, see Fig. 2. If an image contained lesions, their bounding box annotations were converted to a binary mask, ones indicated pixels that belong to a lesion. The binary mask was then resized to the size of the attention map (either 28×28 or 14×14) for comparison. The attention map was rescaled to the range [0, 1] using division by the maximum value of the map, to make it comparable to the binary mask. A pixel-wise mean square error was computed for attention based localization loss, which was then applied for all images that contained bounding boxes, defined as

$$L_{loc} = -\frac{1}{M_1} \sum_{m=1}^{M_1} \frac{1}{n^2} \sum_{i=1}^{n} \sum_{j=1}^{n} [b_{i,j,m} - a'_{i,j,m}]^2 \tag{5}$$

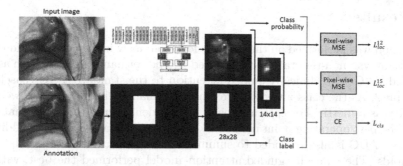

Fig. 2. Illustration of how the loss terms were calculated. MSE = mean square error, CE = cross entropy. Loss terms calculated across batch of images.

where b represents the binary masks, a' represents the rescaled attention maps, n represents both the height and width of the binary masks and attention maps, and M_1 is the number of images that contained bounding boxes. This quantified the difference between the binary masks and attention maps.

The network was then trained end-to-end to minimize a linear combination of classification loss and attention based localization loss, defined as

$$L = \lambda_1 L_{cls} + \lambda_2 L_{loc}^{12} + \lambda_3 L_{loc}^{15} \tag{6}$$

where attention based localization loss was calculated for the attention maps of convolutional layers 12 and 15. Empirically derived balancing parameters (λ) were used to weight the loss terms, with values of $\lambda_1 = 1.0$, $\lambda_2 = 2.5$, $\lambda_3 = 2.5$.

4.3   Technical Details

Backpropagation and stochastic gradient descent (SGD) with momentum was used for training. Images were rescaled to 224×224 pixels. Flipping, scaling, translation, and rotation were used to augment the training data (images/masks).

SGD mini-batch size was 128 images. Classification loss was class weighted to correct for the slight imbalance in the training data. We used a learning rate of 0.001, a momentum of 0.9, and a weight decay of 0.01. Batch normalization and a dropout of 0.5 were used on the first two fully connected layers. The network was trained for 100 epochs until convergence. The model was built on the training set and hyperparameters were derived from performance on the validation set.

A Nvidia GeForce RTX 2080 Ti graphics card with 11 GB memory was used for training. This implementation used Keras and TensorFlow.

5   Results

Evaluation was performed on the test set, for the binary image classification task of 'referral' vs. 'non-referral'. We compared the performance of VGG with attention and VGG with clinically guided attention to the standard VGG (baseline); see Table 2. As the classes were approximately balanced in the test set, we used accuracy as a metric. For each approach, a confidence score threshold that produced the best operating point defined by the accuracy was selected. In addition, the ROC AUC is also provided to summarize the performance across different thresholds. The clinically guided attention model performed the best, with an accuracy of 0.8333 and a ROC AUC of 0.8632. This accuracy corresponds to a sensitivity of 0.8469, a specificity of 0.8208, a precision of 0.8137, and a recall of 0.8469. The precision-recall and ROC curves are provided in Fig. 3.

Table 2. Image classification results.

Model	Accuracy	ROC AUC
Standard VGG	0.8088	0.8238
VGG with attention	0.8186	0.8498
VGG with clinically guided attention	0.8333	0.8632

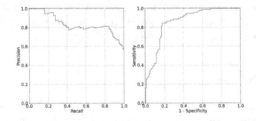

Fig. 3. Precision-Recall curve (AUC = 0.8367) and ROC curve (AUC = 0.8632) for the clinically guided attention model.

Outputs from the three approaches are provided in Fig. 4 to enable a comparison of class predictions and attention maps. Figure 5 expands on further outputs of just the clinically guided attention model, providing a comparison of attention maps from layers 12 and 15.

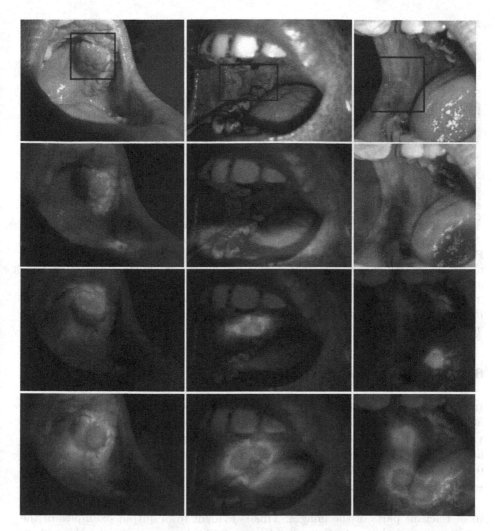

Fig. 4. Output comparison of the three models. Top row: input images, clinically annotated bounding boxes overlaid for visualization, class label (left to right) = ['referral',
'referral', 'referral']. Second row: Standard VGG, no trainable attention so Grad-CAM
used for attention maps, class specific, predicted class and probability (left to right) =
['referral' 0.846, 'referral' 0.969, 'non-referral' 0.773]. Third row: VGG with attention,
layer 15 attention maps, predicted class and probability (left to right) = ['referral'
0.901, 'referral' 0.991, 'non-referral' 0.536]. Bottom row: VGG with clinically guided
attention, layer 15 attention maps, predicted class and probability (left to right) =
['referral' 0.739, 'referral' 0.990, 'referral' 0.662].

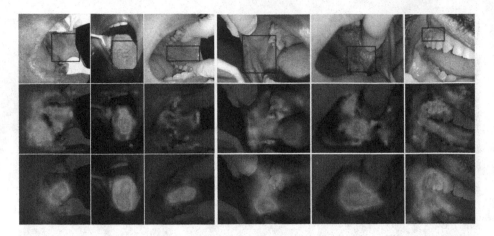

Fig. 5. Correct and incorrect outputs for the clinically guided attention model. Top row: input images, clinically annotated bounding boxes overlaid for visualization, class label (left to right) = ['referral', 'non-referral', 'non-referral', 'referral', 'non-referral', 'referral']. Middle row: layer 12 attention maps. Bottom row: layer 15 attention maps. Predicted class and probability (left to right) = ['referral' 0.738, 'non-referral' 0.836, 'non-referral' 0.929, 'non-referral' 0.687, 'referral' 0.820, 'non-referral' 0.881].

6   Discussion and Conclusion

In this paper, we have demonstrated the performance of multi-scale trainable soft attention which has been clinically guided for the image classification of 'referral' vs. 'non-referral' with respect to oral cancer. The proposed model achieved an accuracy of 0.8333 and a ROC AUC of 0.8632, which is an improvement of 0.0245 and 0.0394 on the baseline model, and 0.0147 and 0.0134 on the model with attention (not clinically guided), respectively.

In image classification, the decision making process may not always use the most relevant parts of the images. This is evident from output examples of the baseline model shown in the second row of Fig. 4, on occasions making decisions without even focussing on the lesions. The attention mechanism offers a much more targeted approach, highlighting areas of interest whilst suppressing irrelevant parts of the image. This is demonstrated in last two rows of Fig. 4, showing the outputs for the two attention models, with lesions being more clearly highlighted. The bottom row does appear superior, with a greater coverage of the lesions, showing that attention benefits from clinical guidance, which also resulted in a higher classification performance.

Multi-scale attention allows complementary focus on different parts of the image at different scales, whilst still receiving clinician guidance on the general region of the lesion. Attention maps from layer 12 appear to attend to part details of lesions/surrounding area and those from layer 15 on whole lesions, apparent from Fig. 5. Incorrect classifications are shown in the last 3 columns of Fig. 5, where localization still performs well (apart from the last column).

Related work reported values of 0.8500 and 0.8875 [10], 0.866 and 0.900 [7], 0.89 and 0.97 [14], and 0.867 and 0.845 [15], for sensitivity and specificity, respectively. Whilst we provide a comparative study of models in Table 2, currently, direct comparisons to related work can be difficult to make because their datasets and consequently their methodologies are designed to tackle different challenges. To drive competition, we aim to release a publicly available dataset.

Performances need to improve before translation into clinical practice. The future plan is to build a larger dataset (only a subset requires annotated bounding boxes), which is key to deep learning in order to improve results. A larger test set will be in place, whereby the generalizability of the model can be properly tested. An alternative is nested k-fold cross validation (keeping model selection in mind). We plan to make use of the metadata as input, and for models to output several of the other clinically assigned labels to gain further benefits of multi-task learning. High quality data will be promoted by putting constraints on what is acceptable and models will be built with a larger input image size.

In conclusion, this paper has demonstrated the use of trainable soft attention for the early detection of oral cancer, with improved performances when that attention was clinically guided. Importantly, the model demonstrates an improved ability to pay attention to the lesions in its decision making. In addition to post-hoc attention maps, trainable attention maps aid model interpretability, helping to strengthen clinical confidence in AI decision making, particularly when maps demonstrate clinical guidance on where to look.

Acknowledgments. We would like to thank the Medical Research Council for providing funding (MR/S013865/1).

References

1. Bray, F., Ferlay, J., Soerjomataram, I., Siegel, R.L.: Global cancer statistics 2018: GLOBOCAN estimates of incidence and mortality worldwide for 36 cancers in 185 countries. CA Cancer J. Clin. **68**(6), 394–424 (2018)
2. Haron, N., et al.: m-Health for early detection of oral cancer in low-and middle-income countries. CA Telemed. e-Health **26**(3), 278–285 (2020)
3. Ribli, D., Horváth, A., Unger, Z., Pollner, P., Csabai, I.: Detecting and classifying lesions in mammograms with deep learning. Sci. Rep. **8**(1), 1–7 (2018)
4. Bahdanau, D., Cho, K., Bengio, Y.: Neural machine translation by jointly learning to align and translate. arXiv preprint arXiv:1409.0473 (2014)
5. Jetley, S., Lord, N.A., Lee, N., Torr, P.H.S.: Learn to pay attention. arXiv preprint arXiv:1804.02391 (2018)
6. Pesce, E., Withey, S.J., Ypsilantis, P.P., Bakewell, R., Goh, V., Montana, G.: Learning to detect chest radiographs containing pulmonary lesions using visual attention networks. Med. Image Anal. **53**, 26–38 (2019)
7. Aubreville, M., et al.: Automatic classification of cancerous tissue in laserendomicroscopy images of the oral cavity using deep learning. Sci. Rep. **7**(1), 1–10 (2017)
8. Szegedy, C., Vanhoucke, V., Ioffe, S., Shlens, J., Wojna, Z.: Rethinking the inception architecture for computer vision. In: Proceedings of the IEEE Conference on Computer Vision and Pattern Recognition, pp. 2818–2826 (2016)

9. Halicek, M., et al.: Head and neck cancer detection in digitized whole-slide histology using convolutional neural networks. Sci. Rep. **9**(1), 1–11 (2019)
10. Uthoff, R.D., et al.: Point-of-care, smartphone-based, dual-modality, dual-view, oral cancer screening device with neural network classification for low-resource communities. PloS One **13**(12), e0207493 (2018)
11. Simonyan, K., Zisserman, A.: Very deep convolutional networks for large-scale image recognition. arXiv preprint arXiv:1409.1556 (2014)
12. Welikala, R.A., et al.: Automated detection and classification of oral lesions using deep learning for early detection of oral cancer. IEEE Access **8**, 132677–132693 (2020)
13. Welikala, R.A., et al.: Fine-tuning deep learning architectures for early detection of oral cancer. In: Bebis, G., Alekseyev, M., Cho, H., Gevertz, J., Rodriguez Martinez, M. (eds.) ISMCO 2020. LNCS, vol. 12508, pp. 25–31. Springer, Cham (2020). https://doi.org/10.1007/978-3-030-64511-3_3
14. Shamim, M.Z.M., Syed, S., Shiblee, M., Usman, M., Ali, S.: Automated detection of oral pre-cancerous tongue lesions using deep learning for early diagnosis of oral cavity cancer. arXiv preprint arXiv:1909.08987 (2019)
15. Jubair, F., Al-karadsheh, O., Malamos, D., Al Mahdi, S., Saad, Y., Hassona, Y.: A novel lightweight deep convolutional neural network for early detection of oral cancer. Oral Diseases (2021)
16. Deng, J., Dong, W., Socher, R., Li, L.J., Li, K., Fei-Fei, L.: ImageNet: a large-scale hierarchical image database. In: IEEE Conference on Computer Vision and Pattern Recognition, pp. 248–255 (2009)

Small and Large Bile Ducts Intrahepatic Cholangiocarcinoma Classification: A Preliminary Feature-Based Study

Chiara Losquadro[1]([✉]), Silvia Conforto[1], Maurizio Schmid[1], Gaetano Giunta[1], Marco Rengo[2], Vincenzo Cardinale[2], Guido Carpino[3], Andrea Laghi[4], Ana Lleo[5,6], Riccardo Muglia[6], Ezio Lanza[6], and Guido Torzilli[5,6]

[1] Department of Engineering, Applied Electronics Section, Roma Tre University, Rome, Italy
chiara.losquadro@uniroma3.it
[2] Department of Medical Surgical Sciences and Biotechnologies, La Sapienza University, Latina, Italy
[3] Division of Health Sciences, Department of Movement, Human and Health Sciences, University of Rome Foro Italico, Rome, Italy
[4] Department of Medical Surgical Sciences and Translational Medicine, La Sapienza University, Rome, Italy
[5] Department of Biomedical Sciences, Humanitas University, Milan, Italy
[6] Humanitas Clinical and Research Center, Milan, Italy

Abstract. Cholangiocarcinoma (CCA) is the second most common liver malignancy and the incidence and mortality rates of this disease are worldwide increasing. This paper deals with the problem of Intrahepatic Cholangiocarcinoma (IH-CCA) classification using Computed Tomography (CT) images. Precisely, a radiomics-based approach is proposed by exploiting abdominal volumetric CT data in order to differentiate large bile duct from small bile duct IH-CCA. The developed method relies on the investigation of intrinsic discriminative properties of CT scans according to feature selection methods. The effectiveness of the proposed method is proved by enrolling in the study a total of 26 patients, including 16 patients with large bile duct and 10 with small bile duct pathological disease, respectively. The conducted tests have shown that our approach is a baseline to provide an efficient classification process with a low computational cost in order to facilitate clinical decision-making procedures.

Keywords: Cholangiocarcinoma (CCA) · Classification · Feature extraction · Computed Tomography (CT) images

1 Introduction

Cholangiocarcinoma (CCA) represents a fatal cancer of the biliary epithelium, arising both within the liver and the extrahepatic bile ducts. According to its

© Springer Nature Switzerland AG 2021
N. Tsapatsoulis et al. (Eds.): CAIP 2021, LNCS 13052, pp. 237–244, 2021.
https://doi.org/10.1007/978-3-030-89128-2_23

origin, it is framed into Intrahepatic Cholangiocarcinoma (IH-CCA) and Extra-hepatic Cholangiocarcinoma (EH-CCA), respectively [10,15]. Globally, CCA is the second most common hepatic malignancy: in particular, recent studies have shown that the incidence and mortality rates of the IH-CCA disease are world-wide increasing [13]. IH-CCAs are pathologically classifiable into conventional IH-CCAs (bile duct IH-CCA), bile ductular IH-CCAs, intraductal neoplasms and several rare variants which are usually diagnosed due to a combination of clinical presentation, laboratory analysis, and radiologic evaluation [5,6]. In particular, Magnetic Resonance Imaging (MRI) and Computed Tomography (CT) are fre-quently utilized to obtain useful radiologic information. Although the mentioned existing diagnostic modalities, the histopathological examination is required for a definitive diagnosis in the great majority of patients [4]. In this respect, to avoid this invasive method, advanced methods of image processing and image texture analysis may represent a useful way to increase the obtainable radiologic information from medical images by considering the inter-relationships of pix-els and the values they could assume, providing then, additional and important information within the frame of IH-CCA pathological condition. As a matter of fact, during the past few decades, artificial intelligence technologies are gaining a great achievement with respect to medical image analysis, with the goal to assist clinicians detecting different diseases, due to the possibility to extract quantita-tive features from image examinations [18]: recent studies have focused on these methodologies in order to provide an effective and automatic way to diagnose also the CCAs disease. In particular, the authors of [9] validate a radiomic model where several features were extracted from CT scans in order to predict the clinical outcome of intrahepatic cholangiocarcinoma; likewise, in [16], radiomics signatures are evaluated starting from ultrasound medicine images in order to develop a CCA classification system. Additionally, the authors of [19] investigate a radiomics-based model for preoperative differentiation of hepatocellular and cholangiocarcinoma (CH-CCA) and intrahepatic cholangiocarcinoma (IH-CCA).

Following the line of reason of these works, in this paper a radiomics-based methodology is proposed and investigated, in order to classify conventional IH-CCAs into small bile duct and large bile duct types. The analyses conducted on challenging CT data have shown the effectiveness of the proposed method providing a baseline classification process characterized by a low computational cost.

The paper is organized as follows: Sect. 2 introduces the problem statement together with the details of the proposed feature selection and extraction algo-rithms. The classification capabilities of the proposed technique are assessed in Sect. 3 where the classification results related to CT images are shown. Finally, Sect. 4 gives some conclusions and some hints for possible future studies.

2   Proposed Classification Algorithm

This section describes an automatic method focused on the exploitation of intrin-sic discriminative properties of abdominal CT images in order to differentiate

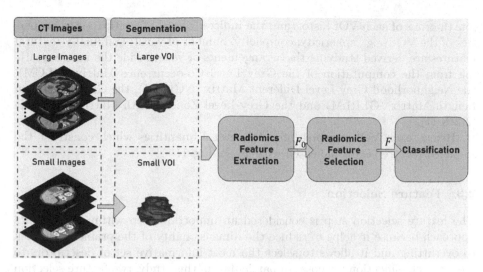

Fig. 1. Block scheme of the proposed CT images feature extraction algorithm.

liver pathologies. Precisely, the proposed solution framed within the context of a radiomics-based approach as a result of which effective classification models can be built. Synthetically, the proposed classification algorithm is pictorially represented in Fig. 1 and it is described in the following lines.

2.1 Tumor Segmentation

In the framework of automatic classification of human lesions, the segmentation process represents a preliminary step of paramount importance in order to achieve accurate classification results [1]. As a matter of fact, tumor segmentation aims to basically separate the pixels corresponding to the diseased tissue from the surrounding anatomy in order to avoid erroneous classification due to the overall anatomy of the patients being observed. For such a purpose, in the proposed study, the volume of interest (VOI) of each CT acquisition has been manually contoured by a radiologist, using the 3D Slicer platform, an open source software package for medical images visualization and computation.

2.2 Feature Extraction

In the attempt to differentiate small from large bile duct IH-CCAs, a feature extraction approach is investigated by exploiting volumetric CT data [12]. In particular, given each patient VOI, several quantitative parameters based on conventional indices (n = 7), first order statistical measures (n = 6), indices from shape (n = 4) and texture analysis (n = 32) are extracted, resulting in a total of 49 radiomics features. Basically, conventional indices reflects native values extracted from the CT images such as quartile, mean, maximum and minimum measures; first order statistical measures consist of the moments up to the

fourth-order of each VOI histogram; the indices from shape return some properties of the VOI (e.g., sphericity, compacity, number of voxel) and finally textural
features are derived studying the arrangements of voxel inside the images starting from the computation of the Grey Level Co-occurrence Matrix (GLCM),
the Neighborhood Grey-Level Different Matrix (NGLDM), the Grey-Level Run
Length Matrix (GLRLM) and the Grey-Level Zone Length Matrix (GLZLM)
[2,3,8,17].

Hence, each VOI is represented by several quantities which construct the
so-called feature vector indicated with F_0 in Fig. 1.

2.3   Feature Selection

The feature selection step is considered an important step within a radiomics
approach because it helps to reduce the dimensionality of the problem, the risk
to overfitting and it allows to select the most informative set of descriptors to
meet the classification purpose. In particular, in this study, two feature selection
methods are compared:

- Principal Component Analysis (PCA), consisting in a feature transformation
  techniques to reduce the dimensionality of the data by transforming data into
  new features [7].
- Relief-Based Algorithm (RBA), consisting in a filter type feature selection
  algorithm which measures feature importance based on the characteristics of
  features, i.e., information, distance, similarity, and statistical measures [11].
  In particular, after this selection procedure, three GLZLM features (Long-
  Zone Emphasis, Long-Zone High Gray-level Emphasis, Long-Zone Low Gray-
  level Emphasis), one GLCM feature (Correlation) and one shape index (Voxel
  Count) were selected for the following classification process.

Before each feature selection step, the feature vector, F_0, is normalized thanks
to the following linear rescaling:

$$F = \frac{F_0 - \mu_{F_0}}{\sigma_{F_0}} \tag{1}$$

meaning μ_{F_0} as the mean and σ_{F_0} as the standard deviation of the feature
vector F_0. This is done to avoid that a very strong feature value could polarize
the performance of the proposed method.

3   Performance Evaluation

This section shows the classification capabilities of the suggested approach in
terms of small versus large IH-CCA VOIs discrimination through the comparison of the feature selection methods described in Subsect. 2.3. Therefore, the
description of the patients dataset is first provided, followed by the details of the
classification procedure together with the discussion of the obtained results.

3.1   Patients Dataset

The analyses are performed on patients affected with either large and small bile duct IH-CCA, which are included in the dataset herein exploited and that has been gathered from the "Humanitas Research Hospital" in Milan. In particular, our dataset contains the examinations of both large and small IH-CCA affected people, acquired using a GE Healthcare EVO CT scanner and stored in a 512×512 DICOM format. For the purpose of this study, a total of 26 patients are enrolled in the study, including respectively 16 large IH-CCA and 10 small IH-CCA examinations. Table 1 summarizes the patients population details in terms of their sex and age (expressed as mean and standard deviation (SD), namely, $\pm SD$).

Table 1. Sex and age details of patients enrolled in the study.

		Large IH-CCA	Small IH-CCA
Sex	*Male*	9	5
	Female	7	5
Age y ($\pm$SD)		69.0 ± 8.51	66.0 ± 11.65

3.2   Classification Procedure and Results

The effectiveness of the devised methodology in discriminating small and large bile ducts pathology is described below.

The classification process is performed dividing the dataset into two non-overlapped groups: the training set, composed by about the 70% of the available data and the test set, composed by the remainder 30% having foresight to exclude all the data involved during the training phase. Furthermore, since the aim of this work is to show the effectiveness of the proposed radiomics-extraction method, we compare several classifiers, precisely the k-nearest neighbour (k-NN, note that in this work the parameter k set equal to 5), the support vector machine (SVM) and the decision tree (DT) because of their low computational burden.

Then, to demonstrate the effectiveness of the proposed algorithm, the classification accuracy, namely A_{cc}, is used as figure of merit, whose analytic expression is given by

$$A_{cc} = \frac{TP + TN}{TP + FP + TN + FN}, \qquad (2)$$

where TP is the total number of true positives, FP is the total number of false positives, TN is the total number of true negatives, and finally FN represents the total number of false negatives.

However, to provide a statistical characterization of the entire classification method, the average classification accuracy, $\bar{A}_{cc}$, is estimated due to a standard Monte Carlo approach. More precisely, for each independent Monte Carlo trial a different selection of the training and test sets is randomly chosen for each class (i.e., large IH-CCA and small IH-CCA). This process reiterates N number of times, and in this study N is set equal to 100, the training and test procedures by independent random extractions.

Figure 2 shows the classification results in terms of $\bar{A}_{cc}$ of PCA and Relief algorithms according to the use of three different classifiers. From Fig. 2 it is evident that the Relief method and the use of SVM classifier is capable of ensuring the maximum $\bar{A}_{cc}$ reaching the 69.67% of average correct classification.

In addition, the classification accuracy obtained in the best case (that is the best one extracted from the 100 different experimental cases) is highlighted according to the confusion matrices depicted in Fig. 3. In particular, the PCA and Relief algorithms allow to reach the maximum accuracy of 88.89% by the use of k-NN and SVM respectively with a very low computational cost.

Finally, the promising results of the proposed method could be surely improved considering for instance the investigation of additional features or the application of neural networks [14].

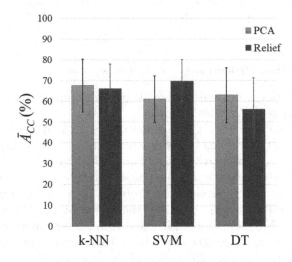

Fig. 2. $\bar{A}_{cc}$ (%) reached through the PCA and Relief algorithms respectively by k-NN, SVM and DT classifiers.

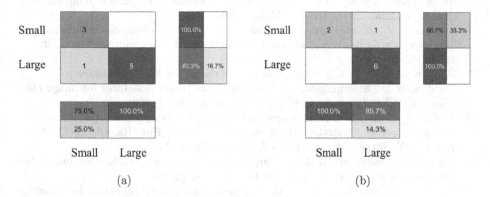

Fig. 3. Confusion matrices, actual class versus predicted class. Subplots refer to (a) confusion matrix of PCA with k-NN classifier, and (b) confusion matrix of Relief algorithm with SVM classifier.

4   Conclusions

This paper has proposed a radiomics extraction procedure to discriminate large from small bile ducts IH-CCA patients. The proposed method has dealt with the exploitation of existing intrinsic discriminative properties of the acquired abdominal CT images. In particular, radiomics features have been extracted starting from patients VOIs and then selected comparing two feature selection methods. The extracted features have finally constituted the input to several classifiers ensuring promising correct classification results.

Possible future research works related to this study might consider a larger sample size dataset, the use of other features to be line-up to those explored in this paper as well as the test of neural networks with the aim to improve the recognition capabilities.

Aknowledgments. Medical datasets were acquired at "Humanitas Research Hospital" in Milan (Italy) with the explicit consent of all patients for research purposes.

References

1. Aggarwal, P., Renu, V., Bhadoria, S., Dethe, C.: Role of segmentation in medical imaging: a comparative study. Int. J. Comput. Appl. **29**(1), 54–61 (2011). https:// doi.org/10.1109/36.739146
2. Ahmed, A., et al.: Radiomic mapping model for prediction of Ki-67 expression in adrenocortical carcinoma. Clin. Radiol. **75**(6), 479-e17 (2020)
3. Amadasun, M., King, R.: Textural features corresponding to textural properties. IEEE Trans. Syst. Man Cybern. **19**(5), 1264–1274 (1989)
4. Bartella, I., Dufour, J.F.: Clinical diagnosis and staging of intrahepatic cholangio-carcinoma. J. Gastrointestin. Liver Dis. **24**(4), 481–489 (2015)

5. Bridgewater, J., et al.: Guidelines for the diagnosis and management of intrahepatic cholangiocarcinoma. J. Hepatol. **60**(6), 1268–1289 (2014)
6. Cardinale, V., et al.: Intra-hepatic and extra-hepatic cholangiocarcinoma: new insight into epidemiology and risk factors. World J. Gastrointest. Oncol. **2**(11), 407 (2010)
7. Dunteman, G.H.: Principal Components Analysis. No. 69, Sage (1989)
8. Haralick, R.M., Shanmugam, K., Dinstein, I.H.: Textural features for image classification. IEEE Trans. Syst. Man Cybern. **6**, 610–621 (1973)
9. Ji, G.-W., et al.: A radiomics approach to predict lymph node metastasis and clinical outcome of intrahepatic cholangiocarcinoma. Eur. Radiol. **29**(7), 3725–3735 (2019). https://doi.org/10.1007/s00330-019-06142-7
10. Khan, S., Toledano, M., Taylor-Robinson, S.: Epidemiology, risk factors, and pathogenesis of cholangiocarcinoma. HPB **10**(2), 77–82 (2008)
11. Kira, K., Rendell, L.A.: A practical approach to feature selection. In: Machine Learning Proceedings 1992, pp. 249–256. Elsevier (1992)
12. Kurani, A.S., Xu, D.H., Furst, J., Raicu, D.S.: Co-occurrence matrices for volumetric data. Heart **27**, 25 (2004)
13. Malhi, H., Gores, G.: Cholangiocarcinoma: modern advances in understanding a deadly old disease. J. Hepatol. **45**, 856–67 (2007). https://doi.org/10.1016/j.jhep.2006.09.001
14. Meng, L., et al.: A deep learning prognosis model help alert for covid-19 patients at high-risk of death: a multi-center study. IEEE J. Biomed. Health Inform. **24**(12), 3576–3584 (2020). https://doi.org/10.1109/JBHI.2020.3034296
15. Patel, T.: Worldwide trends in mortality from biliary tract malignancies. BMC Cancer **2**(1), 1–5 (2002)
16. Peng, Y.T., et al.: Preoperative ultrasound radiomics signatures for noninvasive evaluation of biological characteristics of intrahepatic cholangiocarcinoma. Acad. Radiol. **27**(6), 785–797 (2020)
17. Thibault, G., Angulo, J., Meyer, F.: Advanced statistical matrices for texture characterization: application to cell classification. IEEE Trans. Biomed. Eng. **61**(3), 630–637 (2013)
18. Zhang, J., Xia, Y., Xie, Y., Fulham, M., Feng, D.D.: Classification of medical images in the biomedical literature by jointly using deep and handcrafted visual features. IEEE J. Biomed. Health Inform. **22**(5), 1521–1530 (2018). https://doi.org/10.1109/JBHI.2017.2775662
19. Zhang, J., et al.: Differentiation combined hepatocellular and cholangiocarcinoma from intrahepatic cholangiocarcinoma based on radiomics machine learning. Ann. Transl. Med. **8**(4) (2020)

A Review on Breast Cancer Brain Metastasis: Automated MRI Image Analysis for the Prediction of Primary Cancer Using Radiomics

Vangelis Tzardis[1]([✉]) [iD], Efthyvoulos Kyriacou[2] [iD], Christos P. Loizou[2] [iD], and Anastasia Constantinidou[3,4] [iD]

[1] Frederick University, 3080 Limassol, Cyprus
st019415@stud.frederick.ac.cy
[2] Cyprus University of Technology, 3036 Limassol, Cyprus
christos.loizou@cut.ac.cy
[3] University of Cyprus, 2029 Nicosia, Cyprus
constantinidou.anastasia@ucy.ac.cy
[4] Bank of Cyprus Oncology Centre, 2006 Nicosia, Cyprus

Abstract. Breast cancer brain metastasis (BCBM) still remains a major clinical challenge. Current systemic treatments are often inadequate while diagnosis involves time-consuming series of neuro-imaging acquisitions and dangerous invasive biopsies. Automated image analysis systems for the identification, prediction and follow up of BCBM are therefore required. This review discusses the advancements in the automated MRI brain metastasis (BM) image analysis using radiomic features based classification. Seven BM segmentation studies, and three BCBM identification studies were considered eligible. The latter studies were based on either manual or semi-automated segmentation methods. Almost every fully automated BM segmentation method presented in the literature, reported a maximum dice similarity score (DSC) of 84%, but they resulted in a poor BM segmentation for brain areas less than 5 mm (0.06 ml). The multi-class prediction of BCBM approach, which is more representative for clinical applicability, is based on imaging features and resulted in an area under the curve (AUC) of 60%. Therefore, the need still exists for the development of automated image analysis methods for the identification, follow up and prediction of BCBM. The potential clinical usage of above methods entails further multi-center studies with comprehensive clinical data and multi-class modeling with vast and varying primary and metastatic brain tumors.

Keywords: Magnetic resonance imaging · Breast cancer brain metastasis · Radiomics · Automated image analysis · Tumor segmentation · Tumor classification · Primary tumor origin

© Springer Nature Switzerland AG 2021
N. Tsapatsoulis et al. (Eds.): CAIP 2021, LNCS 13052, pp. 245–255, 2021.
https://doi.org/10.1007/978-3-030-89128-2_24

1 Introduction

This paper provides a literature review of the individual procedures that lead to a fully automated image analysis for the characterization of breast cancer brain metastasis (BCBM) in magnetic resonance images and these are presented in Tables 1 and 2.

Around 80% of the brain metastases (BM) originate from lung, breast, renal cell carcinomas, melanoma and gastrointestinal tract adenocarcinomas. The estimated incidence rate of BCBM varies among studies and is reported to be as high as 30% in autopsy studies [1]. The development of BCBM leads to a low quality of life and a debilitating symptomatology, while few patients live more than one year [2]. Recent studies suggested that, along with systemic therapy, surgery and radiation, BM patients' treatment should also be based on their overall performance status as well as the primary tumor site, its molecular subtype, the number, location and the size of the BM [2, 3].

More specifically, the diagnosis of BCBM is currently performed based on clinical information, magnetic resonance imaging (MRI) modalities and histopathological examination to help discriminate BCBM from primary brain tumors. A summary of fully automated brain metastasis segmentation studies reported in the literature is presented in Table 1. However, conventional assessment of BM using MRI remains descriptive, subjective and mostly qualitative [4]. The evaluation with regards to tumor location and artefacts, among other features, often results in large inter- and intra-reader variabilities. This in turn leads to insufficient diagnostic power, additional imaging, and invasive biopsies [5]. Even though biopsy is considered today as the gold standard for cancer diagnosis [6], it poses several challenges such as low sensitivity and specificity, risk during the biopsy procedure, and relatively long waiting times for the biopsy results.

To support the diagnostic decisions of medical experts, for the past 15 years, a great wealth of literature has been ever more focusing on radiomics features. In Table 2 a summary of BCBM identification methods using radiomic image features analysis, reported in the literature, is presented. More specifically, in oncology, using the big data approach, a large number of quantitative features [4, 5, 7–9] were extracted from digital medical images based on intensity, shape, size, volume and texture. These features in conjunction with clinical data and genomic assays can be correlated with the disease prediction and/or evolution, thus offering a rich source of biomarkers. To this end, image texture analysis, is able to characterize tissues with intrinsic heterogeneous properties, which are usually imperceptible to the human eye. This can be applied to differentiate primary tumors from BCBM and contribute to a more focused primary cancer and subtype detection, thus narrowing down required diagnostic procedures and accelerating therapy initiation [5].

The objective of the current review, was to provide a literature review on the advancements in the automated MRI brain metastasis (BM) image analysis using radiomic features based classification, and furthermore to propose a methodology of radiomics image analysis for the BCBM for prediction and follow up of the evolution of the disease. This methodology involves image acquisition, preprocessing, segmentation of tumors, features extraction, features selection, classification and decision making. Brain tumor segmentation is usually performed either manually by an expert clinician, semi-automatically or fully automatically (using machine learning (ML) and deep learning (DL) approaches) [10] (see also Table 1). Due to the different image acquisition protocols used, MRI images need to be preprocessed before the automated tumor segmentation and features analysis in order to alleviate differences occurring between different acquisitions time-points as well as prior to features extraction in order to improve texture discrimination [9]. Even though DL methods yielded good results in brain tumor segmentation (see also Table 1), their applicability to different datasets in daily clinical practice is still a major challenge. Furthermore, the sensitivity of commonly used radiomic features to the intra- and inter-observer variations of manual and semi-automated segmentations was considerable to 2/3 of the tested features according to studies using MRI phantoms [7]. Lastly, features can be calculated in many ways, therefore the adherence to the Image Biomarker Standardization Initiative (IBSI) guidelines is recommended [11].

2 Literature Review

For contacting the present review, the PubMed and IEEE Explore databases were searched using the following keywords: MRI, brain metastasis/-es, and segmentation. Out of the 37 conference and journal papers found, 7 papers were selected, which investigated automated segmentation in brain MRI images. Furthermore, PubMed and Google Scholar were queried using the keywords: MRI, differentiate | classify | discriminate | identify | predict, brain metastases, breast cancer | breast tumor. Out of the 73 publications found, only the three were devoted to the classification and/or prediction of BCBM and these were selected for the present review. Only the most recent works of each research group were included. The structure of the paper is as follows. In Sect. 2.1 an overview of BM segmentation along with the used preprocessing techniques is documented. Section 2.2 covers the process of BM differentiation based on their primary site origin including BC.

2.1 Image Preprocessing and Brain Metastasis Segmentation

Preprocessing. The images acquired from MRI scanners need to go through different preprocessing steps before being ready for segmentation. Initially, the brain MRI image scans are registered to the same anatomical template [12–15] in order to transform different sets of data into one coordinate system. Images are then resampled to the specific resolutions $0.82 \times 0.82 \times 1.02$ mm^3 [12], $0.50 \times 0.50 \times 1.50$ mm^3 [13], $0.89 \times 0.89 \times 0.89$ mm^3 [14], then the brain is extracted (skull-stripping) [14–16] and the faces are masked out for anonymization [12, 15]. Then, various combinations of techniques are applied on the MRI images, including bias field correction (BFC) in order to correct the inhomogeneities induced from the scanner's magnetic field [12], z-score normalization (ZSN) to transform intensity distributions to $(\mu,\sigma) = (0,1)$ [12–14, 17], as well as intensity saturation (SI) and intensity range histogram normalization (IRN) (e.g. value 0 to 1) as it was shown in [18, 19]. In order to increase the number of images, which were usually provided for most of the studies presented in Table1, new artificial training data were generated such as: random flipping, rotation, intensity scaling, gamma correction (see also Table 1), as well as flipping during inference time (test-time augmentation - TTA) [12, 14, 16]. The majority of the studies performed whole tumor segmentation (WT) [14, 18], whereas in some other studies the proposed methods were trained separately on the necrotic (NCR) and enhancing tumor (ET) sub regions [12], tumor core (TC) and edema (ED) [13].

Segmentation. Liu et al. [16], used post-contrast T1-weighted scans (T1c) and modified the DeepMedic (DM) method [20], by introducing a second sub-path for the local features using larger convolution filters of 5^3 voxels as compared to the 3^3 filter size of the DeepMedic local sub-path. To deal with false positives (FPs), predicted segments were discarded if their sphericity (measure of the resemblance of a structure to a sphere) is smaller than a predefined threshold. Charron *et al.* [12], implemented the original DeepMedic method which was initially trained on gliomas. The team optimized the number of epochs (30) and the patch (24^3 voxels) and batch size (10), to best perform on BMs. Training with 3D T1c, 2D T1-weighted (T1), 2D Fluid-Attenuated Inversion Recovery (FLAIR) and removing <4 mm^3 predictions yielded the best results in terms of a DSC (79%) and FPs (4.4). The 3D T1c and 2D FLAIR training, removal of <4 mm^3 predictions and segmenting NCR and ET separately yielded the best results in terms of sensitivity (98% vs 93%). The addition of virtual patients did not significantly improve the performance. Bousabarah *et al.* [14], demonstrated an ensemble of a conventional U-net [21] (cU-net), a U-net with a weighted multi-scale loss function [22], and a pre-trained cU-net further trained only with BMs <0.4 ml. Ensembling by summation (NetSUM) prevails in sensitivity compared to ensembling by majority voting (NetMV) (0.82 over 0.77), having though much larger false positive rate (0.35 over 0.08). The DSC of NetSUM and NetMV were comparable (0.70 and 0.71 respectively). Consequently, the metrics above indicate a more balanced performance by NetMV. Jalalifar *et al.* [13], proposed two independent one-class Support Vector Machines (OC-SVMs) [23], to segment TC and ED with 2D T1c and FLAIR respectively. In each slice, a 5 × 5 test image patch was compared to all the patch pairs to classify its central pixel. To remove FPs, the predictions were refined by dilation, erosion and connected region analysis (CRA). Despite the satisfying results, the dataset is very small. Xue *et al.* [18], cascaded two 3D Fully Convolutional Networks (FCNs) [24] for bounding box detection and then finer segmentation. The FCNs were trained with a single dataset (largest in this review), and tested with two separate ones. Zhou *et al.* [17], used 3-slice tall T1 spoiled gradient echo images and cascaded two FCNs, similarly with Xue *et al.* [18]. The team experimented among DenseNets [25] and ResNets [26], with various number of layers, a VGG16 [27] and a conventional U-net [21], with the latter yielding the best performance (DSC = 81%, sensitivity = 88%, positive predictive value - PPV = 58%, FPs = 3). For comparison, Zhou *et al.* [17] vs Xue *et al.* [18] showed a DSC of 0.87 vs 0.84 for BMs >6 mm and >5 mm, respectively. Grøvik *et al.* [15], trained and tested their FCN on 3D T1c inversion recovery fast spoiled gradient echo, T1 and T1c fast spin echo and FLAIR scans from two hospitals separately. The team used a modified DeepLab V3 method [28], which uses atrous (dilated) convolutions, inputting 5 2D slices per modality. They simulated a dataset omission by stochastically dropping out 0–3 modalities. For comparison, the original DeepLab V3 network, which was tested without the dropout strategy, vs the Grøvik et al. [15] method showed, calculated on a voxel basis, an AUC of 98.9% both, a DSC of 77.4% vs 79.5%, an intersection over union (IoU) of 49.3% vs 56.1%, a sensitivity of 63.1% vs 67.1%, a PPV of 72.2% vs 79.0%, and 26.3 vs 12.3 FPs per patient, respectively.

2.2 Prediction of Brain Metastasis: Breast Cancer Origin

Preprocessing. It is also beneficial to perform an independent preprocessing of the original images in order to enhance texture discrimination. Ortiz-Ramón et al. [9], normalized the image intensities by clipping in the range (μ, $\pm3\sigma$), isotropically resampled and experimented with different numbers of quantization levels. H. Kniep et al. [5], registered the images to the same anatomical space, isotropically resampled and corrected for field bias. Béresová et al. [8], did not use any preprocessing.

Features Extraction. Ortiz-Ramón et al. [9], used T1-weighted inversion recovery fast-spoiled gradient-echo (T1 IR-SPGR) scans of the same MRI settings and extracted 43 rotation-invariant texture features (TF) based on histogram, gray-level co-occurrence matrix (GLCM), gray-level run-length matrix, gray-level size-zone matrix and neighborhood gray-tone difference matrix, to discriminate between BMs originating from BC, lung cancer (LC) and melanoma (MM). Each feature was computed in 2D brain image slices and 3D stacks using 5 different quantization levels of intensity (number of gray levels - NGL), yielding 10 datasets. H. Kniep et al. [5], used T1, T1c and FLAIR scans of different protocols and resolutions to extract 17 shape, 18 first-order, and 56 TF from the segmented brain lesions, with the latter two categories being also calculated on the basis of 8 wavelet decompositions, totaling up to 1423 features. Béresová et al., [8] used T1c brain images and upon them generated local binary pattern (LBP) maps in order to extract from both images and maps, 7 histogram- and 5 GLCM-based features so as to discriminate BMs originating from BC vs BMs from LC.

Classification and Prediction. Ortiz-Ramón et al. [9], fed a random forest method (RF) with each feature dataset in a nested cross-validation manner to classify brain metastases by their primary site of origin using a radiomics approach based on texture analysis. In each iteration, a model was tested with different subsets of the top-ranked TF. The resulted TF were ranked based on their corresponding p-values and separately on their significance as RF variables. Differentiation of BMs originating from BC vs LC achieved an AUC of 96%, using 4 features and 32 NGL. On the other hand, with the multi-class strategy, BCBM was correctly classified achieving an accuracy of about 59% in total. BM originating from BC vs MM gave poorer results (AUC = 60%). H. Kniep et al. [5], also used an RF, to classify 10 randomly permuted 5-fold cross-validation imaging feature sets. BMs originating from BC, were predicted against small (SCLC) and non-small cell lung cancer (NSCLC), MM and gastrointestinal cancer (GC) (AUC = 61%, p < 0.05). The prediction utilized also the age and sex of the patients alone (AUC = 73%, p-value < 0.01) as well as in conjunction with the image features (AUC = 78%, p < 0.001). For comparison, two neuroradiologists classified BCBM, however with non-significant differences of sensitivities as opposed to the RF classifier. Béresová et al. [8] did not use any unseen testing set for prediction. The team reported statistically significant differentiations (p < 0.004) using the TF contrast, correlation, energy, homogeneity, and entropy. The TF were extracted from the axial 2D (and slice-weighted converted 3D), LBP brain maps. Entropy and energy were calculated on the true 3D LBP maps and on the other hand contrast and correlation were calculated on the 2D T1c images (BM area > 19.35 cm^2), which also indicated statistically significant differences (p < 0.05 and < 0.004 respectively).

Table 1. Summary of fully automated brain metastasis segmentation studies reported in the literature.

Author [Ref]-Year	Dataset(s)	Preprocessing	Network	Data augmentation	Postprocessing	Findings (test data)	Final remarks
Y. Liu [16]-2017	T1c, tr & val/ts: 225/15 pts: 0.67 ml (μ)	-	3D CNN: 3 s-p (local & global patches)	F: sagittal axis	If Sph < Thr: remove	DSC = (0.67 ± 0.03), MSSD = (0.9 ± 0.3)mm, AUC = (0.98 ± 0.01)	S-C. Better DSC than original DM in MM scans (ts: BraTS)
O. Charron [12]-2018	M-M. tr/ts: 146(+62*)/18 pts 412 BMs. 0.5 ml(M)	ZSN, N4BFC	3D CNN: 2 s-p (local & global patches)	Virtual pts*: linear reg/tion by Rt	If vol. < 4 mm³: remove	DSC% (WT)| Se⁺| FP# (79 | 93 | 4.4)C1 (77 | 98 | 7.2)C2	S-C. NS impact of virtual pts. Small ts
K Bousabarah [14]-2020	M-M. tr/ˣts: 469/40 pts. 1140/83 BMs 0.47 ml(M) x	ZSN	En: U-net, U-net w/DS, U-net: tr: BMs < 0.4ml	Random ISc, F, Rt, GC, EDF TTA (3 axes F)	If vol. < 6 mm³: remove	NetMV: DSC| Se⁺|Sp⁺| F|| FPR % 71|77|96|85|8	Unclear calc/tion of FPR. BMs < 0.06 ml undetected
A. Jalalifar [13]-2020	M-M. tr/ts: 35/5 pts	ZSN	2 OC-SVMs	-	D + E, CRA	✦DSC%: 84 ± 6, ✦H95: 1.85 ± 0.48mm	Very small dataset
J. Xue [18]-2020	T1. tr/ts: 1201/231 pts. [0.79, 2.22, 5.42] ml(Q)	SI, IRN	Cc: 3D FCNs	-	-	DSC% (BMs > 5mm)	Se⁺	Sp⁺ 84	95	99	M-C study. 100% BM detection acc/cy. Largest dataset

(continued)

Table 1. (*continued*)

Author [Ref]-Year	Dataset(s)	Preprocessing	Network	Data augmentation	Postprocessing	Findings (test data)	Final remarks
Z. Zhou [17]-2020	T1c. tr/ts: 748/186 pts. 3131/766 BMs	ZSN	Cc: FCNs w/ 3-slice tall input	-	-	(DSCl Se[#]l PPV[#]l%l Fp[#] 81 l 88 l 58 l 3	Most BMs. 1–52 mm BMs w/ 13 primaries. Poor perf/ance: < 6mm
E. Grøvik [15]-2021	M-M. tr/ts: 100/55 pts. 100/65 BMs	-	DeepLab V3 (modded) w/ SD of modalities	-	-	†(DSCIIoUIAUCISelPPV)% 80 l 56 l 99 l 67 l 79 FP[#]: 3.6	M-C study. Missing modality simulation

tr/val/ts: Training/validation/testing set, pts: Patients, M-M: Multimodal, N4BFC: N4 Bias field correction, DS: Deep supervision, s-p: Sub-path, S-D: Stochastic dropout, F: Flipping, ISc: Intensity scaling, Rt: Rotation, EDF: Elastic deformation, GC: Gamma correction, vol.: Volume, Sph: Sphericity, Thr: Threshold, D+E: Dilation/Erosion, H95: Hausdorff distance (95th perc/tile), Se: Sensitivity, Sp: Specificity, FP(R): False positives (rate), En: Ensemble, Cc: Cascade, S-C/M-C: Single-/multi-center study, NS: Non-significant, ♦: Evaluation of TC, (μ)/(M)/(Q): Mean/Median/IQR of BM volume, C1: Three modalities/binary segmentation, C2: Two modalities/NCR-ET segmentation, †: On a voxel basis, ‡: On a BM basis, #: Per patient.

Table 2. Summary of BCBM identification methods using radiomic image features analysis reported in the literature.

Author [ref] Year	Dataset	Preprocessing	Segmentation	Features	Classification	Findings	Final Remarks
M. Béresová [8] 2018	T1c. BMs from BC/LC: 26/32, mean area: 44.58/24.12 cm²	-	1 NR	7 HG- and 5 GLCM-based on T1c and LBP. 2D and 3D	2-sample test. M-W. F-test. Box's test. No unseen testing set	Cn^1, Co^2, Eg^3, Hm^4, Et^5 s.s.d. in: $2D_m$ $(^{1-4})$, $2D_i$ $(^{1-2})$, $3D_m$ $(^{1-5})$	Single-center. Same MRI settings
R. Ortiz-Ramón [9] 2018	T1 IR-SPGR. 38 pts. BMs from BC: 17, LC:27, MM:23	$\mu \pm 3\sigma$, NGL, isotropic resampling	2D: 1 NO (m), 3D: AC (s-a)	FE: 43 TF (2D, 3D, diff NGL). FS: p-value*, RF v.i	RF (250 trees). 100-m ens: tr | ts: 75 | 25%	(4 t-r* 3D TF, NGL:32) BCBM vs LCBM | MBM | all: AUC = (96 | 60)% | Se = 59%	Small dataset. Single-center. Same MRI settings. Different NGL affects acc
H. Kniep [5] 2019	M-M. 189 pts. BMs from BC:143, (N)SCLC: (225)151, GC: 50 MC: 89	atlas registration, N4ITK, WSN	2 NR (s-a)	FE: (Age, Sex)†, (17 SF, 18 FOF*, 56 TF*, 8 WD*,†)‡. FS: Gini	RF (3500 trees). 10 random sets of model-external 5-xCV	BCBM vs all: †AUC = 73%, ‡AUC = 61%, †,‡AUC = 78%	Singe-center. RF prediction diff NS cmp to 2 NR. Variant MRI settings

$2D_m$: 2D features on LBP maps, $2D_i$: 2D features on T1c images, $3D_m$: 3D features on LBP maps, NO: Neurooncologist, m: Manually, s-a: Semiautomatically, AC.: Active contouring, FE: Feature extraction, FS: Feature selection, v.i.: Variable importance metric, ens: Ensemble, tr: Training test, ts: Testing set, t-r: Top-ranked, M-M: Multi-modal, (N)SCLC: (Non-)Small cell lung cancer, GC: Gastrointestinal cancer, N4ITK: Improved N3 bias field correction, WSN: White stripe normalization, NR: Neuroradiologist, SF: Shape features, FOF: First-order features, WD: Wavelet decompositions (calculated on the bases of *, † features), -xCV: -Fold cross validation, HG: Histogram, M-W: Mann-Whitney test, Cn: Contrast, Co: Correlation, Eg: Energy, Hm: Homogeneity, Et: Entropy, s.s.d.: Statistically significant difference, acc.: Accuracy, diff NS cmp: Differs non-significantly compared.

3 Discussion

The task of the automated brain tumor segmentation has been significantly standardized with the contribution of the public NIH Cancer Imaging Archive (TCIA) [29] and the BraTS benchmark [10]. However, regarding brain tumors, both TCIA and BraTS only included gliomas, which constitute to a type of primary brain tumor generally much larger than BMs. Therefore, BM segmentation studies rely on private datasets, thus limiting a widespread experimentation from different research groups.

Segmentations of BM < 0.06 ml (with a diameter size of about 5 mm), were found to be inadequate. Considering data variety and results, the most generalizable BM segmentation studies presented in the literature were single-modal. More specifically, the study by Zhou *et al.* [17], used the most BMs in a test set, BMs with diameter of 1–52 mm, and resulted in a DSC of 84%, and a PPV of 58% per patient. In another multi-center study, Xue *et al.* [18], used the most patients in a test set, and reported a 100% detection accuracy for BMs > 5 mm.

Differentiation of the primary tumor site using imaging features has a long way to go until its clinical applicability. It was shown in [5], that the imaging features are only complementary to the larger predictive power of age and sex data in BCBM. In [9], it was documented that only fair results (AUC = 60%) could be achieved, for BC in multi-class prediction mainly due to misclassifications of BC with MM.

In order to detect new robust imaging texture features, we intend to propose an integrated fully automated segmentation and features analysis system, which will cover both the segmentation, features extraction and classification and decision making procedures. We will utilize therefore a number of techniques as also posed in the literature review presented in this study (see also Tables 1 and 2), which will include image preprocessing, image segmentation, and features extraction. We also intent to use capsule-based networks for segmentation [30] and classification [31], and a new class of features based on amplitude modulation-frequency modulation (AM-FM) methods as it was also proposed in [19].

In general, in order to achieve a BCBM identification through imaging features to be generalizable and reproducible, standardized high-resolution images from multiple centers and comprehensive integration of clinical data are required. Ideally, studies must cover a vast variety of brain lesions, including primary tumors (e.g. glioblastomas) and BMs varying in size, shape, location and origin, including all the different primary sites in addition to BC and different BC subtypes.

References

1. Chamberlain, M.C., Baik, C.S., Gadi, V.K., Bhatia, S., Chow, L.Q.M.: Systemic therapy of brain metastases: non–small cell lung cancer, breast cancer, and melanoma. Neuro. Oncol. **19**(1), i1–i24 (2017). https://doi.org/10.1093/neuonc/now197
2. Hadjipanteli, A., Doolan, P., Kyriacou, E., Constantinidou, A.: Breast cancer brain metastasis: the potential role of MRI beyond current clinical applications. Cancer Manag. Res. **12**, 9953–9964 (2020). https://doi.org/10.2147/CMAR.S252801
3. Mitchell, D., Kwon, H.J., Kubica, P.A., Huff, W.X., O Regan, R., Dey, M.: Brain metastases: an update on multi-disciplinary approach of clinical management. Neurochirurgie (2021). doi: https://doi.org/10.1016/j.neuchi.2021.04.001

4. Zhou, M., Scott, J., Chaudhury, B., Hall, L., Goldgof, D., Yeom, K.W., et al.: Radiomics in brain tumor: Image assessment, quantitative feature descriptors, and machine-learning approaches. AJNR Am. J. Neuroradiol. **39**(2), 208–216 (2018). https://doi.org/10.3174/ajnr. A5391

5. Kniep, H.C., et al.: Radiomics of brain MRI: utility in prediction of metastatic tumor type. Radiology. **290**(2), 479–487 (2019). https://doi.org/10.1148/radiol.2018180946

6. Tandel, G.S., Balestrieri, A., Jujaray, T., Khanna, N.N., Saba, L., Suri, J.S.: Multiclass magnetic resonance imaging brain tumor classification using artificial intelligence paradigm. Comput. Biol. Med. **122**, 103804 (2020). https://doi.org/10.1016/j.compbiomed.2020.103804

7. Baeßler, B., Weiss, K., Pinto dos Santos, D.: Robustness and reproducibility of radiomics in magnetic resonance imaging: a phantom study. Investigative Radiology **54**(4), 221–228 (2019). https://doi.org/10.1097/RLI.0000000000000530

8. Béresová, M., Larroza, A., Arana, E., Varga, J., Balkay, L., Moratal, D.: 2D and 3D texture analysis to differentiate brain metastases on MR images: proceed with caution. Magn. Reson. Mater. Phys., Biol. Med. **31**(2), 285–294 (2017). https://doi.org/10.1007/s10334-017-0653-9

9. Ortiz-Ramón, R., Larroza, A., Ruiz-España, S., Arana, E., Moratal, D.: Classifying brain metastases by their primary site of origin using a radiomics approach based on texture analysis: a feasibility study. Eur. Radiol. **28**(11), 4514–4523 (2018). https://doi.org/10.1007/s00330-018-5463-6

10. Menze, B.H., et al.: The multimodal brain tumor image segmentation benchmark (BRATS). IEEE Trans. Med. Imaging. **34**(10), 1993–2024 (2015). https://doi.org/10.1109/TMI.2014.2377694

11. Zwanenburg, A., Leger, S., Vallières, M., Löck, S.: Image biomarker standardisation initiative (2019). https://arxiv.org/abs/1612.07003

12. Charron, O., Lallement, A., Jarnet, D., Noblet, V., Clavier, J.-B., Meyer, P.: Automatic detection and segmentation of brain metastases on multimodal MR images with a deep convolutional neural network. Comput. Biol. Med. **95**, 43–54 (2018). https://doi.org/10.1016/j.compbiomed.2018.02.004

13. Jalalifar, A., Soliman, H., Ruschin, M., Sahgal, A., Sadeghi-Naini, A.: A brain tumor segmentation framework based on outlier detection using one-class support vector machine. In: 2020 42nd Annual International Conference of the IEEE Engineering in Medicine and Biology Society (EMBC), pp. 1067–1070. IEEE (2020). https://doi.org/10.1109/EMBC44109.2020.9176263

14. Bousabarah, K., Ruge, M., Brand, J.-S., Hoevels, M., Rueß, D., Borggrefe, J., et al.: Deep convolutional neural networks for automated segmentation of brain metastases trained on clinical data. Radiat. Oncol. **15**(1), 87 (2020). https://doi.org/10.1186/s13014-020-01514-6

15. Grøvik, E., Yi, D., Iv, M., Tong, E., Nilsen, L.B., Latysheva, A., et al.: Handling missing MRI sequences in deep learning segmentation of brain metastases: a multicenter study. NPJ Digit. Med. **4**(1), 33 (2021). https://doi.org/10.1038/s41746-021-00398-4

16. Liu, Y., Stojadinovic, S., Hrycushko, B., Wardak, Z., Lau, S., Lu, W., et al.: A deep convolutional neural network-based automatic delineation strategy for multiple brain metastases stereotactic radiosurgery. PLoS ONE **12**(10), e0185844 (2017). https://doi.org/10.1371/journal.pone.0185844

17. Zhou, Z., et al.: MetNet: computer-aided segmentation of brain metastases in post-contrast T1-weighted magnetic resonance imaging. Radiother. Oncol. **153**, 189–196 (2020). https://doi.org/10.1016/j.radonc.2020.09.016

18. Xue, J., Wang, B., Ming, Y., Liu, X., Jiang, Z., Wang, C., et al.: Deep learning-based detection and segmentation-assisted management of brain metastases. Neuro. Oncol. **22**(4), 505–514 (2020). https://doi.org/10.1093/neuonc/noz234

19. Loizou, C.P., Pantzaris, M., Pattichis, C.S.: Normal appearing brain white matter changes in relapsing multiple sclerosis: texture image and classification analysis in serial MRI scans. Magn. Reson. Imaging. **73**, 192–202 (2020). https://doi.org/10.1016/j.mri.2020.08.022

20. Kamnitsas, K., Ledig, C., Newcombe, V., Simpson, J.P., Kane, A.D., Menon, D.K., et al.: Efficient multi-scale 3D CNN with fully connected CRF for accurate brain lesion segmentation. Med. Image Anal. **36**, 61–78 (2017). https://doi.org/10.1016/j.media.2016.10.004

21. Ronneberger, O., Fischer, P., Brox, T.: U-Net: Convolutional Networks for Biomedical Image Segmentation (2015). https://arxiv.org/abs/1505.04597

22. Kickingereder, P., et al.: Automated quantitative tumour response assessment of MRI in neuro-oncology with artificial neural networks: a multicentre, retrospective study. Lancet Oncol. **20**(5), 728–740 (2019). https://doi.org/10.1016/S1470-2045(19)30098-1

23. Cortes, C., Vapnik, V.: Support-vector networks. Mach. Learn. **20**(3), 273–297 (1995). https://doi.org/10.1007/bf00994018

24. Long, J., Shelhamer, E., Darrell, T.: Fully convolutional networks for semantic segmentation. In: 2015 IEEE Conference on Computer Vision and Pattern Recognition (CVPR), pp. 3431–3440. IEEE (2015). https://doi.org/10.1109/cvpr.2015.7298965

25. Huang, G., Liu, Z., van der Maaten, L., Weinberger, K.Q.: Densely Connected Convolution-al Networks (2018). https://arxiv.org/abs/1608.06993

26. He, K., Zhang, X., Ren, S., Sun, J.: Deep residual learning for image recognition (2015). https://arxiv.org/abs/1512.03385

27. Simonyan, K., Zisserman, A.: Very deep convolutional networks for large-scale image recognition (2015). https://arxiv.org/abs/1409.1556

28. Yi, D., et al.: MRI pulse sequence integration for deep-learning based brain metastasis segmentation (2019). https://arxiv.org/abs/1912.08775

29. The cancer imaging archive (TCIA). https://www.cancerimagingarchive.net/. Accessed 03 May 2021

30. LaLonde, R., Xu, Z., Irmakci, I., Jain, S., Bagci, U.: Capsules for biomedical image segmentation. Med. Image Anal. **68**, 101889 (2021). https://doi.org/10.1016/j.media.2020.101889

31. Afshar, P., Mohammadi, A., Plataniotis, K.N.: Brain tumor type classification via capsule networks. In: 2018 25th IEEE International Conference on Image Processing (ICIP), pp. 3129–3133. IEEE (2018). https://doi.org/10.1109/icip.2018.8451379

An Adaptive Semi-automated Integrated System for Multiple Sclerosis Lesion Segmentation in Longitudinal MRI Scans Based on a Convolutional Neural Network

Andreas Georgiou[1]($\boxtimes$) (iD), Christos P. Loizou[1] (iD), Andria Nicolaou[2] (iD),
Marios Pantzaris[3] (iD), and Constantinos S. Pattichis[2] (iD)

[1] Department of Electrical Engineering and Computer Engineering and Informatics,
Cyprus University of Technology, Saripolou Street, 3040 Limassol, Cyprus
andk.georgiou@edu.cut.ac.cy, christos.loizou@cut.ac.cy
[2] Department of Computer Science, University of Cyprus, 1 Panepistimou Avenue, 2109
Aglantzia, Nicosia, Cyprus
{nicolaou.andria,pattichi}@ucy.ac.cy
[3] University of Cyprus School of Molecular Medicine, The Cyprus Institute of Neurology
and Genetics, Nicosia, Cyprus
pantzari@cing.ac.cy

Abstract. This work proposes and evaluates a semi-automated integrated segmentation system for multiple sclerosis (MS) lesions in fluid-attenuated inversion recovery (FLAIR) brain magnetic resonance images (MRI). The proposed system uses an adaptive two-dimensional (2D) full convolutional neural network (CNN) and is applied to each MRI brain slice separately. The system is based on a U-Net architecture and allows manual error corrections by the user. This task produces continuing additional improvements to the accuracy of the segmentation system, which can be adapted and reconfigured interactively based on the data entered by the user of the system. The system was evaluated based on the ISBI dataset, on 20 MRI brain images acquired from 5 MS subjects who repeated their examinations in four consecutive time points (TP_1-TP_4). Manual lesion delineations were provided by two different experts. A Dice Similarity Coefficient (DSC) of 0.76 was achieved using the proposed system which is the highest achieved also by another system. A higher DSC of 0.82 was achieved when the proposed system was evaluated on TP_4 images only. A larger dataset will be analyzed in the future, and new measurement metrics will be suggested.

Keywords: MRI · Multiple sclerosis · Semi-automated lesion segmentation · Convolutional Neural Networks · U-Net

1 Introduction

Multiple Sclerosis (MS) is a chronic, autoimmune, and demyelinating disease of great clinical importance affecting the human central nervous system (CNS). It gradually

© Springer Nature Switzerland AG 2021
N. Tsapatsoulis et al. (Eds.): CAIP 2021, LNCS 13052, pp. 256–265, 2021.
https://doi.org/10.1007/978-3-030-89128-2_25

changes the brain's white matter (BWM) texture, morphology, and structure due to myelin sheath damage [1]. The increase in the subjects' disability in conducting everyday tasks is an important consequence of this chronic condition [2], which rises as the number of BWM lesions increases [1, 2]. The identification and follow up of the MS lesions may quantify the progression of the MS disease [3] and follow up its evolution over time [4]. Segmentation of brain MRI lesions is usually manually performed by experts, which is time consuming and suffers from intra and inter-observer variability [5–7].

A number of semi-automated or automated MS lesion segmentation methods have been proposed in the literature for brain MRI (see also Table 1), which were applied on the ISBI dataset [8]. These were based on Random Forest [9, 10] and Convolutional Neural Networks (CNN) [11–13]. It was also mentioned in [5], that for accurately deriving MS diagnostic criteria, spatial information from the MS lesions is required. Furthermore, automated segmentation may provide an additional tool for following up the appearance, evolution, and development of new brain MS lesions [6]. Accurate segmentation of MS lesions is necessary for the quantitative evaluation of the disease, which is of great value in analyzing disease progression and treatment options.

The first attempts for segmenting MS brain lesions were made by Wicks et al. [5], using a semi-automated segmentation method based on manually selected thresholding. Recently, machine learning approaches for MS lesion segmentation were proposed which were based on manual feature extraction and classification [7, 10]. More specifically, Cabezas et al. [7] introduced a fully semi-automated method, for the segmentation of MS brain lesions in MRI images, using a discriminative classification algorithm based on the Gentle boost classifier. In another study, Maier et al. [10], proposed a random forests classification method, combined with local context intensity features to segment MS lesions in multi-spectral MRI images.

During the last decade, deep learning methods based on CNN, demonstrated exceptional image segmentation performance. More specifically, Afzal et al. [14], proposed an unsupervised 2D patch-wise CNN method for the segmentation of brain MRI lesions. The method incorporated three different routines of MRI images, namely T1, T2 and fluid-attenuated inversion recovery (FLAIR). Suthirth et al. [15], proposed a 3D voxel-wise CNN model using multi-channel 3D patches in MRI volumes as input. By combining the methods presented in [14, 15]. Aslani et al. [13] proposed a CNN model that includes a multi-branch down-sampling path.

This facilitated the network to encode information from multiple modalities separately, where multiscale feature fusion blocks were proposed to combine feature maps from different modalities at different network stages. In the above studies the best Dice Similarity Coefficient (DSC) achieved was 0.76. A high DSC score of 0.86 was recently published by Narayana et al. [16] on brain MRI T_2 MS lesion segmentation on 1008 subjects based on the U-net architecture.

The objective of the present study was to propose and evaluate an adaptive semi-automated integrated system for MS lesions segmentation in FLAIR, brain MRI images. The system, which is presented in Fig. 1, is based on an adaptive 2D full U-net CNN [11] and avails a functionality for manual segmentation corrections, which can be performed by the user of the system. The corrections are taken into consideration by the system

Table 1. MS lesion manual delineations (MD) and semi-automated segmentation methods proposed in the literature based on CNN, applied on the ISBI dataset [8] (N = 5 × 4 Time points = 20 MRI images). The dice coefficient (DSC), the true positive rate (TPR), and the false positive rate (FPR) are shown.

Author	Year	Method	TPR	FPR	DSC	TPR	FPR	DSC
			Validation on R_1			Validation on R_2		
MD_1		R_1	-/-	-/-	-/-	0.65	0.17	0.73
MD_2		R_2	0.83	0.36	0.73/-	-/-	-/-	-/-
Jesson [9]	2015	RF	0.61/-	0.135/-	0.70/-	0.501/-	0.127/-	0.68/-
Oskar [10]	2015	RF	0.53/0.56	0.49/0.49	0.70/0.70	0.38/0.39	0.44/0.44	0.66/.066
Olaf [11]	2015	U-Net	0.69 /0.7	0.27/0.23	0.71/0.70	0.70/0.70	0.27/0.21	0.69/0.71
Brosch [12]	2016	3D CNN	0.75/0.78	0.55/0.65	0.68/0.63	0.63/0.69	0.53/0.62	0.64/0.66
Aslani [13]	2019	MB Res-Net	0.67/0.7	0.12/0.2	0.76/0.76	0.54/0.57	0.12/0.19	0.70/0.71
Proposed system	2021	U-Net (MECM)	0.745/-	0.22/-	0.76/-	-/-	-/-	

R_1: Rater 1, R_2: Rater 2, MD1, MD2: Manual delineation for R_1 (-/) and R_2 (/-), DSC: Dice similarity coefficient, TPR: Lesion true positive rate, FPR: Lesion false positive rate, RF: Random forest, CNN: Convolutional neural network, MB-Res-Net: Multi Branch Res-Net, MECM: Manual Error Correction Module (see Subsect. 2.5).

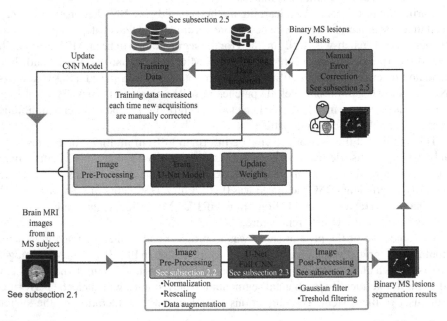

Fig. 1. Flow diagram of the CNN MRI lesion segmentation system proposed in this work.

and are used as an input to the next training procedure of the CNN in order to improve the final segmentation.

2 Materials and Methods

2.1 Acquisition of Brain MRI Images

Table 2 presents the demographics for the subjects investigated in this study obtained from the ISBI Dataset [8]. Blinded manual segmentations were provided by two different experts with 4 and 10 years of clinical experience, respectively. Each MRI acquisition consistent of 181 longitudinal slices and acquired with a 3 T MRI scanner (Philips Medical System, Best, The Netherlands), using four different types of MRI sequence scans (T_1, T_2, Proton density (PD) and FLAIR). The dataset provided two versions of MRI images, an original and a denoised image using a bias correction filter and a nonuniform intensity normalization (N4) algorithm [17], skull stripped [18], and dura stripped [19]. The inversion time (T1) was 835 ms, the time echo (TE) was 68 ms, and the field of view consisted of a 181 × 217 pixels window and a voxel size of 0.82 × 0.82 × 2.2 mm^3.

Table 2. MS Patient demographics of the ISBI dataset [8] used in this study (N = 5).

Data set	N (M/F)	Timepoints	Age at baseline
		Mean(±SD) [Years]	Mean(±SD) [Years]
RR	4 (1/3)	4.5(±0.50)	40.0(±7.55)
PP	1 (0/1)	4.0	57.9
Total training	5 (1/4)	4.4(±0.55)	43.5(±10.3)
Total slices	3822		

MS: Multiple Sclerosis, RR: Relapsing Remitting, PP: Primary Progressive, M/F: Male/Female.

2.2 Image Pre-processing

Before automated lesion segmentation the MRI images were uniformly intensity normalized [20] using MinMaxScaler method from Sklearn library, setting minimum and maximum pixels intensities values from 0 to 1 respectively. Images were then rescaled from 181 × 217 pixels size to a fixed size of 256 × 256 pixels, which corresponds to the input size of the first CNN layer. The image size was selected through experimentation that showed best segmentation results. During each training iteration, a number of randomly deformed transformations of the training images was implemented using ImageDataGenerator method of Keras library, enabling horizontal, vertical flipping and shifting image, combined with a random rotation (in range of 0°–180°) and center scaling transformations(in range of a scaling factor 0.8–1.2) [21].

2.3   CNN Model Architecture (U-Net)

A CNN based U-Net model [11] was used to implement the proposed segmentation system as presented in Fig. 1. Its overall architecture is illustrated in Fig. 2. The U-Net model provides the ability of pixel-wise predictions by keeping the exact size of the input image. The model consists of six consecutive down-sampled blocks followed by six consecutive up-sampled blocks. For the activation function, a rectified linear unit (RELU) for nonlinearity was used, and a kernel size of 3×3 pixels was selected across the network. All down-sampling convolutions had a 2×2 max pooling operation with a stride of two for down-sampling where the image size was kept constant by using the same padding.

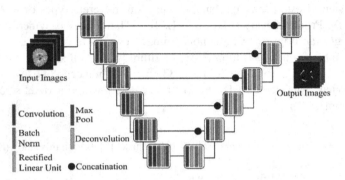

Fig. 2. U-Net model [11] architecture of the proposed MRI lesion segmentation system.

2.4   Image Post-processing

A Gaussian filter for filtering additive noise was applied on the MRI images with a kernel size of 7×7 pixels. The kernel size was empirically derived through experimentation, selected through a training procedure. The Gaussian filter decreases additive noise and removes small false-positive regions [22]. Additionally, a threshold of 0.43 was applied, after image denoising. The threshold value was defined so that the maximum DSC was achieved and the value was estimated through experimentation.

2.5   Proposed CNN System

The proposed CNN integrated system is illustrated in Fig. 1, having a progressive evolution through the training process. More specifically, a manual error correction module was incorporated in Fig. 1. This improves the model's segmentation accuracy by incorporating the user's input. The training procedure is not executed at once, but it is repeated at regular intervals making the training process part of the automated segmentation process (see also Fig. 1). Each time new MRI data are entered as an input to the system (i.e. at each time point (TP)), the expert can perform manual corrections, which help in improving the output segmentation of the proposed CNN model. The manually corrected images are included in the training data.

2.6 Software Implementation

The proposed system, was implemented in Python language[1] using the Tensor Flow[2] method, backend through the Kera's[3] application programming interface. The system was tested on an Intel computer with a 6-core CPU, a processor with a frequency of 3.6 GHz, and a RAM workstation capacity of 32 GB. It was also necessary to define certain hyper-parameters of the model through the training procedures, which were needed for the model's smooth tuning. All models used in this work were trained for 500 iterations (epochs).

2.7 Evaluation Metrics

To evaluate the performance, the following evaluation metrics were used:

1. The TPR, is calculated when the expert detects a lesion (when a lesion is present), and the semi-automated computerized method identifies it as so [23]:

$$TPR = \frac{TP}{TP + FN} \tag{1}$$

where TP denotes the number of similar pixels estimated for the manual and the semi-automated segmentation, and FN denotes the number of pixels that were estimated true by the user but not by the prosed segmentation method.

2. The FPR, is calculated when the expert detects no lesion and the semi-automated method incorrectly detects that there is a lesion present [23]:

$$FPR = \frac{FP}{FP + TN} \tag{2}$$

where FP denotes the number of pixels that incorrectly overlaps the manual and the semi-automated segmentation masks.

3. The DSC, which is used to compare the agreement between the manual and the semi-automated segmentations is defined as follows [24]:

$$DSC = \frac{2 \times TP}{2 \times TP + FP + FN} \tag{3}$$

4. The Precision is considered as the false discovery rate or lesions false positive rate (FPR) between the automatically segmented lesions and the manually annotated lesions and is expressed as precision as follows [25]:

$$Precision = \frac{TP}{TP + FP} \tag{4}$$

---

[1] https://www.python.org/;
[2] https://www.tensorflow.org/.
[3] https://keras.io/.

3   Experimental Results

3.1   Manual and Semi-automated Lesions Segmentation on Brain MRI Images

Figure 3a) illustrates manual (performed by R_1), while Fig. 3b) shows the semi-automated lesion segmentations (performed by the proposed semi-automated segmentation system), of brain MRI FLAIR images for an MS subject (subject 3), at TP_4.

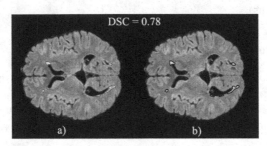

Fig. 3. a) Manual (red color) and b) semi-automated (blue color) segmentations on brain MRI FLAIR image for slice 83 (horizontal plane, from bottom to the top), from the ISBI dataset. (Color figure online)

3.2   Evaluation of the Proposed System

Table 3 presents a comparison of the proposed system based on TPR, FPR, DSC and Precision when compared to a full standard U-Net CNN method (excluding the module presented in Subsect. 2.5), trained with images of time points TP_1-TP_3 (N = 19) and evaluated on the last timepoint (TP_4) [11] (N = 1) with a cross validation procedure repeated for five times (for the TP4 images).

Table 3. Comparison of the proposed segmentation system vs a U-Net CNN architecture evaluated on the last time-point TP_4

Method type	TPR	FPR	DSC	Precision
U-Net	0.65	0.24	0.69	0.76
Proposed system	0.81	0.17	0.82	0.83

TPR: True positive rate, FPR: False positive rate, DSC: Dice similarity coefficient.

The performance of the system proposed in this work was also investigated based on the DSC. Manual corrections on the lesion segmentations performed by the user were inputted back into the system (see also Fig. 1), so that the system is adapted to the new data. More specifically, the improvement of DSC is shown in Fig. 4 over the four different timepoints (TP_1-TP_4). It is shown that by incorporating the manual correction, DSC improved considerably. Furthermore, the slope (0.0525) of the DSC can be estimated, which might be used to assess the DSC rate improvement.

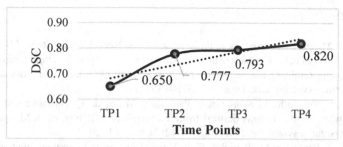

Fig. 4. Improvement of the DSC using the proposed segmentation system (manual error correction module) for each TP.

4 Discussion

The objective of the present study was to propose and evaluate an adaptive semi-automated integrated segmentation system for MS lesions segmentation in FLAIR brain MRI images. The system, which was compared with other CNN segmentation methods, presented in Table 1, was based on an adaptive 2D full U-net CNN [11] and avails a functionality for manual segmentation correction which is performed by the user of system. The corrections were taken into consideration by the system and were used as an input to the next training procedure of the CNN in order to improve the final segmentation. The proposed system achieved a higher DSC score when compared to previously proposed deep learning network approaches in [11, 12] but a similar performance to [13] (see also Table 1).

The incorporation of the manual error corrections which is implemented as a new feature in the proposed segmentation system, produces gradual improvements to the system which were presented in Fig. 4. More specifically, the diagram of Fig. 4 illustrates an increase of the DSC score at $TP_1 = 0.65$ to $TP_4 = 0.82$. Table 3 also showed that the proposed system performed better to previous segmentation methods presented in Table 1 in terms of all evaluation metrics used in this study (TPR, FPR, DSC and Precision) when evaluated only at TP_4. An excellent DSC score equal to 0.86 was recently published by Narayana et al. [16] on brain MRI T_2 MS lesion segmentation on 1008 subjects based on the U-net architecture.

5 Concluding Remarks

A classic U-Net network architecture has been presented in this work for the semi-automated segmentation of MS lesions in MRI images. A DSC of 0.76 was achieved using the proposed system which is the highest achieved also by another system. A higher DSC of 0.82 was achieved when the proposed system was evaluated on TP_4 images only. Manual segmentation error corrections performed by the expert using the system were taken into consideration in order to improve the system's segmentation performance. Ongoing work includes the integration of lesion segmentation into a three-dimensional reconstructed brain MRI system [26] where lesion texture analysis in the form of rule display [27] will be incorporated. Furthermore, the proposed system needs to be evaluated in a larger number of subjects.

References

1. Trip, S.A., Miller, D.H.: Imaging in multiple sclerosis. Neurol. Pract. **76**(3), 11–19 (2005)
2. Gross, H.J., Watson, C.: Characteristics, burden of illness, and physical functioning of patients with relapsing-remitting and secondary progressive multiple sclerosis: a cross-sectional US survey. Neuropsychiatr. Dis. Treat. **13**, 1349–1357 (2017)
3. Loizou, C.P., Petroudi, S., Seimenis, I., Pantziaris, M., et al.: Quantitative texture analysis of brain white matter lesions derived from T2-weighted MR images in MS patients with clinically isolated syndrome. J Neuroradiol **42**(2), 99–114 (2015)
4. Loizou, C.P., Pantzaris, M., Pattichis, C.S.: Normal appearing brain white matter changes in relapsing multiple sclerosis: texture image and classification analysis in serial MRI scans. Magn. Reson. Imaging **73**(August), 192–202 (2020)
5. Wicks, D.A.G., Tofts, P.S., Miller, D.H., de Boulay, G.H., et al.: Volume measurement of multiple sclerosis lesions with magnetic resonance images - a preliminary study. Neuroradiology **34**(6), 475–479 (1992)
6. Polman, C.H., Reingold, S.C., Banwell, B., Clanet, M., et al.: Diagnostic criteria for multiple sclerosis: 2010 revisions to the McDonald criteria. Ann Neurol **69**(2), 292–302 (2011)
7. Cabezas, M., Oliver, A., Valverde, S., Beltran, B., et al.: BOOST: a supervised approach for multiple sclerosis lesion segmentation. J. Neurosci. Methods **237**, 108–117 (2014)
8. Carass, A., Roy, S., Jog, A., Cuzzocreo, J.L., et al.: Longitudinal multiple sclerosis lesion segmentation: resource and challenge. Neuroimage **148**, 77–102 (2017)
9. Jesson, A., Arbel, T.: Hierarchical MRF and random forest segmentation of MS lesions and healthy tissues in brain MRI (2015)
10. Maier, O., Handels, H.: MS lesion segmentation in MRI with random forests. In: Proc 2015 Longitudinal Multiple Sclerosis Lesion Segmentation Challenge, pp. 5–6 (2015)
11. Weng, W., Zhu, X.: INet: convolutional networks for biomedical image segmentation. IEEE Access **9**, 16591–16603 (2021)
12. Brosch, T., Tang, L.Y.W.W., Yoo, Y., Li, D.K.B.B., et al.: Deep 3D convolutional encoder networks with shortcuts for multiscale feature integration applied to multiple sclerosis lesion segmentation. IEEE Trans. Med. Imaging **35**(5), 1229–1239 (2016)
13. Aslani, S., Dayan, M., Storelli, L., Filippi, M., et al.: Multi-branch convolutional neural network for multiple sclerosis lesion segmentation. Neuroimage **196**(March), 1–15 (2019)
14. Afzal, H.M.R., Luo, S., Ramadan, S., Lechner-Scott, J., et al.: Automatic and robust segmentation of multiple sclerosis lesions with convolutional neural networks. Comput. Mater. Contin. **66**(1), 977–991 (2021)
15. Vaidya, S., Chunduru, A., Muthuganapathy, R., Krishnamurthi, G.: Longitudinal multiple sclerosis lesion segmentation using 3D convolutional neural networks (2015)
16. Narayana, P.A., Coronado, I., Sujit, S.J., Wolinsky, J.S., Lublin, F.D., Gabr, R.E.: Deep-learning-based neural tissue segmentation of MRI in multiple sclerosis: effect of training set size. J. Magn. Reson. Imaging **51**(5), 1487–1496 (2019). https://doi.org/10.1002/jmri.26959
17. Tustison, N.J., Avants, B.B., Cook, P.A., Zheng, Y., et al.: N4ITK: improved N3 bias correction. IEEE Trans. Med. Imaging **29**(6), 1310–1320 (2010)
18. Carass, A., Wheeler, M.B., Cuzzocreo, J., Bazin, P., et al.: Image analysis and communications laboratory, electrical and computer engineering. In: Division of Psychiatric Neuroimaging, Psychiatry and Behavioral Sciences, MedIC, Neuroradiology Division, Radiology and Radiological Science, The Johns Hopkins University Library (London), pp. 656–659 (2007)
19. Shiee, N., Bazin, P.-L.L., Cuzzocreo, J.L., Ye, C., et al.: Reconstruction of the human cerebral cortex robust to white matter lesions: method and validation. Hum. Brain Mapp **35**(7), 3385–3401 (2014)
20. Rafael C. González, R.E.W.: Digital Image Processing. Prentice Hall (2007)

21. Wang, J., Perez, L.: The Effectiveness of data augmentation in image classification using deep learning (2017)
22. Sweeney, E.M., Shinohara, R.T., Reich, D.S., Crainiceanu, C.M., et al.: Automatic lesion incidence estimation and detection in multiple sclerosis using. AJNR Am. J. Neuroradiol. **34**(1), 68–73 (2013)
23. Styner, M., Lee, J., Chin, B., Chin, M.S., et al.: 3D Segmentation in the clinic: a grand challenge II: MS lesion segmentation (2008)
24. Zijdenbos, A.P., Dawant, B.M., Margolin, R.A., Palmer, A.C.: Morphometric analysis of white matter lesions in MR images: method and validation. IEEE Trans. Med. Imaging **13**(4), 716–724 (1994)
25. Molyneux, P.D.: Precision and reliability for measurement of change in MRI lesion volume in multiple sclerosis: a comparison of two computer assisted techniques. J. Neurol. Neurosurg. Psychiat. **65**(1), 42–47 (1998)
26. Gregoriou C., Loizou, C.P., Georgiou A., Pantzaris M., Pattichis, C.S.: A Three-dimensional reconstruction integrated system for brain multiple sclerosis lesions. In: Proceedings of Computer Analysis of Images and Patterns, 19th International Conference, CAIP 2021, This volume (2021)
27. Nicolaou A., Loizou, C.P., Pantzaris M., Kakas A., Pattichis, C.S.: Rule extraction in the assessment of brain MRI lesions in multiple sclerosis: preliminary findings. In: Proceedings of Computer Analysis of Images and Patterns, 19th International Conference, CAIP 2021, This volume (2021)

A Three-Dimensional Reconstruction Integrated System for Brain Multiple Sclerosis Lesions

Charalambos Gregoriou[1]([envelope]) [iD], Christos P. Loizou[1] [iD], Andreas Georgiou[1] [iD], Marios Pantzaris[2] [iD], and Constantinos S. Pattichis[3] [iD]

[1] Department of Electrical Engineering and Computer Engineering and Informatics, Cyprus University of Technology, 3040 Limassol, Cyprus
{chg.grigoriou,andk.georgiou}@edu.cut.ac.cy,
christos.loizou@cut.ac.cy
[2] Cyprus School of Molecular Medicine, The Cyprus Institute of Neurology and Genetics, Nicosia, Cyprus
pantzari@cing.ac.cy
[3] Department of Computer Science, University of Cyprus, 1 Panepistimou Avenue, 2109 Aglantzia, Nicosia, Cyprus
pattichi@ucy.ac.cy

Abstract. In the course of a human brain acquisition, which is acquired by a magnetic resonance imager (MRI), two-dimensional (2D) slices of the brain are captured. These have to be aligned and reconstructed to a three-dimensional (3D) volume, which will better assist the doctor in following up the development of the disease. In this study, a 3D reconstruction integrated system for MRI brain multiple sclerosis (MS) lesion visualization is proposed. Brain MRI images from 5 MS subjects were acquired at four diffident consecutive time points (TP_1-TP_4) with an interval of 6–12 months. MS lesions were manually segmented by an expert neurologist and semi-automatically by a system and reconstructed in a brain volume. The proposed system assists the doctor in following up the MS disease progression and provides support to better manage the disease. The proposed system includes a 5-stage investigation (pre-processing, lesion segmentation, 3D reconstruction, volume estimation and method evaluation), as well as a module for the quantitative evaluation of the method. Twenty MRI images of the brain were used to evaluate the proposed system. Results show that the 3D reconstruction method proposed in this work, can be used to differentiate brain tissues and recognize MS lesions by providing improved 3D visualization. These preliminary results provide evidence that the proposed system could be applied in the future in clinical practice given that it is further evaluated on more subjects.

Keywords: Magnetic resonance imaging · Multiple sclerosis · 3D lesions reconstruction · Integrated system

1 Introduction

Brain image acquisitions are gradually becoming more and more information-rich thanks to the increasing resolution and versatility of brain imaging methods [1, 2]. Multiple

© Springer Nature Switzerland AG 2021
N. Tsapatsoulis et al. (Eds.): CAIP 2021, LNCS 13052, pp. 266–276, 2021.
https://doi.org/10.1007/978-3-030-89128-2_26

Sclerosis (MS) is a chronic central nervous system disease that results in loss of sensory and motor function [2]. It is one of the most common diseases that affects the neurological abilities of young adults [1]. Neurological disabilities affect the immune system and induce inflammation to the neurons, demyelination, and axonal damage [1].

Three-Dimensional (3D) visualization of brain magnetic resonance images (MRI), has been a developing part of medical imaging for many years and the literature surrounding it, is growing rapidly ever since. It aims in providing experts with both qualitative and quantitative information on the subject of study and help comprehend it in its full dimensionality [3]. Despite the massive research that has been carried out in 3D visualization in MRI, there are only a few other studies found in the current literature, that focused on the 3D reconstruction and visualization of the brain in MS subjects [4–6]. More specifically, in [4] the spherical harmonics method was used to estimate the 3D shape of the MS lesions and to determine the lesions' volume. In [5], the 3D Slicer tool was introduced, which was used to reconstruct brain MRI images and lesions. Furthermore, in order to estimate the loss of brain volume over time and asses the evolution of the MS disease, the FreeSurfer [7] and MIPAV [8] software tools were used. It was also recently proposed [6], that 3D texture analysis and 3D lesion visualization may be useful in following up the progression and development of the MS disease. Furthermore, in [6], researchers used the isosurface rendering method and texture mapping techniques to reconstruct the MRI images and estimated also the volume of the lesions in two different time points (TP_1-TP_2).

The objective of this work is to propose and evaluate a 3D reconstruction integrated system for brain MS lesions visualization, which will be able to reconstruct 2D MRI MS acquisitions into a 3D volume. The system is composed out of the following functions: 1) Brain and lesions re-slicing and visualization in three different planes (sagittal, coronal, transverse); 2) Visualization and comparison of four different brain MRI acquisition time points (TP_1-TP_4); 3) Comparison of the manual and the automated [9] lesion segmentations, which are embedded into a 3D volume; 4) Volume estimation [mm^3] of each lesion; 5) Approximation for the total volume of all lesions at TP_1 to TP_4; 6) Comparison and quantification of the MS disease; and 7) Follow up the development of the MS disease.

To the best of our knowledge, there are no other studies reported in the current literature, where 3D reconstruction of brain images and lesions from MS subjects at TP_1-TP_4 was developed, validated, or investigated.

2 Material and Methods

The 3D reconstruction integrated system proposed in this work was developed using the VisPy python library [10], which is used for interactive and high-level visualization. The system is also combined with a simple user interface developed with PyQT5 and other supplementary libraries and tools [11–13]. Figure 1 illustrates the flow diagram depicting the steps followed for reconstructing the 3D brain MRI images and MS lesions into a 3D volume, which will be described below.

2.1   Acquisition of Brain MRI Images

The brain MRI images used in this work were available from the International Sympo-sium on Biomedical Imaging (ISBI) 2015 contest [14] which provided 19 participants with four different MRI scan sequences (T_1-weighted (T_{1w}), magnetization prepared rapid gradient echo (PD_w), T_2-weighted (T_{2w}) and T_{2w} fluid attenuated inversion recov-ery (T_{2w} FLAIR)) [14]. In this work only T_{2w} FLAIR images were used (Nr = 5). A 3.0 T MRI scanner was used to generate the images with an inversion recovery time of 823 ms, an echo time of 68 ms, and a voxel size of $0.82 \times 0.82 \times 2.2$ mm^3 [14]. The MRI images were inhomogeneity corrected using the N4 method, a non-uniform inten-sity normalization [15] and then rigidly registered using a baseline (TP_1) as described in [14]. In addition, additive noise filtering was applied using skull [16] and dura [17] stripped masks [14]. Finally, an inhomogeneity correction N4 was performed once again.

The entire dataset consisted of 19 subjects out of which only the five were used in this study (Nr = 5, Age = 43.5 ± 10.3 (mean $\pm$ std) years old)). The four of them had brain acquisitions from four consecutive time points (T_1-T_4) and one of them from five consecutive time points (T_1-T_5) separated by an average acquisition time of one year [14]. The MRI brain scans consisted of 181 longitudinal slices of size 217×181 pixels. Manual blinded delineations were performed by two raters (R_1, R_2) with four and ten years of clinical experience in manual delineation respectively [14].

2.2   MRI MS Lesion Preprocessing and Semi-automated Segmentation

Before the automated lesion segmentation [9] the MRI images were intensity normalized between 0 and 4095 as shown in (1) (see also Fig. 1, Step 3). The lesion segmentation was performed using a Convolutional Neural Network (CNN) U-Net system [9]. A CNN architecture that is symmetric and it uses an encoder and a decoder to extract spatial features from the images and to construct segmentation maps [18]. A basic CNN architecture repeats a sequence of two 3×3 convolution networks followed by a max-pooling operation with a pooling size of 2×2 and a stride of two [18]. The detailed architecture and additional information for the CNN model and the MRI lesion segmen-tation is documented in [9]. The model in [18] is also embedded in the tool proposed in this work, giving the opportunity to the user to segment new MRI scans.

2.3   Image Normalization and Contour Lesion Generation

The original MRI image intensities were histogram normalized (see Fig. 1, step 3), using the MinMaxScaler function from Scikit-Image library, to map the intensity values in the range of 0 and 255 as follows:

$$N_i = \frac{X_i - min(X)}{max(X) - min(X)} * (T - D) + D \tag{1}$$

where X_i and N_i are the intensity and the normalized intensity of each pixel respectively, and X the entire image array, T is the new maximum intensity and D the new minimum intensity.

The segmented lesions estimated by the system in [9] were in a form of a binary array mask that can be used to estimate the lesions contour coordinates. The contour coordinates were estimated using the OpenCV's method findContours [12] (see Fig. 1, step 4), where a list with the (x, y) coordinates of all the contour points were provided. The coordinates were then saved in a dictionary using their slice number as the key.

2.4 3D Lesion Volume Estimation

In the next step, the segmented masks were given to the SciPy module which labels all the connected components of the array according to a $3 \times 3 \times 3$ pixels cube structuring element filled with ones[1] (see Fig. 1, step 5). The labeling method assumes that all the pixels with zero intensities form the background of the image and are not included in the lesions mask. The structuring element estimates the outline of the lesions by assuming connecting components and labels. After iterating through all the segmented masks, the labeling method returns an array of the same shape as the input array and an integer. The returned array is labeled with all the connected pixels having the same integer as the label, and the integer returned is the number of 3D lesions found in the masks. To estimate each lesion's volume the NumPy method count nonzero is called for each of the above estimated labels (see also Fig. 1, step 5).

2.5 3D Reconstruction

Prior to the 3D reconstruction, the user loads the json file, which contains the data for visualization (see Fig. 1, step 6). The proposed system then automatically draws the lesion contours on the original MRI brain images, using the OpenCV method drawContours [12] (see Fig. 1, step 7). Adding on, the tool automatically multiplies the segmented masks with the original images. This operation isolates the lesions and provides separate visualization (see also Figs. 2c) and 2d)). In the following, the OpenGL, NumPy and VisPy libraries were used to improve interactive 3D visualization [10]. VisPy offers the ability to render the MS lesion volumes with the direct volume rendering (DVR) translucent method. DVR renders the lesions volume by casting virtual rays through the volume and projecting the result as a 2D image on a plane. It retains the interior of the volume unchanged and provides spatial information between different structures [19]. This is a key factor for the visualization and the validity as indicated in [1, 2, 20]. The Vertex shader (see Fig. 1, step 9), is a code written in GLSL[2] programming language that is used to calculate the position of the vertices on the screen. It is also used to calculate how the virtual rays will transfer through the volume and decides which parts of the volume are visible and have to be drawn. The fragment shader (see Fig. 1, step 10), assigns the color of each pixel and its opacity, it is also written in GLSL. VisPy additionally, offers the ability to modify the volume by changing the shaders or attaching codes in the shaders [10], and to re-slice the brain volume (see Fig. 1, step 13). All of the above tool properties offer the ability to the user to remove and replace slices in three

---

[1] https://docs.scipy.org/doc/scipy/reference/tutorial/ndimage.html.

[2] GLSL stand for OpenGL Shading Language it is a programming language similar to C and it runs directly on the graphics processor unit.

different directions (coronal, sagittal, transverse). Furthermore, VisPy's canvas supply basic interactive keys such as rotation, zooming, and changing the center of rotation.

Eventually, one of the challenges of 3D reconstruction is to maintain the texture of the MRI image. In order to visualize the texture of the MRI as representative as possible, a GLSL code is attached on the fragment shader (see Fig. 1, step 11), which changes the colormap and the opacity of each pixel. Specifically, a color map of RGBA values using the image intensities was generated in such a way so that the brain volume can be illustrated in original colors. As a result, the pixels with zero intensity (black color) are fully transparent, while as the pixel intensity gets higher, the transparency of the pixels is depressed.

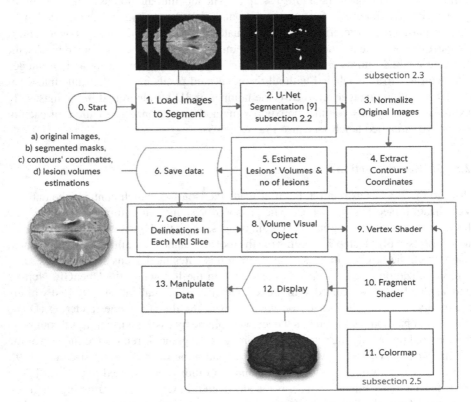

Fig. 1. Flow diagram of the proposed 3D reconstruction system in MRI brain images showing all processing steps. Given that MS lesions are whiter in a FLAIR image this results in MS lesion volumes that are more opaque and bright than other brain tissues.

2.6 Evaluation Metrics

The following evaluation metrics were used to assess the proposed 3D reconstruction method: a) Lesion load (L), b) Volume (V), c) The correlation coefficient, and d) The t-test p-value. Furthermore, the average difference of the total number of lesions (LAD) and the total volume (VAD) in mm^3 between the two different raters and the automated system [9] for all time points (TP_1–TP_4) were calculated as follows:

$$LAD_{j,k} = \frac{\sum |L_j(TP_i) - L_k(TP_i)|}{\sum TP} \tag{2}$$

$$VAD_{j,k} = \frac{\sum |V_j(TP_i) - V_k(TP_i)|}{\sum TP} \tag{3}$$

where L and V are the total lesion volume and the number of lesions respectively, j and k represent R_1 or R_2 or the semi-automated segmentation system and TP_i the time point of the given subject under evaluation.

3 Results

Figure 2 illustrates the 3D brain reconstruction from an MS patient aged 43.5 years in two consecutive time points (TP_1, TP_2), in the left and right columns, in Figs. 2a) and 2b) respectively. In Figs. 2c) and 2d) re-slicing of the brain in three different directions after rotation (−34.0 x-axis, −39.5 y-axis) is shown. The rotations were performed to better visualize the interior of the brain. Figures 2e) and 2f) show the 3D reconstructed lesions for TP_1 and TP_2 after the extraction of the lesions (Number of lesions at TP_1/TP_2: 35/44, Volume [mm^3]: 31746/33382). It is noted that the 3D reconstructed lesions shown in Figs. 2e) and 2f) can be rotated and re-sliced according to the expert's input.

Table 1 presents a comparison of all MS lesions investigated in this study based on the number of segmented lesions and volume performed by the two experts versus the semi-automated system proposed in [9], for all subjects investigated (Nr = 5) at TP_1–TP_5. The upper part of Table 1 tabulates the mean (± std) of the number of lesions and the total lesion volume (mm^3) for different time points (TP_1-TP_5) for the segmentations performed by R_1 and R_2 and the semi-automated CNN U-Net model. The middle section of Table 1 shows the correlation coefficient and the t test p-value of lesion segmentations and volume per time point. The t test p-value indicates the level of statically significance difference between manual and automated segmentations. Lastly, the lower section of Table 1, indicates the average difference of lesion load and the total lesion volume in mm^3 between R_1 vs R_2, R_1 vs the semi-automated system, and R_2 vs the semi-automated system for each different TP.

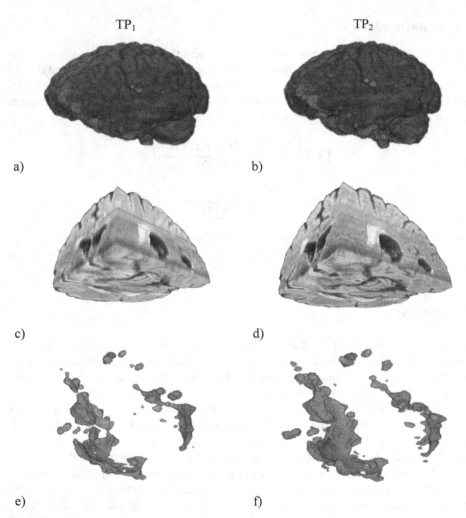

Fig. 2. MRI brain 3D reconstruction of an MS subject at TP_1 and TP_2 in the left and right columns, respectively. b) Re-slicing of the 3D reconstructed brain in 3 different planes (coronal, sagittal, transverse) (Number of slices shown/out of a total number of slices: 152/256, 165/256, 116/181). e), f) 3D reconstruction of the MRI MS lesions (Total number of MS lesions (TP_1/TP_2): 35/44.

4 Discussion

The objective of this work was to provide MS clinicians, a tool that can be used to estimate, follow up and compare MS lesions progression, by reconstructing 2D MRI scans into a 3D volume. The tool can reconstruct and visualize up to four different TPs, which can be resliced, rotated and switched between showing the entire brain or the lesions. Also, the tool provides an estimation of the lesion load at each TP, as well as an estimation of each lesion's volume in mm^3. The results presented in this study, showed that the proposed 3D reconstruction method can be used to differentiate brain tissue and distinguish MS lesions by providing an improved 3D visualization. The method preserves the texture of the original MRI scans to an acceptable level (see also Fig. 2). It can also be observed from the upper part of Table 1 that lesion load varies between different TPs. More specifically the std for the mean number of lesions and the lesions volume varies a lot between R1, R2 vs the automated system. As opposed to mean and std, Pearson correlation between R1 and R2 (see middle part of Table 1), shows a good correlation of above 0.5 for generally all different TPs. For the TP1 correlation is very high ($\rho = 0.91$), regarding the number of lesions. The correlation coefficient for the number of lesions between R_1 and R_2 and the automated system at TP_1 and TP_2 is 0.86 and 0.81, respectively.

Different methods were proposed in the current literature for the 3D reconstruction of MS brain lesions and or quantification estimation. More specifically, in [4] a method for estimating the shape and volume of MS lesions was presented. The 3D surface shape of the MS lesions was approximated by 2D contours, using the spherical harmonics method. The volume estimation was finally calculated by virtually slicing the 3D lesion, summing the areas of each slice, and multiplying them by the virtual slice thickness. Cordovez M. et al. [5], visually and quantitatively compared two different time points in 11 MS patients using the FreeSurfer [7] and MIPAV [8] software. The 3D slicer was used for the brain reconstruction, where a good 3D visualization of the lesions was achieved. A study for the 3D reconstruction of MS MRI brain scans was proposed in [6], where the surface of the brain was reconstructed using the isosurface method combined with texture mapping. In all of the above-mentioned studies, the drawbacks in the 3D reconstructions are twofold. More specifically, the texture and the interior of the MRI scans was not maintained, and none of the studies mentioned interactive visualizations such as re-slicing of the brain. In addition, no other studies were found in the literature where brain visualization at four different time points, as it was proposed in this study, for comparing the evolution of the MS disease were investigated. The software proposed in this study will be available to download at https://drive.google.com/drive/folders/1z7595UBTcsOzY7Hp9kCYsJ6tnHYQEfDM?usp=sharing. A number of other studies were also reported recently for the reconstruction of MRI using deep neural networks [22, 23]. In [22] a deep neural network was proposed for 4D reconstruction of the artic flow MRI data, while in [23] a deep neural recurrent network was presented to reduce MRI image acquisition time and improve the MRI reconstruction.

Table 1. Comparison of MS lesions based on the number of segmented number of lesions and volume performed by the two experts (R_1, R_2) vs the semi-automated segmentation system [9], for all subjects investigated in this study (Nr = 5) at TP_1-TP_5. The mean($\pm$std) of the total number of lesions (-/) and the total volume in mm^3 (/-) are shown for all subjects (see upper part of the table). Pearson correlation (t Test p-value) of the total number of lesions (-/) and the total volume in mm^3 (/-) are shown (see middle part of the table). The average difference of the total number of lesions (-/) and the total volume in mm^3 (/-) are shown (see lower part of the table).

Rater	TP_1	TP_2	TP_3	TP_4	TP_5
	Mean number of lesions (-/) and lesion volume (/-) in mm^3 of all lesions investigated [mean ($\pm$std)]				
R_1	26.4(8.8)/16761(15391)	28(4.1)/17257(17615)	24.8(5.5)/17944(18230)	21.8(7.9)/17762(18422)	25/8808
R_2	37.6(29)/20727(17267)	32(8.1)/17624(15311)	33(6.6)/19374(19663)	33(13.4)/19825(19690)	26/12198
Semi-automated	31.4(10.5)/22561(20615)	31.6(7.8)/23742(22368)	29.4(4.8)/23046 (17267)	34.8(4.9)/20727(21780)	34/13765
	Pearson correlation (t test with p-values for significantly and non-significantly different at p < 0.05) for the mean number of lesions (-/) and the lesion volume (/-)				
R_1 vs R_2	0.91(0.15)/0.99(0.03)	0.51(0.13)/0.98(0.42)	0.60(0.01)/0.99(0.13)	0.75(0.03)/0.99(0.04)	-/-
R_1 vs Semi-automated	0.86(0.05)/0.99(0.04)	0.81(0.09)/0.99(0.02)	−0.55(0.16)/0.99(0.02)	−0.11(0.02)/0.99(0.03)	-/-
R_2 vs Semi-automated	0.83(0.27)/0.98(0.24)	−0.08(0.47)/0.99(0.06)	−0.28(0.21)/0.99(0.02)	0.39(0.38)/0.99(0.05)	-/-
	Average difference for the number of lesions (-/) and the lesion volume (/-) in mm^3				
R_1 vs R_2	13.2/4141	5.6/2901	8.6/2097	11.2/2326	1/3005
R_1 vs Semi-automated	5.4/5800	4.4/6485	7.8/5103	13/5896	9/4395
R_2 vs Semi-automated	11.8/3643	8.8/6118	6/3958	9.8/3834	8/1389

5 Conclusions and Future Trends

The preliminary results presented in this study, provide evidence that the proposed 3D reconstruction system could be applied in the future in clinical practice. It was shown that the system may provide reliable, quantitative information for the neurologists for following up the evolution of the MS disease [21]. However, results reported in this study, (number of lesions and lesions volume [mm^3]), should be further validated in a future work in a larger number of subjects, as well as compared with additional studies from the literature. Moreover, lesion texture analysis needs to be incorporated in the proposed system in the form of rule display [24], aiding the neurologists to better understand the MS pathogenesis as it was shown in [21].

References

1. Karussis, D.: The diagnosis of multiple sclerosis and the various related demyelinating syndromes: a critical review. J. Autoimmun. **48–49**, 134–142 (2014)
2. Dobson, R., Giovannoni, G.: Multiple sclerosis–a review. Eur. J. Neurol. **26**(1), 27–40 (2019)
3. Eickhoff, S.B., Yeo, B.T.T., Genon, S.: Imaging-based parcellations of the human brain. Nat. Rev. Neurosci. **19**(11), 672–686 (2018)
4. Goldberg-Zimring, D., Azhari, H., Miron, S., Achiron, A.: 3-D surface reconstruction of multiple sclerosis lesions using spherical harmonics. Magn. Reson. Med. **46**(4), 756–766 (2001)
5. Cordovez, M.J., Gálvez, G.M., Rojas, C.G., Bravo, C.C., Cerda, E.C.: Use of lesion volumes and loads for monitoring patients with multiple sclerosis. Local experience and literature review. Rev. Chil. Radiol. **19**(4), 156–164 (2013)
6. Loizou, C.P., Papacharalambous, C., Samaras, G., Kyriacou, E., et al.: brain image and lesions registration and 3d reconstruction in dicom mri images. In: Proceedings of 30th IEEE Symposium Computer Based Medical Systems, pp. 419–422 (2017)
7. Fischl, B.: FreeSurfer. Neuroimage **62**(2), 774–781 (2012)
8. McAuliffe, M.J., Lalonde, F.M., McGarry, D., Gandler, W., et al.: Medical image processing, analysis & visualization in clinical research. In: Proceedings of 14th IEEE Symposium Computer Based Medical System, pp. 381–388 (2001)
9. Georgiou, A., Loizou, C.P., Nicolaou A., Pantzaris M., Pattichis, C.S.: An adaptive semi-automated integrated system for multiple sclerosis lesion segmentation in longitudinal mri scans based on a convolutional neural network. In Proceedings of Computer Analysis of Images and Patterns, 19th International Conference, CAIP 2021, This volume, (2021)
10. Campagnola, L., Klein, A., Larson, E., Rossant, C., et al.: VisPy: harnessing the GPU for fast, high-level visualization. In: Proceedings of 14th Python in Science Conference, pp. 93–99. HAL, Texas (2015)
11. Harris, C.R., Millman, K.J., van der Walt, S.J., Gommers, R., et al.: Array programming with NumPy. Nature **585**(7825), 357–362 (2020)
12. Kaehler, A., Bradski, G.: Learning OpenCV3: Computer Vision in C++ with the OpenCV library. Sebastopol (2016)
13. Virtanen, P., et al.: SciPy 1.0: Fundamental algorithms for scientific computing in Python. Nat. Methods **17**(3), 261–272 (2020). https://doi.org/10.1038/s41592-019-0686-2
14. Carass, A., Roy, S., Jog, A., Cuzzocreo, J.L., et al.: Longitudinal multiple sclerosis lesion segmentation: resource and challenge. Neuroimage **148**, 77–102 (2017)
15. Tustison, N.J., Avants, B.B., Cook, P.A., Zheng, Y., et al.: N4ITK: improved N3 bias correction. IEEE Trans. Med. Imag. **29**(6), 1310–1320 (2010)

16. Shiee, N., Bazin, P.-L., Cuzzocreo, J.L., Ye, C., et al.: Reconstruction of the human cerebral cortex robust to white matter lesions: method & validation. Hum. Brain Mapp. **35**(7), 3385–3401 (2014)
17. Carass, A., Wheeler, M.B., Cuzzocreo, J., Bazin, P.L., et al.: A joint registration and segmentation approach to skull stripping. In: Proceedings of 4th IEEE International Symposium Biomedical Imaging, pp. 656–659 (2007)
18. Ibtehaz, N., Rahman, M.S.: MultiResUNet: rethinking the U-Net architecture for multimodal biomedical image segmentation. Neur. Networks **121**, 74–87 (2020)
19. Dougherty, G. (ed.): Medical Image Processing. Springer, New York (2011). https://doi.org/10.1007/978-1-4419-9779-1
20. Filippi, M., Rocca, M.A., Ciccarelli, O., De Stefano, N., et al.: MRI criteria for the diagnosis of multiple sclerosis: MAGNIMS consensus guidelines. Lancet Neurol. **15**(3), 292–303 (2016)
21. Loizou, C.P., Pantzaris, M., Pattichis, C.S.: Normal appearing brain white matter changes in relapsing multiple sclerosis: texture image and classification analysis in serial MRI scans. Magn. Reson. Imaging **73**, 192–202 (2020)
22. Vishnevskiy, V., Walheim, J., Korke, S.: Deep variational network for rapid 4D flow MRI reconstruction. Nat. Mach. Intell. **2**(4), 228–235 (2020)
23. Zhou, B., Zhou, K.S.: DuDoRNet: learning a dual-domain recurrent network for fast MRI reconstruction with deep T1 prior. In: Proceedings of the IEEE/CVF Conference on Computer Vision and Pattern Recognition, pp. 4273–4282 (2020)
24. Nicolaou A., Loizou, C.P., Pantzaris M., Kakas A., Pattichis, C.S.: Rule extraction in the assessment of brain MRI lesions in multiple sclerosis: preliminary findings. In Proceeding of Computer Analysis of Images and Patterns, 19th International Conference, CAIP 2021, This volume, (2021)

Rule Extraction in the Assessment of Brain MRI Lesions in Multiple Sclerosis: Preliminary Findings

Andria Nicolaou[1]([✉]) [iD], Christos P. Loizou[2] [iD], Marios Pantzaris[3] [iD],
Antonis Kakas[1] [iD], and Constantinos S. Pattichis[1] [iD]

[1] Department of Computer Science, University of Cyprus, Nicosia, Cyprus
{nicolaou.andria,antonis,pattichi}@ucy.ac.cy
[2] Department of Electrical Engineering, Computer Engineering and Informatics,
Cyprus University of Technology, Limassol, Cyprus
christos.loizou@cut.ac.cy
[3] Cyprus School of Molecular Medicine, Cyprus Institute of Neurology and Genetics,
Nicosia, Cyprus
pantzari@cing.ac.cy

Abstract. Various artificial intelligence (AI) algorithms have been proposed in the
literature, that are used as medical assistants in clinical diagnostic tasks. Explainability methods are lighting the black-box nature of these algorithms. The objective of this study was the extraction of rules for the assessment of brain magnetic resonance imaging (MRI) lesions in Multiple Sclerosis (MS) subjects based on texture features. Rule extraction of lesion features was used to explain and provide information on the disease diagnosis and progression. A dataset of 38 subjects diagnosed with a clinically isolated syndrome (CIS) of MS and MRI detectable brain lesions were scanned twice with an interval of 6–12 months. MS lesions were manually segmented by an experienced neurologist. Features were extracted from the segmented MS lesions and were correlated with the expanded disability status scale (EDSS) ten years after the initial diagnosis in order to quantify future disability progression. The subjects were separated into two different groups, G_1: EDSS $\leq$ 3.5 and G_2: EDSS > 3.5. Classification models were implemented on the KNIME analytics platform using decision trees (DT), to estimate the models with high accuracy and extract the best rules. The results of this study show the effectiveness of rule extraction as it can differentiate MS subjects with benign course of the disease (G_1: EDSS $\leq$ 3.5) and subjects with advanced accumulating disability (G_2: EDSS > 3.5) using texture features. Further work is currently in progress to incorporate argumentation modeling to enable rule combination as well as better explainability. The proposed methodology should also be evaluated on more subjects.

Keywords: Multiple Sclerosis · Brain MRI · Lesions · Texture features · Classification · Rule extraction · Explainability

© Springer Nature Switzerland AG 2021
N. Tsapatsoulis et al. (Eds.): CAIP 2021, LNCS 13052, pp. 277–286, 2021.
https://doi.org/10.1007/978-3-030-89128-2_27

1 Introduction

Multiple Sclerosis (MS) is a chronic autoimmune disease of the central nervous system (CNS), damaging tissues of the brain and spinal cord [1]. The pathological characteristics of MS are perivenular inflammatory lesions, leading to demyelinating plaques [2]. As the clinical outcome differs within patients, expert neurologists perform visual MS lesion evaluation based on clinical, imaging, and laboratory evidence, following the McDonald criteria, to diagnose the disease [3, 4]. Magnetic resonance imaging (MRI) plays a fundamental role in the diagnosis of MS as it allows structural damage to be quantified in the patients' tissue and to be followed in time [1]. In addition, MRI can be used to provide additional information about treatment options. Texture feature analysis extracts quantitative data from an image based on mathematical computations. The texture of images refers to the appearance, structure, and arrangement of the parts of an object within the image. Mathematical parameters may be computed from the distribution of pixels that could characterize the underlying structure of the objects shown in the MRI image [5].

Several researchers have studied the textural characteristics of lesions in MS. More specifically, in [6], it was shown that texture analysis characterizes MS activity and progression, and can be further tested as a potential tool for evaluating treatment benefits for MS patients. In [7], an MRI texture analysis protocol with fixed imaging sequence and anatomical levels of interest was proposed as a robust quantitative clinical means for evaluating MS lesions. Furthermore, in [8] it was shown that texture features may provide some prognostic evidence regarding the future disability of patients. In another study [9], texture features provided complementary information for following up the development of the MS disease. Finally, in [10] it was suggested that texture features extracted from MS lesions may differentiate chronic active from chronic stable lesions.

Deep learning methods [11] have been very effective for medical diagnosis, but the black-box nature of the algorithms has restricted their clinical use. However, explainability studies aim to illustrate specific features that could influence the decision of a model [11]. Recent medical research [12] proposed explainable artificial intelligence (XAI) systems, which target to be transparent, and to provide understandable decisions. In [13], an explainable model for basic brain tumor detection using convolutional neural networks on MRI slices was implemented. Moreover, in [14], explainable assessments of Alzheimer's disease were extracted using machine learning on brain MRI imaging data, whereas in [15] an explainable AI model for stroke prediction using integrating machine learning with symbolic reasoning was built. However, to the best of our knowledge, no study investigated MS lesion imaging explainability assessment.

The objective of this study was the extraction of explainable information in the form of rules for the assessment of brain MRI lesions and their interrelation to disability in MS subjects based on texture features. Machine learning models were developed using a decision tree (DT) classifier, in order to determine the models with high accuracy and extract the best lesion texture features' rules. These rules may then be used to differentiate subjects with benign course of the disease and subjects with advanced accumulating disability. The patient's images were acquired and analyzed at the initial stage of the disease ($Time_0$) and after 6–12 months ($Time_{6-12}$). The lesion feature findings were correlated with the expanded disability status scale (EDSS), which is used for evaluating the degree of neurologic impairment in MS and ranges from 0 (normal) to 10 (death due to MS) [16].

2 Materials and Methods

The flow diagram of Fig. 1 illustrates the different processing steps for rule extraction investigated in this study, for the assessment of brain MRI lesions in MS subjects. Further analysis for each step is given herein below.

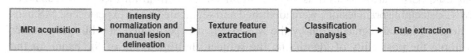

Fig. 1. Flow diagram for the assessment of brain MRI lesions in MS and rule extraction.

2.1 Study Group and MRI Acquisition

A dataset of 38 subjects (17 males, and 21 females) with a clinically isolated syndrome (CIS) of MS and MRI detectable brain lesions were scanned twice with an interval of 6–12 months. The transverse MR images used for analysis were obtained using a T2-weighted turbo spin-echo pulse sequence (repetition time = 4408 ms, echo time = 100 ms, echo spacing = 10.8 ms). The reconstructed image had a slice thickness of 5 mm and a field of view of 230 mm with a pixel resolution of 2.226 pixels per mm. Standardized planning procedures were followed during each MRI examination. The MRI images were acquired using a 1.5 T whole-body Philips ACS NT MR imager (Philips Medical Systems, Best, the Netherlands). A built-in quadrature radiofrequency body coil and a quadrature radiofrequency head coil were used for proton excitation and signal detection respectively. MRI acquisition information was also given in [8, 9].

An expert MS neurologist (co-author, M. Pantzaris), performed the clinical evaluation, referred subjects for a baseline MRI (at $Time_0$), and followed them for over ten years. All patients were subjected to an EDSS evaluation, ten years after the initial diagnosis to quantify future disability progression. The subjects were separated into two different groups (i.e. G_1: EDSS $\leq$ 3.5 and G_2: EDSS > 3.5). The reason for selecting an EDSS cutoff point of 3.5 is that for EDSS > 3.5, the physician can assess neurological signs, meaning that the patient starts accumulating disability. Thus, any patient having an EDSS $\leq$ 3.5 at ten years after the initial MRI scan can be regarded as having a rather benign course of the disease.

2.2 Intensity Normalization and Manual Lesion Delineation

Before the manual lesion segmentation, all MRI images were intensity normalized between the grayscale values of 0 and 255 as documented in [8, 9] where all additional details about the algorithm may be found. A normalization algorithm adjusted distributions of each follow up scan ($Time_{6-12}$), to match those of the chosen baseline

(Time$_0$) scan in order to improve image compatibility, reduce partial volume effects and facilitate MR image comparability between serial MR scans [17], such as those obtained from the MS group of this work. Before features extraction, the MS lesions were visually identified and manually segmented by a neurologist (co-author M. Pantzaris) using a software tool developed by our group [8].

The expert neurologist manually delineated the brain lesions by selecting consecutive points at the visually defined borders between the lesions and the adjacent normal-appearing white matter (NAWM) on the acquired transverse T$_2$W sections. For the purpose of intensity normalization, the neurologist manually segmented cerebrovascular fluid (CSF) areas as well as areas with air (sinuses) from all MS brain scans. All MRI-detectable brain lesions were identified and segmented by an experienced MS neurologist. Manual segmentation by the MS expert was performed in a blinded manner without the possibility of identifying the subject, the time-point of the exam, or the clinical findings. The selected points and delineations were saved to be used for texture feature extraction and analysis.

2.3 Texture Feature Extraction

Texture features were extracted from all manually segmented MS lesions, detected from brain MRI scans, and were then estimated by averaging the corresponding values for all lesions for each group of patients (G$_1$, G$_2$). The following groups of features were extracted [8, 9]: 5 Statistical Features (SF), 12 Spatial Gray Level Dependence Matrices (SGLDM), 4 Gy Level Difference Statistics (GLDS), 5 Neighborhood Gray Tone Difference Matrix (NGTDM), 4 Statistical Feature Matrix (SFM), 6 Laws Texture Energy Measures (LTEM), 4 Fractal Dimension Texture Analysis (FDTA), 2 Fourier Power Spectrum (FPS), and 5 Shape Parameters (SP).

2.4 Classification Analysis

Texture classification modeling was used to predict EDSS score at year 10. Models to predict subjects with EDSS $\leq$ 3.5 (G$_1$) versus those with EDSS > 3.5 (G$_2$) based on texture features, and lesion images were developed. The classification analysis performance metrics documented in this study were based on the evaluation set. Overfitting of data by cross-validation was avoided. The classification models were implemented on the KNIME analytics platform using decision trees (DT). Specifically, the Decision Tree Learner node was used for the model training and the Decision Tree Predictor node for the model evaluation [18, 19]. The synthetic minority over-sampling technique was applied during the model training, using the SMOTE node, to improve the performance of the model and avoid overfitting. SMOTE creates new samples for the minority group of the model (G$_2$) with the same statistical properties [18]. The Parameter Optimization nodes were also added, to check the minimum numbers of records per node in the DT and return the one with high accuracy [19].

As shown in Table 1, the initial dataset of 38 subjects was separated into the two aforementioned groups (G$_1$, G$_2$). In order to build the proposed models, data were split into a training and an evaluation group, using 80% for the training and 20% for the evaluation step. Each model was run for both 10-fold and 20-fold cross-validation sense.

Applying the over-sampling technique on the training set, G_1 and G_2 were generated having the same number of subjects. Thus, the models were trained by 54 subjects and evaluated by 4 subjects in each iteration.

Table 1. MS subject distribution for training and evaluation of the classification models.

Subject/ Data sets	All	G_1_EDSS $\leq$ 3.5	G_2_EDSS $>$ 3.5
Initial	38	29	9
Training	34	27	7
Over-sample training	54	27	27
Evaluation	4	2	2

2.5 Rule Extraction

The extraction of the lesion texture features' rules was achieved using the Decision Tree to Ruleset KNIME node which converts the DT to a set of rules [14]. For each rule, the following evaluation metrics [20] were used:

$$Accuracy = \frac{TP + TN}{TP + FP + TN + FN} \tag{1}$$

$$Sensitivity = \frac{TP}{TP + FN} \tag{2}$$

$$Specificity = \frac{TN}{TN + FP} \tag{3}$$

$$Precision = \frac{TP}{TP + FP} \tag{4}$$

$$Recall = \frac{TP}{TP + TN} \tag{5}$$

where TP and TN denote the number of positive and negative instances that are correctly classified, and FP and FN indicate the number of misclassified negative and positive instances, respectively [20].

3 Results

The models were trained and evaluated for all the lesion texture features at the initial stage of the disease ($Time_0$) as well as at $Time_0 + Time_{6-12}$.

Table 2 illustrates the results of the evaluation metrics for the model classification. It is shown that the texture feature groups can be used to differentiate subjects with benign

Table 2. Model analysis of brain MRI lesion texture features extracted from MS subjects (n = 38) at $Time_0$ and $Time_0$ + $Time_{6-12}$ in the left and right columns, averaged at 10 (-/) and 20 iterations (/-) respectively.

Feature groups	$Time_0$					$Time_0$ + $Time_{6-12}$				
	Accuracy	Sensitivity	Specificity	Precision	Recall	Accuracy	Sensitivity	Specificity	Precision	Recall
SF	0.60/0.51	0.60/0.51	0.60/0.51	0.65/0.52	0.60/0.51	0.80/0.71	0.80/0.71	0.80/0.71	0.81/0.77	0.80/0.71
SGLDM	0.63/0.60	0.63/0.60	0.63/0.60	0.66/0.62	0.63/0.60	0.63/0.60	0.63/0.60	0.63/0.60	0.67/0.62	0.63/0.60
GLDS	0.75/0.70	0.75/0.70	0.75/0.70	0.78/0.73	0.75/0.70	0.65/0.60	0.65/0.60	0.65/0.60	0.69/0.63	0.65/0.60
NGTDM	0.80/0.71	0.80/0.71	0.80/0.71	0.85/0.76	0.80/0.71	0.83/0.73	0.83/0.73	0.83/0.73	0.85/0.77	0.83/0.73
SFM	0.60/0.56	0.60/0.56	0.60/0.56	0.64/0.59	0.60/0.56	0.63/0.61	0.63/0.61	0.63/0.61	0.66/0.65	0.63/0.61
LTEM	0.68/0.64	0.68/0.64	0.68/0.64	0.73/0.66	0.68/0.64	0.65/0.60	0.65/0.60	0.65/0.60	0.69/0.63	0.65/0.60
FDTA	0.63/0.60	0.63/0.60	0.63/0.60	0.67/0.66	0.63/0.60	0.63/0.60	0.63/0.60	0.63/0.60	0.68/0.63	0.63/0.60
FPS	0.75/0.65	0.75/0.65	0.75/0.65	0.81/0.67	0.75/0.65	0.75/0.69	0.75/0.69	0.75/0.69	0.80/0.72	0.75/0.69
SP	0.65/0.63	0.65/0.63	0.65/0.63	0.70/0.66	0.65/0.63	0.75/0.60	0.75/0.60	0.75/0.60	0.78/0.63	0.75/0.60

SF: Statistical Features, SGLDM: Spatial Gray Level Dependence Matrices, GLDS: Gray Level Difference Statistics, NGTDM: Neighborhood Gray Tone Difference Matrix, SFM: Statistical Feature Matrix, LTEM: Laws Texture Energy Measures, FDTA: Fractal Dimension Texture Analysis, FPS: Fourier Power Spectrum, SP: Shape Parameters.

course of the disease (EDSS $\leq$ 3.5) and subjects with advanced accumulating disability (EDSS > 3.5). At $Time_0$ (see left part of Table 2), the texture feature groups GLDS, NGTDM, LTEM, FPS, and SP may be used to separate subjects into the two different groups (G_1 vs G_2). When including all lesions at $Time_0$ and $Time_{6-12}$ (see right part of Table 2), the texture feature groups SF, GLDS, NGTDM, LTEM, FPS, and SP may be used to separate subjects into the two different groups (G_1 vs G_2). The selection of the texture feature groups was achieved using models with an accuracy $\geq$ 0.65.

Table 3. Summary findings of the rules extracted for the proposed models with Accuracy $\geq$ 0.65 based on Table 2.

	$Time_0$			$Time_0 + Time_{6-12}$		
Feature groups		$G_1.EDSS \leq 3.5$	$G_2.EDSS > 3.5$		$G_1.EDSS \leq 3.5$	$G_2.EDSS > 3.5$
SF	–	–	–	$Median_{6-12}$	≤105	>105
GLDS	Mean	>6.779	≤6.779	$Mean_0$	>6.687	≤6.687
	Angular second Moment	≤0.070	>0.070	–	–	–
NGTDM	Strength	≤27284	>27284	$Strength_0$	≤27284	>27284
	–	–	–	$Coarseness_0$	≤12.72	>12.72
LTEM	SS	>175	≤175	LL_0	≤44319	>44319
	LL	≤43719	>43719	–	–	–
FPS	Radial Sum	≤1808	>1808	Radial Sum_{6-12}	≤1346	>1346
SP	Perimeter [mm]	≤61	>61	$Perimeter_{6-12}$ [mm]	≤53	>53
	–	–	–	$Area_{6-12}$ [mm^2]	≤208	>208

SF: Statistical Features, GLDS: Gray Level Difference Statistics, NGTDM: Neighborhood Gray Tone Difference Matrix, LTEM: Laws Texture Energy Measures, FPS: Fourier Power Spectrum, SP: Shape Parameters, SS: texture energy from SS kernel, LL: texture energy from LL kernel.

Table 3 indicates the summary findings of the rules that were extracted from the proposed models (see also Table 2). It is shown from Table 3 that when evaluating and selecting the models with an accuracy $\geq$ 0.65, this may provide the best rules to assess the brain MRI lesion texture features in MS. It is worth mentioning that some of the texture features, e.g. Strength of the NGTDM group, are strong enough to differentiate subjects into the two different groups (G_1 vs G_2), as both at $Time_0$ and $Time_0 + Time_{6-12}$ have exactly the same rule.

4 Discussion

The objective of this study was the extraction of explainable information in the form of rules for the assessment of brain MRI lesions and their interrelation to disability in MS subjects based on texture features. Classification models were developed using DT, to estimate the models with high accuracy and extract the best lesion texture features' rules that can be used to differentiate subjects with a benign course (EDSS ≤ 3.5) of the disease and subjects with an advanced accumulating disability (EDSS > 3.5).

The main findings of this study can be summarized as follows:

1) Some texture feature groups (e.g. GLDS, LTEM, FPS) at $Time_0$ can achieve very good accuracy, with no additional increase in accuracy observed when combining features at $Time_0 + Time_{6-12}$ (see also Table 2).
2) Simple rules including even one texture feature group (e.g. FPS) can achieve an accuracy that is greater than 0.7 (see also Table 3).

To the best of our knowledge, we found no other studies reported in the literature where rules were extracted for the assessment of brain MRI lesions in MS subjects based on lesions' texture features. There were a few other studies reported in the literature, where texture feature analysis on MS brain MRI lesions was investigated for assessing the evolution of the disease. More specifically, in [7] a texture analysis protocol classified the MS plaques from white matter and normal-appearing white matter. It was shown that the MRI texture analysis protocol with fixed imaging sequence and anatomical levels of interest, is a robust quantitative clinical means for evaluating MS lesions. In [8], several texture features of MRI brain white matter and MS lesions were extracted. It was shown that these may be used to differentiate between brain lesions normal white matter and normal-appearing white matter, that lead to minimal (EDSS ≤ 2) and mild clinical signs (EDSS > 2). Moreover, in [9], classification models were implemented to notice the normal-appearing white matter changes in relapsing MS using texture features. It was shown in [10] that MS texture features extracted from MPRAGE images indicate higher intralesional heterogeneity, however, they demonstrate only a fair accuracy to differentiate chronic active from chronic stable MS lesions. Further similarities of this study with the study in [10], can be found in texture features and classification analysis.

A number of recent studies proposed the use of XAI in the medical domain. In [14], rule extraction was implemented to explain the assessment of Alzheimer's disease (AD) based on MRI hippocampus and entorhinal cortex texture analysis using decision trees and random forests integrated with argumentation modeling, achieving an overall accuracy of 77% and 74%, respectively. Similarly, in [15], rule extraction using random forests, followed by argumentation modeling were used to build an explainable AI model for stroke risk prediction based on carotid ultrasound imaging plaque texture, shape, and stenosis features. The best models achieved an overall accuracy of 78% in differentiating asymptomatic versus symptomatic subjects.

5   Conclusion

MS lesion texture features extracted from brain MRI scans can be analyzed to provide explainable information in the form of rules for the diagnosis and differentiation of the disease progression. The main findings of this study can be summarized as follows: (i) some texture feature groups (e.g. GLDS, LTEM, FPS) at $Time_0$ can achieve very good accuracy, and (ii) simple rules including even one texture feature group (e.g. FPS) can achieve an accuracy that is greater than 0.7.

Future work will incorporate an argumentation-based reasoning framework to build XAI modeling as it was demonstrated in [14, 15]. It is expected that this new model will provide further information on MS diagnosis and progression. Moreover, future work will include the integration of lesion texture rule display together with a visualization of the three-dimensional reconstructed brain MRI [21] where MS lesions were segmented based on a semi-automated Convolutional Neural Network system [22]. Furthermore, a larger number of subjects is required to improve the validation of results and establish the clinical use of the proposed system.

References

1. Filippi, M., Rocca, M.A.: MRI evidence for multiple sclerosis as a diffuse disease of the central nervous system. J. Neurol. **252**, 16–24 (2005)
2. Dobson, R., Giovannoni, G.: Multiple sclerosis-a review. Eur. J. Neurol. **26**, 27–40 (2019)
3. Thompson, A.J., Banwell, B.L., Barkhof, F., Carroll, W.M., et al.: Diagnosis of multiple sclerosis: 2017 revisions of the McDonald criteria. Lancet Neurol. **17**, 162–173 (2018)
4. McDonald, W.I., Compston, A., Edan, G., Goodkin, D., et al.: Recommended diagnostic criteria for multiple sclerosis: guidelines from the international panel on the diagnosis of multiple sclerosis. Ann. Neurol. **50**, 121–127 (2001)
5. Castellano, G., Bonilha, L., Li, L.M., Cendes, F.: Texture analysis of medical images. Clin. Radiol. **59**, 1061–1069 (2004)
6. Zhang, Y.: MRI texture analysis in multiple sclerosis. Int. J. Biomed. Imaging **2012**, 1–7 (2012)
7. Harrison, L.C.V., et al.: MRI texture analysis in multiple sclerosis: toward a clinical analysis protocol. Acad. Radiol. **17**, 696–707 (2010)
8. Loizou, C.P., Petroudi, S., Seimenis, I., Pantziaris, M., Pattichis, C.S.: Quantitative texture analysis of brain white matter lesions derived from T2-weighted MR images in MS patients with clinically isolated syndrome. J. Neuroradiol. **42**, 99–114 (2015)
9. Loizou, C.P., Pantzaris, M., Pattichis, C.S.: Normal appearing brain white matter changes in relapsing multiple sclerosis: texture image and classification analysis in serial MRI scans. Magn. Reson. Imaging **73**, 192–202 (2020)
10. Weber, C.E., Wittayer, M., Kraemer, M., Dabringhaus, A., et al.: Quantitative MRI texture analysis in chronic active multiple sclerosis lesions. Magn. Reson. Imaging **79**, 97–102 (2021)
11. Singh, A., Sengupta, S., Lakshminarayanan, V.: Explainable deep learning models in medical image analysis. J. Imaging **6**, 1–19 (2020)
12. Holzinger, A., Biemann, C., Pattichis, C.S., Kell, D.B.: What do we need to build explainable AI systems for the medical domain? arXiv, 1–28 (2017)
13. Windisch, P., et al.: Implementation of model explainability for a basic brain tumor detection using convolutional neural networks on MRI slices. Neuroradiology **62**(11), 1515–1518 (2020). https://doi.org/10.1007/s00234-020-02465-1
14. Achilleos, K.G., Leandrou, S., Prentzas, N., Kyriacou, P.A. et al.: Extracting explainable assessments of Alzheimer's disease via machine learning on brain MRI imaging data. In: 2020 IEEE 20th International Conference of Bioinformatics and Bioengineering BIBE, pp. 1036–1041 (2020)
15. Prentzas, N., Nicolaides, A., Kyriacou, E., Kakas, A., Pattichis C.: Integrating machine learning with symbolic reasoning to build an explainable AI model for stroke prediction. In: 2019 IEEE 19th International Conference of Bioinformatics and Bioengineering BIBE, pp. 817–821 (2019)

16. Kurtzke, J.F.: Rating neurologic impairment in multiple sclerosis: an expanded disability status scale (EDSS). Neurology **33**, 1444–1452 (1983)
17. Meier, D.S., Guttmann, C.R.G.: Time-series analysis of MRI intensity patterns in multiple sclerosis. Neuroimage **20**, 1193–1209 (2003)
18. Widmann, M., Roccato, A.: From modeling to model evaluation, 1st edn. KNIME Press, Switzerland (2021)
19. Silipo, R.: Practicing data science, 3rd edn. KNIME Press, Switzerland (2021)
20. Hossin, M., Sulaiman, M.N.: A review on evaluation metrics for data classification evaluations. Int. J. Data Min. Knowl. Manag. Process **5**(2), 1–11 (2015)
21. Gregoriou, C., Loizou, C.P., Georgiou, A., Pantzaris, M., Pattichis, C.S.: A three-dimensional reconstruction integrated system for brain multiple sclerosis lesions. In: 19th International Conference of Computation Analysis of Images Patterns CAIP (2021)
22. Georgiou, A., Loizou, C.P., Nicolaou, A., Pantzaris, M., Pattichis, C.S.: An adaptive semi-automated integrated system for multiple sclerosis lesion segmentation in longitudinal mri scans based on a convolutional neural network. In: 19th International Conference of Computation Analysis of Images Patterns CAIP (2021)

Invariant Moments, Textural and Deep Features for Diagnostic MR and CT Image Retrieval

Lorenzo Putzu[1]([✉])[iD], Andrea Loddo[1,2][iD], and Cecilia Di Ruberto[1][iD]

[1] Department of Mathematics and Computer Science, University of Cagliari,
via Ospedale 72, 09124 Cagliari, Italy
{lorenzo.putzu,andrea.loddo,dirubert}@unica.it
[2] Department of Electrical and Electronic Engineering, University of Cagliari,
piazza d'Armi, 09123 Cagliari, Italy

Abstract. Image analysis in the medical field aims to offer tools for the diagnosis and detection of life-threatening illness. This study means to propose a novel content-based image retrieval system oriented to medical diagnosis. In particular, we exploit several classic and deep image descriptors together with different similarity measures on three different data set, containing computed tomography and magnetic resonance images. Experiments show that feature selection can bring benefit if applied to deep and texture features, contrary to what observed for invariant moments. Moreover, the cityblock distance emerged to be quite suitable overall in this domain, although some other distances also exhibit satisfying robustness.

Keywords: Biomedical image retrieval · CBIR · Hand-crafted features · Deep features

1 Introduction

Image retrieval faces the problem of searching for semantically related or identical images in an extensive image gallery by analysing their visual content, provided a query image that represents a specific scenario. In particular, it is a long-standing research topic in the computer vision environment [1] and several authors tried to propose their own *Content-Based Image Retrieval* (CBIR) systems in order to face the depicted issue. Considering the ever-increasing amount and availability of image and video files, it became paramount to develop appropriate information systems that can effectively handle such huge data sets with the most suitable relevance scores for the tasks [2], also considering how image searching has become one of the most convenient techniques. We can broadly group CBIR techniques based on the level of retrieval, i.e. instance level and category level. In the first case, the system receives a query image of a specific object or scene to locate images of the same typology [3]. On the contrary, category-level image retrieval aims to find images of the same class as the query one [4].

© Springer Nature Switzerland AG 2021
N. Tsapatsoulis et al. (Eds.): CAIP 2021, LNCS 13052, pp. 287–297, 2021.
https://doi.org/10.1007/978-3-030-89128-2_28

In both cases, the image features are the basis for CBIR, representing specific properties of the images. They can be either global for the entire image or local for a reduced group of pixels. There exist several possible CBIR applications [5] and medical image search [6] is one of them. In this paper, we focus on how to exploit several classic and deep image descriptors to propose a new medical diagnosis-related CBIR system in such a way that a pathologist can input an unseen query image to obtain similar images e.g. belonging to a specific disease. On the contrary, several works focus on retrieval based solely on visual similarity, without considering whether an image represents a particular pathology [7–9]. Concretely, we performed several evaluations on a lung Computed Tomography (CT) data set and two brain Magnetic Resonance (MR) data sets. We also pinpoint that we used deep features without explicitly training or fine-tuning new learners. Our contribution is fourfold: *i)* we examined and compared several image descriptors to face a category-level image retrieval problem, oriented to medical diagnosis; *ii)* we tested several similarity measures to discover the most appropriate for this task; *iii)* we investigated the extent to which feature selection may be beneficial to lead to an increase in retrieval performance; *iv)* we combined heterogeneous features to discover if it can improve the produced retrieval results.

The remainder of this work is organised as follows. Section 2 discusses the role of image retrieval and its applications in the medical field, while the components of our image retrieval system are described in Sect. 3. In Sect. 4 we illustrate our experimental evaluation, including the used data set and the experimental setup. The results are presented and discussed in Sect. 5, and finally, in Sect. 6 we draw the conclusions and directions for future works.

2   Related Works

Image analysis and processing in the medical image field aim to offer tools for diagnosis, lesion detection, or early detection of life-threatening illness. In addition to them, several authors also investigate CBIR systems because they can have impactful clinical applications. A considerable number of works on medical image retrieval concerns radiographic images, such as mammograms, plain X-rays, CT, Positron Emission Tomography (PET), and MR images. In this field, traditional features, e.g., shape [5], invariant moments [10] and texture [11], or Scale-Invariant Feature Transform (SIFT) are highly representative and used [12], especially before the Convolutional Neural Network (CNN) AlexNet outbreak [13]. Since 2012, in fact, also deep learning has impacted image retrieval applications due to its powerful feature representations [14].

Speaking about CT and MR, Ma et al. [15] proposed a CBIR system for CT images built on SVM classifier trained with low-level features, including the bag-of-visual-words (BoVW) based on the HOG, wavelet features, Local Binary Pattern (LBP) and histogram of CT values. Kumar et al. [9] proposed the use of Zernike moments for CT and MR images retrieval and compared its performance with standard benchmark approaches, like LBP and derivates. The authors in [16,17] proposed region-based CBIR systems and new distance metric methods. In particular, the first one takes into account different region of interest (ROI) patches from every single MR image. Its purpose is to extract discriminative BoVW features from the key regions of the images to distinguish among the different categories. The second one also works on ROIs of MR with tumours. Moreover, the authors proposed a margin information descriptor (MID) to characterise intensity variation of the tumour or surrounding region and perform a brain category retrieval task.

Concerning the deep features, the authors in [8] proposed to detect irrelevant image blocks in each medical image class via analysing the error histogram of an autoencoder to reduce the dimensionality of the features for image retrieval. At the same time, Mbilinyi et al. [18] analysed the effectiveness of the deep features extracted in retrieving similar medical images by examining different CNNs pretrained on natural images and CNNs fine-tuned and fully-trained from scratch on medical images, from a collection with different medical image modalities.

In this scenario, we also pinpoint the extensive use of radiography, particularly chest X-rays (CXR), for COVID-19 patient triaging, diagnosing, and monitoring to face the public health crisis of the novel coronavirus disease 2019 (COVID-19) pandemic. In particular, a recent effort in this sense has been realised by Zhong et al. [19]. They proposed a metric learning-based CBIR model for analysing CXR images, founded on Resnet50 [20] as the backbone architecture for the feature extraction.

3   Our Image Retrieval System

This section describes the components used to create our image retrieval system: feature extraction, feature selection, and image ranking.

3.1   Feature Extraction

We evaluated different feature sets that we grouped in three important classes: invariant moments, texture features and deep features.

As **Invariant Moments** we computed the Hu, Zernike, Legendre and Chebyshev moments. *Hu moments* [21] are invariant to scale, translation and rotation changes. *Legendre moments* are orthogonal moments first introduced by Teague [22]. Also, *Zernike moments* are orthogonal to each other and represent image properties without redundancy [23]. *Chebyshev moments* are derived from Chebyshev polynomials and can extract global features in an image by varying moment order [24]. Here we extracted moments of first (Chebyshev 1) and second kind (Chebyshev 2). The order of the moments is equal to 5 for all the mentioned invariant moments. A higher-order would have decreased system performance, adding features representative of irrelevant details or noise [24]. As **Texture Features**, we computed the rotation invariant Gray Level Co-occurrence Matrix (GLCM) features as proposed in [25] and the rotation invariant LBP features [26]. In both cases we focused in fine textures, thus we computed four GLCMs with $d = 1$ and $\theta = [0°, 45°, 90°, 135°]$ and the LBP map in the neighbourhood identified by r and n equal to 1 and 8 respectively. From the GLCMs we extracted thirteen features [27] and converted into rotationally invariant ones Har_{ri} (for more details see [25]). The LBP map is converted into a rotationally invariant one, and its histogram is used as a feature vector LBP_{ri} [26]. Finally, we extracted the **CNN features** from four different well-known network architectures: AlexNet, Vgg19Net, ResNet50 and GoogleNet. AlexNet [28] and Vgg19 [29] are the shallowest among the tested architectures, being composed of 8 and 19 layers, respectively. In both cases, we extracted the features from the second last fully connected layer (fc7) for a feature vector of size $h = 4096$. ResNet50 [20] and GoogleNet [30] are much more deeper, being composed of 50 and 100. In both cases, we extracted the features from the one fully connected layer for a feature vector of size $h = 1000$. We extracted features directly from the CNNs without a fine-tuning process [31], since they have already learnt a sufficient representational power and generalisation ability to perform different visual recognition tasks [31,32].

3.2 Feature Selection

The feature selection step has been implemented with the *ReliefF* [33] algorithm. It returns a vector containing all the k selected features sorted according to their relevance. ReliefF algorithm penalises the features that differentiate neighbours of the same class or group neighbours of different classes and rewards features that group neighbours of the same class or differentiate neighbours of different classes.

3.3 Image Ranking

To rank the images according to the similarity to a given query, we evaluated several distance measures: Euclidean, cityblock, Chebyshev, cosine, correlation, Spearman, Canberra [34], D1 [35], χ^2, and KLDiv (Kullback-Leibler or Jensen-Shannon divergence) [36]. Given two images, I_i and I_j, and their respective feature vectors, F_i and F_j, of size L, the distance D_{ij} between them can be calculated using the mentioned metrics as listed in Table 1. Obviously, being based on distances, the images are ranked in ascending order, from the least to the most distant to a given query.

4 Experimental Evaluation

This section describes the experimental evaluation carried out to assess the performances of our system for MR and CT medical image retrieval. In the following, we first describe the data set used and then the experimental set-up.

Table 1. Equations of the considered distance measures.

Euclidean	$D_{ij} = \sqrt{\sum_{k=1}^{L}(F_i^k - F_j^k)^2}$		cityblock	$D_{ij} = \sum_{k=1}^{L} |F_i^k - F_j^k|$
KLDiv	$D_{ij} = \sum_{k=1}^{L} F_i^k \times log(F_i^k/F_j^k)$		cosine	$D_{ij} = 1 - \frac{F_i F_j'}{\sqrt{(F_i F_i')(F_j F_j')}}$
correlation	$D_{ij} = 1 - \frac{(F_i - \overline{F_i})(F_j - \overline{F_j})'}{\sqrt{(F_i - \overline{F_i})(F_i - \overline{F_i})'}\sqrt{(F_j - \overline{F_j})(F_j - \overline{F_j})'}}$,		Canberra	$D_{ij} = \sum_{k=1}^{L} \frac{|F_i^k - F_j^k|}{|F_i^k| + |F_j^k|}$
where	$\overline{F_i} = \frac{1}{L}\sum_{k=1}^{L} F_i^k, \overline{F_j} = \frac{1}{L}\sum_{k=1}^{L} F_j^k$		D1	$D_{ij} = \sum_{k=1}^{L} |\frac{F_i^k - F_j^k}{1 + F_i^k + F_j^k}|$
Spearman	$D_{ij} = \frac{1+r}{2}$, where $r = \frac{S_{ij}}{\sqrt{S_i S_j}}$,		χ^2	$D_{ij} = \sum_{k=1}^{L} \frac{(F_i^k - F_j^k)^2}{F_i^k + F_j^k}$
	$S_i = \sum_{k=1}^{L}(F_i - \overline{F_i})^2, S_j = \sum_{k=1}^{L}(F_j - \overline{F_j})^2,$		Chebyshev	$D_{ij} = max_k|F_i^k - F_j^k|$
	$S_{ij} = \sum_{k=1}^{L}(F_i - \overline{F_i})(F_j - \overline{F_j})$			

4.1 Data Sets

We now describe the three benchmark data sets, illustrated in Fig. 1.

Emphysema-CT (EMP) [37] is a CT database created to diagnose diseases related to emphysema. The images data set has been created by selecting 168 square patches (size 61×61) that the physicians have manually annotated into three main categories: Normal Tissue (NT - 59 observations), Centri Lobular Emphysema (CLE - 50) and Para Septal Emphysema (PSE - 59). **Brain-Tumor-MRI** (BRAIN) [38] is a brain MR database containing different images section from 233 patients. Here, we used the axial sections subset containing 870 images (size 512×512), including 143 meningiomas, 436 gliomas, and 291 pituitary tumours. **OASIS-MRI** (OASIS) [39] also contains brain MRI data but collected to study the Clinical Dementia Rating (CDR). Again we focused on the axial section images only, for a total of 436 images, divided into 336 representing cognitively healthy individuals and 100 with subtypes of dementia (70 with very mild, 28 with mild, and 2 with moderate).

4.2 Experimental Setup

To make the experiments repeatable and objective, we automated the retrieval process by considering the images belonging to the same class of the query image as being relevant. Although in real cases the users/physicians would like to examine a reduced number of images k, we used the following values; $k = 1, 10, 20, 50, 100$ to make an exhaustive comparison, however. In addition, to obtain the best parameters in terms of feature set and similarity measure, we performed a number of queries equal to the size of the data set, n. Thus, iteratively, we extracted an image from the data set and used it as a query Q on the remaining $n - 1$ data set images. Finally, we evaluated the retrieval performances using tree well known metrics, that are: *Precision* (P), *Recall* (R) and *mean Average Precision* (mAP) (for more details see [31]).

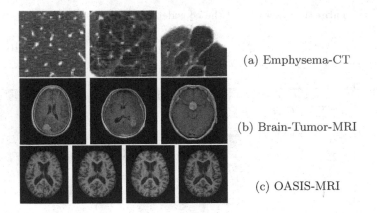

(a) Emphysema-CT

(b) Brain-Tumor-MRI

(c) OASIS-MRI

Fig. 1. Illustrations of the data sets classes. From left to right, (a) shows NT, CLE and PSE; in (b): meningioma, glioma and pituitary tumours. Finally, (c) represents four classes of dementia: absent, very mild, mild and moderate.

5   Results

We performed several experiments with the previously mentioned data sets, mainly devoted to evaluating the performances of the image retrieval system exploiting different feature sets. However, firstly we performed an analysis focused on the similarity measures and the percentage of feature selected. In Fig. 2 we report the mAP trends of each similarity distance against feature selection percentage. It appears clearly how the globally more performing similarity measure is the cityblock, although some others also exhibit satisfactory robustness. Also, it is quite evident that the best selection amount is 20% for both texture and deep features, while, in general, the invariant moments do not benefit from feature selection.

Once established the appropriate distance measure and the selection percentage for each descriptor, we performed an accurate comparison between single descriptors and combinations of heterogeneous descriptors to assess their retrieval performances. For the sake of brevity, in Table 2 we reported the performances obtained on each data set by single descriptors (one for each category) and the combination of descriptors that brought to the best results. As it can be observed, with few exceptions (particularly for the invariant moments), the best performing features are almost the same for all the tested data set. In particular, HAR_{ri} appears for every data set, and ResNet50 features appear both individually for two out of three data sets and among the best feature combination for all the data sets. On the contrary, for the invariant moments' category, none descriptor outperformed the others, but rather, for each data set, a different descriptor appear on top, even though with performance close to the others. For what concerns values, we observe that there is not a clear gap between the different descriptors. For Emphysema-CT, only the deep features outperform the others significantly, while in the other cases, the performance

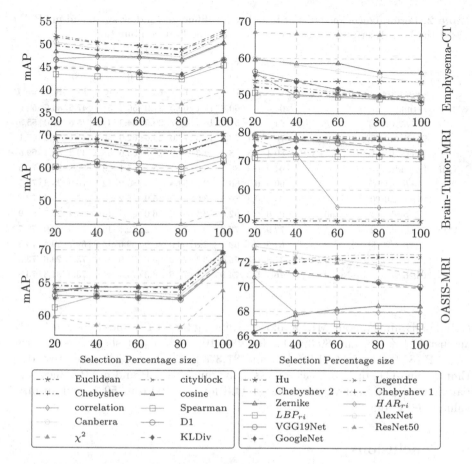

Fig. 2. mAP trends with different percentages of feature selection. On the left, we compare different similarity measures averaged for all the descriptors. On the right, we report every single descriptor computed with the cityblock measure.

values are almost comparable, especially for the mAP value. Nevertheless, for small k values, the deep features, even more in combination with other features, show their effectiveness, making them the most suitable for a retrieval system for diagnosis support. These results confirm what reported in [18], i.e. that deep features extracted from ResNet50 produce excellent performances in this task, even though, in contrast to [18], we did not perform any fine-tuning strategy on the used data sets.

Unfortunately, we can make few comparisons with the state-of-the-art since most of the existing system focused on retrieving organs/sections rather than diseases. Others instead used a more complex pipeline that exploits a semi-automated patch extraction or image segmentation to extract the features from the region of interest only [16,17]. However, our approach proved to be competitive, even if less structured and more generic than other works. Indeed, when

Table 2. Performances in terms of P and R for different window sizes, and mAP. We report the best descriptor for each category, and the best combination.

		Windows size									
		1		10		20		50		100	
	Descriptor	P	R	P	R	P	R	P	R	P	R	mAP
EMP	Zernike	64.3	1.2	60.8	11.1	60.0	21.9	56.4	51.4	47.6	86.6	59.6
	HAR_{ri}(20%)	64.3	1.2	56.3	10.2	56.7	20.5	54.3	49.2	45.6	83.0	55.9
	ResNet50(20%)	76.2	1.4	73.0	13.3	70.0	25.6	62.0	56.6	49.1	89.2	67.1
	Chebyshev 1-LBP_{ri}-ResNet50	80.4	1.5	78.4	14.3	75.2	27.6	63.6	58.2	49.2	89.4	69.6
BRAIN	Legendre	94.0	0.3	88.5	2.9	87.1	5.7	85.1	13.7	82.8	26.3	78.6
	HAR_{ri}(20%)	90.9	0.3	86.7	2.8	84.5	5.4	80.9	12.7	76.9	23.8	72.5
	ResNet50(20%)	95.5	0.3	89.8	3.0	88.0	5.9	85.2	14.1	83.0	26.9	79.1
	Chebyshev 1-ResNet50	98.6	0.3	92.0	3.1	90.2	6.0	87.8	14.4	85.4	27.5	81.9
OASIS	Chebyshev 2	72.9	0.3	73.4	3.0	72.6	6.2	72.1	15.2	71.9	29.9	72.7
	HAR_{ri}(20%)	67.9	0.3	69.1	2.8	69.6	5.9	69.6	14.9	69.7	29.4	70.8
	AlexNet(20%)	77.3	0.3	74.5	3.1	74.2	6.0	73.9	15.1	73.4	29.6	73.3
	Chebyshev 1-ResNet50	76.4	0.3	74.9	3.1	74.6	6.3	74.2	16.1	73.9	31.3	73.8

tested on Brain-Tumour-MRI it allows achieving better P values for small k (92% against 87.5% [16] and 89.3% [17] for $k = 10$) with a just slight loss in terms of mAP (81.9% against 91.0% [16] and 87.3% [17]). Last of all, when tested on Emphysema-CT, our approach allows achieving better P and mAP values compared to [9] that achieve a P value of 44.8 for $k = 100$ (our 59.2) and an mAP value of 60.2 (our 69.6).

6   Conclusions

In this work, we designed a CBIR system oriented to medical diagnosis, specifically CT and MR image retrieval. We evaluated different descriptors, including hand-crafted (invariant moments and texture) and deep features. It is important to notice that we used deep features without explicitly training or fine-tuning new learners, which make our system not specific for a single target. We also exploited a feature selection step to evaluate which features are more informative or present redundant data or noise. After a systematic evaluation of three image data sets, which included: *i)* comparison of different distance measures, *ii)* investigation of different percentage of feature selection, *iii)* comparison of single descriptors, and *iv)* combination of heterogeneous descriptors, we determined the optimal components for our CBIR system. In particular, we defined the appropriate distance measure, the cityblock, and the appropriate selection percentage for each descriptor, in the quantity of 20% for deep and texture features and 100% (none selection) for invariant moments. Finally, we established the most performing single descriptor and combination of descriptors, that in both cases include deep features, specifically from ResNet50. Indeed the obtained results demonstrated the effectiveness of such descriptors, even compared with other state-of-the-art approaches exploiting a more complex pipeline. As future work, we plan to extend our study to analyse different medical images, such as microscopic or X-ray images. We also aim to explore distance metric learning methods to further improve our proposed CBIR system. An additional extension of the proposed method could be a patch-based image representation in order to handle large images with mainly normal tissue regions.

References

1. Lew, M.S., Sebe, N., Djeraba, C., et al.: Content-based multimedia information retrieval: state of the art and challenges. ACM TOMM **2**(1), 1–19 (2006)
2. Putzu, L., Piras, L., Giacinto, G.: Ten years of relevance score for content based image retrieval. In: Perner, P. (ed.) MLDM 2018. LNCS (LNAI), vol. 10935, pp. 117–131. Springer, Cham (2018). https://doi.org/10.1007/978-3-319-96133-0_9
3. Zheng, L., Yang, Y., Tian, Q.: Sift meets CNN: a decade survey of instance retrieval. IEEE Trans. PAMI **40**(5), 1224–1244 (2017)
4. Guillaumin, M., Mensink, T., Verbeek, J., et al.: TagProp: discriminative metric learning in nearest neighbor models for image auto-annotation. In: IEEE ICCV, pp. 309–316 (2009)
5. Di Ruberto, C., Morgera, A.: Moment-based techniques for image retrieval. In: 2008 19th International Workshop on Database and Expert Systems Applications, pp. 155–159 (2008)
6. Cao, Y., et al.: Medical image retrieval: a multimodal approach. Cancer Inf. 13s3, CIN.S14053 (2014)
7. De Oliveira, J.E., Machado, A.M., Chavez, G.C., Lopes, A.P.B., Deserno, T.M., Araújo, A.D.A.: MammoSys: a content-based image retrieval system using breast density patterns. Comp. Meth. Prog. Biom. **99**(3), 289–297 (2010)
8. Çamlica, Z., Tizhoosh, H., Khalvati, F.: Autoencoding the retrieval relevance of medical images. In: International Conference IPTA, pp. 550–555 (2015)
9. Kumar, Y., Aggarwal, A., Tiwari, S., Singh, K.: An efficient and robust approach for biomedical image retrieval using Zernike moments. Biom. Sig. Proc. Cont. **39**, 459–473 (2018)
10. Karakasis, E., Amanatiadis, A., Gasteratos, A., Chatzichristofis, S.: Image moment invariants as local features for content based image retrieval using the bag-of-visual-words model. Patt. Recog. Lett. **55**, 22–27 (2015)
11. Di Ruberto, C., Loddo, A., Putzu, L.: Histological image analysis by invariant descriptors. In: International Conference ICIAP, vol. 10484, pp. 345–356 (2017)
12. Shi, M., Avrithis, Y., Jégou, H.: Early burst detection for memory-efficient image retrieval. In: IEEE International Conference CVPR, pp. 605–613 (2015)
13. Krizhevsky, A., Sutskever, I., Hinton, G.E.: ImageNet classification with deep convolutional neural networks. In: NeurIPS, pp. 1097–1105 (2012)
14. Wan, J., et al.: Deep learning for content-based image retrieval: a comprehensive study. In: ACM MM, pp. 157–166 (2014)
15. Ma, L., Liu, X., Gao, Y., et al.: A new method of content based medical image retrieval and its applications to CT imaging sign retrieval. J. Biom. Inf. **66**, 148–158 (2017)
16. Huang, M., Yang, W., Wu, Y., et al.: Content-based image retrieval using spatial layout information in brain tumor T1-weighted contrast-enhanced MR images. PLoS ONE **9**(7), 1–13 (2014)

17. Yang, W., Feng, Q., Yu, M., et al.: Content-based retrieval of brain tumor in contrast-enhanced MRI images using tumor margin information and learned distance metric. Med. Phys. **39**(11), 6929–42 (2012)

18. Mbilinyi, A., Schuldt, H.: Cross-modality medical image retrieval with deep features. In: IEEE International Conference BIBM, pp. 2632–2639. IEEE Computer Society (2020)

19. Zhong, A., Li, X., Wu, D., et al.: Deep metric learning-based image retrieval system for chest radiograph and its clinical applications in covid-19. Med. Im. Anal. **70**, 101993 (2021)

20. He, K., Zhang, X., Ren, S., Sun, J.: Deep residual learning for image recognition. In: IEEE International Conference CVPR, pp. 770–778 (2016)

21. Hu, M.K.: Visual pattern recognition by moment invariants. IRE Trans. Inf. Theory **8**(2), 179–187 (1962)

22. Teague, M.R.: Image analysis via the general theory of moments. J. Opt. Soc. Am. **70**(8), 920–930 (1980)

23. Oujaoura, M., Minaoui, B., Fakir, M.: Image annotation by moments. In: Moments and Moment Invariants - Theory and Applications, vol. 1, pp. 227–252 (2014)

24. Di Ruberto, C., Putzu, L., Rodriguez, G.: Fast and accurate computation of orthogonal moments for texture analysis. Patt. Recogn. **83**, 498–510 (2018)

25. Putzu, L., Di Ruberto, C.: Rotation invariant co-occurrence matrix features. In: International Conference, ICIAP, vol. 10484, pp. 391–401 (2017)

26. Ojala, T., Pietikäinen, M.T.M.: Multiresolution gray-scale and rotation invariant texture classification with local binary pattern. IEEE Trans. PAMI **24**(7), 971–987 (2002)

27. Haralick, R.M., Shanmugam, K., Dinstein, I.H.: Textural features for image classification. IEEE Trans. Syst. Man Cybern. SMC-**3**(6), 610–621 (1973)

28. Krizhevsky, A., Sutskever, I., Hinton, G.E.: ImageNet classification with deep convolutional neural networks. In: International Conference, NIPS. vol. 1, pp. 1097–1105 (2012)

29. Simonyan, K., Zisserman, A.: Very deep convolutional networks for large-scale image recognition. In: International Conference ICLR (2015)

30. Szegedy, C., Liu, W., Jia, Y., et al.: Going deeper with convolutions. In: IEEE Conference CVPR, pp. 1–9 (2015)

31. Putzu, L., Piras, L., Giacinto, G.: Convolutional neural networks for relevance feedback in content based image retrieval. Mult. Tools Appl. **79**, 26995–27021 (2020)

32. Donahue, J., Jia, Y., Vinyals, O., et al.: DeCAF: a deep convolutional activation feature for generic visual recognition. In: International Conference ML, pp. 647–655 (2014)

33. Robnik-Sikonja, M., Kononenko, I.: Theoretical and empirical analysis of ReliefF and RReliefF. Mach. Learn. **53**, 23–69 (2003)

34. Jurman, G., Riccadonna, S., Visintainer, R., Furlanello, C.: Canberra distance on ranked lists. In: International Workshop Advances in Ranking - NIPS, pp. 22–27 (2009)

35. Dubay, S., Singh, S., Singh, R.: Local diagonal extrema pattern: a new and efficient feature descriptor for CT image retrieval. IEEE Sign. Proc. Lett. **22**(9), 1215–1219 (2015)

36. Rubner, Y., Tomasi, C., Guibas, L.: The earth mover's distance as a metric for image retrieval. Int. J. CV **40**(2), 99–121 (2000)

37. Sørensen, L., Shaker, S.B., de Bruijne, M.: Quantitative analysis of pulmonary emphysema using local binary patterns. IEEE Trans. Med. Imag. **29**(2), 559–569 (2010)
38. Cheng, J., et al.: Enhanced performance of brain tumor classification via tumor region augmentation and partition. PloS One **10**(12), 1–13 (2015)
39. Marcus, D.S., Wang, T.H., Parker, J., et al.: Open access series of imaging studies (OASIS): cross-sectional MRI data in young, middle aged, nondemented, and demented older adults. J. Cogn. Neuros. **19**(9), 1498–1507 (2007)

Toward Multiwavelet Haar-Schauder Entropy for Biomedical Signal Reconstruction

Malika Jallouli[1](✉), Wafa Belhadj Khalifa[1](✉), Anouar Ben Mabrouk[2,3](✉), and Mohamed Ali Mahjoub[1](✉)

[1] Université de Sousse, Ecole Nationale d'Ingénieurs de Sousse, LATIS Laboratory of Advanced Technology and Intelligent Systems, 4023 Sousse, Tunisia
`wafa.belhajkhelifa@isamm.uma.tn`, `mohamedali.mahjoub@eniso.rnu.tn`
[2] Department of Mathematics, Higher Institute of Applied Mathematics and Computer Science, University of Kairouan, 3100 Kairouan, Tunisia
`anouar.benmabrouk@fsm.rnu.tn`
[3] Department of Mathematics, Faculty of Science, University of Tabuk, Tabuk, Saudi Arabia

Abstract. In this paper, a wavelet/multiwavelet approach is proposed for biosignal reconstruction based on a correlation between the two well known Haar and Schauder wavelets called Haar-Schauder multiwavelet. A multiwavelet entropy is then proposed in order to optimize and evaluate the order/disorder of the reconstructed signals. Finally, an experimentation step is carried on ECG and EMG biosignals to validate the proposed approaches.

Keywords: Biosignals · ECG · EMG · Wavelet · Mutliwavelet · Entropy

1 Introduction

The term biosignal is often used to refer to bioelectrical signals. The usual understanding is to refer only to time-varying signals, although spatial parameter variations (e.g. the nucleotide sequence determining the genetic code) are sometimes subsumedas well. Thus, among the best-known bioelectrical signals are: Electrocardiogram (ECG) and Electromyogram (EMG). The ECG signal is a quick exam that takes only a few minutes, it is painless and non-invasive. However, its interpretation remains complex and requires methodical analysis and clinical experience. EMG signal describes the muscle electrical signals. It is sometimes referred to a myoelectric activity. In fact, muscle tissue conducts electrical potentials similar to the nerves way.

These very specific signals constitute an important source of information for practitioners. They are graphic representation of the functional behavior of several organs (such as the heart or the brain). However, the amount of data produced makes their interpretation rather cumbersome and requires considerable time and experience.

© Springer Nature Switzerland AG 2021
N. Tsapatsoulis et al. (Eds.): CAIP 2021, LNCS 13052, pp. 298–307, 2021.
https://doi.org/10.1007/978-3-030-89128-2_29

Therefore, it is important to find a process to model these biosignals. The obtained biosignal models should be faithful to the initial signal, compact and relevant in order to facilitate their interpretation and help practitioners to make a precise diagnosis based on numerical parameters.

Classical techniques, in general, analyse the signal over long periods thus they are not adequate to model impulse events. High variability and the necessity to combine features temporally well localized with others well localized in frequency remain the most important challenges in biomedical signal modeling ([1, 3, 4]). Wavelet transform provides the ability to localize the information in the time/frequency plane ([2, 16]).

Multiwavelets' concept is introduced to more ameliorate wavelet tools and to achieve more efficiency a multi camera wavelet system. Theoretically speaking, it is possible to build multiwavelets that have simultaneously desirable properties such as orthogonality, compact support and symmetry [7, 10, 12]. In almost all these existing studies, there is quietly one way to construct them.

In this work, we aim to propose a recently developed optimized modeling process for biosignals in order to reconstruct them via a compact set of representative parameters. One of our aims is to develop a concrete modeling process based on multiwavelets defined from two single wavelets such as Haar and Schauder.

The main problem of signal reconstruction is the non-possibility to fix a priory optimal reconstruction order. The abstraction of Shannon entropy as a measure of randomness provides a framework for calculating the minimum amount of data required to describe an image without loss of information ([18]). Ruppert-Felsot et al. [17] and [8] provided a clear description of entropy as a tool for searching a basis in discrete wavelet packets that carries substantial information about the turbulent flow fields. The wavelet entropy proposed there provided a measure of the homogeneity of the spectral distribution of a signal from discrete wavelet transform.

In this paper, we propose a multiwavelet associated entropy inspired from Shannon entropy to define in a precise way the end of the modeling process of biomedical signals with Haar-Schauder (HSCH) multiwavelets.

The results of the reconstruction method, based on Haar-Schauder multiwavelet and optimized by HSCH entropy, are compared with those obtained by the classical Haar and Schauder wavelets.

This paper is organized as follow. A theoretical description of the proposed method based on a new modeling process of biomedical signals using HSCH multiwavelet associated entropy is provided in Sect. 2. In Sect. 3, experimentations are conducted especially on ECG and EMG signals to prove the effectiveness of the proposed theory. We conclude afterward.

2   HSCH Multiwavelet Modeling and Associated Entropy

The proposed method in the present paper is based on a combination of two different scaling functions (Haar and Schauder). The generated multiscale function is a new multiwavelet called HSCH multiwavelet. Its first component is the

Haar scaling function $\varphi_1 = \chi_{[0\ 1[}$, which describes well the piece-wise constant signals. The second one is the Schauder scaling function $\varphi_2 = (1 - |x|)\chi_{[-1\ 1[}$ which is more adapted to linear-wise signals. The choice of these basic wavelets is motivated by their closeness to discrete signals, which are usually represented as piece-wise linear functions, and their histograms which look like Haar wavelet series. Furthermore, Haar and Schauder wavelets have several interesting features ([14]). They are both compactly supported, so as the associated multi-wavelet. They are both explicit and their computational applications are fast and easy. In fact, computing HSCH multiwavelet coefficients may be performed by recursive averaging and differentiating coefficients. Haar and Schauder's nonzero recursion coefficients are reduced (2 or 3) and adequate for any type of signals. HSCH multiscaling function has exactly three nonzero low-pass filter elements H_k, $k = -1, 0, 1$ and thus three nonzero elements for the associated high-pass filter H_k, $k = -1, 0, 1$. The computation of these filters is easy and provided in [19,20]. The HSCH multiscaling function is simply and explicitly the vector valued-function $\Phi_{HSCH} = \left(\varphi_1\ \varphi_2 \right)^T$.

The next concept to be applied here is the well-known entropy measure. The stating idea is due to Shannon entropy which measures the uncertainty of a random process [18], and which is classically defined by

$$S = -\sum_{i=1}^{L} p_i \ln(p_i), \tag{1}$$

where p_i is the probability of occurrence of the $i - th$ event.

The evolution of the signal during the reconstruction process can be considered as being a phenomenon whose stability can be determined by a stationary value of a quantity proposed in this article as a multiwavelet variant of the entropy, and which will be called multiwavelet entropy (MWEnt). This is based on the multiwavelet decomposition of signals into multiwavelet series.

The decomposition of the signal S at the level J using HSCH multiwavelet will be obtained by [11,20]

$$S_{HSCH}^{J} = \sum_{k} A_{J,k}^{HSCH} \Phi_{J,k} + \sum_{j=0}^{J} \sum_{k} D_{j,k}^{HSCH} \Psi_{j,k}, \tag{2}$$

where $A_{J,k}^{HSCH}$ is the approximation coefficient of the signal at the level J and the position k, and $D_{j,k}^{HSCH}$ is the detail coefficient of the signal at the level j and the position k. However, it is important to determine in a precise and efficient way the optimal order J which makes it possible to have a reconstructed signal that is compact and faithful to the initial signal S. This is one motivation to introduce the concept of MWEnt. In fact, a stabilization of the value of MWEnt over several iterations of the reconstruction process allows to deduce an optimal order of a stable reconstruction. Indeed, as the entropy in its general definition as well as mathematical meaning is a type of dimension, so it should be somehow a global measure of invariance for our reconstruction process.

A special expression of entropy by means of multiwavelets is proposed. For a signal S, the multiwavelet entropy will be as follows (see also [19,20]). For J fixed as level, we denote the J-level MwEnt or simple the J-MwEnt,

$$MwEnt(J) = -\sum_{j=0}^{J} p_j^{mw} \log(p_j^{mw}), \quad L \in \mathbb{N} \tag{3}$$

where p_j^{mw} is the j-proportion of multiwavelet energy at the level j expressed by means of the multiwavelet coefficients

$$p_j^{mw} = \frac{MwEnr_j}{MWEnr_{tot}}, \tag{4}$$

where the $MwEnr_j$ is the energy at the level j defined by means of the multiwavelet approximation coefficients contributing at the level j as

$$MWE_j = \sum_k \|A_{j,k}^{HSCH}\|^2 \tag{5}$$

and where $MwEnr_{tot}$ is the total energy of the signal,

$$MwEnr_{tot} = \sum_{j,k} \|A_{j,k}^{HSCH}\|^2. \tag{6}$$

The norms $\|A_{j,k}^{HSCH}\|$ used above are the usual matrix norms. The proposed reconstruction process in the present paper is an iterative one. It is based on HSCH multiwavelets modeling ([19,20]). At each iteration, multiwavelet associated entropy is applied in order to reach an optimal reconstruction order. The organigram in Fig. 1 illustrates the proposed method.

Applying HSCH multiwavelet in the processing resides for a level J of decomposition a number of positions k such that $0 \leq k \leq 10.2^J$. In this case the decomposition of ECG and EMG for a level J is

$$S_{HSCH}^J = AS_J^{HSCH} + \sum_{j=0}^{J} DS_j^{HSCH}, \tag{7}$$

where AS_J^{HSCH} is the projection of the signal on the approximation space V_J^{HSCH} corresponding to the HSCH multiscaling function, and the DS_j^{HSCH} are respectively the projections of the signal on the detail spaces W_j^{HSCH} relative to HSCH multiwavelets respectively.

To illustrate the closeness of S_{HSCH}^J to the original signal S, we computed a Normalized Average Quadratic Error defined on a grid of N points as

$$Er_J = \frac{\sum_{i=1}^{N}(S(i) - S_{HSCH}^J(i))^2}{\sum_{i=1}^{N} S^2(i)}. \tag{8}$$

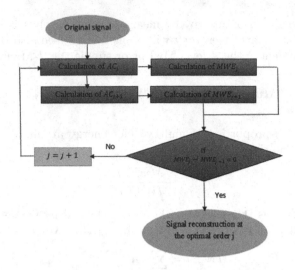

Fig. 1. Biosignal modeling steps using the MwEnt.

3   Experimentations

In order to validate the proposed approach, we applied the new multiwavelet for the decomposition of the well known ECG and EMG signals. In this experimental part, a reconstruction process is developed. Next, the multiwavelet entropy is computed to evaluate the most appropriate reconstruction order for each signal considered. The ECG and EMG signals used in this section are obtained from the physionet MIT-BIH Arrhythmia database ([9]).

3.1   Biosignal Reconstruction

In this section, we start with a comparison between results obtained on the ECG signal and ones obtained by classical simple Haar and Schauder wavelets. Then we continue by exposing experimentation on EMG biosignal.

ECG Signal Reconstruction. Figure 2 illustrates an example of an ECG signal. We propose to apply first a reconstruction process based on single Haar and Schauder wavelets. Then, the proposed HSCH multiwavelet method is used in order to evaluate the contribution of the new method compared to the classical ones.

Figure 3 illustrates the decomposition of the ECG signal by means of Haar wavelet, Schauder wavelet and HSCH multiwavelet for different levels. This makes it possible to visualize the evolution of the reconstruction process according to the classical methods due to Haar wavelet and schauder wavelet transform,

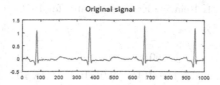

Fig. 2. Initial ECG biosignal.

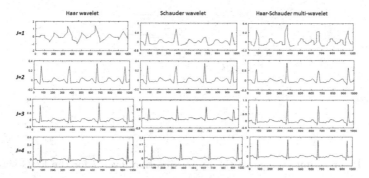

Fig. 3. Reconsruction of the ECG signal by Haar wavelet, Schauder wavelet and HSCH multi-wavelet

then according to the proposed method based on HSCH multiwavelet transform. We notice that, for the three methods, the reconstructed signal fits more and more the original one as the level J increases. However, the difference in reconstruction resides in the precision of the reconstructed signal compared with the initial signal. In this direction, the HSCH multiwavelet method shows a reconstructed signal more closer to the initial one regarding classical methods for the same order J.

Table 1 shows the estimations of the Normalised Average Quadratic Error between the real ECG signal and the reconstructed one obtained by the use of Haar wavelet, Schauder wavelet and HSCH multiwavelet decomposition at different levels.

We notice the improvement of the error as the level J increases even if the increase is not as fast for both wavelet transforms. However, classical Haar wavelet presents values of NAQE greater than the multiwavelet even the same level J. Numerical tests show that the level $J = 4$ is the best one for HSCH multiwavelet method. For $J > 4$, we noticed that perturbations yield a small increase in the NAQE.

Figure 4 illustrates different curves that represent the evolution of the value of NAQE as a function of the reconstruction level for Haar wavelet, Schauder wavelet ans HSCH multiwavelet. It is clear that the curve resulting from the multiwavelet method is clearly closer than the other curves, which shows that the reconstructed signal with HSCH multiwavelet is more faithful to the initial signal.

Table 1. Relative Error estimates for ECG signal.

Level	$J = 1$	$J = 2$	$J = 3$	$J = 4$
H-wavelet	0.0012	$8.7\ 10^{-4}$	$5.46\ 10^{-4}$	$5.01\ 10^{-4}$
Sch-wavelet	0.0014	$8.95\ 10^{-4}$	$6.18\ 10^{-4}$	$3.7\ 10^{-4}$
HSCH multiwavelet	$9.8\ 10^{-4}$	$7.4\ 10^{-4}$	$4.37\ 10^{-4}$	$1.09\ 10^{-4}$

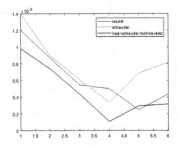

Fig. 4. Error estimates against the approximation level J for the ECG signal.

EMG Signal Reconstruction. We now propose to apply our HSCH multi-wavelet for EMG biosignals to prove that such reconstruction process can be generalized to other types of biosignals.

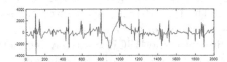

Fig. 5. Original EMG signal.

Figure 6 represents the evolution of the reconstructed signal as a function of the reconstruction order J. It is clear that the more the value of J increases, the more the reconstructed signal approaches the initial signal.

Table 2 shows the variation of the NAQE error according to the level of reconstruction. The optimum error is reached for the level $J = 6$.

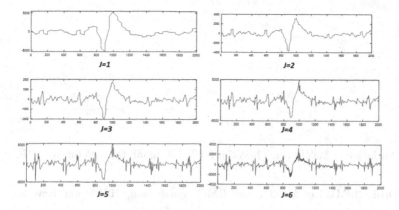

Fig. 6. EMG reconstructed with Haar-Schauder Multiwavelet.

Table 2. NAQE estimates for EMG with HSCH multiwavelet.

J	1–2	3	4	5	6
Er_J	10^{-2}	$2.08\ 10^{-4}$	$1.68\ 10^{-4}$	$1.7\ 10^{-5}$	$4.69\ 10^{-7}$

3.2   Multiwavelet Entropy

We now propose to apply our multiwavelets entropy ([19,20]) for the ECG and EMG signals to prove that such entropy may be an automatic black box allowing to reach the optimal reconstruction level fastly and accurately.

Assessment of the multiwavelet entropy allows to determine in a precise way the optimal level of reconstruction. Indeed, as entropy in its general form and definition as well as mathematical/physical meaning is a type of dimension. So, it should be a somehow global measure of invariance for the studied system. Its value should therefore be stationary or quietly constant as the multiresolution level increases.

Figure 7 illustrates the variation of the multiwavelet entropy according to the order of reconstruction for ECG and EMG signals shown in Fig. 2 and 5.

It is clear that the value of MWEnt becomes stationary for the three signals at a very precise value. This value defines the stabilization of the reconstructed signal. The optimal order is thus obtained.

In Table 3 the values of MWEnt are presented during the reconstruction process. This proves the stabilization of this value as soon as the optimal order is reached for the different types of signals.

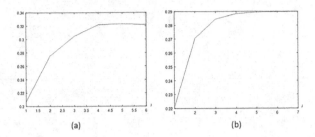

Fig. 7. Variation of MWEnt: (a) MWEnt for ECG signal, (b) MWEnt for EMG signal.

Table 3. MWEnt for ECG and EMG signals using HSCH multiwavelet.

J	1	2	3	4	5	6
MWEnt for ECG	0.21	0.27	0.30	0.31	0.31	0.31
MWEnt for EMG	0.17	0.25	0.27	0.29	0.296	0.299

4 Conclusion

The present paper proposes to apply a new concept of wavelet/multiwavelet entropy for biosignals reconstruction. Haar-Schauder multiwavelets have been examined for the purpose of biomedical signal reconstruction such as ECG and EMG. Next, a modified variant of Shannon entropy, based on multiwavelet decomposition, has been acted in order to obtain the optimal reconstruction order in a precise and efficient way. This processing optimized the reconstruction process by determining in advance the number of iterations required for the reconstruction. An experimental step is carried out using the physionet MIT-BIH Arrhythmia database. A comparison between classical Haar and Schauder single wavelets and the proposed HSCH multiwavelet shows better results with more precise and faithful reconstructed models for the last one. Several perspectives can be envisaged such as the application of the MwEnt to more complex signals and proposing other types of multiwavelets which could be more efficient in reconstruction. Some recent constructions of efficient wavelets/multiwavelets have been developed in [5,6] and have been also tested for biomedical purposes. These may be starting steps for an eventual future works by applying generally the concept of Clifford algebras in computer science.

References

1. Adison, P.S., Walker, J.S., Rodrigo, G.: Time-Frequency analysis of biosignals. JIEEE Eng. Med. Biol. Mag. **28**(5), 14–29 (2009)
2. Adison, P.S., Watson, J.N., Clegg, J.R., Olze, M., Sterz, F., Robertson, C.E.: Evaluation arrhythmias in ECG signals using wavelet transforms. IEEE Eng. Med. Biol. **19**(5), 104–109 (2000)

3. Alqudah, A.M.: An enhanced method for real-time modeling of cardiac related biosignals unsing Gaussian mixtures. J. Med. Eng. Technol. **41**(8), 600–611 (2017)
4. Andrea, R.V., Dorizzi, B., Boudy, J.: ECG analysis using hidden Markov models. IEEE Trans. Biomed. Eng. **53**(8), 1541–1549 (2006)
5. Arfaoui, S., Ben Mabrouk, A., Cattani, C.: New type of gegenbauer-hermite monogenic polynomials and associated clifford wavelets. J. Math. Imaging Vis. **62**(1), 73–97 (2020)
6. Arfaoui, S., Ben Mabrouk, A., Cattani, C.: New type of gegenbauer-jacobi-hermite monogenic polynomials and associated continuous clifford wavelet transform. Acta Appl. Math. **170**(1), 1–35 (2020). https://doi.org/10.1007/s10440-020-00322-0
7. Calandra, R., Brazile, B.S.: Multivariate Multiresolution Multiwevelets, Thesis in Mathematics. Texas Tech University (2009)
8. Fischer, P.: Multiresolution analysis for 2D turbulence Part 1: Wavelets vs cosince packets, a comparative study. Discret. Contin. Dyn. Syst. B. **5**(3), 659 (2005)
9. Goldberger, A., et al.: PhysioBank, physiotoolkit, and physionet: components of a new research resource for complex physiologic signals. Circulation [Online] **23**, e215–e220 (2000)
10. Hardin, D.P., Roach, D.W.: Multiwavelet prefiltersi: orthogonal prefilters preserving approximation order p-2. Analog and digital signal processing. IEEE Trans. Circ. Syst. - II **45**(8), 1106–1112 (1998)
11. Jallouli, M., Zemni, M., Ben Mabrouk, A., Mahjoub, M.A.: Towards New multiwavelets: associated filters and algorithms. part 1: Theoretical Framework and Investigation of Biomedical Signals, ECG and Coronavirus Cases. Soft Computing (2021)
12. Keinert, F.: Wavelets and Multiwavelets. Chapman (2004)
13. Labat, D.: Recent Advances in Wavelet Analyses: Part 1. A review of Concepts J, Hydrol (2005)
14. Mallat, S.: A Wavelet Tour of Signal Processing. The Sparse Way (2015)
15. Nicolis, O., Mateu, J.: 2D Anisotropic Wavelet Entropy with an Application to Earthquakes in Chile. Entropy (2009)
16. Rafiee, J., Rafiee, M.A., Prause, N., Schoen, M.P.: Wavelet basis functions in biomedical signal processing. Expert Syst. Appl. **38**(5), 6190–6201 (2016)
17. Ruppert-Felsot, J., Praud, O., Sharon, E., Swinney, H.L.: Extraction of coherent structures in a rotating turbulent ow experiment. Phys. Rev. **72**(1), 016311 (2005)
18. Shannon, C.: A mathematical theory of communication. Bell Syst. Tech. J. **27**(3), 379–423 (1948)
19. Zemni, M., Jallouli, M., Ben Mabrouk, A., Mahjoub, M.A.: Explicit Haar-Schauder multi-wavelet filters and algorithms. Part II: Relative entropy-based estimation for optimal modeling of biomedical signals. Int. J. Wavelets Multiresolution Inf. Process. **17**(5), 1950038 (2019) https://doi.org/10.1142/S0219691319500383
20. Zemni, M., Jallouli, M., Ben Mabrouk A., Mahjoub, M.A.: ECG signal processing with haar-schauder multiwavelet. In: Proceedings of the 9th International Conference on Information Systems and Technologies - Icist (2019). https://doi.org/10.1145/3361570.3361611

Machine Learning

Handling Missing Observations with an RNN-based Prediction-Update Cycle

Stefan Becker[1]([✉])([iD]), Ronny Hug[1]([iD]), Wolfgang Huebner[1]([iD]), Michael Arens[1]([iD]), and Brendan T. Morris[2]([iD])

[1] Fraunhofer IOSB, Ettlingen, Germany
{stefan.becker,ronny.hug,wolfgang.huebner,
michael.arens}@iosb.fraunhofer.de
[2] University of Nevada, Las Vegas, USA
brendan.morris@unlv.edu
http://www.iosb.fraunhofer.de

Abstract. In tasks such as tracking, time-series data inevitably carry missing observations. While traditional tracking approaches can handle missing observations, *recurrent neural networks* (RNNs) are designed to receive input data in every step. Furthermore, current solutions for RNNs, like omitting the missing data or data imputation, are not sufficient to account for the resulting increased uncertainty. Towards this end, this paper introduces an RNN-based approach that provides a full temporal filtering cycle for motion state estimation. The Kalman filter inspired approach enables to deal with missing observations and outliers. For providing a full temporal filtering cycle, a basic RNN is extended to take observations and the associated belief about its accuracy into account for updating the current state. An RNN prediction model, which generates a parametrized distribution to capture the predicted states, is combined with an RNN update model, which relies on the prediction model output and the current observation. By providing the model with masking information, binary-encoded missing events, the model can overcome limitations of standard techniques for dealing with missing input values. The model abilities are demonstrated on synthetic data reflecting prototypical pedestrian tracking scenarios.

Keywords: Recurrent Neural Networks (RNNs) · Trajectory data · Missing input data · Outliers · Filtering

1 Introduction and Related Work

One important task for autonomous systems is estimating pedestrians' motion states based on observations. After the success of RNNs in a variety of sequence processing tasks, like speech recognition [9,13] and caption generation [11,36],

Fraunhofer IOSB is a member of the Fraunhofer Center for Machine Learning.

© Springer Nature Switzerland AG 2021
N. Tsapatsoulis et al. (Eds.): CAIP 2021, LNCS 13052, pp. 311–321, 2021.
https://doi.org/10.1007/978-3-030-89128-2_30

these models are also successfully applied to pedestrian trajectory prediction (see for example [1,15,17,33]). While tracking approaches based on Bayesian formalization explicitly model the increase in the prediction uncertainty when an observation is missing, RNN-based models are designed to receive input data in every step. The two main ways to address missing values in time series are data imputation and omitting the missing data [31]. Data imputation means to substitute the missing values with methods like interpolation [20] or spline fitting [10]. Nonetheless, various imputation methods estimate better missing data, which results in a process where imputation and prediction models are separated [8]. Since the model does not effectively explore the missing pattern, only suboptimal results are achieved. The simplest omitting strategy is to remove samples in which a value is missing. This may work for training but cannot be applied during inference. Alternatively, and in particular for RNNs, the problem can be modeled with marked missing values. A missing value can be masked and explicitly excluded, or the model can be encouraged to learn that a specific value represents the missing observation [7].

Most approaches are for healthcare applications [35] or in the field of speech recognition [24]. More recently, Che et al.[8] customized an RNN model to incorporate the patterns of missingness for time series classification. Also, for classifying time series, Lipton et al. [22] treated the pattern of missing data as a feature to diagnose clinical data collected from a pediatric intensive care unit.

This paper introduces an RNN-based full temporal filtering cycle for motion state estimation to better deal with missing observations. The approach is intended to serve as a module for single object motion filtering in a multi-object deep learning trajectory prediction pipeline. In trajectory prediction applications, deep learning-based approaches are increasingly replacing classic approaches due to their ability to better capture contextual cues from the static (e.g. obstacles) or dynamic environment (e.g. other objects in the scene) [29]. Although there exist variants relying on *generative adversarial networks* (GANs) [2,14], *temporal convolution networks* (TCNs) [3,23], and *transformers* [12,30], for encoding object motion, the most popular basis is RNNs. The proposed approach can partly be adapted to the other deep learning approaches but is then essentially limited to the additional masking information. It should be noted that due to the positional encoding and the attention mechanism applied in *transformers*, these models can deal with missing observations by exploiting the remaining observations [12]. The positional encoding extends to unseen lengths, but it is primarily designed for a fixed input length. Thus, it is not clear how well this approach generalizes to variable input lengths, and our proposed approach is designed for varying input lengths. For a comprehensive overview of current deep learning-based approaches for trajectory prediction, the reader is referred to these surveys [19,27,29].

For providing a full temporal filtering cycle, two RNNs are combined to recursively infer the prior and posterior motion states. Thereby, an RNN update model (Update-RNN) relies on the output of an RNN prediction model (Prediction-RNN) in addition to the current observation for inferring the current state. The Prediction-RNN generated a parametrized distribution to capture future states and their prediction uncertainties. Both networks are additionally provided with masking information to enable the networks to learn a representation for missing observations. Thereby, the Prediction-RNN can capture the increased uncertainty when observations are missing, and the Update-RNN can learn to trust in the prior states in these situations. The evaluation is done on synthetically generated data reflecting prototypical pedestrian tracking scenarios.

In the following, a brief formalization of the problem and a description of the RNN-based Prediction-Update-Cycle are provided in Sect. 2. The achieved results are presented in Sect. 3. Finally, a conclusion is given in Sect. 4.

2   RNN-based Prediction-Update-Cycle

The goal is to devise a model that can successfully infer motion states of tracked objects and deal with missing observations. In the context of RNNs, trajectory prediction is formally stated as the problem of inferring trajectories of objects (e.g. pedestrians), conditioned on their track history. Given an input sequence $\mathcal{Z}$ of consecutive observed positions $\vec{z}^k = (x^k, y^k)$ at time step k along a trajectory, the task is to filter the current position $\hat{\vec{z}}^k = (x^k, y^k)$ and to generate predictions for future positions $\{\vec{z}^{k+1}, \vec{z}^{k+2}, \ldots\}$. Almost all deep-learning-based trajectory prediction models conditioned solely on positions ignore that the observed positional data includes uncertainties [12]. Conditioning is done under the assumption that a noise-free, full input trajectory is provided. We combine two RNNs in a Kalman filter-like Prediction-Update cycle to deal with the included uncertainties in the observations.

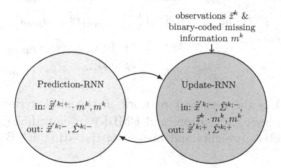

Fig. 1. Visualization of the proposed *prediction-update* cycle. The Update-RNN estimates the unknown system state $\hat{\vec{x}}'^k$ from the observations $\vec{z}^k$ and estimated prior state $\hat{\vec{x}}'^{k-1}$ or rather $\hat{\vec{x}}'^{k;-}$ provided by the Prediction-RNN.

Prediction Network: The Prediction-RNN generates the distribution over the next position $\vec{z}^{k+1}$, the density of the predicted state $p^-((\vec{z} = \vec{x}')^{k+1}) \triangleq p(\vec{x}'^{k+1}|\vec{z}^{0:k})$. Compared to Bayesian filtering, $\vec{x}'^k$ is not the full dynamical state $\vec{x}^k$, but the state $\vec{z}^k$ can be interpreted as observable state by mapping the RNN state $\vec{h}$ to the observation space [5]. For generating the distribution $p^-(\cdot)$ over the next positions, the model parametrizes a mixture density network (MDN) [6].

For reflecting the increased prediction uncertainty in case of a missing observation, which would result in a changed possible position distribution, we propose to extend the input sequence with masking information in the form of a binary-coded indicator variable $m^k \in \{0, 1\}$, which marks an observation as missing. The binary-coded masking is used to incorporate a replacement value as missing. In the case of dealing with missing last k_{miss} observation, the model generates a distribution over the $1 + k_{miss}$ next steps. In practice, only conditioned on the information from the $k - k_{miss}$ observed time steps. The model can be trained by maximizing the likelihood of the data given the output Gaussian mixture parameters. The loss function $\mathcal{L}_{pred}$ of the Prediction-RNN using one mixture component is given by $\mathcal{L}_{pred}(\mathcal{Z}') = \sum_{k=1}^{K} -\log\left\{\mathcal{N}(\vec{z}^{k+1+k_{miss}} | \vec{x}'^k, \Sigma^k)\right\}$. Thus, for $k_{miss} = 0$ this is the default loss for a next step prediction RNN-MDN. $\mathcal{Z}'$ is the combination of estimated states by the Update-RNN multiplied with the masking information and concatenation of the masking information $\{\hat{\vec{x}}'^k \cdot m^k, m^k\}$. Although the Update-RNN estimates the current state $\vec{x}'^k$ for every time step, the estimates are replaced with the missing placeholder for conditioning. Note that the Prediction-RNN is not used for long-term prediction but for providing the Update-RNN with a prior state with uncertainty $\hat{\vec{x}}'^{k;-}, \hat{\Sigma}^{k;-}$ together with observations. Since RNNs are only capable of generating conditional predictions for one time step at a time, we can create a next step prediction with increased uncertainty using the missing placeholders. With an embedding of the inputs, the Prediction-RNN can be defined as follows:

$$\vec{e}^k_{pred} = \text{EMB}(\hat{\vec{x}}'^{k;+} \cdot m^k, m^k; \vec{\Theta}_{epred}),$$
$$\vec{h}^k_{pred} = \text{RNN}(\vec{h}^{k-1}_{pred}, \vec{e}^k_{pred}; \vec{\Theta}_{RNNpred}),$$
$$\hat{\vec{z}}^{k+1+k_{miss}}, \hat{\Sigma}^{k+1+k_{miss}} = \text{MLP}(\vec{h}^k_{pred}; \vec{\Theta}_{MLPpred})$$
$$\hat{\vec{x}}'^{k;-}, \hat{\Sigma}^{k;-} = \hat{\vec{z}}^{k+1+k_{miss}}, \hat{\Sigma}^{k+1+k_{miss}} \tag{1}$$

Here, $\text{RNN}(\cdot)$ is the recurrent network, $\vec{h}$ the hidden state of the RNN, $\text{MLP}(\cdot)$ the multilayer perceptron, and $\text{EMB}(\cdot)$ an embedding layer. $\vec{\Theta}$ represents the parameters (weights and biases) of the MLP, EMB or respectively RNN.

Update Network: The Update-RNN is used for generating the posterior $p^+(\vec{x}'^k)$. The posterior is the probability distribution over $\vec{x}'^k$ conditioned on all past observations $\vec{z}^{0:k}$. It is important to note, that the $\hat{\vec{x}}'^{k;+}, \hat{\Sigma}^{k;+}$ and $\hat{\vec{x}}'^{k;-}, \hat{\Sigma}^{k;-}$ depend on the whole history of inputs in contrast to the Markov assumption of the Kalman filter. In case of a missing observation, the corresponding $\vec{z}^k$ is replaced with a placeholder value by multiplying with the masking value m^k. Here, we used zero as placeholder values for missing. Besides the observations, the output of the Prediction-RNN is also used as input for the Update-RNN. Although a division is not defined for matrices (Kalman gain multiplies prior uncertainty with the inverse observation uncertainty), we can think of the Kalman gain as a ratio that controls the influence of a new observation on the updated (posterior) state estimate. Following a Kalman filter, the Update-RNN learns to weight both inputs in order to generate the parameter of an MDN for representing the posterior. The weighting factors K ($K_{pred} = (1 - K_{obs})$) can therefore be seen as a pseudo-Kalman gain. The Update-RNN with an embedding of the inputs is given by:

$$\vec{e}_{up}^k = \mathrm{EMB}(\hat{\tilde{x}}^{'k;-}, \hat{\Sigma}^{k;-}, \tilde{z}^k \cdot m^k, m^k; \vec{\Theta}_{eup}),$$
$$\vec{h}_{up}^k = \mathrm{RNN}(\vec{h}_{up}^{k-1}, \vec{e}_{up}^k; \vec{\Theta}_{RNNup}),$$
$$\hat{\tilde{z}}^k = K_{pred} \cdot \hat{\tilde{x}}^{'k;-} + K_{obs} \cdot \tilde{z}^k, \hat{\Sigma}^k = \mathrm{MLP}(\vec{h}_{up}^k; \vec{\Theta}_{MLPup})$$
$$\hat{\tilde{x}}^{'k;+}, \hat{\Sigma}^{k;+} = \hat{\tilde{z}}^k, \hat{\Sigma}^k \qquad (2)$$

Here, $\tilde{z}^k$ is an actual, noisy observation, a realization of $\tilde{z}^k$ despite the inputs of an RNN being deterministic. The pseudo-Kalman gain can be realized with every activation function keeping the output between zero and one, switching between trusting the prior or the current observations. Here, K is generated with a *softplus* activation function. The Update-RNN can learn when to rely on predictions instead of observations due to provided prediction uncertainty and the masking information. Similar to Kalman filtering, the information of both RNN is exchanged iteratively. The Update-RNN is trained by minimizing the filtering loss $\mathcal{L}_{up}$ in the form of the negative log-likelihood of the ground truth current position under the filtered position. By combining both models, we get a full Prediction-Update cycle to filter noisy observations and handle missing observations from variable input sequences. In Fig. 1 the RNN-based Prediction-Update cycle is visualized.

3 Data Generation and Evaluation

This section consists of a brief evaluation of the proposed Prediction-Update-RNN cycle. The evaluation is concerned with verifying the approach's overall viability in situations with missing observations and outliers from tracking maneuvering pedestrians. For initial results, synthetic generated data is used due to the fact that current pedestrian trajectory data sets do not consider aspects like motion smoothness (see for example *TrajNet++* [19], *UCY* [21], *ETH* [26], SDD [28]) despite RNNs can generalize to deal with noisy inputs. Further, problems such as limited training samples are avoided. Although using synthetic data, we make use of a real-world dataset with maneuvering pedestrians to capture similar conditions (*Daimler Path Prediction* dataset [32]). For generating synthetic trajectories of a basic maneuvering pedestrian on a ground plane, random agents are sampled from a Gaussian distribution according to a preferred pedestrian walking speed [34] ($\mathcal{N}(1, 38m/s, (0.37m/s)^2)$). The frame rate is set to $16fps$. During a single trajectory simulation, the agents can perform a turning maneuver. The heading change is sampled from a uniform distribution between

Table 1. Results for a comparison between the proposed RNN-based Prediction-Update cycle compared to two variants of RNN-MDNs (1to1 and encoder). The displacement error is shown for different observation noise levels and for varying probabilities of outliers and missing observations.

	Fully observed $Ber_{miss}(0.0, 1.0)$			
	No outlier $Ber_{outl}(0.0, 1.0)$		With outlier $(Ber_{outl}(0.1, 0.9); \sigma_{outl} = 0.5)$	
Approach	$\sigma_w = 0.01$	$\sigma_w = 0.05$	$\sigma_w = 0.01$	$\sigma_w = 0.05$
	ADE/m σ_{ADE}/m	ADE/m σ_{ADE}/m	ADE/m σ_{ADE}/m	ADE/m σ_{ADE}/m
Prediction-Update-RNN	0.011 0.012	0.051 0.027	0.038 0.097	0.083 0.106
RNN-(1to1)-MDN	0.018 0.039	0.053 0.064	0.076 0.084	0.094 0.096
RNN-(encoder)-MDN	0.028 0.016	0.067 0.035	0.112 0.214	0.135 0.196

	Missing observations $Ber_{miss}(0.1, 0.9)$			
	No outlier $Ber_{outl}(0.0, 1.0)$		With outlier $(Ber_{outl}(0.1, 0.9); \sigma_{outl} = 0.5)$	
Approach	$\sigma_w = 0.01$	$\sigma_w = 0.05$	$\sigma_w = 0.01$	$\sigma_w = 0.05$
	ADE/m σ_{ADE}/m	ADE/m σ_{ADE}/m	ADE/m σ_{ADE}/m	ADE/m σ_{ADE}/m
Prediction-Update-RNN	0.021 0.057	0.060 0.058	0.039 0.096	0.090 0.114
RNN-(1to1)-MDN (imputation)	0.031 0.056	0.065 0.080	0.087 0.110	0.101 0.120
RNN-(encoder)-MDN (imputation)	0.040 0.029	0.069 0.037	0.104 0.194	0.138 0.195

$45°$ and $100°$. The duration of the turning event is sampled from a Gaussian distribution based on the mean sojourn time estimated from the ground truth sequences ($\mathcal{N}(1.83s, (0.29s)^2)$). The positional observation noise is assumed to follow a bimodal Gaussian mixture model for considering outliers. The outlier observation noise is set to $\sigma_{outl} = 0.5m$ and the standard observation noise is varied ($\sigma_w = 0.05m$ and $\sigma_w = 0.01m$). Outlier and missing events are drawn from a Bernoulli distribution $Ber(\cdot, \cdot)$.

Implementation Details: The models have been implemented using *Pytorch* [25]. The Prediction-RNN is pre-trained for 100 epochs on noise-free trajectory data and then for 100 epochs on noisy trajectory data. After that, both models are jointly trained for 400 epochs. In the joint training, the estimated states are iteratively exchanged over the sequence length, whereas in pre-training, the Prediction-RNN is conditioned directly on the observations. For training, the ADAM optimizer [18] with a learning rate of 0.001 is used. As RNN variant, the *long short-term memory* (LSTM) [16] is utilized. Both models use an embedding dimension and a hidden state dimension of 64.

Results and Analysis: For every experiment, 1000 noisy trajectories are synthetically generated with a ratio of using 80% for training and 20% for evaluation. The results are summarized in Table 1. For comparison, the average displacement error (ADE) is calculated as the average L2 distance between the estimated positions and the ground truth positions. Further, the probability of a missing observation and outlier are varied ($Ber(0.0, 1.0)$ and $Ber(0.1, 0.9)$). As reference models, a one-to-one RNN-MDN (RNN-(1to1)-MDN), which estimates the true positions stepwise, and an RNN-encoder with an MDN on top (RNN-(encoder)-MDN), which first fully observes the input sequence, are used to generalize from the noisy inputs. The sequence length varies between 8 and 20 time steps. In case observations are missing, the reference models receive the predictions from the Prediction-RNN for data imputation. Compared to linear interpolation, a better performance is achieved in the experiments and better comparability to the Prediction-Update cycle is guaranteed. These results show that the proposed Prediction-Update-RNN can better handle outliers and missing observations. Even in the experiments without outliers, the achieved result is better. Due to provided binary-coded masking patterns, the approach learns to ignore the placeholder inputs and to fully trust the predictions. Missing observations lead to increasing prediction uncertainties. Thus the model corrects the position estimates by relying more strongly on the new observations. These effects are visualized in Fig. 2 and 3.

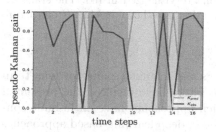 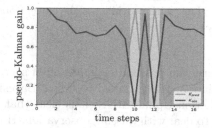

Fig. 2. Visualization of the pseudo-Kalmam gain for low observation noise sequences. The weighting towards trusting the prediction is visualized with dark yellow and towards observations with dark blue. Time steps with missing observations are highlighted with a yellow background. (Color figure online)

Figure 2 shows the pseudo-Kalman gain for a low observation noise sequence. The weighting towards trusting the prediction is visualized with dark yellow ($K_{pred} = (1 - K_{obs})$) and correspondingly K_{obs} with dark blue. Following this color scheme, time steps with missing observations are highlighted with a dark yellow background. It is clearly visible how the approach relies only on the predicted position to estimate the posterior position when the observation is replaced with a placeholder value. The ability of the Prediction-RNN to capture the increased prediction uncertainty when observations are missing is depicted in Fig. 3 for an example trajectory of the *ETH* dataset [26]. The ground truth trajectory for conditioning is shown in green. Missing observations are marked with a cross. The covariance ellipses capture the 3σ area around the predicted position. The predicted position varies reasonably around the prediction from the fully observed trajectory. Besides demonstrating the Prediction-RNN ability, this example shows the noise present in the ground truth data. Since RNNs can generalize to produce smooth trajectories, neglecting such noise levels in order to focus on prediction comparison, seems, on the one hand, to be reasonable. On the other hand, evaluating the filtering performance by not considering the noise or discretization artifacts in the underlying trajectories data is not adequate. Further, the position estimation error naturally influences prediction performance

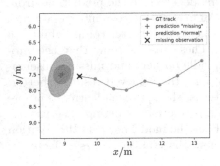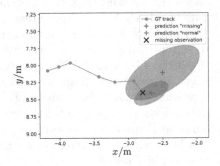

Fig. 3. Example predictions for two trajectories from the *ETH* dataset [26]. The prediction uncertainties with a missing observation are visualized in red. The standard predictions, with providing of the current observation, are shown in blue. (Color figure online)

because the position estimate often serves as the reference point for long-term prediction. Since classic recursive Bayesian filters have inspired the presented Prediction-Update-RNN, it is clear that such approaches have an in-built concept to deal with missing observation. However, deep learning-based approaches are increasingly replacing classic approaches due to their ability to capture better contextual cues from the static or dynamic environment [29]. Moreover, in order to deal with outliers, classic approaches require strategies such as gating. Gating is a hard decision made about which observations are considered valid. The Prediction-Update-RNN learns a conditioned decision on the influence of new observations instead of relying on a fixed gating threshold. For a comparison of deep learning-based approaches with several Kalman filter variants for capturing pedestrian maneuvers based on fully observed trajectories, the reader is referred to the following works [4,5].

4   Conclusion

In this paper, an RNN-based Prediction-Update cycle has been presented. The model enables improved handling of missing observations and outliers present in time-series data. The model abilities were shown on synthetic data reflecting prototypical pedestrian maneuvers. By iteratively exchanging the estimates of two separated RNNs and providing a binary-coded missing pattern, the model can learn to trust the prior estimates or rely more strongly on the current observations. Whereas data imputation only fills missing values, the proposed Prediction-Update-RNN provides information about the included uncertainties in the filling values. Compared to RNN-based reference models using data imputation, the model achieved better performance in terms of the average displacement error in the experiments.

References

1. Alahi, A., Goel, K., Ramanathan, V., Robicquet, A., Fei-Fei, L., Savarese, S.: Social LSTM: human trajectory prediction in crowded spaces. In: Conference on Computer Vision and Pattern Recognition (CVPR), pp. 961–971 (2016)
2. Amirian, J., Hayet, J.B., Pettre, J.: Social ways: learning multi-modal distributions of pedestrian trajectories with GANs. In: Conference on Computer Vision and Pattern Recognition Workshops (CVPRW), pp. 2964–2972 (2019)
3. Becker, S., Hug, R., Hübner, W., Arens, M.: RED: a simple but effective baseline predictor for the *TrajNet* benchmark. In: Leal-Taixé, L., Roth, S. (eds.) ECCV 2018. LNCS, vol. 11131, pp. 138–153. Springer, Cham (2019). https://doi.org/10.1007/978-3-030-11015-4_13
4. Becker, S., Hug, R., Hübner, W., Arens, M.: An RNN-based IMM filter surrogate. In: Felsberg, M., Forssén, P.-E., Sintorn, I.-M., Unger, J. (eds.) SCIA 2019. LNCS, vol. 11482, pp. 387–398. Springer, Cham (2019). https://doi.org/10.1007/978-3-030-20205-7_32
5. Becker, S.: Dynamic Switching State Systems for Visual Tracking. Ph.D. thesis, Karlsruher Institut für Technologie (KIT) (2020)
6. Bishop, C.M.: Mixture Density Networks. Technical report, Microsoft Research (1994)
7. Brownlee, J.: Introduction to time series forecasting with python: how to prepare data and develop models to predict the future. Machine Learning Mastery (2017)
8. Che, Z., Purushotham, S., Cho, K., Sontag, D., Liu, Y.: Recurrent neural networks for multivariate time series with missing values. Sci. Rep. (SREP) **8**, 6085 (2018)
9. Chung, J., Kastner, K., Dinh, L., Goel, K., Courville, A., Bengio, Y.: A recurrent latent variable model for sequential data. In: Advances in Neural Information Processing Systems (NeurIPS), vol. 28, pp. 2980–2988 (2015)
10. De Boor, C.: A practical guide to splines; rev. ed. Applied mathematical sciences, Springer, Berlin (2001)
11. Donahue, J., et al.: Long-term recurrent convolutional networks for visual recognition and description. In: Conference on Computer Vision and Pattern Recognition (CVPR), pp. 2625–2634 (2015)
12. Giuliari, F., Hasan, I., Cristani, M., Galasso, F.: Transformer networks for trajectory forecasting. In: International Conference on Pattern Recognition (ICPR), pp. 10335–10342 (2021)
13. Graves, A., Mohamed, A., Hinton, G.: Speech recognition with deep recurrent neural networks. In: International Conference on Acoustics, Speech and Signal Processing, pp. 6645–6649 (2013)
14. Gupta, A., Johnson, J., Fei-Fei, L., Savarese, S., Alahi, A.: Social GAN: socially acceptable trajectories with generative adversarial networks. In: Conference on Computer Vision and Pattern Recognition (CVPR), pp. 2255–2264 (2018)
15. Hasan, I., Setti, F., Tsesmelis, T., Bue, A.D., Galasso, F., Cristani, M.: MX-LSTM: mixing tracklets and vislets to jointly forecast trajectories and head poses. In: Conference on Computer Vision and Pattern Recognition (CVPR), pp. 6067–6076 (2018)
16. Hochreiter, S., Schmidhuber, J.: Long short-term memory. Neural Comput. **9**(8), 1735–1780 (1997)
17. Hug, R., Becker, S., Hübner, W., Arens, M.: On the reliability of LSTM-MDL models for pedestrian trajectory prediction. In: Chen, L., Ben Amor, B., Ghorbel, F. (eds.) RFMI 2017. CCIS, vol. 842, pp. 20–34. Springer, Cham (2019). https://doi.org/10.1007/978-3-030-19816-9_2

18. Kingma, D., Ba, J.: Adam: a method for stochastic optimization. In: International Conference on Learning Representations (ICLR) (2015)
19. Kothari, P., Kreiss, S., Alahi, A.: Human trajectory forecasting in crowds: a deep learning perspective. IEEE Transactions on Intelligent Transportation Systems, pp. 1–15 (2021)
20. Kreindler, D., Lumsden, C.J.: The effects of the irregular sample and missing data in time series analysis. Nonlinear Dyn. Psychol. Life Sci. **10**(2), 187–214 (2006)
21. Lerner, A., Chrysanthou, Y., Lischinski, D.: Crowds by example. computer graphic. Forum **26**(3), 655–664 (2007)
22. Lipton, Z.C., Kale, D., Wetzel, R.: Directly modeling missing data in sequences with Rnns: improved classification of clinical time series. In: Proceedings of the 1st Machine Learning for Healthcare Conference, vol. 56, pp. 253–270. PMLR, Children's Hospital LA, Los Angeles, CA, USA (2016)
23. Nikhil, N., Morris, B.T.: Convolutional neural network for trajectory prediction. In: Leal-Taixé, L., Roth, S. (eds.) ECCV 2018. LNCS, vol. 11131, pp. 186–196. Springer, Cham (2019). https://doi.org/10.1007/978-3-030-11015-4_16
24. Parveen, S., Green, P.: Speech recognition with missing data using recurrent neural nets. In: Advances in Neural Information Processing Systems (NeurIPS), pp. 1189–1195. MIT Press (2002)
25. Paszke, A., et al.: PyTorch: an imperative style, high-performance deep learning library. In: Advances in Neural Information Processing Systems (NeurIPS), pp. 8024–8035. Curran Associates, Inc. (2019)
26. Pellegrini, S., Ess, A., Schindler, K., van Gool, L.: You'll never walk alone: modeling social behavior for multi-target tracking. In: International Conference on Computer Vision (ICCV), pp. 261–268 (2009)
27. Rasouli, A.: Deep Learning for Vision-based Prediction: A Survey. arXiv abs/2007.00095 (2020)
28. Robicquet, A., Sadeghian, A., Alahi, A., Savarese, S.: Learning social etiquette: human trajectory understanding in crowded scenes. In: Leibe, B., Matas, J., Sebe, N., Welling, M. (eds.) ECCV 2016. LNCS, vol. 9912, pp. 549–565. Springer, Cham (2016). https://doi.org/10.1007/978-3-319-46484-8_33
29. Rudenko, A., Palmieri, L., Herman, M., Kitani, K.M., Gavrila, D.M., Arras, K.O.: Human motion trajectory prediction: a survey. Int. J. Robot. Res. **39**, 895–935 (2020)
30. Saleh, K.: Pedestrian Trajectory Prediction using Context-Augmented Transformer Networks. arXiv abs/2012.01757 (2020)
31. Schafer, J.L., Graham, J.W.: Missing data: our view of the state of the art. Psychol. Methods **7**(2), 147–177 (2002)
32. Schneider, N., Gavrila, D.M.: Pedestrian path prediction with recursive bayesian filters: a comparative study. In: Weickert, J., Hein, M., Schiele, B. (eds.) GCPR 2013. LNCS, vol. 8142, pp. 174–183. Springer, Heidelberg (2013). https://doi.org/10.1007/978-3-642-40602-7_18
33. Syed, A., Morris, B.T., et al.: CNN, segmentation or semantic embeddings: evaluating scene context for trajectory prediction. In: George, B. (ed.) ISVC 2020. LNCS, vol. 12510, pp. 706–717. Springer, Cham (2020). https://doi.org/10.1007/978-3-030-64559-5_56
34. Teknom, K.: Microscopic Pedestrian Flow Characteristics: Development of an Image Processing Data Collection and Simulation Model. Ph.D. thesis, Tohoku University (2002)

35. Tresp, V., Briegel, T.: A solution for missing data in recurrent neural networks with an application to blood glucose prediction. In: International Conference on Neural Information Processing Systems (NeurIPS), pp. 971–977. MIT Press, Cambridge, MA, USA (1997)
36. Xu, K., et al.: Show, attend and tell: neural image caption generation with visual attention. In: International Conference on Machine Learning (ICML), vol. 37, pp. 2048–2057. PMLR (2015)

eGAN: Unsupervised Approach to Class Imbalance Using Transfer Learning

Ademola Okerinde[(✉)], William Hsu, Tom Theis, Nasik Nafi, and Lior Shamir

Kansas State University, 2164 Engineering Hall, Manhattan, KS 43017-6221, USA
{okerinde,bhsu,theis,nnafi,lshamir}@ksu.edu

Abstract. Class imbalance is an inherent problem in many machine learning classification tasks. This often leads to learned models that are unusable for any practical purpose. In this study, we explore an unsupervised approach to address class imbalance by leveraging transfer learning from pre-trained image classification models. To this end, an encoder-based Generative Adversarial Network (eGAN) is proposed which modifies the generator of a GAN by introducing an encoder module and adopts the GAN loss function to directly classify the majority and minority class. To the best of our knowledge, this is the first work to tackle this problem using GAN-based loss function rather than augmenting the dataset with synthesized fake images. Our approach eliminates the epistemic uncertainty in the model predictions, as $P(minority)$ and $P(majority)$ need not sum up to 1. The impact of transfer learning and combinations of different pre-trained image classification models at the generator and the discriminator level is also explored. Best result of 0.69 F1-score was obtained on CIFAR-10 classification task with an enforced imbalance ratio of 1:2500. Our implementation code is available at - https://github.com/demolakstate/eGAN_addressing_class_imbalance_with_transfer_learning_on_GAN.git.

Keywords: Class imbalance · Transfer learning · GAN · Nash equilibrium

1 Introduction

A dataset is considered imbalanced when there is a significant, or in some cases, extreme disproportion between the number of samples of the different classes in the dataset. The class or classes with large number of samples are called the majority, while the class with few examples are denoted as the minority. In many cases, the machine learning model is required to correctly classify the minority class while minimizing the misclassification of the majority class. However, the skewness in the data often leads machine learning classification methods to favour the majority class.

Class imbalance problem in computer vision is normally approached either at the data level or algorithm level. Using data augmentation, a class with a small number of samples can be expanded into a class with a much larger number of

© Springer Nature Switzerland AG 2021
N. Tsapatsoulis et al. (Eds.): CAIP 2021, LNCS 13052, pp. 322–331, 2021.
https://doi.org/10.1007/978-3-030-89128-2_31

samples. Earlier data augmentation was achieved simply by transforming images via scaling, cropping, flipping, padding, rotation, brightness, contrast, saturation level etc. [15]. Now-a-days, synthetic images can also be generated using generative models such as VAE, GAN [4,8]. As a result, a humongous image dataset can be created from the images of the minority class.

At the algorithm level, the objective function is tweaked to heavily penalize the network for mis-classifying the minority class [10,13]. The most popular is cost-sensitive approach. Here, the classifier is modified to incorporate varying penalty for each of considered groups of examples. By assigning a higher cost to the less represented set of samples its importance is boosted during training.

Transfer learning has been known to help improve the performance of machine learning models [14]. By fine-tuning varying number of layers in the pre-trained image classification model, the pre-trained model can serve as a feature extractor, while adding a classifier head for more specific feature learning for the current task.

In this work, we compared the performance of various pre-trained image classification models for the task of unsupervised image classification with varying imbalance ratios. Our architecture, named eGAN, is developed to serve as a basis for this comparison. Using GAN [4], we reparameterise the task of the discriminator as a classifier which outputs a positive score for majority samples and a negative score for the minority ones. We integrate an encoder module to the GAN network that encodes the minority samples into a latent code from which the generator learns. While most GAN-based architectures focus on the output of the generator, in our proposed approach, as we intuitively adapted the vanilla GAN network and the corresponding loss function to directly classify the majority and the minority class, we are more concerned about the performance of the discriminator.

2   Related Work

There have been a lot of work in the last few decades to address class imbalance. Earlier approaches include deliberate undersampling of the majority class or oversampling of the minority class by mere copying [3]. However, for image data the earlier approach leads to loss of useful data information while the latter approach causes overfitting [11]. Data augmentation via rotation, scaling, cropping etc. can be considered as a variant of oversampling which copy the same data, however with little modification [15]. VAE and GAN enables the generation of completely new data [4,8]. In recent years, GAN-based approaches have gained much popularity than others and a good number of variants of vanilla GAN have been proposed to address class imbalance [1,12,16].

In [6], an ensemble method was proposed based on advanced generative adversarial network to generate new samples for the minority class to restore balance. Our opinion is that the computational demand of such approach is enormous, and many low-income countries of the world do not have access to such computation power. Deep Cascading (DC) with a long sequence of decision trees could help to handle unbalanced data [2]. A DC is a sequence of n classifiers where each sample x passes to the next classifier only if the current one classifies it as positive according to a high-sensitivity decision threshold. However, this works well with foreground-background imbalance unlike the classification task. Transfer learning with GAN was used to generate images from limited data in [14]. Their result showed that knowledge from pre-trained networks can ensure faster convergence and significantly improve the quality of generated images.

3 Methodology and Experimental Design

In this section, we discuss our proposed approach and the various testbeds that were used in our experiments.

3.1 Addressing Class Imbalance with eGAN

The proposed architecture is based on adaptation of existing Deep Convolutional Generative Adversarial Network (DCGAN)[12] by incorporating an encoder module. This module encodes minority samples in latent space needed by the generator G to generate minority samples that are capable of fooling the discriminator D. On the other hand, the discriminator D is fed with data samples drawn from majority distribution and the generated output of the generator G. D and G are simultaneously optimized through the following two-player minimax game with value function V(G,D) in 1.

$$\underset{G}{min}\ \underset{D}{max}\ V(D,G) = E_{X_{ma} \sim P_{ma}}[\log D(X_{ma})] + E_{X_{mi} \sim P_{mi}}[\log(1 - D(G(X_{mi})))] \tag{1}$$

where X_{ma} and X_{mi} are majority and minority sample distributions respectively.

Over the course of iteration, the discriminator D is optimized to assign a negative score to the minority data distribution and a positive score to the majority data distribution. This enables the discriminator D to act as a classifier.

Encoder-Generator Module. Our latent space is composed of 128 units vector. Rather than feeding the generator with random noise as is typical of most GAN implementation, we added an encoder module that forces the generator to learn from known distribution (minority distribution). The encoder part consists of the pre-trained DenseNet121 followed by global average pooling layer and latent dimension space. The generator part has two transposed convolutional layers. We use LeakyRelu activation function with alpha set to 0.2; batch normalization and Sigmoid function at the final layer.

Discriminator Module. The pre-trained discriminator has 7,038,529 parameters out of which only 39,937 are trainable. A layer of global average pooling follows the pre-trained DenseNet121. We use a dropout of 0.2 followed by the final one unit dense layer. The overall architecture of our encoder-based generative adversarial network is shown in Fig. 1.

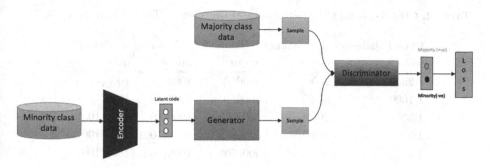

Fig. 1. Encoder-based Generative Adversarial Network (eGAN) architecture

3.2   Selection of Pre-trained Image Classification Weights

We perform experiments on VGG16, VGG19, EfficientNetB2, ResNet101 and DenseNet121 pre-trained classification models on ImageNet dataset. Here, we fine-tuned only top five layers at each of the pre-trained models. Table 1 shows the maximum precision, recall and F1-score obtained on CIFAR-100 with imabalance ratio 1:50 by using different combinations of pre-trained models.

Table 1. Comparative analysis of different pre-trained models configuration on Generator and Discriminator using CIFAR-100 dataset

Discriminator pre-trained	Generator pre-trained	Precision	Recall	F1
ResNet101	VGG19	0.72	1.0	0.78
VGG19	ResNet101	0.73	1.0	0.69
EfficientNetB2	VGG19	1.0	0.22	0.32
VGG19	EfficientNetB2	0.7	0.86	0.71
ResNet101	VGG16	1.0	0.17	0.27

DenseNet121 [5] was used for pretraining our eGAN. After experimenting different pre-trained architectures and different layers of fine-tuning, we obtained best result with fine-tuning only top 5-layer out of 427 layers of DenseNet121.

3.3   Dataset

Several commonly used datasets were used in this study. In order to model the real-world scenario of heavy imbalance, we used only few samples of the minority class as input to the encoder module. Detail overview is shown in Table 2.

The CIFAR-10 dataset [9] consists of 60,000 32 × 32 colour images in 10 classes, with 6000 images per class. There are 50,000 training images and 10,000 test images. Here we use *airplane* as minority and *automobile* as majority.

Table 2. CIFAR-10 and CIFAR-100 dataset overview - CIFAR-100 in parenthesis

Class imbalance ratio	# Training minor	# Training major	# Testing minor	# Testing major
1:2500	2(-)	5000(-)	1000(-)	1000(-)
1:1000	5(-)	5000(-)	1000(-)	1000(-)
1:500	10(1)	5000(500)	1000(100)	1000(100)
1:50	100(10)	5000(500)	1000(100)	1000(100)
1:1	500(500)	500(500)	1000(100)	1000(100)
*1:1	−(1)	−(1)	−(100)	−(100)

CIFAR-100 [9] is similar to CIFAR-10, except it has 100 classes containing 600 images each. Each class has 500 training images and 100 testing images. The 100 classes in the CIFAR-100 are grouped into 20 superclasses. Each image comes with a "fine" label (the class in which it belongs) and a "coarse" label (the superclass). We also use the pneumonia subset of Stanford CheXpert dataset [7] for experimenting on an inherent imbalanced dataset. The dataset contains 4576 and 167407 minority and majority samples, respectively.

4   Results and Discussion

All discriminator scores that are less than zero are classified as minority, otherwise they are classified as majority class. Table 1 shows the result obtained by combining various pre-trained models. Adam optimizer with a learning rate of 1e-4 was used to train all models for 100 epochs.

As can be seen in Fig. 2, the model achieved a Nash equilibrium on test data at around 10 epochs. Here, we perform inference on test data at every epoch and plot the number of samples correctly classified. Five layers of DenseNet121 were fine-tuned at each generator and discriminator module, while 422 layers' weight were kept fixed. At Nash, the discriminator correctly classified roughly 700/1000 of each of the minority and majority test data. This result convinces us that transfer learning with GAN can be used to overcome the challenge of highly imbalanced dataset, owing to the fact that we train only with 10 samples of the minority class and 5000 samples of the majority class. Similar performance is observed in CIFAR-100 with imbalance ratio 1:50.

Without pre-training the discriminator, the effect of the high imbalance in the training set is revealed, as the discriminator is skewed towards the majority class in the training set, thereby missing all the minority samples in the test data. This can be seen in Fig. 3 on CIFAR-100. This behaviour pattern is observed on CIFAR-10 as well. We experiment with no pre-training at all, neither in the discriminator nor generator, and observed exact same pattern. Therefore, we can safely conclude that the use of transfer learning helps unsupervised image classification in a highly imbalanced domain.

Training can be stopped as soon as Nash equilibrium is reached, as this point gives the model best performance on the minority and majority class.

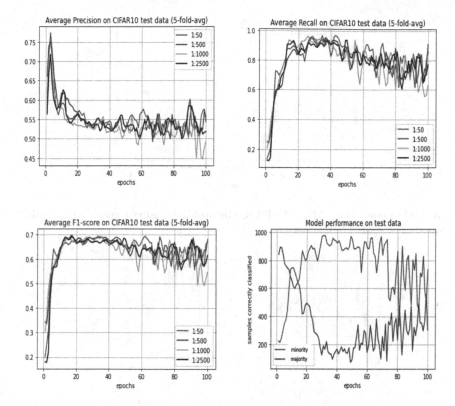

Fig. 2. eGAN's average precision (top left), average recall (top right), average F1-score (bottom left), and average performance (bottom right) on CIFAR-10 test dataset using DenseNet121.

An acceptable threshold can also be set for the absolute difference of the number of correctly classified samples of both classes. For instance, if the | correctly_classified_minority - correctly_classified_majority | $\leq$ 20. The precision, recall and F1-score curves on CIFAR-10 averaged over five folds at different imbalance ratios are shown in Fig. 2.

We observed that at the early training epochs, typically between 1 and 40 epochs, the generator tries to achieve its objective of fooling the discriminator by generating samples from the majority class fed into the discriminator. That results in more of the minority samples being mis-classified as the discriminator "knows" the distribution of the majority too well. A drastic change occurs when the generator start generating samples from latent vector, which can fool the discriminator as seen in generator and discriminator loss shown in Fig. 4.

4.1 Imbalance Ratios

To eliminate bias in model performance, we conducted 5-fold-cross-validation on the minority samples and average the result.

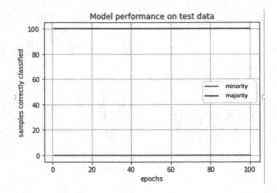

Fig. 3. Performance of eGAN on **CIFAR-100** test data using DenseNet121. Only the generator network is pre-trained

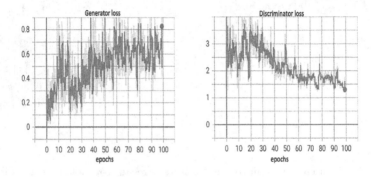

Fig. 4. Discriminator and Generator loss.

Class Ratio of 1:2500. We experiment on CIFAR-10 dataset by deliberately using an unbalanced subset of the training set. At the 4th epoch, our model correctly identifies 257 minority and 821 majority samples out of 1000 each. At epoch 5, a sharp change occurred that led to 821 minority samples being correctly classified, while only correctly classifying 233 majority samples as shown in Table 3. We also observed that at epoch 72 the performance of the network on the majority and minority classes reached a nash equilibrium with a threshold difference of less than or equal to 20.

Class Ratio of 1:1000. At epoch 81 on CIFAR-10 dataset, nash equilibrium was reached. At this epoch, 532 and 525 minority and majority test data respectively were correctly classified. We observed that the classifier had another major shift between epoch 5 and 6. At epoch 5, best result was obtained. The network was able to classify 863 majority tests and 585 minority tests correctly out of the 1000 samples. At epoch 6, 634 majority and 810 minority tests were classified correctly as swown in Table 4. Maximum precision, F1-score and recall of 0.88, 0.74 and 0.99 were obtained at epochs 3, 6 and 37, respectively.

Table 3. Confusion matrix of imbalance ratio 1:2500

	Predicted minority	Predicted majority
Actually minority	821	179
Actually majority	767	233

Table 4. Confusion matrix of imbalance ratio 1:1000

	Predicted minority	Predicted majority
Actually minority	810	190
Actually majority	366	634

Class Ratio of 1:500. Both CIFAR-10 and CIFAR-100 were used to experiment imbalance ratio 1:500. On CIFAR-10, maximum precision, recall and F1-score on averaging 5-fold-cross-validation are 0.75, 0.95 and 0.70 respectively as shown in Table 5. The maximum precision is slightly lower on CIFAR-100 with 0.60. However, the recall and F1-score which are 0.96 and 0.68 are roughly the same.

Class Ratio of 1:50. We demonstrate our model performance on imbalance ratio 1:50 using CIFAR-100 and CIFAR-10. For CIFAR-100, a sudden change occurred between epoch 69 and 70 as follows majority: 57, minority: 59; and majority: 55, minority: 59. A nash equilibrium is attained at epoch 68, with 56 correctly classified minority as well as majority class. At epoch 97 maximum F1-score and recall of 0.66 and 0.77 were obtained respectively, while maximum precision of 0.6 was obtained at epoch 74.

Table 5. Maximum precision, recall and F1-score on CIFAR-10 and CIFAR-100 (avg. 5-fold) - CIFAR-100 in parenthesis

Ratio	Precision	Recall	F1
1:2500	0.72(-)	0.94(-)	0.69(-)
1:1000	0.72(-)	0.96(-)	0.69(-)
1:500	0.75(0.60)	0.95(0.96)	0.70(0.68)
1:50	0.78(0.7)	0.97(0.86)	0.69(0.71)
1:1	**0.82**(0.72)	0.98(**1.0**)	0.72(**0.78**)
*1:1	−(0.53)	−(0.86)	−(0.63)

Class Ratio of 1:1. We use CIFAR-100 to demonstrate the performance of eGAN on a balanced dataset. We notice that the experimental performance follow the same pattern as imbalanced dataset. Training starts with mostly all the majority correctly classified and all the minority mis-classified. At epoch 48,

Table 6. Precision, recall and F1-score on pneumonia subset of CheXpert dataset

| Model | #Training | | #Testing | | | | | |
	Minor	Major	Minor	Major	Imbalance ratio	Precision	Recall	F1
EGAN	30	1080	1000	1000	1:36	0.51	0.97	0.67
Baseline	30	1080	1000	1000	1:36	0.5	1.0	0.67

a nash equilibrium (with threshold less than or equal to 5) is achieved, with 76 and 71 of minority and majority correctly classified respectively. The maximum F1-score of 0.78 is reached at epoch 53 as shown in Table 5. Instead of using 500 samples each of minority and majority class, by training on a single instance of minority and majority sample (*1:1) of CIFAR-100, we obtained an F1-score of 0.63. This demonstrates the impact of transfer learning on the training.

Class Ratio of 1:36. For pneumonia subset of CheXpert dataset with imbalance ratio 1:36, the best performed model achieves 0.51, 0.97, and 0.67 precision, recall, and F1-score respectively. The results shown in Table 6 is evaluated on 1000 of each minority and majority test set. As can be seen in the table, our approach did not beat the baseline classification model because this task is more of an anomaly detection task rather than a classification problem. Also, the pre-trained image classification model source dataset (i.e. ImageNet) is different from the medical domain. Exploring more variants of complex GAN architectures like BigGAN, StyleGAN and ProGAN could possibly help.

5   Conclusion

In this work, we demonstrates the capability of a GAN-based unsupervised technique to address class imbalance using pre-trained models. We conducts experiment with varying levels of imbalance ratios in the training dataset. Instead of synthesizing artificial images with the generator for data augmentation, we employ the discriminator as a classifier and formulate the loss function accordingly. Experimental results reveal that transfer learning plays a significant role in the model performance. The performance measure of interest plays a significant role in deciding the trained model from which epoch to deploy in production as the model at different epochs favour different evaluation metrics for example sensitivity. Future work will focus on the usage of this approach for anomaly detection task where the distinguishing features between normal (majority) and abnormal (minority) are less profound. Our work can be further explored in object detection tasks in case of imbalance between foreground and background.

References

1. Antoniou, A., Storkey, A., Edwards, H.: Data augmentation generative adversarial networks. arXiv preprint arXiv:1711.04340 (2017)

2. Bria, A., Marrocco, C., Tortorella, F.: Addressing class imbalance in deep learning for small lesion detection on medical images. Comput. Biol. Med. **120**, 103735 (2020)
3. Drummond, C., Holte, R.C., et al.: C4. 5, class imbalance, and cost sensitivity: why under-sampling beats over-sampling. In: Workshop on Learning from Imbalanced Datasets II, vol. 11, pp. 1–8. Citeseer (2003)
4. Goodfellow, I.J., et al.: Generative adversarial networks. arXiv preprint arXiv:1406.2661 (2014)
5. Huang, G., Liu, Z., Van Der Maaten, L., Weinberger, K.Q.: Densely connected convolutional networks. In: Proceedings of the IEEE Conference on Computer Vision and Pattern Recognition, pp. 4700–4708 (2017)
6. Huang, Y., Jin, Y., Li, Y., Lin, Z.: Towards imbalanced image classification: a generative adversarial network ensemble learning method. IEEE Access **8**, 88399–88409 (2020)
7. Irvin, J., et al.: Chexpert: a large chest radiograph dataset with uncertainty labels and expert comparison. In: Proceedings of the AAAI Conference on Artificial Intelligence, vol. 33, pp. 590–597 (2019)
8. Kingma, D.P., Welling, M.: Auto-encoding variational bayes. arXiv preprint arXiv:1312.6114 (2013)
9. Krizhevsky, A., Hinton, G., et al.: Learning multiple layers of features from tiny images (2009)
10. Ling, C.X., Sheng, V.S.: Cost-sensitive learning and the class imbalance problem. Encyclopedia Mach. Learn. **2011**, 231–235 (2008)
11. Nafi, N.M., Hsu, W.H.: Addressing class imbalance in image-based plant disease detection: Deep generative vs. sampling-based approaches. In: 2020 International Conference on Systems, Signals and Image Processing (IWSSIP), pp. 243–248. IEEE (2020)
12. Radford, A., Metz, L., Chintala, S.: Unsupervised representation learning with deep convolutional generative adversarial networks. arXiv preprint arXiv:1511.06434 (2015)
13. Sun, Y., Kamel, M.S., Wong, A.K., Wang, Y.: Cost-sensitive boosting for classification of imbalanced data. Pattern Recogn. **40**(12), 3358–3378 (2007)
14. Wang, Y., Wu, C., Herranz, L., van de Weijer, J., Gonzalez-Garcia, A., Raducanu, B.: Transferring gans: generating images from limited data. In: Proceedings of the European Conference on Computer Vision, pp. 218–234 (2018)
15. Zhang, C., Zhou, P., Li, C., Liu, L.: A convolutional neural network for leaves recognition using data augmentation. In: 2015 IEEE International Conference on Computer and Information Technology; Ubiquitous Computing and Communications; Dependable, Autonomic and Secure Computing; Pervasive Intelligence and Computing, pp. 2143–2150. IEEE (2015)
16. Zhu, J.Y., Park, T., Isola, P., Efros, A.A.: Unpaired image-to-image translation using cycle-consistent adversarial networks. In: Proceedings of the IEEE International Conference on Computer Vision, pp. 2223–2232 (2017)

Progressive Contextual Excitation
for Smart Farming Application

Chia-Hung Bai[✉], Setya Widyawan Prakosa[✉], He-Yen Hsieh[✉],
Jenq-Shiou Leu[✉], and Wen-Hsien Fang[✉]

National Taiwan University of Science and Technology, Taipei, Taiwan
{m10502219,d10702804,m10502103,jsleu,whf}@mail.ntust.edu.tw

Abstract. This paper attempts to address the issue of smart farming application, which targets discriminating distinct cocoa bean categories. In smart farming application, one critical issue is how to distinguish little difference among all categories. Our proposed scheme is designed to construct a more robust representation to better leverage textual information. The key concept is to adaptively accumulate contextual representations to obtain the contextual channel attention. Specifically, we introduce a contextual memory cell to progressively select the contextual channel-wise statistics. The accumulated contextual statistics are then used to explore the channel-wise relationship which implicitly correlates contextual channel states. Accordingly, we propose the progressive contextual excitation (PCE) module employing channel-attention-based architecture to simultaneously correlate the contextual channel-wise relationships. The progressive manner via the contextual memory cell demonstrates efficiently to guide high-level representation by keeping more detailed information, which benefits to discriminate small variations in tackling the smart farming application task. We evaluate our model on the cocoa beans dataset which comprises fine-grained cocoa bean categories. The experiments show a significant boost compared with existing approaches.

Keywords: Deep learning · Progressive contextual excitation · Smart farming

1 Introduction

Image classification serves as an active research topic in computer vision tasks. Benefited from the effectiveness of convolutional neural networks (CNNs), a significant improvement has been demonstrated in a wide range of vision-related works. With rapid development in information technology, an artificial-intelligence-based (AI-based) method has been applied to the area of agriculture on a large scale. Several researchers [3,8,10,11,15,18,21,22] attempt to optimize the farming process with an AI-based strategy in the smart farming application (SFA). Wayan et al. [18] propose to improve the classification rate of cocoa beans based on a support vector machine equipped with image processing technology. Tan et al. [15] propose an electronic-nose-based system to assist farmers

© Springer Nature Switzerland AG 2021
N. Tsapatsoulis et al. (Eds.): CAIP 2021, LNCS 13052, pp. 332–340, 2021.
https://doi.org/10.1007/978-3-030-89128-2_32

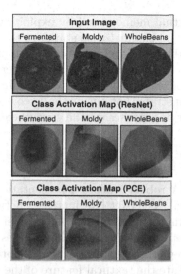

Fig. 1. Progressive Contextual Excitation (PCE): The proposed PCE module progressively leverages contextual representations to excite high-level representation. Specifically, PCE selects the contextual representations to obtain channel attention, enabling high-level representation to involve rich detailed information from low-level representations, which benefits to discriminates small variations.

to determine the fermentation rate and quality of cocoa beans. Adhitya et al. [2] improve the classification rate of cocoa beans with the additional gray level co-occurrence matrix (GLCM) textural features. These approaches [2,18] attempt to enhance the feature representations for the smart farming application task. However, the task of smart farming application faces a more challenging issue as little difference presented among distinct categories shown in Fig. 1. We adopt a ResNet-based model [7] to serve as our baseline and apply it to the SFA task. While, the ResNet-based model is unable to discriminate the little difference shown in the second row of Fig. 1. Inspired by the *squeeze-and-excitation* mechanism [9], we focus on the exploration of contextual channel-wise statistics through multi-level representations. In this paper, we propose a Progressive Contextual Excitation (PCE) model to adaptively accumulate multi-level representations via a gate mechanism. A contextual representation containing rich information is then retrieved. We then explore the channel-wise relationship with the accumulated contextual representation to obtain the channel attention, which is called *contextual channel attention*. Therefore, a high-level representation is guided by the contextual channel attention to keep rich detailed information. Consequently, we show a significant improvement by exploiting contextual channel attention in our experiments. Figure 2 illustrates our smart farming application model, called as Progressive Contextual Excitation (PCE). The main contributions of our model tackling to smart farming application task as follows:

1. We introduce a contextual memory cell by exploiting multi-level representations to obtain contextual channel-wise statistics to address the SFA task.
2. We suggest enriching the high-level representation guided by contextual representations concerning contextual channel-wise attention which involves rich detailed information for effectively discriminating fine-grained categories.
3. The experiments demonstrate that our method achieves superior performance in comparison to other approaches on the cocoa bean dataset adopted in [2].

2 Related Work

Feature Representation. A robust feature representation is vital for the image classification task [7,13,14]. VGGNet [13] increases network depth to retrieve a powerful representation. Inception network [14] considers multi-scale contexts to strengthen the representations. ResNet [7] manipulates the identity-based connections to construct a deeper network for obtaining more robust representations. Adhitya et al. [2] incorporate the textural feature of the gray level co-occurrence matrix (GLCM) [6] into an off-the-shelf network to tackle the smart farming application task. This paper adopts the ResNet [7] as the visual extractor to retrieve the visual representations and further leverages contextual representations.

Attention Mechanism. The attention mechanism has demonstrated significant advance in many fields, such as vision related tasks [5,9,16] and language-processing tasks [4,17]. These methods indicates the attention mechanism as a useful method to explore element-wise relationship [17] or channel-wise relationship [9]. With the powerful feature enhancement strategy, the attention mechanism has been successfully applied in the image captioning [19], image question answering [20], and referring image segmentation [12] task. Vaswani et al. [17] adopt the self-attention mechanism to explore the data relationship for adjusting the representations. Hu et al. [9] manipulate the channel-wise relationship to enhance the features within a deep network. In this paper, we propose to accumulate contextual representations and further explore the channel-wise relationship with the accumulated information. Hence, the obtained channel attention incorporates multi-level cues to better guide the high-level representation keep detailed information. The proposed progressive contextual excitation (PCE) model correlates the channel-wise statistics among multi-level representations and achieves significant improvement compared to other approaches in addressing the smart farming application task.

3 Method

This section illustrates the proposed key components of our approach to tackle the smart farming application task. Figure 2 sketches the overall scheme of the proposed Progressive Contextual Excitation (PCE) model. Our model is based on a visual extractor, where we adopt ResNet-50 [7] as the backbone, to tackle

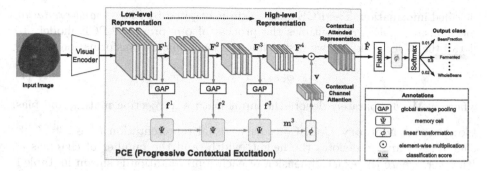

Fig. 2. The framework of our PCE model. Given an input image, we employ a visual encoder to extract the visual representations. The PCE model accumulates multi-level representations through a gating selection mechanism. We further explore the channel-wise relationship with the accumulated information concerning contextual representations. Next, the contextual channel attention is obtained via a linear transformation. Therefore, high-level representation leverages the contextual channel attention for better discriminating fine-grained categories.

the SFA task. Given an image belonging to one of the cocoa bean categories, we aim to predict a correct class. As expected, a classifier discriminates distinct cocoa bean types based on visual representations. Here, we illustrate the workflow of our model. The first process begins with a visual extractor to generate contextual representations. To efficiently leverage context information, a PCE model is employed to progressively select contextual features through a gate mechanism. One linear transformation layer is employed to obtain contextual channel attention followed by a sigmoid operation. We then multiply high-level representation with the contextual channel attention. Finally, a linear transformation layer can efficiently determine the belonging category although a small variation exists.

3.1   Feature Extraction

We employ the ResNet-50 [7] as our visual extractor. Given an image $I \in \mathbb{R}^{H^I \times W^I \times 3}$, we extract the multi-level representations $\left\{ \mathbf{F}^l \right\}_{l=1}^{4}$ via the extractor's four different layers. For a representation in a lower level, the spatial semantics comprise rich detailed information. A high-level representation contains abstract information that enables the network to discriminate distinct object types.

3.2   Progressive Contextual Excitation

Our proposed Progressive Contextual Excitation (PCE) aims to obtain channel attention from multi-level cues for guiding high-level representation to keep more detailed information. The PCE mechanism is inspired by *squeeze-and-excitation* mechanism [9]. Figure 2 outlines the process of our proposed PCE model. To detail the PCE workflow, we first define the basic linear transformation of ϕ by

$$\phi(x) = x\mathbf{W} + b, \tag{1}$$

where x, $\mathbf{W}$, b separately denote the input feature, projection matrix, and bias.

Contextual Memory Cell. Consider a visual representation $F^l \in \mathbb{R}^{h^l \times w^l \times c^l}$ that h^l, w^l, and c^l denotes the height, width, and the number of channels at lth layer, where the exact dimension of each representation is shown in Table 1. We employ a *global-average-pooling* mechanism to obtain channel-wise statistics $f^l \in \mathbb{R}^{c^l}$. Next, we adopt a memory cell to adaptively select input information and integrate the contextual channel-wise statistics. The memory cell Ψ can be defined by

$$i^l = \sigma(\phi(f^l) + \phi(m^{l-1})), \tag{2}$$
$$o^l = \sigma(\phi(f^l) + \phi(m^{l-1})), \tag{3}$$
$$s^l = tanh(\phi(f^l) + \phi(i_l \cdot m^{l-1})), \tag{4}$$
$$m^l = o^l \cdot m^{l-1} + (1 - o^l) \cdot s^l, \tag{5}$$

where f^l, i^l, o^l, s^l, σ, $tanh$, m^l indicate the channel-wise statistics, input gate, output gate, candidate state, sigmoid activation function, hyperbolic tangent function, and memorized cue, respectively. Here, the memorized cue m^l incorporates contextual channel-wise representations to involve fine-grained visual information.

Contextual Channel Attention. Consider the final contextual channel-wise representations $m^3 \in \mathbb{R}^{c^4/r}$ in Fig. 2 that the dimension of m^3 is c^4 divided by a reduction factor r, where $c^4 = 2048$ and r is set to 16 empirically. We explore contextual channel-wise dependence through a linear transformation to learn the channel-wise relationship. The contextual channel attention $\mathbf{v}$ is thus determined by $\mathbf{v} = \sigma(\phi(m^3))$, where $\mathbf{v} \in \mathbb{R}^{c^4}$. Next, we define the contextual attended representation concerning the contextual channel-wise statistics as

$$\hat{\mathbf{F}}_j = \mathbf{F}_j^4 \cdot \mathbf{v}_j, \tag{6}$$

where $\mathbf{F}_j^4$ and $\mathbf{v}_j$ represent the high-level representation and contextual channel attention at the jth channel. The contextual attended representation $\hat{\mathbf{F}}$ concerns low-level visual representations which involve fine-grained information and serves as the input to the following linear transformation for determining the category.

4   Experiments

4.1   Dataset

The cocoa beans dataset [1] is an available benchmark for fine-grained cocoa bean classification and smart farming application task. The dataset contains 7482 images comprising the following categories: (1) *whole beans* indicate a cocoa bean without any fracture covered by a whole seed coat; (2) *beans fractions* contain fractions less than half of the whole cocoa bean. (3) *skin-damaged beans* lack part of bean shell which the corresponding size is less than half of the whole cocoa bean. (4) *fermented beans* indicate a cocoa bean after curing process including washed or dried procedure serves as the final product. (5) *unfermented beans* comprise more than half of observable sliced grayish chips' surface. (6) *moldy beans* contain visible mold and fungus inside. We randomly divide the dataset into 75% for training, 15% as validation split, and the rest 10% for testing split.

4.2   Implementation Details

We adopt ResNet-50 [7] as our visual encoder to obtain multi-level visual representations. The visual encoder is trained from scratch without using any pre-trained weight of other datasets. The input image is resized to 224×224 pixels through the bilinear interpolation. We use the SGD optimizer to train our model with a batch size of 128. The learning rate is set to $1e^{-2}$ initially and decayed by $5e^{-4}$. Table 1 represents the exact dimension for each representation at the lth layer.

Table 1. The dimensions of multi-level representations. Each visual representation is applied by the global average pooling (GAP) to obtain the channel-wise statistics. Subsequently, we adopt a linear transformation ϕ to coordinate the shape of contextual channel-wise statistics. The dimension is sequenced by *height* $\times$ *width* $\times$ *channel*. The symbol ϕ in (6) denotes the linear transformation.

Representation	Dimension	GAP	ϕ	Projected dimension
$\mathbf{F}^1$	$256 \times 56 \times 56$	Yes	Yes	2048
$\mathbf{F}^2$	$512 \times 28 \times 28$	Yes	Yes	2048
$\mathbf{F}^3$	$1024 \times 14 \times 14$	Yes	Yes	2048
$\mathbf{F}^4$	$2048 \times 7 \times 7$	No	No	-

4.3   Comparison with Other Approaches

Table 2 compares our approach with other methods on the cocoa beans testing splits with the top-1 accuracy metric. The accuracy metric is defined as $\frac{TP+TN}{TP+TN+FP+FN}$, where TP, TN, FP, and FN separately represent true positive, true negative, false positive, and false negative. The comparison shows that our approach achieves a better performance with a significant improvement at least of 3.38%. Without any post-processing of the gray level co-occurrence matrix (GLCM) [6], our model still achieves superior performance compared to Adhitya's model [2]. Concerning the involvement of contextual channel-wise statistics, our proposed PCE outperforms the ResNet-50 based model, which means that high-level visual representation considering contextual representations to leverage detailed information is beneficial for discriminating the fine-grained cocoa beans categories.

Table 2. Results on the cocoa beans testing split. The notation "*" indicates the model applying GLCM [6] for post-processing to enhance the visual features.

Model	Post-processing	Top-1 Accu
Adhitya's model [2] (SVM)	No	59.14
Adhitya's model [2] (XGBoost)	No	56.99
Adhitya's model [2] (SVM*)	Yes	61.04
Adhitya's model [2] (XGBoost*)	Yes	65.08
ResNet-50	No	82.71
PCE	No	**86.09**

4.4   Visualization

To elaborate the effectiveness of our proposed PCE model, we employ the class activation mapping (CAM) [23] mechanism to analyze the spatial regions being focused during the fine-grained category prediction. We visualize the two types of cocoa beans, namely *fermented beans* and *whole beans* on the first row and second row in Fig. 3, respectively. The second column indicates the class activation mapping generated from the ResNet-50 model. The third column shows the result of our PCE model. With our contextual channel-wise attention, the model concentrates more on the accurate spatial regions compared to the ResNet-50.

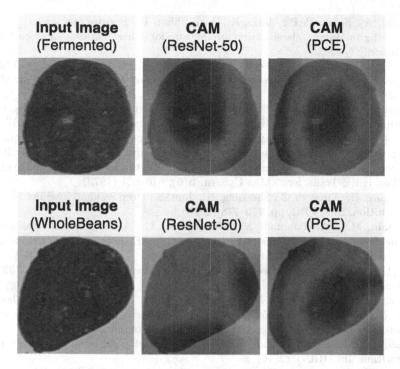

Fig. 3. Class Activation Mapping (CAM). The first row visualizes the CAM from the ResNet-50 and our proposed PCE model on the *fermented beans* category. And the second row shows the results of the type of the *whole beans*. Our model (the third column) captures more precise spatial regions compared to the ResNet-50 model. We attribute the effectiveness to the advantage of our contextual channel-wise attention which efficiently distinguishes fine-grained categories.

5   Conclusion

We have demonstrated the effectiveness of our PCE model and further boost the performance to tackle the task of the smart farming application with fine-grained categories. The proposed PCE model correlates the contextual channel-wise statistics. Furthermore, the contextual channel attention enables the high-level representation to efficiently manipulate detailed information for discriminating categories with small variations. As a result, our model significantly improves the performance compared with other approaches on the cocoa beans benchmark.

Acknowledgments. The authors gratefully acknowledge the support by the Ministry of Science and Technology, Taiwan, under grant MOST-110-2221-E-011-013 -

References

1. Badan standardisasi nasional (bsn). In Biji kakao SNI 2323:2008 ICS 1.67.140.30 Kakao.; Badan Standardisasi Nasional: Jakarta, Indonesia (2008)

2. Yudhi, A., Setya, W.P., Mario, K., Jenq-Shiou, L.: Feature extraction for cocoa bean digital image classification prediction for smart farming application. In: Agronomy (2020)
3. Bacco, M., Barsocchi, P., Ferro, E., Gotta, A., Ruggeri, M.: The digitisation of agriculture: a survey of research activities on smart farming. Array **3–4**, 100009 (2019)
4. Dzmitry, B., Kyunghyun, C., Yoshua, B.: Neural machine translation by jointly learning to align and translate. In: Yoshua, B., Yann, L., (eds.) ICLR (2015)
5. Denil, M., Bazzani, L., Larochelle, H., de Freitas, N.: Learning where to attend with deep architectures for image tracking. Neural Comput. **24**(8), 2151–2184 (2012)
6. Robert, M., Haralick, K., Sam, S., Its'hak, D.: Textural features for image classification. IEEE Trans. Syst. Man Cybern. **3**(6), 610–621 (1973)
7. Kaiming, H., Xiangyu, Z., Shaoqing, R., Jian, S.: Deep residual learning for image recognition. In: CVPR, pp. 770–778 (2016)
8. Hossain, M.S., Al-Hammadi, M., Muhammad, G.: Automatic fruit classification using deep learning for industrial applications. In: IEEE Transactions on Industrial Informatics, pp. 1027–1034 (2019)
9. Jie, H., Li, S., Gang, S.: Squeeze-and-excitation networks. In: CVPR, pp. 7132–7141 (2018)
10. Behera, S.K., Rath, A.K., Mahapatra, A., Sethy, P.K.: Identification, classification & grading of fruits using machine learning & computer intelligence: a review. Journal of Ambient Intelligence and Humanized Computing (2020)
11. Mahajan, M.S.: Optimization and classification of fruit using machine learning algorithm. In: IJIRST (2016)
12. Shi, H., Li, H., Meng, F., Wu, Q.: Key-word-aware network for referring expression image segmentation. In: Ferrari, V., Hebert, M., Sminchisescu, C., Weiss, Y. (eds.) ECCV 2018. LNCS, vol. 11210, pp. 38–54. Springer, Cham (2018). https://doi.org/10.1007/978-3-030-01231-1_3
13. Karen, S., Andrew, Z.: Very deep convolutional networks for large-scale image recognition. In: ICLR (2015)
14. Christian, S., et al.: Going deeper with convolutions. In: CVPR, pp. 1–9 (2015)
15. Juzhong, T., Balu, B., Darin, S., Saila, R., Pathmanathan, U.: Sensing fermentation degree of cocoa (theobroma cacao l.) beans by machine learning classification models based electronic nose system. Journal of Food Process Engineering (2019)
16. Yichuan, T., Nitish, S., Ruslan, S.: Learning generative models with visual attention. In: NIPS, pp. 1808–1816 (2014)
17. Ashish, V., et al.: Attention is all you need. In: NIPS, pp. 5998–6008 (2017)
18. Astika, I.W., Solahudin, M., Kurniawan, A., Wulandari, Y.: Determination of cocoa bean quality with image processing and artificial neural network. In: AFITA (2010)
19. Kelvin, X., et al.: Show, attend and tell: neural image caption generation with visual attention. In: ICML, vol. 37, pp. 2048–2057 (2015)
20. Zichao, Y., Xiaodong, H., Jianfeng, G., Li, D., Alexander, J.S.: Stacked attention networks for image question answering. In: CVPR, pp. 21–29 (2016)
21. Aubain, Y., Camille, E., Kidiyo, K.: Cocoa beans fermentation degree assessment for quality control using machine vision and multiclass svm classifier. In: International Journal of Innovation and Applied Studies, pp. 1711–1717 (2018)
22. Zawbaa, H.M., Maryam, H., Mona, A., Hassanien, A.E.: Automatic fruit classification using random forest algorithm. In: HIS, pp. 164–168 (2014)
23. Bolei, Z., Aditya, K., Àgata, L., Aude, O., Antonio, T.: Learning deep features for discriminative localization. In: CVPR, pp. 2921–2929 (2016)

Fine-Grained Image Classification
for Pollen Grain Microscope Images

Francesca Trenta$^{(\boxtimes)}$ ⓘ, Alessandro Ortis ⓘ, and Sebastiano Battiato ⓘ

IPLAB, University of Catania, Catania 95125, Italy
francesca.trenta@unict.it, {ortis,battiato}@dmi.unict.it

Abstract. Pollen classification is an important task in many fields, including allergology, archaeobotany and biodiversity conservation. However, the visual classification of pollen grains is a major challenge due to the difficulty in identifying the subtle variations between the subcategories of objects. The pollen image analysis process is often time-consuming and require expert evaluations. Even simple tasks, such as image classification or segmentation requires significant efforts from experts in aerobiology. Hence, there is a strong need to develop automatic solutions for microscopy image analysis. These considerations underline the effort to study and develop new efficient algorithms. With the growing interest in Deep Learning (DL), much research efforts have been spent to the development of several approaches to accomplish this task. Hence, this study covers the application of effective Deep Learning methods in combination with Fine-Grained Visual Classification (FGVC) approaches, comparing them with other Deep Learning-based methods from the state-of-art. All experiments were conducted using the dataset Pollen13K, composed of more than 13,000 pollen objects subdivided in 4 classes. The results of experiments confirmed the effectiveness of our proposed pipeline that reached over 97% in terms of accuracy and F1-score.

Keywords: Pollen classification · Fine-grained visualization · Machine learning

1 Introduction

With the rapid development of technologies in the field of Artificial Intelligence (AI), image data analysis has attracted much research attention over the last few years. In particular, typical problems in Computer Vision and Machine Learning field are related to image classification tasks. Indeed, image classification embraces several issues including discriminative feature extraction. The rapid emergence in developing such innovative pipeline to solve image classification has led to the spread of AI methods for extracting features from images. In this regard, Deep Learning (DL) approaches, provided a remarkable contribution. In fact, the main advantage of DL methods is the capacity to automatically

© Springer Nature Switzerland AG 2021
N. Tsapatsoulis et al. (Eds.): CAIP 2021, LNCS 13052, pp. 341–351, 2021.
https://doi.org/10.1007/978-3-030-89128-2_33

learn meaning features from high volume of data, rather than traditional ML solutions which involve the design of hand crafted features. The most valuable techniques include the use of neural networks such as AlexNet [13], ResNet [10], EfficientNet [17], which achieved effective results in a large variety of classification problems. In this work, we proposed an effective pipeline to perform image classification, making use of promising solutions which have reached state-of-art results in wide range of applications. Specifically, this study investigated the problem of classifying pollen grains having similar appearance. The dataset used for the experiments is Pollen13K[1] [2], composed of more than 13,000 pollen objects. In particular, the Pollen13K dataset includes 5 categories of objects. However, we considered the 4 classes and the train/test data splitting used during the International Pollen Grain Classification Challenge 2020. The dataset is publicly available, however, due to the nature of the competition, details about the employed methods are missing [3]. The classification of pollen objects has become a hot research topic in the field of aerobiology. Hence, the automation of pollen classification that could operate largely independently of a human operator would be of great benefit. Motivated by these considerations, we defined an innovative pipeline to improve pollen grains classification by using a Fine-Grained Visual Classification (FGVC) based approach [6]. The methods consists of a progressive training step and the application of a jigsaw patches generator in order to extract information from images at different granularity. We also implemented a Test-Time Augmentation (TTA) method to improve object classification predictions.

The paper is organised as follows. In Sect. 2, we report the most interesting work regarding image classification, outlining the most promising approaches to solve this task. In Sect. 3, we report the pipeline used to classify pollen images, detailing the approaches for improving classification predictions. Section 4 details the experiments, giving an overview of the methods used, comparing them to other state-of-the-art methods. Section 5 reports the experiments details regarding other DL approaches used to perform a benchmarking evaluation. In Sect. 6, we discuss the results of the experiments. Finally, Sect. 7 outlines the conclusions.

2   Related Works

In recent years, there has been a growing interest in implementing effective methods for object classification. In this regard, we provide a brief survey of the recent advances in Deep Learning, outlining the most significant contributions to solving image classification, these approaches can be summarized into three categories, reported as follows.

Training Data Augmentation. A number of data augmentation strategies have been proposed over the years [15]. The most used techniques encompass the application of simple geometric transformations, such as horizontal flipping,

---

[1] more details are available on the dataset website: https://iplab.dmi.unict.it/pollengraindataset/dataset.

color space augmentations, and random cropping, devoting to increasing the amount of training data for neural networks [13]. Recent works have shown that forming new artificial samples by combining two or more images from training data can lead to significant improvements in the performance of neural networks. Modern works include approaches such as MixUp [21], CutMix [20], and CutOcclusion [9]. In [21], the authors propose an innovative approach for creating a new example by performing a weighted linear interpolation of two existing images. In [20], the authors implemented a method to encourage the model to focus on less prominent parts of an image. The strategy is based on replacing a region from an image with patches from another one. The added patches further improve localization capability of the model by identifying the object considering a partial view. Penghui et al. [9] define a novel method for data training augmentation forcing the neural network to pay more attention to the surrounding area of a given object by including an occlusion region into an image. Specifically, the proposed approach was designed for the classification of pollen grains. In particular, the authors demonstrated that the most discriminating parts rely on the surrounding area of pollen object. Therefore, they introduced black patches around the center point of image in order to force the network to learn information from pollen wall and aperture area.

Fine-Grained Visual Classification. In recent years, the most promising solutions were devoted to the analysis of the granularity of images. Although neural networks such as AlexNet [13], ResNet [18], etc. have achieved remarkable results in image classification task, these models often fail to discriminate objects presenting a limited intra-class variation. For this reason, the key for improvement is represented by FGVC-based approaches. For example, Chen et al. [5], defined a method to "destruct" and "reconstruct" images. With regard to the "destruction" stage, the authors subdivided input images into k patches. Then, they shuffled them in order to create a new sample. For "construction", the authors implemented a region alignment mechanism to force the model to restore the spatial layout of image regions. The main advantage of this method is the capability of the model to pay more attention on local parts of the images than global features.

Tricks for Improving Classification Predictions. One of the most used techniques for improving class predictions is Test Time Augmentation (TTA). This approach has been applied to several works including [16], where the authors proposed a test augmentation by applying horizontal flipping to input images, and [12], where the authors propose a test time augmentation method based on dynamically selecting transformations according to the loss function. Basically, the idea behind TTA method is based on performing a data augmentation on the test set in order to create different variants of the same image and perform the prediction on them. In general, a system of soft voting is implemented to determine which prediction is the most voted in order to assign a certain label. Several works propose the average of the resulting predictions or the sum of the probabilities to determine the confidence of the model.

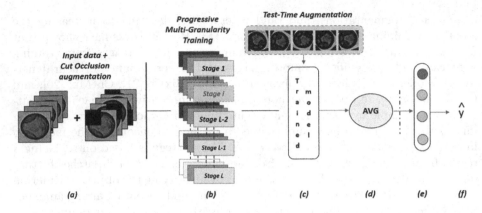

Fig. 1. The overall pipeline. (a) Input data consisting of pollen grains from Pollen13K and augmented dataset with Cut Occlusion. (b) Training performed using progressive multi-granularity strategy. (c) Test-time augmentation. (d) Average calculation of predictions. (e) Max value of each predictions. (f) Predicted label.

3   Method and Materials

3.1   *CutOcclusion*: Training Data Augmentation

In [9], data augmentation is performed by operating Cut Occlusion strategy. As mentioned previously, the strategy was shown to be effective for pollen object classification. Inspired by the results of Penghui et al. [9], we reproduce the Cut Occlusion strategy in order to create new instances of the training data. The main advantage of this approach consists in avoiding some parts of the images by means of black patches in order to help the model to extract discriminative features from pollen wall. This strategy can bring substantial improvements for the pollen classification task, where extracting discriminative features from aperture area of pollen grain instance seems to be more important than concentrating on the center area of the pollen object.

3.2   Pipeline

In Fig. 1, the full pipeline is depicted. The architecture design was firstly introduced in [6], which tackles the image classification task by introducing a novel approach based on Progressive Multi-Granularity training strategy (PMG). As discussed, pollen objects, from the Pollen13K dataset, belonging to different classes, present a similar appearance. In addition, objects in the same category could report a varying appearance. Therefore, applying a fine-grained visual classification approach, taking advantage of local features information, could lead to remarkable improvements. The framework consists of two main components: (a) a progressive training method to add new layers during training process in order to extract discriminative features from images with different granularities. Hence, the process starts at low stage and progressively include new layers. (b) a jigsaw patches generator [19] to capture local information from images. In [6], the authors used ResNet50 as backbone. In our study, we use ResNet101 [10] as feature extractor. For each layer $\mathcal{L}$ of the feature extractor, a new convolution layer is added taking as input the feature maps from the output of intermediate layers that is transformed into a vector representation. Then, classification modules are added to calculate the probability distribution between classes. Finally, the outputs of the last levels are concatenated.

Progressive Training. This technique allows to train the model starting from the low stage and then adding new layers. The advantage of this technique consists in forcing the model to learn discriminative information from local details rather than focusing on global information. The loss cross entropy function $\mathcal{L}_{CE}$ is applied to the output of each stage and the output of the concatenated features.

Jigsaw Puzzle Generator. To train the PMG model, we define a set of jigsaw puzzle permutations. This approach has been widely employed to find the multi-granularities of the images during training stage. Given an image x, it can be subdivided into k patches. Then, the patches are shuffled randomly and merged together into a new image x'.

3.3   Test Time Augmentation.

In order to boost the prediction accuracy, we implemented a strategy called Test Time Augmentation (TTA), which consists of applying data augmentation techniques to the test set in order to improve the prediction of a given class of objects. For this reason, we create several variants of the same image, applying a horizontal or vertical flip, standard color augmentation, or other geometric transformations. We computed a prediction for each of these images. Then, we average these predictions and calculate the max value in order to obtain which prediction has the highest confidence score. Finally, we computed the predicted class for the analysed object. By applying this strategy, we avoid the uncertain of the model by averaging the predictions and averaging the error. In this context, we created 5 different variants for each single image of the test set by applying a horizontal flip, a vertical flip and a random rotation. The rotation angle ranges from $-90°$ to $90°$. In addition, as in the training set, we performed an image resize of 550×550 pixels and a centre crop of 448×448 pixels. Moreover we normalized data setting with a mean and a standard deviation of 0.5.

4   Dataset

The dataset Pollen13K [2] is composed of $13,416$ objects divided into 5 categories, respectively: *Coryllus Avellana* (well-developed), *Coryllus Avellana* (anomalous), Alnus, *Cuprissaceae*, and Debris. However, considering the small number of observations related to *Cupressaceae* class (43), we did not include them in the dataset used for the experiments. Hence, the dataset is composed of images depicting one pollen type among the 4 mentioned categories, these patches have been manually labelled by experts in the field of aerobiology. The dataset includes: (1) 84 × 84 RGB images for each segmented object, for each of the four categories; (2) binary masks for single object segmentation (84 × 84 resolution); (3) segmented versions of the patches obtained by applying the segmentation mask and padding the background with all green pixels (84 × 84 resolution).

5   Experiments

To evaluate the performance of the proposed pipeline, rigorous experiments are performed on two image datasets: Pollen13K [2] and Augmented Pollen13K. We implemented the proposed approaches as well as several state-of-the-art approaches on these datasets. The experimental settings are given in the following subsections.

Proposed Pipeline: CutOcclusion + PMG + TTA. In order to boost the predictions, we performed dataset augmentation by using Cut Occlusion strategy [9]. We inserted occlusions around the center of pollen area. In particular, we created 4 different variants for each image of the training set. With regard to PMG method [6], all experiments were conducted using PyTorch [14] over a cluster of GPU NVIDIA® T4. We employed ResNet101 [10] as backbone. All setting are indicated in [6], where $S = 3$, $\alpha = 1$, and $\beta = 2$. In addition, the input images were resized to 550×550 pixels and randomly cropped by 448×448 pixels. A random horizontal flipping is applied for data augmentation for training data. We use Stochastic Gradient Descent (SGD) [11] optimizer and batch normalization as the regularizer. We train the model for 100 epochs. The batch size was set to 16. Moreover, we used a weight decay of 0.0005 and a momentum of 0.9. Finally, we performed TTA algorithm to improve the performance of our proposed pipeline.

Other Approaches. We use the Pytorch [14] Deep Learning library for performing the experiments related to other advanced DL networks: ResNet101 and Residual Attention Network (ResAttNet). We resized input images by 256×256 pixels. A 224×224 center crop is sampled from an augment image, applying geometric transformations. The network is trained using Stochastic Gradient Descent (SGD) with a momentum of 0.9. We set initial learning rate to 0.0001, decaying learning rate by a factor of 0.1 every 7 epochs. We set the number of epochs to 100. All the experiments use a batch size of 16. With regard to the methods based on CutMix strategy [20], we set hyperparameters values to $\beta = 1.0$ and *cutmix probability* to 0.5.

WRS Method. In this study, we evaluate the performance of the aforementioned algorithms using the Pollen13K dataset [2]. To the best of our knowledge, it represents the public dataset with the largest number of pollen objects, with

Table 1. Comparison between DL approaches. On the top part of the table, we reported evaluation results without applying TTA. On the bottom part of the table, we reported results by applying TTA method.

Method	Accuracy	F1 (weighted)	F1 (macro)
WRS + ResNet101	92.667 %	92.899 %	88.233 %
WRS + ResAttNet	83.776 %	84.497 %	77.627 %
WRS + CutMix + ResNet101	93.922 %	94.046 %	90.097 %
PMG [6]	**96.384 %**	**96.349 %**	**93.585 %**
Method + TTA	Accuracy	F1 (weighted)	F1 (macro)
WRS + ResNet101 + TTA	94.626 %	94.702 %	91.266 %
WRS + ResAttNet + TTA	83.626 %	84.748 %	80.416 %
WRS + CutMix + ResNet101 + TTA	95.228 %	95.280 %	92.581 %
CutOcclusion + PMG + TTA (Proposed)	**97.087 %**	**97.050 %**	**94.726 %**

more than 13,000 objects. However, the Pollen13K dataset consists of imbalanced classes since the largest class consists of the objects from class Alnus (8,216 objects in the train set). Other classes include a total number of objects less than 1,600. Motivated by these issues, we provided an effective solution by implementing the Weighted Random Sampler (WRS) function to deal with imbalanced dataset and preventing overfitting problems. Hence, one of the proposed solution is to oversample minority classes [4]. By applying this technique, we balanced batches of data. As a result, during training time, the model will not concentrate significantly on one class over another and risks of overfitting are reduced. Basically, the WRS method uses the array of weights which corresponds to weights given to each class. The goal is to assign a higher weight to the minor class, providing a more robust classification. Finally, we evaluate the performance of each classifier by also using the weighted and macro F1 score, which represent two more reliable performance metrics than accuracy. The weighted F1 score function calculates the F1 metrics for each class, and their average weighted by support (i.e., the number of true instances for each class). The F1 macro score computes the F1 for each label and returns the average.

6 Results and Discussion

This section presents the results obtained for a pool of Deep Learning algorithms, providing also a benchmarking evaluation of the performance of the techniques herein proposed for pollen grains classification. In Table 1, results show that PMG [6] method lead to a boost of the prediction accuracy than other methods. The main advantage of this approach include the analysis of images with different granularities. Basically, it forces each stage of the network to focus on local features rather than concentrating on global information. Furthermore, a jigsaw generator perform an image splitting into several patches during the training phase, providing discriminative information at the specific granularity level. Although the method based on Residual Attention Network (ResAttNet) produces remarkable results, it fails to outperform the other proposed methods. In both cases, we observed that the CutMix-based approach leads to better classification results than the pre-trained model (ResNet101) without using this strategy as data augmentation. According to these results, the method that yields to good results, both in terms of accuracy and F1 scores, is the approach based on progressive training and the use of jigsaw generator, i.e., PMG. With the attempt to further improve performance of the PMG model, we defined a data augmentation technique, based on Cut Occlusion, and a Test Time Augmentation to achieve higher accuracy during inference. The proposed pipeline yield to better results than other methods, confirming the effectiveness of the proposed framework. This strategy tends to improve the classification results and obtain results consistent with state-of-the-art. We reported the achieved results in Table 1. As observed, the TTA strategy provides reliable results in terms of accuracy and F1-score (weighted and macro) than other methods where this strategy was not applied to. With regard to ResAttNet, the accuracy value is decreased compared to the value from previous experiment. Instead, the F1-score metrics improve their value. In general, we observe that Deep Learning methods in combination with TTA strategy yields to better results than methods not using Test Time Augmentation strategy.

Misclassification: In Fig. 2, we report some examples of misclassification obtained by the standard PMG [6] algorithm without using training augmentation and TTA approach. As observed, the standard PMG approach [6] misclassifies objects of class 1 with objects of class 3 and vice versa. In fact, the object from these classes present similar characteristics, leading to a challenging image classification task. Furthermore, objects belonging to class 4 are indicated as class 3 objects. Probably, it depends on the presence of other objects within the image which fools the model, forcing it to extract their features and leading to a misclassification. However, the implementation of Cut Occlusion strategy allows us to avoid these objects, encouraging the network to focus on the object depicted at the center of the image, reporting better classification predictions.

Comparisons with Previous Studies. We also reported comparison between our proposed approach and previous studies for the classification of pollen grains. Fang et al. [7] propose a blending strategy consisting of a Destruction and Construction Learning architecture [5] and DenseNAS [8] output vectors to be used as the input of a Random Forest Classifier, which performs the final classification. Penghui Gui et al. [9] generated a number of images by applying the cut occlusion approach. The trained model is based on ResNet101. In our previous study [1], we investigated the performance of several Machine Learning approaches, such as AlexNet, SmallerVGGNet, etc. Fang et al. [7] leads to the best results in terms of accuracy (97.539) and F1-score (97.510), whereas Penghui Gui et al. achieved an accuracy of 97.290 and an F1-score of 97.260. Our previous method [1] achieved an accuracy of 89.730 % and an F1-score of 89.140%. In our experiments by using the proposed pipeline, we achieved an accuracy of

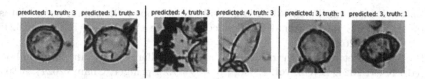

Fig. 2. Example of bad classification performed by PMG. These objects are classified accurately by PMG with training augmentation and TTA.

97.087 % and an F1-score (weighted) of 97.050 %, providing results similar to [7] and [9]. We also performed cross-validation, achieving good results in terms of accuracy (96.5%) and F1-score (96%). Although the experiments suggest that the algorithm we proposed provides better results, an accurate validation was not performed. On the contrary, our 3-fold cross-validation method has proved to provide more robust results.

7 Conclusions

In this paper, we tackled the problem of the classification of pollen grains by designing an innovative pipeline, consisting of a progressive training strategy and a jigsaw generator to extract information about image granularity. We further applied a data augmentation method to input images of the training set by forming 4 variants of each image. In addition, a Test Time Augmentation (TTA) method was implemented by applying simple geometric transformations to test images, creating 5 variants of each image in order to provide better classification results. The results show that our proposed pipeline has obtained a robust and consistent results with respect to state-of-the-art methods for image classification. One explanation of the performance improvement of the pipeline could depend on the dataset augmentation for both train and test which generally leads to a better generalization of the model, improving also its performance.

8 Future Works

As future development, we plan to collect more data with the aim of improving the effectiveness of the proposed approach. Specifically, we will address further application in the aerobiology field with special focus to designing more advanced DL pipelines to perform pollen object classification.

References

1. Battiato, S., et al.: Detection and classification of pollen grain microscope images. In: Proceedings of the IEEE/CVF Conference on Computer Vision and Pattern Recognition Workshops, pp. 980–981 (2020)

2. Battiato, S., et al.: Pollen13k: a large scale microscope pollen grain image dataset. In: 2020 IEEE International Conference on Image Processing (ICIP), pp. 2456–2460. IEEE (2020)
3. Battiato, S., et al.: Pollen grain classification challenge 2020. In: Del Bimbo, A. (ed.) ICPR 2021. LNCS, vol. 12668, pp. 469–479. Springer, Cham (2021). https://doi.org/10.1007/978-3-030-68793-9_34
4. Buda, M., et al.: A systematic study of the class imbalance problem in convolutional neural networks. Neural Netw. **106**, 249–259 (2018)
5. Chen, Y., et al.: Destruction and construction learning for fine-grained image recognition. In: Proceedings of the IEEE/CVF Conference on Computer Vision and Pattern Recognition, pp. 5157–5166 (2019)
6. Du, R., Chang, D., Bhunia, A.K., Xie, J., Ma, Z., Song, Y.-Z., Guo, J.: Fine-grained visual classification via progressive multi-granularity training of jigsaw patches. In: Vedaldi, A., Bischof, H., Brox, T., Frahm, J.-M. (eds.) ECCV 2020. LNCS, vol. 12365, pp. 153–168. Springer, Cham (2020). https://doi.org/10.1007/978-3-030-58565-5_10
7. Fang, C., Hu, Y., Zhang, B., Doermann, D., et al.: The fusion of neural architecture search and destruction and construction learning. In: Del Bimbo, A. (ed.) ICPR 2021. LNCS, vol. 12668, pp. 480–489. Springer, Cham (2021). https://doi.org/10.1007/978-3-030-68793-9_35
8. Fang, J., et al.: Densely connected search space for more flexible neural architecture search. In: Proceedings of the IEEE/CVF Conference on Computer Vision and Pattern Recognition, pp. 10628–10637 (2020)
9. Gui, P., Wang, R., Zhu, Z., Zhu, F., Zhao, Q., et al.: Improved data augmentation of deep convolutional neural network for pollen grains classification. In: Del Bimbo, A. (ed.) ICPR 2021. LNCS, vol. 12668, pp. 490–500. Springer, Cham (2021). https://doi.org/10.1007/978-3-030-68793-9_36
10. He, K., et al.: Deep residual learning for image recognition. In: Proceedings of the IEEE Conference on Computer Vision and Pattern Recognition, pp. 770–778 (2016)
11. Kiefer, J., et al.: Stochastic estimation of the maximum of a regression function. Ann. Math. Stat. **23**(3), 462–466 (1952)
12. Kim, I., et al.: Learning loss for test-time augmentation. arXiv preprint arXiv:2010.11422 (2020)
13. Krizhevsky, A., et al.: Imagenet classification with deep convolutional neural networks. Adv. Neural Inf. Process. Syst. **25**, 1097–1105 (2012)
14. Paszke, A., et al.: Automatic differentiation in pytorch (2017)
15. Simard, P.Y., LeCun, Y.A., Denker, J.S., Victorri, B.: Transformation invariance in pattern recognition – tangent distance and tangent propagation. In: Montavon, G., Orr, G.B., Müller, K.-R. (eds.) Neural Networks: Tricks of the Trade. LNCS, vol. 7700, pp. 235–269. Springer, Heidelberg (2012). https://doi.org/10.1007/978-3-642-35289-8_17
16. Simonyan, K., Zisserman, A.: Very deep convolutional networks for large-scale image recognition. arXiv preprint arXiv:1409.1556 (2014)
17. Tan, M., Le, Q.: Efficientnet: rethinking model scaling for convolutional neural networks. In: International Conference on Machine Learning, pp. 6105–6114. PMLR (2019)
18. Wang, F., et al.: Residual attention network for image classification. In: Proceedings of the IEEE Conference on Computer Vision and Pattern Recognition, pp. 3156–3164 (2017)

19. Wei, C., et al.: Iterative reorganization with weak spatial constraints: solving arbitrary jigsaw puzzles for unsupervised representation learning. In: Proceedings of the IEEE/CVF Conference on Computer Vision and Pattern Recognition, pp. 1910–1919 (2019)
20. Yun, S., et al.: Cutmix: regularization strategy to train strong classifiers with localizable features. In: Proceedings of the IEEE/CVF International Conference on Computer Vision, pp. 6023–6032 (2019)
21. Zhang, H., et al.: Mixup: beyond empirical risk minimization. arXiv preprint arXiv:1710.09412 (2017)

Adaptive Style Transfer Using SISR

Anindita Das[✉] [iD], Prithwish Sen[iD], and Nilkanta Sahu[iD]

Indian Institute of Information Technology Guwahati, Guwahati, India
anindita.das@iiitg.ac.in

Abstract. Style transfer is the process that aims to recreate a given image (target image) with the style of another image (style image). In this work, a new style transfer scheme is proposed that uses a single-image super resolution (SISR) network to increase the resolution of the given target image as well as the style image and perform the transformation process using the pre-trained VGG19 model. The Combination of perceptual loss and total variation loss is used which results in more photo-realistic output. With the change in content weight, the output image contains different semantic information and precise structure of the target image resulting in visually distinguishable results. The generated outputs can be altered accordingly by the user from artistic style to photo-realistic style by changing the weights. Detailed experimentation is done with different target image and style image pairs. The subjective quality of the stylised images is measured. Experimental results show that the quality of the generated image is better than the state of the art existing schemes. This proposed scheme preserves more information from the target image and creates less distortion for all combinations of different types of images. For more effective comparison, the contour of the stylizing images are extracted and also similarity is measured. This experiment shows that the result images have contour closer to the target images, also measured similarity is found maximum which indicates more preservation of semantic information than other existing schemes.

Keywords: Style transfer · SISR · NST · Artistic style · Photorealistic style

1 Introduction

Humans had been allured by the art of painting from the very beginning of civilization. Vincent van Gogh, Pablo Picasso and Leonardo da Vinci are few among many who enriched the art form with their unique style and skill. They are mastered in this artistic skill to make different visual representations by fusing the target and the styled representation together. Since 1990 [7], the theories after the appealing artworks had not only fascinated the artist but also many eminent researchers. Redrawing a painting with a particular style needs an experienced artist as well as a lot of time. Studies were made to explore techniques to automatically render images into synthetic artforms. A recent computer vision

© Springer Nature Switzerland AG 2021
N. Tsapatsoulis et al. (Eds.): CAIP 2021, LNCS 13052, pp. 352–361, 2021.
https://doi.org/10.1007/978-3-030-89128-2_34

technique, known as Style transfer [7], was used for the reconstruction of the target image with the style of another image. Input target and style images are blended into a reconstructed output image. This output includes the content of the target image, which seems to be "painted" with the style of the styled image. Some of the applications of style transfer include Photo and video editors, Commercial paintings, Gaming, VR and many more. At first, researchers started with Supervised Learning methods which employ input (Style and target pair images) to output mapping [9,17,19]. But the problem is, performance does not increase with the increase in data, also to create/find such a huge image pair dataset is impractical. Recently, Neural Style Transfer (NST) [7] has given a new direction to the problem at hand. In NST based approaches, neural networks extract statistical features of images (both target image and style image) related to content and style. Style features of the input image are changed with the help of features from the style image. With this upgraded approach, only one style reference is sufficient rather than using a pair of target and style images. NST can render both artistic and photorealistic images with varying amounts of content information. In the context of style transfer, we try to accomplish two competing objectives i.e. achieving transformation in local features and preserving geometric features simultaneously. Another challenge is the scene complexity of the real world while considering realism.

Thus, our goal is to synthesize styles in an image that can be made applicable to any input images creating artistic or photorealistic stylized images. In this proposed work, a new pipeline is proposed which can render the output image both in artistic style and photorealistic style by varying the amount of content information. The network must produce results with better visual quality, preserve more information from the target image, create less distortion for all combinations of different types of images, and distinguishable outputs with different weights.

2   Related Work

Since the success of deep learning based approaches, computer vision applications had progressed with rapid speed. The domain NST also had drawn a notable amount of attention. Although visual effect and performance had been drastically improvised, still the existing algorithms are insufficient to map the true visual interpretation among the target image and stylized image.

In the year 2015, Gatys et al. [7] introduced a pipeline that includes a pretrained VGG-19 architecture on ImageNet to extract the features representing semantic content and style. Capturing the artistic style of a painting with a scalable deep network and its construction was investigated in 2016 by Vincent Dumoulin et al. [5]. Some other works proposed like Park et al. [16] introduced a completely unique style-attentional network (SANet) which blends the local style properties based on the semantic spatial distribution of the target image. Nicholas Kolkin et al. [11] proposed a Style Transfer by Relaxed Optimal Transport and Self-Similarity that allows user-specific point-to-point for better visual output.

Li et al. [14] analyzed the system of transfer network and divided it into three major steps i.e. style transfer, image reconstruction, and increasing resolution. They generated a super-resolution style transfer network (SRSTN) and used it for output image. Chang et al.[3] and Dmytro Kotovenko et al. [12] presented two different networks, spatial relation-augmented VGG (SRVGG) and content-and style-aware stylization network respectively. They both mainly focused on paintings. Along with this, Dmytro Kotovenko et al. also presented a normalization layer for high resolution image synthesis. Assembling self-attention mechanism [22] into a style-agnostic reconstruction autoencoder framework and introduced multi-scale style swap and a flexible stroke fusion plan to adaptively blend many style patterns into the output image.

Huan Wang et al. [21] presented a contemporary knowledge distillation method (Collaborative Distillation) using encoder-decoder based neural style transfer. Another work by Chiu et al. [4], where the authors proposed a framework of an autoencoder and bottleneck feature transformation. Li et al. [13] used the shape of the transformation matrix theoretically and presented an arbitrary style transfer method to learn the transformation matrix with a feed-forward network.

Works such as [1,2,23] made in the field of photo-realistic style transfer are worth noting. The authors used a wavelet based approach, a high-resolution daytime translation (HiDT) model and a two-staged method respectively. Gao, Wei, et al. [6] in 2020, introduced a video multi-style transfer (VMST) framework. They used a combination of four different loss functions i.e. Perceptual loss, Total variation loss and other two to solve the temporal flickering issue. Another work by Yihuai et al. [15] in which medical image transformation problems is considered. They proposed a method using CycleGAN with a mixture of perceptual loss and total variation loss function.

3   Proposed Scheme

It is observed that maintaining the curves and structures of the target image after performing the style transfer seems to be a challenging task. To address this problem, the proposed pipeline consists of a three-folded contribution as shown in Fig. 1. Firstly, with SISR the content details are amplified. Secondly, with the traditional style transfer [7] adaptive content weights are introduced. Thus these adaptive weights can be tuned as per requirements. Finally, reconstruction of the output stylized image is done by fusing perceptual loss function and total variation loss function.

3.1   SISR Network

In CNN, it has been observed that higher layer feature maps represent the global content/semantic of the target image more closely than lower layers. But the higher layer features of the image are not suitable to capture the minute details

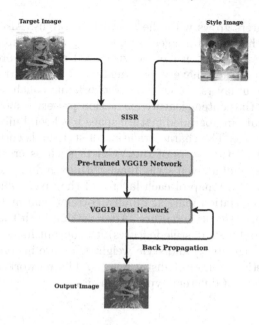

Fig. 1. Schematic diagram

and local structures. As we know that smoothing or low pass filtering operation is analogous to downsampling. Conversely, we can assume that finding a super resolution of an image will strengthen its high-frequency component or semantic contents. So, the use of a SISR before extracting semantic features will help in capturing global semantics along with local structures. This SISR network is inspired from the paper [20]. The task of this network is to convert the input images to high-resolution images. The network has two beneficial outcomes in comparison to other networks. Firstly, both high and low-resolution subnets are connected in parallel, unlike conventional networks where the connection is in series. Secondly, it performs repeated multi-scale fusion with low-resolution representations of the same depth and indistinguishable levels to enhance high-resolution representation. This network involves the fusion between different feature maps and we concatenate these feature maps received just like the inception module. This enables high-resolution subnets to possess both high-resolution and low-resolution feature maps information.

3.2 Style Transfer Network

This image style transfer approach uses the traditional pre-trained VGG-19 architecture on ImageNet [18] to extract the feature and texture image that represent semantic target and style images respectively as shown in Fig. 2. The super resolution images are fed to the network. Given input images representing a set of filtered images at each processing step in the VGG-19 network. The count of various filters grows with the network hierarchy. Also, the dimension-ality reduction of the filtered images is done using a downsampling mechanism (e.g. pooling). The information can be visualised at every processing layer of the network. In [8] authors have experimented with the extraction of content features from different layers of the VGG-Network and concluded that using the 'Conv4_1' layer for the content feature extraction preserves more high-level con-tents. On top of that, an upgraded feature space has been built that records the style of an input image. The characterization of style finds correlations between various features at all the layers of the VGG-19 such as creating images that match with the design of a given style image with excluding meaningful content of the scene. The feature map of each layer and the pixel values are multiplied and resulted in a correlation matrix. Finally, correlational matrices of all layers are combined to form the style feature. After getting both the ingredients (i.e. semantic features and texture style features), the output image is generated. An adaptive content weight (C_w) and style weight (S_w) are introduced to alter the outputs from artistic to photo-realistic images. The network is checked for all possible combinations of different types of images.

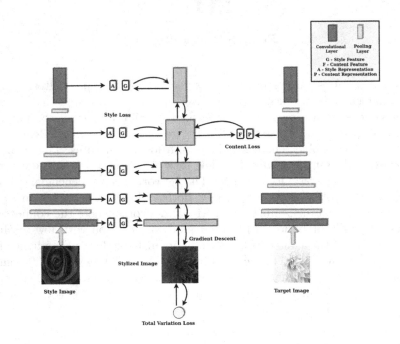

Fig. 2. Style transformation network

3.3   Loss Function

In style transfer, to achieve a good result correct choice of loss function[10] is essential. In our proposed work, we have used total variation loss [6] along with content loss and style loss to generate a better visual experience. It is very important for the target image and the output image to have the similar semantics. For content loss, Euclidean distance is used as shown by the Eq. 1.

$$l_c^{\phi,i}(y,\hat{y}) = \frac{1}{C_i, H_i, W_i}||\phi_i(\hat{y}) - \phi_i(y)||^2 \tag{1}$$

Here, $\hat{y}$ is the styled output image and y is the target image. As i is the convolutional layer, so $\phi_i(y)$ is the feature map of size shape $C_i \times H_i \times W_i$. The style loss is calculated as the squared L-2 norm of the difference between the Gram matrices of the output image and style image. Mathematically, Gram Matrix is computed by the given Equation-2.

$$G_i^{\phi}(y)_{c,\hat{c}} = \frac{1}{C_i, H_i, W_i}\sum_{h=1}^{H_i}\sum_{w=1}^{W_i}\phi_i(y)_{h,w,c}\phi_i(y)_{h,w,\hat{c}} \tag{2}$$

The style loss equation where $G_i^{\phi}(y)$ and $G_i^{\phi}(\hat{y})$ is the gram matrix of style image and output image respectively as shown below:

$$l_s^{\phi,i}(y,\hat{y}) = ||G_i^{\phi}(y) - G_i^{\phi}(\hat{y})||^2 \tag{3}$$

Total variation loss is a measure of variation of an image with respect to its spatial variation. It also prevents excessive contrast and encourages the spatial denoising in the stylized image. The total variation loss is calculated in one dimension array as $l_{TV}^{\phi,i}(y)$ shown below can be easily generalized to two dimensional images where β is generally taken as 1 or 2 as number of dimensions of the image.

$$l_{TV}^{\phi,i}(y) = \sum|y_{i+1} - y_i|^{\beta} \tag{4}$$

$$\text{Total Loss} = \underbrace{l_c^{\phi,i}(y,\hat{y})}_{\text{Content Loss}} + \underbrace{l_s^{\phi,i}(y,\hat{y})}_{\text{Style Loss}} + \underbrace{l_{TV}^{\phi,i}(y)}_{\text{Total variation loss}} \tag{5}$$

Style image y_s and target image y_c is taken in the i-th layer to perform feature mapping, texture extraction and style reconstruction. Here C_w is the content weight and S_w is the style weight and λ_R is the parameter of Regularizer which plays a critical role in smoothing process. TV_w weight is the total variation weight in the total variation loss function which is half the content loss in next iterations in the training. If TV_w loss is very high then resultant image generated will be of poor quality. The output image $\hat{y}$ that is generated during the style transfer [15] in the shown Equation-6.

$$\hat{y} = argminC_w l_c^{\phi,i}(y,y_c) + S_w l_s^{\phi,i}(y,y_s) + TV_w l_{TV}^{\phi,i}(y) + \lambda_R l_R(y) \tag{6}$$

4 Experiments

4.1 Experimental Setup

The whole experiment is carried out with NVIDIA Geforce GTX 1650 Max-Q GPU. The input and output images are of size (500,500) and containing 3 channels. We start our experiment by setting up SISR and pre-trained VGG19 architecture followed by perceptual loss and total variation loss evaluation.

4.2 Results

In the proposed scheme, the SISR network increases the resolution of the content and style image so that more detailed feature and texture images can be extracted during style transfer with the VGG19 network. It is found that our result preserves the finer structure and creates less distortion when compared with other existing approaches. The resultant image has a more uniform color distribution, which makes it more realistic than others. The proposed scheme is tested in all combinations with different types of images, night-to-day and vice versa with varied range (as shown in Table 1) of content weight keeping style weight constant shown in Figs. 3 and 4.

Target   Style   Cw=0.5

Cw=10   Cw=70   Cw=100

Fig. 3. Result of image style transfer using SISR

The regeneration of the stylized image is done by updating the parameters by backpropagation until the total loss becomes minimal. An empirical study is conducted to compare the best visual effects. 20 volunteers had been invited from different departments to rate the successful output images. They were asked to rate the generated images from 1 to 5 according to the visual quality of the images. The lowest and highest quality are represented by 1 and 5 respectively.

Fig. 4. Results of image style transfer using SISR

Table 1. Results with different weights

Result	Content weight(C_w)	Style weight(S_w)
Artistic	0.5–10	1
Photorealistic	70–100	1

In the case of artistic images, this method produces a result similar to the distortion of the painting but better than the existing methods. Below are the Table 2 of mean opinion scores of this scheme with other's results. For more efficient comparisons as shown in Fig. 5, contour extracted for the stylizing images with image graying followed by sobel operator and similarity measure is done. This experiment shows that resultant images have contour closer to the target image and with maximum similarity.

Table 2. MOS comparison of artistic image and photorealistic image with other's results

Criteria	Gayts's [8]	Artistic result	Li's [14]	Photorealistic result
Style information	26.65%	39.04%	46.83%	45.3%
Content information	10.83%	55.96%	25.42%	66.89%
Visual effect	15.20%	73.66%	20.27%	80.81%

Target   Style   Gatys's [8]   Artistic   Li's [14]   Realistic

Fig. 5. Comparison result with other approach using grayscale and contour images

4.3   Conclusion

The traditional NST algorithm along with the SISR network achieves good results in both artistic and photo-realistic style with a finer structure and less distortion. The reconstruction of the output stylized image by the combination of the perceptual and total variation loss function results in distinguishable outputs with different weights. An empirical study is conducted to evaluate the approach. It has been inferred from experimental results that the stylized output image has better visual effects in comparison to other models. The contour of stylised images and target images are compared. This experiment shows that the resultant images have contour closer to the target images and indicates more preservation of semantic information.

References

1. An, J., Xiong, H., Huan, J., Luo, J.: Ultrafast photorealistic style transfer via neural architecture search. In: AAAI, pp. 10443–10450 (2020)
2. Anokhin, I., et al.: High-resolution daytime translation without domain labels. In: Proceedings of the IEEE/CVF Conference on Computer Vision and Pattern Recognition, pp. 7488–7497 (2020)
3. Chang, J.R., Chen, Y.S.: Exploiting spatial relation for reducing distortion in style transfer. In: Proceedings of the IEEE/CVF Winter Conference on Applications of Computer Vision, pp. 1209–1217 (2021)
4. Chiu, T.-Y., Gurari, D.: Iterative feature transformation for fast and versatile universal style transfer. In: Vedaldi, A., Bischof, H., Brox, T., Frahm, J.-M. (eds.) ECCV 2020. LNCS, vol. 12364, pp. 169–184. Springer, Cham (2020). https://doi.org/10.1007/978-3-030-58529-7_11
5. Dumoulin, V., Shlens, J., Kudlur, M.: A learned representation for artistic style. arXiv preprint arXiv:1610.07629 (2016)

6. Gao, W., Li, Y., Yin, Y., Yang, M.H.: Fast video multi-style transfer. In: Proceedings of the IEEE/CVF Winter Conference on Applications of Computer Vision, pp. 3222–3230 (2020)
7. Gatys, L.A., Ecker, A.S., Bethge, M.: A neural algorithm of artistic style. arXiv preprint arXiv:1508.06576 (2015)
8. Gatys, L.A., Ecker, A.S., Bethge, M.: Image style transfer using convolutional neural networks. In: Proceedings of the IEEE Conference on Computer Vision and Pattern Recognition, pp. 2414–2423 (2016)
9. Gooch, B., Gooch, A.: Non-photorealistic Rendering. CRC Press, New York (2001)
10. Johnson, J., Alahi, A., Fei-Fei, L.: Perceptual losses for real-time style transfer and super-resolution. In: Leibe, B., Matas, J., Sebe, N., Welling, M. (eds.) ECCV 2016. LNCS, vol. 9906, pp. 694–711. Springer, Cham (2016). https://doi.org/10.1007/978-3-319-46475-6_43
11. Kolkin, N., Salavon, J., Shakhnarovich, G.: Style transfer by relaxed optimal transport and self-similarity. In: Proceedings of the IEEE Conference on Computer Vision and Pattern Recognition, pp. 10051–10060 (2019)
12. Kotovenko, D., Sanakoyeu, A., Ma, P., Lang, S., Ommer, B.: A content transformation block for image style transfer. In: Proceedings of the IEEE Conference on Computer Vision and Pattern Recognition, pp. 10032–10041 (2019)
13. Li, X., Liu, S., Kautz, J., Yang, M.H.: Learning linear transformations for fast arbitrary style transfer. arXiv preprint arXiv:1808.04537 (2018)
14. Li, Z., Zhou, F., Yang, L., Li, X., Li, J.: Accelerate neural style transfer with super-resolution. Multimed. Tools Appl. 79(7), 4347–4364 (2020)
15. Liang, Y., Lee, D., Li, Y., Shin, B.S.: Unpaired medical image colorization using generative adversarial network. Multimed. Tools Appl. 1–15 (2021)
16. Park, D.Y., Lee, K.H.: Arbitrary style transfer with style-attentional networks. In: Proceedings of the IEEE Conference on Computer Vision and Pattern Recognition, pp. 5880–5888 (2019)
17. Rosin, P., Collomosse, J.: Image and Video-based Artistic Stylisation, vol. 42. Springer, Heidelberg (2012). https://doi.org/10.1007/978-1-4471-4519-6
18. Simonyan, K., Zisserman, A.: Very deep convolutional networks for large-scale image recognition (2015)
19. Strothotte, T., Schlechtweg, S.: Non-photorealistic Computer Graphics: Modeling, Rendering, and Animation. Morgan Kaufmann, New York (2002)
20. Sun, K., Xiao, B., Liu, D., Wang, J.: Deep high-resolution representation learning for human pose estimation. In: Proceedings of the IEEE Conference on Computer Vision and Pattern Recognition, pp. 5693–5703 (2019)
21. Wang, H., Li, Y., Wang, Y., Hu, H., Yang, M.H.: Collaborative distillation for ultra-resolution universal style transfer. In: Proceedings of the IEEE/CVF Conference on Computer Vision and Pattern Recognition, pp. 1860–1869 (2020)
22. Yao, Y., Ren, J., Xie, X., Liu, W., Liu, Y.J., Wang, J.: Attention-aware multi-stroke style transfer. In: Proceedings of the IEEE Conference on Computer Vision and Pattern Recognition, pp. 1467–1475 (2019)
23. Yoo, J., Uh, Y., Chun, S., Kang, B., Ha, J.W.: Photorealistic style transfer via wavelet transforms. In: Proceedings of the IEEE International Conference on Computer Vision, pp. 9036–9045 (2019)

Object-Centric Anomaly Detection
Using Memory Augmentation

Jacob Velling Dueholm[1][(✉)] [iD], Kamal Nasrollahi[1,2] [iD],
and Thomas Baltzer Moeslund[1] [iD]

[1] Visual Analysis and Perception Laboratory, Aalborg University, Aalborg, Denmark
{jvdu,kn,tbm}@create.aau.dk
[2] Research Department, Milestone Systems A/S, Brøndby, Denmark

Abstract. Video anomaly detection is becoming of increased interest as surveillance is becoming more widespread. We propose an object-centric method with memory augmentation (ObjMemAE) for video anomaly detection. Recently, object-centric approaches is seen at the top of the leaderboards, where we take the novel approach of combining an object-centric approach with memory augmentation using a long term memory bank storing prototypical objects. The memory module also allows the use of additional object-centric features. The proposed method is shown to outperform the baseline by 4.5%, with an AUC score of 98.3% on the UCSD-Ped2 dataset achieving state-of-the-art.

Keywords: Anomaly detection · Object-centric · Memory augmentation · Autoencoder · Unsupervised learning

1 Introduction

The need for intelligent video analysis is in increasing demand as video surveillance is becoming increasingly widespread with an expected growth. Manually looking through vast amount of video is tedious, time consuming, and human observers are known to be error prone and subject to fatigue. A sought-after feature of intelligent video analysis is to only extract sequences of interest, defined as rare cases which deviates from the norm, also known as anomaly or abnormality detection.

In this work we target the surveillance application specified by a single scene, where behavioural patterns are to be learned for each unique scene. What is to be considered normal in one scene, might be considered abnormal in another. A typical approach to video anomaly detection is based on a reconstruction idea where a model is trained on normal data only, to learn to reconstruct the input. In the test case of an unseen anomalous input the system will not be able to accurately reconstruct the input, resulting in a high reconstruction error that can be used to determine between regularity and abnormality.

We further focus on object-centric methods [4,23] and datasets, as these have shown promising results recently both in terms of accuracy and localization

© Springer Nature Switzerland AG 2021
N. Tsapatsoulis et al. (Eds.): CAIP 2021, LNCS 13052, pp. 362–371, 2021.
https://doi.org/10.1007/978-3-030-89128-2_35

ability. Object-centric methods employ a generic pre-trained object detector to the full frames. This intuitively helps the reconstruction based methods by not having to reconstruct the entire frame, but focus on the objects of interest which form the majority of anomalies, at least for the current datasets used in the research community. To further aid the unsupervised anomaly detection we also take the approach of utilizing a memory module [1,14] storing the prototypical events, helping to learn whats normal. Using an object-centric approach also opens up the ability to use object-centric features, which is yet to be exploited by these recent object-centric anomaly detection methods.

The main contributions are of this work are twofold:

- Introducing memory augmentation into object-centric anomaly detection.
- Query of object-centric features through a memory module.

2   Related Work

For an overview of the anomaly detection field we refer to two recent surveys [6,17].

2.1   Reconstruction, Prediction and Hybrid Approaches

Reconstruction based approaches seek to learn normalcy, where the expectation is that anomalous activity will have a large reconstruction error, comparing the input with its reconstruction. This approach has shown promise due to the era of deep learning and specifically the convolutional autoencoder (CAE) and the generative adversarial network (GAN) [2]. Hasan et al. [3] is the first example of applying CAE to compare hand-crafted features Histograms of Oriented Gradients and Histograms of Optical Flows, showing the potential of learned representations. A similar approach is seen using GANs [13,18]. A special comparison approach was seen by Ravanbakhsh et al. [18], not calculating the reconstruction error on the pixel level, but instead applied a pre-trained AlexNet trained on ImageNet for a semantic difference.

A variation of the reconstruction based approach is the prediction based approach, arguing anomalous actions are naturally harder to predict. This approach was pioneered by Liu et al. [8], using a sliding time window to predict the future frame. The future prediction is compared to the actual input in the same manner as with reconstruction.

2.2   Memory Augmentation

Memory augmentation have previously been used in similar computer vision fields to anomaly detection, such as the work done by Wu et al. [21], augmenting 3D CNNs with long-term feature banks and testing its use on spatial-temporal action localization, action classification and video classification tasks. Specifically for anomaly detection, Gong et al. [1] where the first to test their memory augmented 3D Conv-AE. Here a memory bank of 2000 items is used as the authors show diminishing returns for a larger bank on the UCSD-Ped2 dataset. Yang et al. [22] extended this framework to multiple scales, resizing the input to different scales, with each their decoder. Features are fused using a multi-scale attention fuser before the final reconstructing using a single decoder. Park et al. [14] also used a similar approach proposing both a prediction based method using the U-Net architecture [19], and a reconstruction method using the same U-Net without the skip connections. As input for the decoder they not only use the memory updated features, but also but also concatenate the original decoded features. The memory bank only consist of 10 items, but is also updating during testing to account for drift.

2.3   Object-Centric Anomaly Detection

For state-of-the-art performance we look towards object-centric methods i.e. methods utilizing pre-trained object detectors to reconstruct detected objects instead of the traditional full frame. Firstly introduced by Ionesc et al. [4] using a single-shot object detector based on feature pyramid network together with three autoencoders for appearance and motion features. The latent features of the autoencoders are used in a k-means clustering from which support vector machine binary classifiers are trained. In Yu et al. [23] the object detector is being assisted by a motion based background subtraction to combat the "closed world" problem of unknown classes of the detector. Spatial-temporal cubes are constructed and using visual cloze test to sequential leave one out of the collection for the reconstruction.

3   ObjMemAE Method

The proposed system, object-centric autoencoder with memory augmentation (ObjMemAE), is depicted in Fig. 1 and described in detail the following sections. A pre-trained object detector is used on the full frames, where each detected object serves as input to an autoencoder to get a reconstruction. The autoencoder contains a memory module storing the features of prototypical detections. Additional object-centric features are extracted for each detection for an improved reconstruction. The reconstruction is compared to the original detection to get the reconstruction error, which is used as the anomaly score.

3.1   Object Detection

For object detection, the Scaled-YOLOv4-CSP [20] is applied using a COCO [7] pre-trained model. The COCO dataset consists of 80 classes, where only a few are deemed relevant to the surveillance setting, hence only the following classes are used: *person, bicycle, car, motorcycle, airplane, bus, train* and *truck*. To filter out small and noisy detections a confidence threshold of 0.3 is used for both training and testing.

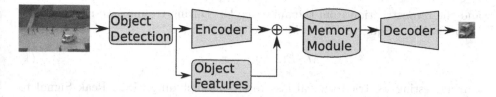

Fig. 1. Overview of the ObjMemAE framework. A pre-trained object detector is used, where objects are input to an encoder functioning as a bottleneck. The latent features are concatenated with object centric features and updated by a query into the memory module of normal objects. Lastly the updated features are decoded into the final reconstruction, where the reconstruction error functions as a mean of anomaly score.

3.2 Autoencoder: Encoder - Memory Module - Decoder

Each detection is used as a crop, converted to grayscale, resized to 64×64 pixels, and lastly normalized. The object detections are stacked into temporal cubes serving as input to the 3D convolution autoencoder with an encoder-memory module-decoder structure, similar to the architecture proposed in [1], with the two difference being the reduced stacking of 8 frames for improved runtime at marginal loss of accuracy compared to the original 16 frames, and the additional object centric features. Temporal cubes are formed by stacking 8 frames surrounding the detection to decode the spatial-temporal information. The encoder performs 3D convolutions in a bottleneck resulting in a compressed representation. The autoencoder use the same architecture as Gong et al. [1], consisting of 3 layers of 3D convolutions with batch normalization and leaky ReLu activations, and mirrored for the decoders deconvolution to obtain the original input dimensions. The decoded features queries into the memory module updating the features towards the stored features learned during training on normal data only. During testing, in case of an anomalous input, the encoded memory updated features towards normalcy will intuitively result in a larger reconstruction error. The memory module will store 2000 prototypical object detections, each with a feature length consisting of the encoded representation and the additional object-centric features. The object-centric features are based on the detections, and includes the class output and confidence score from the object detector; the position of the bounding box related to the full image; and the ratio of the bounding box before resizing. The object-centric features are concatenated with the decoding features as input to the memory module, to give a stronger query when updating the features. The object-centric features are removed before the decoding reconstruction.

3.3 Anomaly Score

To obtain the anomaly score for a frame we calculate the anomaly score for each of the detections in the frame, and using the max score as the frame score. The anomaly score is based on the reconstruction error between the detection input and the reconstruction from the autoencoder. During training we minimize the L2-norm Mean Squared Error (MSE) as seen in Eq. 1.

$$MSE(x, \hat{x}) = \| x - \hat{x} \|_2^2 \tag{1}$$

During testing we tried several loss functions including MSE, Peak Signal to Noise Ratio (PSNR) and found the custom loss function, see Eq. 2, to yield slightly better results.

$$SPLoss(x, \hat{x}) = \sqrt{\sum \| x - \hat{x} \|^2} \tag{2}$$

The frame anomaly scores over time for an entire sequence is especially noisy for the object-centric methods. Hence we apply Gaussian filtering as in Ioenescu et al. [4], to get a smooth signal as seen in Fig. 3.

4   Experiments

4.1   Dataset

The experiments are conducted on the UCSD-Ped2 [12] anomaly detection dataset. We exclude UCSD-Ped1 due to lower usage in the field, and the lower image resolution of only 238×158 pixels. UCSD-Ped2 have a resolution of 360×240 pixels and consists of 16 training and 12 test videos, corresponding to 2550 training frames and 2010 test frames.

4.2   Evaluation Metric

Ground truth is used to calculate the frame level area under the curve (AUC). AUC in the literature is calculated in two different ways: by averaging the score for each video (macro), or if all videos are concatenated and then computing the AUC (micro). Note the macro AUC is only calculated and averaged over scenes where both normal and anomalous frames occur, hence only evaluated on 8 scenes for the UCSD-ped2 dataset. We argue micro AUC is the one to be used as this best reflects the real use case of an anomaly detection system, as this corresponds to setting only one threshold for the system for all scenes, to determine between regular and abnormal frames. For the remainder of the paper AUC refers to micro AUC.

4.3   Implementation Details

The framework is implemented in PyTorch [15]. The model is trained for 80 epochs using a batch size of 64 and the Adam optimizer [5] with a multistep scheme starting with a learning rate of 1e-3 and lowering to 1e-4 after 40 epochs. The model is trained in less than two hours on a NVIDIA GTX 2080TI with pre-computed detections.

5 Results and Discussion

5.1 Quantitative Results

The AUC score for our proposed method is listed in Table 1, and compared to various state-of-the-art methods evaluated on the UCSD-Ped2 dataset. Here it is shown the proposed object-centric memory augmented framework with object-centric features is able to achieve state-of-the-art performance on the UCSD-Ped2 dataset with its anomalous objects typically found in a surveillance scene.

Table 1. Comparing against state-of-the-art. AUC scores is listed in %.

Input	Year	Publ.	Method	UCSD-Ped2	
				Micro	Macro
Full frame	2017	ICME	[10] ConvLSTM-AE	88.1	
	2016	CVPR	[3] Conv-AE	90.0	
	2017	CVPR	[11] Sparse coding revisit	92.2	
	2017	ICIP	[18] GAN	93.5	
	2020	WACV	[16] Siamese network	94.0	
	2019	ICCV	[1] MemAE	94.1	
	2018	CVPR	[8] Future Frame Prediction	95.4	
	2020	CVPR	[14] MNAD-recon.	90.2	
	2020	CVPR	[14] MNAD-pred.	97.0	
Object-centric	2020	ACM	[23] VEC-A	96.9	
	2020	ACM	[23] VEC-AM	97.3	
	2019	CVPR	[4] Object-centric auto-encoders		97.8
			ObjMemAE (ours)	**98.3**	**99.6**

5.2 Ablation Study

We compare the results of the original proposed framework MemAE [1] with our object-centric counter part using various combination of object features. Table 2 shows the transition from the original full frame methods to the object-centric methods with a lower AUC score. The object-centric method allows the use of object-centric features, where the best results is achieved using a combination of bounding box and class information for a AUC score of 98.3% on the UCSD-Ped2 dataset, which is a state-of-the-art result.

5.3 Qualitative Results

A few reconstruction examples are shown in Fig. 2, confirming the principle in being unable to accurately reconstruct anomalous objects. A success and a failure cases is shown in Figs. 3 and 4, respectively. In the success case, the

Table 2. Ablation study on UCSD-Ped2 of the various combinations of object centric features. LF: latent features from the decoder. bbox: bounding box top left position and width and height. conf: confidence score. Ratio: ratio of the width and height of the bounding box.

Method	Input	Stack	Memory	AUC
[1] AE conv3d	Frames	16	N/A	91.7
[1] MemAE conv3d	Frames	16	LF	94.1
MemAE conv3d	Objects	8	LF+bbox	98.0
MemAE conv3d	Objects	8	LF+class	96.5
MemAE conv3d	Objects	8	LF+conf	95.5
MemAE conv3d	Objects	8	LF+ratio	98.2
MemAE conv3d	Objects	8	LF+bbox+class	**98.3**
MemAE conv3d	Objects	8	LF+conf+ratio	97.5
MemAE conv3d	Objects	8	LF+bbox+class+conf+ratio	96.5

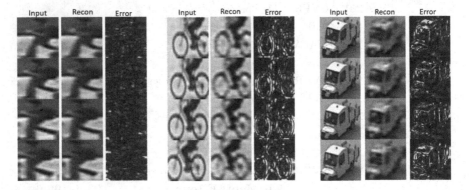

Fig. 2. Reconstruction examples showing a larger reconstruction error for anomalous objects not found in the training data, such as bike and cars.

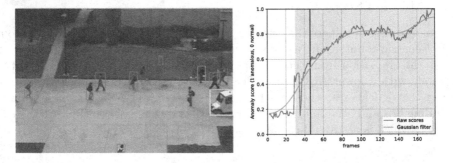

Fig. 3. Success case of the proposed method on USCD-Ped 4 of a car entring the scene. Left: Input full frame with objects represented by bounding boxes color coded with the reconstruction error from blue (low error) to red (high error). Right: full scene overview with anomalous frames marked in red. (Color figure online)

entering car is successfully early identified by a spike in the anomaly score, while in the failure case, the skater is missed. Two approaches can be found to improve on special cases as the skater. The first being extending the object detector to include skateboards. This is less general and challenged by the low resolution of surveillance footage. Another approach is looking at the temporal aspects of the skater being slightly faster than the surrounding pedestrians. The proposed method utilizes 3D convolutions on 8 consecutive frames, but stronger temporal features is needed to pick up on these subtle differences in velocity.

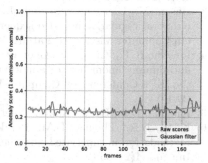

Fig. 4. Failure case of the proposed method on USCD-Ped 12 missing the skater marked with an arrow. Left: Input full frame with objects represented by bounding boxes color coded with the reconstruction error from blue (low error) to red (high error). Right: full scene overview with anomalous frames marked in red. (Color figure online)

This method was also tested on the Avenue dataset [9]. In this case no improvements over the baseline was found with proposed object-centric method with object-centric features. This is expected as our method focuses on objects and their features, while anomalies in the Avenue dataset are more related to behaviours such as running and dancing. For a better performance on such a dataset focusing on temporal information is needed and is therefore considered as future work.

6   Conclusion

We propose an object-centric anomaly detection framework utilizing memory augmentation. The object-centric approach allows the autoencoder to focus on reconstructing the objects of interest instead of the entire frame, while the memory module aids in storing whats normal in the unsupervised setting. Furthermore, object-centric features are added for an improved reconstruction. The method is tested on UCSD-Ped2 anomaly detection dataset, achieving state-of-the-art performance with an AUC of 98.3%. Future work include incorporating additional temporal features to combat anomaly detection datasets focusing on temporal anomalies.

Acknowledgments. This work was supported by the Milestone Research Programme at Aalborg University.

References

1. Gong, D., et al.: Memorizing normality to detect anomaly: memory-augmented deep autoencoder for unsupervised anomaly detection. In: Proceedings of the IEEE International Conference on Computer Vision, pp. 1705–1714 (2019)
2. Goodfellow, I.J., et al.: Generative adversarial networks. arXiv preprint arXiv:1406.2661 (2014)
3. Hasan, M., Choi, J., Neumann, J., Roy-Chowdhury, A.K., Davis, L.S.: Learning temporal regularity in video sequences. In: Proceedings of the IEEE Conference On Computer Vision and Pattern Recognition, pp. 733–742 (2016)
4. Ionescu, R.T., Khan, F.S., Georgescu, M.I., Shao, L.: Object-centric auto-encoders and dummy anomalies for abnormal event detection in video. In: Proceedings of the IEEE Conference on Computer Vision and Pattern Recognition, pp. 7842–7851 (2019)
5. Kingma, D.P., Ba, J.: Adam: a method for stochastic optimization (2017)
6. Kiran, B.R., Thomas, D.M., Parakkal, R.: An overview of deep learning based methods for unsupervised and semi-supervised anomaly detection in videos. J. Imaging **4**(2), 36 (2018)
7. Lin, T.Y., et al.: Microsoft coco: common objects in context (2015)
8. Liu, W., Luo, W., Lian, D., Gao, S.: Future frame prediction for anomaly detection-a new baseline. In: Proceedings of the IEEE Conference on Computer Vision and Pattern Recognition, pp. 6536–6545 (2018)
9. Lu, C., Shi, J., Jia, J.: Abnormal event detection at 150 fps in Matlab. In: Proceedings of the IEEE International Conference on Computer Vision, pp. 2720–2727 (2013)
10. Luo, W., Liu, W., Gao, S.: Remembering history with convolutional LSTM for anomaly detection. In: 2017 IEEE International Conference on Multimedia and Expo (ICME), pp. 439–444. IEEE (2017)
11. Luo, W., Liu, W., Gao, S.: A revisit of sparse coding based anomaly detection in stacked RNN framework. In: Proceedings of the IEEE International Conference on Computer Vision, pp. 341–349 (2017)
12. Mahadevan, V., Li, W., Bhalodia, V., Vasconcelos, N.: Anomaly detection in crowded scenes. In: 2010 IEEE Computer Society Conference on Computer Vision and Pattern Recognition, pp. 1975–1981. IEEE (2010)
13. Nguyen, T.N., Meunier, J.: Anomaly detection in video sequence with appearance-motion correspondence. In: Proceedings of the IEEE International Conference on Computer Vision, pp. 1273–1283 (2019)
14. Park, H., Noh, J., Ham, B.: Learning memory-guided normality for anomaly detection. In: Proceedings of the IEEE/CVF Conference on Computer Vision and Pattern Recognition, pp. 14372–14381 (2020)
15. Paszke, A., et al.: Pytorch: an imperative style, high-performance deep learning library. In: Wallach, H., Larochelle, H., Beygelzimer, A., d' Alché-Buc, F., Fox, E., Garnett, R. (eds.) Advances in Neural Information Processing Systems, vol. 32, pp. 8024–8035 (2019)
16. Ramachandra, B., Jones, M., Vatsavai, R.: Learning a distance function with a Siamese network to localize anomalies in videos. In: The IEEE Winter Conference on Applications of Computer Vision, pp. 2598–2607 (2020)

17. Ramachandra, B., Jones, M., Vatsavai, R.R.: A survey of single-scene video anomaly detection. IEEE Trans. Pattern Anal. Mach. Intell. (2020). https://ieeexplore.ieee.org/document/9271895
18. Ravanbakhsh, M., Nabi, M., Sangineto, E., Marcenaro, L., Regazzoni, C., Sebe, N.: Abnormal event detection in videos using generative adversarial nets. In: 2017 IEEE International Conference on Image Processing (ICIP), pp. 1577–1581. IEEE (2017)
19. Ronneberger, O., Fischer, P., Brox, T.: U-Net: convolutional networks for biomedical image segmentation. In: Navab, N., Hornegger, J., Wells, W.M., Frangi, A.F. (eds.) MICCAI 2015. LNCS, vol. 9351, pp. 234–241. Springer, Cham (2015). https://doi.org/10.1007/978-3-319-24574-4_28
20. Wang, C.Y., Bochkovskiy, A., Liao, H.Y.M.: Scaled-YOLOv4: scaling cross stage partial network (2021)
21. Wu, C.Y., Feichtenhofer, C., Fan, H., He, K., Krahenbuhl, P., Girshick, R.: Long-term feature banks for detailed video understanding. In: Proceedings of the IEEE/CVF Conference on Computer Vision and Pattern Recognition, pp. 284–293 (2019)
22. Yang, Y., Xiang, S., Zhang, R.: Improving unsupervised anomaly localization by applying multi-scale memories to autoencoders. arXiv preprint arXiv:2012.11113 (2020)
23. Yu, G., et al.: Cloze test helps: effective video anomaly detection via learning to complete video events. In: Proceedings of the 28th ACM International Conference on Multimedia, pp. 583–591 (2020)

Document Language Classification: Hierarchical Model with Deep Learning Approach

Sarathi Shah[✉] and M. V. Joshi

Dhirubhai Ambani Institute of Information and Communication Technology,
Gandhinagar 382007, Gujarat, India
{201911019,mv_joshi}@daiict.ac.in

Abstract. Optical character recognition (OCR) refers to the task of recognizing the characters or text from digital document images. OCR is a widely researched area for the past many years due to its applications in various fields. It helps in the natural language processing of the documents, convert the document text to speech, semantic analysis of the text, searching in the documents etc. Multilingual OCR works with documents having more than one language. Different OCR models have been created and optimized for a particular language. However, while dealing with multiple languages or translation of documents, one needs to detect the language of the document first and then give it as input to a model-specific to that language. Most of the researched work in this area focuses on identifying scripts, but considering that the Convolutional Neural Network (CNN) can learn appropriate features, our work focuses on language detection using learned features. We use a hierarchical based method in which a binary classification followed by the multiclass classification is used to improve detection accuracy. Largely, the current approaches do not use hierarchy and hence fail to identify the language correctly. The proposed hierarchical approach is used to detect six Indian languages namely: Tamil, Telugu, Kannada, Hindi, Marathi, Gujarati, using the CNN from printed documents based on the text content in a page. Experiments are performed on scanned government documents, and results indicate that the proposed approach performs better than the other similar methods. Advantage of our approach is that it is based on features extracted from the entire page rather than the words or characters, and it can also be applied to handwritten documents.

Keywords: Document analysis · Optical character recognition · Script language identification · Convolutional neural network · Language classification · Indian languages

1 Introduction

Language detection in documents contains two wide research areas where one research is to identify multiple languages/scripts contained inside a single document page, and another is the classification of multiple documents written in

© Springer Nature Switzerland AG 2021
N. Tsapatsoulis et al. (Eds.): CAIP 2021, LNCS 13052, pp. 372–381, 2021.
https://doi.org/10.1007/978-3-030-89128-2_36

different languages/scripts. Script detection task can be carried out in different ways on printed documents. Two of the common approaches that are used for script identification are: handcrafted feature based methods and deep learning based approaches. Both these approaches extract features at four different levels as per the classification requirements that correspond to page-level, text-block level, word-level and character-level.

Feature-based methods extract different types of features from segmented images, and these features are then given to classification methods for identification. The features include both global and local. Local features contain statistical, structural or geometric, and template matching features whereas global features contain gradient features and wavelet transform like features [23]. Major work in this area is on feature extraction using various filters and then applying classification techniques to classify the scripts. These extracted features are flattened out and stacked together to create a complete feature vector. Some of the classification techniques include Support Vector Machine (SVM), Gaussian Mixture Model (GMM) based classification, decision tree, linear discriminant based classification etc. Though, these methods work fine and have achieved good accuracy, when it comes to particular language detection rather than just script detection, they are not effective. They classify two different languages of the same script as belonging to the same class.

Self learned feature extraction methods based on deep learning techniques perform better in classification. These models are now being experimented in different type of document analysis tasks, including script identification. The benefit of using these approaches is that we do not need to pre-decide what types of features we need to extract for the selected set of languages.

The paper structure is as follows. In the following Sect. 1.1, we explain the literature survey on different approaches. Section 2 explains the proposed approach and the architecture of the CNN used. Section 3 contains experimental results that has details of the dataset used and the performance of the proposed approach. Section 4 concludes the paper with a summary of the work, drawbacks and possible future work.

1.1 Literature Review

Much of the literature of script identification on indic scripts is word or text line level classification. These methods segment lines or words from the pages and train models according to it. For these experiments, less document pages are required because words or lines segmented from those will be much higher in numbers. The line and word level methods are useful for multi script classification. Chanda et al. [3] classified English, Devanagari and Bangla scripts with two stage feature extraction. In the first stage, 64-dimensional chain-code histogram features were extracted and in the second stage 400-dimensional gradient features were extracted. These features were classified using SVM with Gaussian kernel. Singh et al. [20] classified Bangla, Devanagari, Gurumukhi, Malayalam, Oriya, Telugu and the Roman handwritten scripts with 82 features designed using elliptical and polygonal methods at page level. They used Multi Layer

Perceptron (MLP) as classifier. In [5], Thai and English scripts were classified using character level features, overlapping information of components, topological features and water reservoir based features. The authors in [15], used structural and mathematical morphological features to detect the script by making use of a simple logistic regression. They used page level detection of printed documents with Bangla, Urdu, Oriya, Gujarati, Telugu, Kannada and Malayalam scripts. Line based detection by extracting stroke based features for English and Devanagari scripts was performed by Pal and Chaudhuri [17]. Chanda et al. [4] used structural features, topological features and water reservoir based features with SVM classifier for the classification of Sinhala, Tamil, English scripts at the word-level. Note that all these approaches require proper feature selection for classification. For example, considering the Gujarati language, one can extract curve and stroke-based features. In the Hindi language, we can extract the horizontal profile based features. Hence, the selection of features varies with requirements and the input data. Instead of this, deep learning based approaches use the neural network such as CNN or combination of convolutional models with Recurrent Neural Networks (RNN), Long Short-Term Memory Networks (LSTM) etc. [8] to self learn the features and then classify using these learnt features. In the approaches based on CNN, features are obtained by computing weighted average at every pixel in a convolution operation. Hence, it avoids the task of manual feature selection for classification. These methods have given comparable accuracy with previous approaches.

Shi et al. [19] classified Arabic, Chinese, Thai and other languages with the CNN at word level. Rashid et al. [18] did bilingual classification of printed documents containing Greek and Latin languages using the CNN at connected component level. Naz et al. [14] classified Urdu script using multidimensional LSTM network at line level. Bhunia et al. [2] applied attention based convolutional-LSTM network on publicly available script identification datasets. Few papers Fujii et al. [9], Mei et al. [10] used RNN for the classification of multiple scripts. Lu et al. [13] extracted global and local features of natural scene image using the CNN models and used adaboost algorithm to combine both the features for classification. All these methods focus on script identification and do not distinguish between languages, whereas our research focuses on language identification. Most of these methods are not based on hierarchy whereas we have applied hierarchy based two level classification.

2   Proposed Approach

In this work, we propose a hierarchical classification approach to classify six languages Gujarati, Hindi, Marathi, Tamil, Telugu and Kannada, using the Convolutional Neural Networks (CNN) on printed documents.

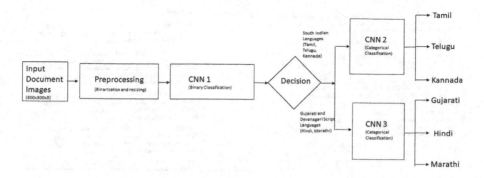

Fig. 1. Language identification using hierarchical approach

Given the documents as images, we first perform preprocessing on images to remove color distortion and resize the images to meet the input shape requirement of the CNN models. As the input dataset contains corrupted old document images, it is necessary to remove colour distortions to achieve better performance. The proposed approach is illustrated by the block schematic as shown in Fig. 1. The model takes printed documents of six different languages as input, pre-processes the input and performs classification using two level classification approach. In the first level, it classifies input images into two classes having significant dissimilarities in writing styles, i.e., South Indian languages and Gujarati and languages with Devanagari script. After this primary classification, categorical classification of each language is done with the CNN in the second level. Both use the same CNN model architecture for classification. CNN extracts features in convolutional layers and the final dense layers contribute to classification. It may be mentioned here that the time taken by the CNN for training and testing is less when compared to that taken by proper manual feature extraction approaches.

2.1 Preprocessing

The collected dataset for the experiments contains the scanned government documents, both old and new. Old documents often have noise in them, and as seen in Fig. 3, they also are corrupted in colour. Due to these, the first step as preprocessing is to minimize the noise and do binarize images. All the document images are converted to have two pixel levels i.e., black and white using Otsu's global thresholding method [16]. After this step, all the images are resized to 300 × 300, which represents input for training the CNNs. Figure 2 shows the output after applying binarization to an old document image.

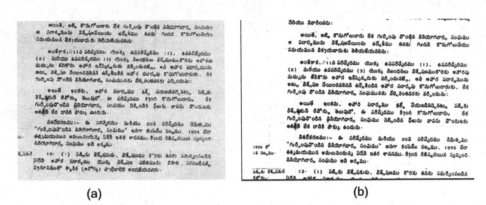

(a)　　　　　　　　　　　　　　　　(b)

Fig. 2. One of the images in the dataset. (a) Original image, (b) Binarized image after preprocessing step.

2.2　Convolutional Neural Network Model

The CNN architectures used in the first and second stage are similar with few differences. As already mentioned, all input images are resized to have the same size, and every image has only two color levels (binary). The CNN includes three-layers of a layer-set in which two of them are convolutional layers, followed by a maxpool layer. Two convolutional layers follow these starting three layers-set, for feature extraction. The last two layers constitute dense layers, and the output layer contains sigmoid activation function in the first level of binary classification. In the second level, the output layer has softmax activation function since it outputs more than two classes. Loss for both the models is calculated with the cross entropy function. The dense layers accept a flattened feature map of length 150 as input and perform classification. We consider auxiliary loss [22] to improve the accuracy by assuring the proper weight update.

3　Experimental Results

Two types of experiments were carried out on the selected dataset where one was to classify using the non-hierarchical model, and the other was using a hierarchical model. The results of both the experiments are discussed in Sect. 3.2.

3.1　Dataset

The dataset used for these experiments contains images of Marathi, Gujarati, Hindi, Tamil, Telugu, and Kannada languages. 300 images of each language are considered for experiments. All the documents used for experiments are printed documents. All the pages are scanned constitutional law document pages available free on the internet [1] and total 1800 pages are considered for the experiments. Each document page contains regional language with few non-regional language words in them, as most of the government documents are printed in regional languages with few non-regional or English words, numbers in it. Example of dataset document images are shown in Fig. 3.

Fig. 3. Dataset document images. (a) Telugu, (b) Tamil, (c) Kannada, (d) Marathi, (e) Hindi, (f) Gujarati.

3.2 Performance of the Proposed Approach

The dataset is split into training and testing with 70:30 ratio. For both the levels, Adam optimizer [12] is used for obtaining the convergence by using a learning rate of 0.00001. All CNNs are trained for 100 epochs and input training data is further split into training and validation data with the ratio of 80:20. Following Table 1 shows the output accuracy of different separate models and overall accuracy.

General CNN model is a direct classification of six languages with a single CNN architecture, same as our 2nd level CNNs. This refers to classifying all six languages using a single CNN architecture i.e., non-hierarchical. This gives an

Table 1. Accuracy measures of different models.

Model description	Accuracy (in %)
General CNN (Non-hierarchical)	94.66
1st level CNN Model (For binary classification)	98.94
Overall accuracy	98.5

accuracy of 94.66%, whereas the hierarchical approach gives an overall accuracy of 98.5%, which proves higher performance of the hierarchical model over the non-hierarchical general model.

The above results show that the CNN achieves excellent performance by self-updating weights to fit the data. It even fits and classifies Hindi and Marathi languages. Although, both the languages have similar writing styles and use the same script, still it extracts the features and correctly classifies between these two language documents. The model achieves this accuracy even if document contains few words or numbers of other non regional languages. Our model gives better accuracy when compared to similar previous approaches as shown in Table 2 for page level/text-block level classification for printed documents. Note that the classification done by [15] achieves greater accuracy but the model is based on script identification and hence fails to classify similar script languages.

Table 2. Comparison with similar approaches.

Method	Script/Language set	Accuracy (in %)
Spitz et al. [21]	Latin-based 23 languages	90
Dhandra et al. [7]	English, Hindi, Kannada, Urdu	97
Joshi et al. [11]	Gujarati, Devanagari, Kannad, Tamil and six other Indian scripts	97.11
Proposed approach	Hindi, Gujarati, Marathi, Tamil and two other south Indian languages	98.5
Obaidullah et al. [15]	Bangla, Devnagari, Roman, Gujarati and six other Indian scripts	98.9

We would like to mention here that we cannot compare our results with some of those currently existing methods because our model classifies languages using the CNN, whereas most of the work published focuses on scripts identification [23,24]. All the previous works consider Hindi and Marathi as a single class as Devanagari, but here our model classifies both as different languages although they have similar script. The CNN model at page-level, classifies Marathi and Hindi with proper features extraction. We cannot compare our model with few hierarchical models similar to us [6] as their works are based mainly at line or word level which means one has to segment out the words separately and perform feature extraction prior to classification. In our approach, the CNN integrates both the tasks and achieves good performance. Few automated models also cannot be compared as they classify at the line or word level [8].

4 Conclusion

Experiments show that CNNs are very powerful models that can easily extract features from an image, can be trained to fit the training set and, easily and accurately they can perform classification. Although, we have used printed documents, it can also be applied to handwritten documents. It is better to go hierarchical by first classifying based on script or similarities in scripts and then perform individual language classification. Further, the classification can be improved combining page-level and line-level or line-level and word-level to increase the accuracy on texts with similar scripts, i.e., Hindi and Marathi.

If we want to translate government documents from one language to another, we must first identify the language. Current approaches and OCR models recognize characters based on scripts, but in this case, we need to do language detection at Natural Language Processing (NLP) level if we want to translate the languages. Rather than doing this, one can directly use the CNN to achieve language detection at the page level. However, the accuracy of the models may suffer if we go for more refined classification i.e., word level or line level, since in this case we fail to capture the global characteristics of similar script languages. Our model also suffers in detecting multiple language texts in a single page as it is not doing multiscript classification, which is a drawback of our model.

References

1. https://legislative.gov.in/
2. Bhunia, A.K., Konwer, A., Bhunia, A.K., Bhowmick, A., Roy, P.P., Pal, U.: Script identification in natural scene image and video frames using an attention based convolutional-LSTM network. Pattern Recogn. **85**, 172–184 (2019)
3. Chanda, S., Pal, S., Franke, K., Pal, U.: Two-stage approach for word-wise script identification. In: 2009 10th International Conference on Document Analysis and Recognition, pp. 926–930 (2009). https://doi.org/10.1109/ICDAR.2009.239
4. Chanda, S., Pal, S., Pal, U.: Word-wise Sinhala Tamil and English script identification using Gaussian kernel SVM. In: 2008 19th International Conference on Pattern Recognition, pp. 1–4 (2008). https://doi.org/10.1109/ICPR.2008.4761823
5. Chanda, S., Terrades, O.R., Pal, U.: SVM based scheme for Thai and English script identification. In: Ninth International Conference on Document Analysis and Recognition (ICDAR 2007), vol. 1, pp. 551–555 (2007). https://doi.org/10.1109/ICDAR.2007.4378770
6. Chanda, S., Pal, S., Franke, K., Pal, U.: Two-stage approach for word-wise script identification. In: Proceedings of the 2009 10th International Conference on Document Analysis and Recognition, pp. 926–930, ICDAR 2009. IEEE Computer Society, USA (2009). https://doi.org/10.1109/ICDAR.2009.239

7. Dhandra, B.V., Nagabhushan, P., Hangarge, M., Hegadi, R., Malemath, V.S.: Script identification based on morphological reconstruction in document images. In: 18th International Conference on Pattern Recognition (ICPR 2006), vol. 2, pp. 950–953 (2006). https://doi.org/10.1109/ICPR.2006.1030

8. Donda, M.V., Prajapati, H.B., Dabhi, V.K.: Survey on automatic script identification techniques. In: 2019 IEEE 5th International Conference for Convergence in Technology (I2CT), pp. 1–5 (2019). https://doi.org/10.1109/I2CT45611.2019.9033660

9. Fujii, Y., Driesen, K., Baccash, J., Hurst, A., Popat, A.C.: Sequence-to-label script identification for multilingual OCR. In: 2017 14th IAPR International Conference on Document Analysis and Recognition (ICDAR), vol. 01, pp. 161–168 (2017). https://doi.org/10.1109/ICDAR.2017.35

10. Mei, J., Dai, L., Shi, B., Bai, X.: Scene text script identification with convolutional recurrent neural networks. In: 2016 23rd International Conference on Pattern Recognition (ICPR), pp. 4053–4058 (2016). https://doi.org/10.1109/ICPR.2016.7900268

11. Joshi, G.D., Garg, S., Sivaswamy, J.: Script identification from Indian documents. In: Bunke, H., Spitz, A.L. (eds.) Doc. Anal. Syst. VII, pp. 255–267. Springer, Heidelberg (2006)

12. Kingma, D.P., Ba, J.: Adam: a method for stochastic optimization (2017)

13. Lu, L., Yi, Y., Huang, F., Wang, K., Wang, Q.: Integrating local CNN and global CNN for script identification in natural scene images. IEEE Access **7**, 52669–52679 (2019). https://doi.org/10.1109/ACCESS.2019.2911964

14. Naz, S., et al.: Urdu nastaliq recognition using convolutional-recursive deep learning. Neurocomputing **243**, 80–87 (2017)

15. Obaidullah, S.M., Mondal, A., Roy, K.: Structural feature based approach for script identification from printed Indian document. In: 2014 International Conference on Signal Processing and Integrated Networks (SPIN), pp. 120–124 (2014). https://doi.org/10.1109/SPIN.2014.6776933

16. Otsu, N.: A threshold selection method from gray-level histograms. IEEE Trans. Syst. Man Cybern. **9**(1), 62–66 (1979). https://doi.org/10.1109/TSMC.1979.4310076

17. Pal, U., Chaudhuri, B.B.: Script line separation from Indian multi-script documents. In: Proceedings of the Fifth International Conference on Document Analysis and Recognition, ICDAR 1999 (Cat. No. PR00318), pp. 406–409 (1999). https://doi.org/10.1109/ICDAR.1999.791810

18. Rashid, S.F., Shafait, F., Breuel, T.: Connected component level multiscript identification from ancient document images (2010)

19. Shi, B., Yao, C., Zhang, C., Guo, X., Huang, F., Bai, X.: Automatic script identification in the wild. In: 2015 13th International Conference on Document Analysis and Recognition (ICDAR), pp. 531–535 (2015). https://doi.org/10.1109/ICDAR.2015.7333818

20. Singh, P.K., Dalal, S.K., Sarkar, R., Nasipuri, M.: Page-level script identification from multi-script handwritten documents. In: Proceedings of the 2015 Third International Conference on Computer, Communication, Control and Information Technology (C3IT), pp. 1–6 (2015). https://doi.org/10.1109/C3IT.2015.7060113

21. Spitz, A.L.: Determination of the script and language content of document images. IEEE Trans. Pattern Anal. Mach. Intell. **19**(3), 235–245 (1997). https://doi.org/10.1109/34.584100

22. Szegedy, C., et al.: Going deeper with convolutions (2014)

23. Ubul, K., Tursun, G., Aysa, A., Impedovo, D., Pirlo, G., Yibulayin, T.: Script identification of multi-script documents: a survey. IEEE Access **5**, 6546–6559 (2017). https://doi.org/10.1109/ACCESS.2017.2689159

24. Zakarde, S., Rojatkar, D.: A review on south-east and south-west Asian script identification. In: 2019 Innovations in Power and Advanced Computing Technologies (i-PACT), vol. 1, pp. 1–4 (2019). https://doi.org/10.1109/i-PACT44901.2019.8960081

Parsing Digitized Vietnamese Paper Documents

Linh Truong Dieu[1,2](✉), Thuan Trong Nguyen[1,2](✉), Nguyen D. Vo[1,2](✉),
Tam V. Nguyen[3](✉), and Khang Nguyen[1,2](✉)

[1] University of Information Technology, Ho Chi Minh, Vietnam
{17520691,18521471}@gm.uit.edu.vn,
{nguyenvd,khangnttm}@uit.edu.vn
[2] Vietnam National University, Ho Chi Minh, Vietnam
[3] University of Dayton, Dayton, USA
tamnguyen@udayton.edu

Abstract. In recent years, the need to exploit digitized document data
has been increasing. In this paper, we address the problem of parsing
digitized Vietnamese paper documents. The digitized Vietnamese documents are mainly in the form of scanned images with diverse layouts
and special characters introducing many challenges. To this end, we first
collect the UIT-DODV dataset, a novel Vietnamese document image
dataset that includes scientific papers in Vietnamese derived from different scientific conferences. We compile both images that were converted
from PDF and scanned by a smartphone in addition a physical scanner that poses many new challenges. Additionally, we further leverage
the state-of-the-art object detector along with the fused loss function
to efficiently parse the Vietnamese paper documents. Extensive experiments conducted on the UIT-DODV dataset provide a comprehensive
evaluation and insightful analysis.

Keywords: Object detection · Page object detection · Deep learning ·
Convolutional neural network

1 Introduction

The COVID19 pandemic has been changing our lives, which requires us to have
a proactive approach toward accessing future technologies for manufacturing
processes. With digital transformation, paper documents are also gradually converted and replaced by electronic documents for storage on the Cloud Storage, convenient for accessing and searching. The paper documents are stored in
images or PDF files format depending on each organization, which leads to many
challenges to extract necessary information. This requires a good enough detector
model as the foundation for extracting information tasks. The problem's input is
a document image with objects on a possible page: Caption, Table, Figure, and
Formula. The output is an image containing the position of the objects expressed
by bounding boxes and their labels (as shown in Fig. 1).

© Springer Nature Switzerland AG 2021
N. Tsapatsoulis et al. (Eds.): CAIP 2021, LNCS 13052, pp. 382–392, 2021.
https://doi.org/10.1007/978-3-030-89128-2_37

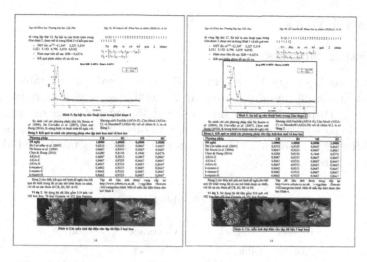

Fig. 1. The problem of detecting objects in Vietnamese document images. The input (left) is the document image. The output (right) is the formula position (yellow), figure (blue), caption (red), table (green). [View better in colored version] (Color figure online)

Most of datasets for object detection on the document page are only converted from PDF, Latex, Word documents. In recent years, with smartphone development, documents are stored in image format using scan apps. Therefore, in this paper, we have introduced the UIT-DODV dataset, including images from PDF and scanned from both scanner and smartphone with more challenges and more practical, suitable for present and future trends. The first reason for choosing the Vietnamese scientific documents is the lack of current datasets. Secondly, the semantic complexity is also a more challenging part of the dataset, promising cause complex problem for the "state of the art" (SOTA) method. We hope that our dataset will play an important role in future works such as OCR (Optical Character Recognition), VQA (Visual Question Answering) on Vietnamese documents. The main contributions to this paper include:

- To the best of our knowledge, UIT-DODV[1] is the first Vietnamese scientific documents dataset.
- We explored the performance of four SOTA models: CascadeTabNet, Faster-RCNN, YOLOv4, YOLOv4x-mish to evaluate challenges encountering from the dataset.
- We proposed a fused loss function for this task. We believe this work is a cornerstone for the development of future algorithms for the given problem.

---

[1] UIT-DODV published at https://uit-together.github.io/datasets/.

2   Related Work

2.1   Existing Datasets

Marmot consists of 2,000 pages in PDF format used for the table detection task. **TableBank** [9] contains more than 278,000 images with more than 47,000 table objects. **PubLayNet** [17] included 358,353 images from research papers and scientific papers in the medical fields. **cTDaR 2019** [4] is a dataset used in the ICDAR2019 competition. The details of the datasets mentioned above are described in Table 1.

Table 1. The statistic of publicly available datasets.

Dataset	Images	Categories	Coverage	Source	Year
Marmot[5]	2,000	Table	Chinese, English	Founder Apabi Library and Citeseer website	2012
TableBank [9]	417,234	Table	English, Chinese, Japanese, Arabic	Miscellaneous	2019
PublayNet [17]	358,353	Title, Text, List, Table, Figure	Medical	PubMed Central	2019
cTDaR2019 Modern Dataset [4]	840	Table	Printed documents	Miscellaneous	2019
UIT-DODV (Ours)	**2,394**	**Table, Figure, Formula, Caption**	**Vietnames research papers**	**National Conference, Can Tho University**	**2021**

[5]https://www.icst.pku.edu.cn/cpdp/sjzy/index.htm

2.2   Parsing Digitized Paper Documents Problem

Traditional Approaches: Traditional methods mainly consist of shape-based methods [6] and texture-based [3,14]. In 2002, Cesarini et al. [2], proposed to detect horizontal and vertical lines, then defining the area surrounded by these lines. Gatos et al. [5] improved it by adding intersection detection in 2005. Overall, the traditional detection methods made significant progress.

Deep Learning Approaches: Li et al. [10] used the layout analysis methods to identify some candidate tablespaces, then applied a conditional random field (CRF) and CNN to classify. Vo et al. [16] combined the region proposals from Fast-RCNN and Faster-RCNN before applying bounding box regression to boost performance. Later, Huang et al. [7] proposed a method of object detection table based on YOLOv3. Sun et al. [15] proposed a method by combining Faster R-CNN [13] and corner locating. Recently, Prasad et al. [12] proposed CascadeTabNet that based on Cascade Mask R-CNN HRNet.

3   UIT-DODV Dataset

3.1   Dataset Collection

We collected the data from the scientific paper available on the website of Can Tho University (CTU). In addition, we used the physical scanner and the scanning app on smartphone to scan the hard-copy of National Conference "Selected issues of Information Technology and Communications" held organized by the Institute of Information Technology (Vietnam Academy of Science and Technology) and universities, from the following publications:

- The XXI edition with the topic "Internet of Things" was held from July 27–28, 2018, at Hong Duc University, Thanh Hoa Province.
- The XXII edition with the topic "Transforming the number of socio-economic operating in the Industry 4.0" was held from June 28 to 29, 2019, at Thai Binh University, Thai Binh Province.

The images in UIT-DODV are created by converting paper documents to digital images, using document conversion program, physical scanners and scanning app on smartphone. The purpose of collecting data from multiple sources is to create diversity for data in terms of layout, presentation form and data domain is also expanded with scanned images instead of just using the transferred image convert from PDF. After collecting the desired data, we proceeded to label attachments. Instead of tagging the data from where, we used the pre-train model from the PAA [8] method to predict the bounding box for the object before manually editing those objects. Figure 2 visualizes some exemplary samples in our dataset.

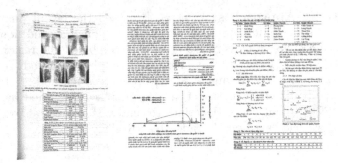

Fig. 2. Exemplary samples in UIT-DODV dataset.

3.2   Category Selection

We followed other image-based document datasets to select the categories. In particular, the first selected object is **Table** - appearing in most published datasets. **Figure** is the simplest and most effective way to turn complex ideas into a concise form, which can be a statistical graph that helps visualize the results of research. Shapes include natural sceneries, graphs, charts, layout designs, block diagrams, or maps. Likewise, **Formula** is equally important to describe relationships between concepts and objects concretely and efficiently. The formulas are usually numbered and may occupy several text lines. In addition, the formula object that contains equations and non-math text in a math region leads to the challenge of this object. Besides, with the desire to build a dataset that can be used for many different tasks such as OCR or VQA, we chose to add a new label, **Caption** - presenting a brief and yet complete explanation of the figure or table. The caption for a figure usually appears below the graphic; for a table, above.

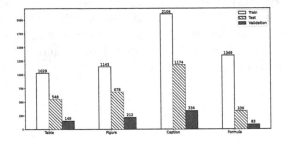

Fig. 3. Statistical of experimental data.

3.3   Dataset Description

UIT-DODV is the first Vietnamese document image dataset, including 2,394 images with four classes: Table, Figure, Caption, Formula. UIT-DODV converted 1,696 images from PDF with size 1,654 × 2,338, 247 images scanned from the physical scanner and expanded with 451 images scanned from the smartphone.

UIT-DODV has the following highlights: **(1) Variety of images:** images in our dataset are of two types, with images converted from PDF as complete documents and images. Scan images often have lower resolutions depending on the scanning angle as well as the lighting conditions that can cause the document page to be blurred, distorted, skewed, or obscured. **(2) Variety of layout:** data collected from other scientific conferences/journals, a common feature of these conferences/journals is that they often use their templates (typically document pages can represent document pages in the form of one column or two columns). **(3) The challenge comes from data classes:** with the simultaneous use of two formula objects (Formula) and Caption creates a challenge for our dataset as well. As in building detection models for these objects. The vast majority of a document page is represented as text, so spotting these objects quickly is very difficult.

4   Computational Model

4.1   Object Detector

Faster-RCNN [13] Faster R-CNN is revamped from its precursors R-CNN and Fast R-CNN and achieves near real-time speeds when skipping the time spent on regional suggestions with Selective Search (SS). Faster R-CNN used pre-trained models to create feature maps. By using attention mechanism as discussed in [11], the feature is fed to the Region proposal network (RPN) to find the proposed regions, from there generating anchor boxes, anchor boxes continue to be classified. Finally, Non-maximum suppression (NMS) algorithm filters out overlapping anchors.

CascadeTabNet [12] is a new approach that uses a single CNN. The CNN HRNetV2p W32 backbone is responsible for converting the input image into a feature map. "RPN Head" (Dense Head) predicts preliminary object proposals for this feature map. "BBox Heads" takes RoI features as input and conducts RoI-wise predictions. The output for each section includes two predictions, classification score, and box regression points. "Mask Head" predicts masks for objects. CascadeTabNet uses Cascade R-CNN's late-stage segmentation branching strategy. "Bbox Heads" performed object detection then complemented with segmentation masks implemented by "Mask Head" for all detected objects.

YOLOv4 [1] is proposed by Bochkovskiy, Wang, and Liao, which is considered the best version in the YOLO family in both accuracy and speed. Thanks to combining CSPNet architecture with Darknet-53 (YOLOv3) for the backbone and adding 2 SPP and PANet modules. At the same time, the BoF and BoS techniques are utilized to help YOLOv4 achieve an AP score of 43.5% on the COCO dataset with an execution speed of 65 FPS.

YOLOv4x-mish is a different version with some changes from YOLOv4. The YOLOv4x-mish change first CSPDarknet stage of backbone to original Darknet-53 balances speed and accuracy. A CSP-ize PAN block replaces a PAN architecture in the neck, decreasing about 40% in computational volume. In this CSPPAN block, the SPP model stays the same as YOLOv4 to increase the receptive field. In the Head section, YOLOv4x-mish change activation function Leaky ReLU to Mish is a non-monotonic activation function, which helps improve accuracy for the model prediction.

4.2   Loss Function

The idea behind CE loss is to penalize the wrong predictions more than to reward the right predictions. The cross-entropy loss function is defined as follows.

$$\mathcal{L}_{CE}(p_t) = -log(p_t) \tag{1}$$

where p_t is the probability for the class t. Recently, the focal loss (FL) was proposed as an version of CE loss that handles the class imbalance problem by assigning more weights to hard or easily misclassified examples. The focal loss function is defined as:

$$\mathcal{L}_{FL}(p_t) = -\alpha(1 - p_t)^\gamma log(p_t), \tag{2}$$

where α is a balanced form for Focal Loss, defaults to 0.25; the gamma (γ) for calculating the modulating factor, defaults to 2.0.

In this work, we argue that each loss function has its own advantages and drawbacks. In the given problem of parsing digitized Vietnamese paper documents, we propose combining different loss functions in order to further improve the performance. Here, the fused loss function is defined as below.

$$\mathcal{L}_{fused}(p_t) = \lambda\mathcal{L}_{CE}(p_t) + (1 - \lambda)\mathcal{L}_{FL}(p_t), \tag{3}$$

where the effect of each individual loss function is decided by the weight λ. In our implementation, we set λ as 0.6 to emphasize the cross entropy loss.

Table 2. Experimental results of different object detection methods with default configuration. The best performance is marked in boldface.

Method	Table	Figure	Formula	Caption	AP_{50}	AP_{75}	AP
Faster-RCNN	91.60	79.70	45.60	57.70	86.20	76.20	68.70
CascadeTabNet	**95.70**	**83.40**	**48.10**	**67.40**	**89.00**	**80.20**	**73.60**
YOLOv4	84.20	78.00	40.20	60.80	90.20	75.20	65.80
YOLOv4x_mish	82.00	75.70	45.20	61.30	90.70	77.70	66.10

Table 3. Experimental results of different loss functions. The best performance is marked in boldface.

Method	Cls Loss	Table	Figure	Formula	Caption	AP_{50}	AP_{75}	AP
Faster R-CNN	Cross-entropy	91.60	79.70	45.60	57.70	86.20	76.20	68.70
Faster R-CNN	Focal	91.60	79.30	37.60	64.30	87.90	74.50	68.20
Faster R-CNN	Fused loss function	92.40	79.70	45.50	66.40	89.00	77.90	71.00
CascadeTabNet	Cross-entropy	**95.70**	**83.40**	**48.10**	67.40	89.00	80.20	73.60
CascadeTabNet	Focal	95.50	82.40	44.20	65.20	87.80	77.60	71.80
CascadeTabNet	Fused loss function	94.30	83.00	47.50	**73.30**	**89.10**	**81.60**	**74.50**

5 Experimental Results and Discussion

5.1 Experimental Setting

We divided the dataset into three subsets: training (1,440), validation (234), and testing (720) set, as shown in Fig. 3. The entire experiment was conducted on a GeForce RTX 2080 Ti GPU with 11019MiB. Both Faster R-CNN and CascadeTabNet are trained on MMDetection framework *V2.10.0*. For Faster R-CNN, we trained on 24 epochs with default configure using backbone *X-101-64x4d-FPN* and CascadeTabNet with configuration provided by Prasad et al.[2]. For YOLOv4 and YOLOv4x-mish, we implemented it on the Darknet framework. The mAP is used to evaluate the model's performance.

5.2 Analysis Results

The best weights on the validation set are used to evaluate the testing set and reported in Table 2. We found that the two-stage methods gave high accuracy for object detection. CascadeTabNet gave the best results when achieved AP_{75} and mAP is 80.20%, 73.60% respectively. Faster-RCNN gives quite good results in object detection. However, it still missed Caption - the object that accounts for the largest distribution of the dataset with AP is 57.70%, that lower than the other three methods. The two one-stage methods gave the best results on AP_{50}. However, when increasing the IoU threshold to 0.75, the score for YOLOv4 (75.20%) and YOLOv4x-mish (77.70%) were lower than that of the two-stage methods. After visualizing the result shown in Fig. 4, we notice that YOLOv4, YOLOv4x-mish have difficulty creating a perfect bounding box compared to the two two-stage methods. The Table and Figure objects also create a challenge to distinguish, and there are also many cases of overlapping bounding boxes. We further conducted the experiment on different loss functions for the top-2 methods, i.e., Faster-RCNN and CascadeTabNet. As shown in Table 3, the fused loss function yields the best performance in terms of AP. The fused loss function also achieves the best performance in the "Caption" semantic class. This clearly demonstrates the need of using the fused loss function in the problem of parsing digitized Vietnamese paper documents.

---

[2] https://github.com/DevashishPrasad/CascadeTabNet.

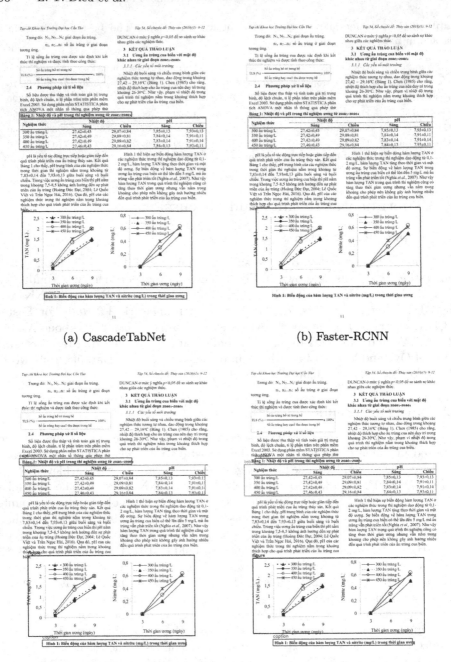

(a) CascadeTabNet

(b) Faster-RCNN

(c) YOLOv4

(d) YOLOv4x-mish

Fig. 4. Visualize results for the four object detection methods with four classes: formula (yellow), caption (red), table (green), figure (blue) [View better in colored version] (Color figure online)

6   Conclusion and Future Work

In this paper, we have introduced the first Vietnamese scientific document image dataset - UIT-DODV - with four critical objects of a research paper: Table, Figure, Caption, and Formula with a total of 2,394 images. We conducted experiments on SOTA object detection methods: Faster-RCNN, YOLOv4, YOLOv4x-mish, and a method was applied on the table detection problem - CascadeTab-Net. Based on the preliminary experimental results, we proposed a fused loss function. The final result with CascadeTabNet achieved the highest mAP result with **74.50%** with a fused loss function. In the future, we plan to build a mobile application that identifies elements in an image document page. In addition, we continue to expand the UIT-DODV dataset in both quantity and diversity in the structure of the documents to serve larger problems in the document image understanding field such as OCR or VQA.

Acknowledgment. The research team would like to express our sincere thanks to the Multimedia Communications Laboratory (MMLab) - University of Information Technology, VNU-HCM for supporting this research. We want to thank Can Tho University Journal of Science for the assistance in the data collection. This project is partially funded under National Science Foundation (NSF) under Grant No. 2025234 and Vietnam National University Ho Chi Minh City (VNU-HCM) under grant number DSC2021-26-03.

References

1. Bochkovskiy, A., Wang, C.-Y., Liao, H.-Y.M.: Yolov4: optimal speed and accuracy of object detection (2020)
2. Cesarini, F., Marinai, S., Sarti, L., Soda, G.: Trainable table location in document images. In: Object Recognition Supported by User Interaction for Service Robots, vol. 3, pp. 236–240 (2002)
3. Etemad, K., Doermann, D., Chellappa, R.: Multiscale segmentation of unstructured document pages using soft decision integration. IEEE Trans. Pattern Anal. Mach. Intell. **19**(1), 92–96 (1997)
4. Gao, L., et al.: ICDAR 2019 competition on table detection and recognition (ctdar). In: 2019 International Conference on Document Analysis and Recognition (ICDAR), pp. 1510–1515 (2019)
5. Gatos, B., Danatsas, D., Pratikakis, I., Perantonis, S.J.: Automatic table detection in document images. In: Singh, S., Singh, M., Apte, C., Perner, P. (eds.) ICAPR 2005, Part I. LNCS, vol. 3686, pp. 609–618. Springer, Heidelberg (2005). https://doi.org/10.1007/11551188_67
6. Ha, J., Phillips, I., Haralick, R.: Document page decomposition using bounding boxes of connected components of black pixels. In: Proceedings of SPIE - The International Society for Optical Engineering (March 1995)
7. Huang, Y., et al.: A YOLO-based table detection method. In: 2019 International Conference on Document Analysis and Recognition (ICDAR), pp. 813–818. IEEE (2019)

8. Kim, K., Lee, H.S.: Probabilistic anchor assignment with IoU prediction for object detection. In: Vedaldi, A., Bischof, H., Brox, T., Frahm, J.-M. (eds.) ECCV 2020, Part XXV. LNCS, vol. 12370, pp. 355–371. Springer, Cham (2020). https://doi.org/10.1007/978-3-030-58595-2_22

9. Li, M., Cui, L., Huang, S., Wei, F., Zhou, M., Li, Z.: Tablebank: table benchmark for image-based table detection and recognition. In: Proceedings of The 12th Language Resources and Evaluation Conference, pp. 1918–1925 (2020)

10. Li, X., Yin, F., Liu, C.: Page object detection from pdf document images by deep structured prediction and supervised clustering. In: 2018 24th International Conference on Pattern Recognition (ICPR), pp. 3627–3632 (2018)

11. Nguyen, T.V., Zhao, Q., Yan, S.: Attentive systems: a survey. Int. J. Comput. Vis. **126**(1), 86–110 (2018)

12. Prasad, D., Gadpal, A., Kapadni, K., Visave, M., Sultanpure, K.: CascadeTabNet: an approach for end to end table detection and structure recognition from image-based documents (2020)

13. Ren, S., He, K., Girshick, R., Sun, J.: Towards real-time object detection with region proposal networks. Faster R-CNN (2016)

14. Sauvola, J., Pietikäinen, M.: Page segmentation and classification using fast feature extraction and connectivity analysis, vol. 2, pp. 1127–1131 (September 1995). ISBN 0-8186-7128-9

15. Sun, N., Zhu, Y., Hu, X., et al.: Table detection using boundary refining via corner locating. In: Lin, Z. (ed.) PRCV 2019, Part I. LNCS, vol. 11857, pp. 135–146. Springer, Cham (2019). https://doi.org/10.1007/978-3-030-31654-9_12

16. Vo, N.D., Nguyen, K., Nguyen, T.V., Nguyen, K.: Ensemble of deep object detectors for page object detection. In: Proceedings of the 12th International Conference on Ubiquitous Information Management and Communication, pp. 1–6 (2018)

17. Zhong, X., Tang, J., Jimeno Yepes, A.: Publaynet: largest dataset ever for document layout analysis. In: 2019 International Conference on Document Analysis and Recognition (ICDAR), pp. 1015–1022 (2019)

EnGraf-Net: Multiple Granularity Branch Network with Fine-Coarse Graft Grained for Classification Task

Riccardo La Grassa[✉][iD], Ignazio Gallo[iD], and Nicola Landro[iD]

Department of Theoretical and Applied Sciences, University of Insubria, Varese, Italy
rlagrassa@uninsubria.it

Abstract. Fine-Grained classification models can expressly focus on the relevant details useful to distinguish highly similar classes typically when the intra-class variance is high and the inter-class variance is low given a dataset. Most of these models use part annotations as bounding box, location part, text attributes to enhance the performance of classification and other models use sophisticated techniques to extract an attention map automatically. We assume that part-based approaches as the automatic cropping method suffers from a missing representation of local features, which are fundamental to distinguish similar objects. While Fine-Grained classification endeavours to recognize the leaf of a graph, humans recognize an object trying also to make a semantic association. In this paper, we use the semantic association structured as a hierarchy (taxonomy) as supervised signals and used them in an end-to-end deep neural network model termed as EnGraf-Net. Extensive experiments on three well-known datasets: Cifar-100, CUB-200-2011 and FGVC-Aircraft prove the superiority of EnGraf-Net over many Fine-Grained models and it is competitive with the most recent best models without using any cropping technique or manual annotations.

Keywords: Fine-grained classification · Hierarchical classification

1 Introduction

In Neuroscience, pattern separation is a process defined as the capability to discriminate a set of similar patterns into less-similar sets of outputs patterns. In [28], the authors show the evidence of the capability of the pattern separation in the Dentate Gyrus (DG) neurons and the pattern completion (a complementary process of pattern separation) in the CA3 neurons. DG and CA3 area of the hippocampus have been long hypothesized to be responsible for these processes and [7,27,28] provides strong empirical support for this functional dissociation. In [29], entitled *CA3 Sees the Big Picture while Dentate Gyrus Splits Hairs*, the authors support the same idea and provide furthermore result to this conclusion. Again, in [25], theoretical models suggest the DG performs pattern separation of cortical inputs before sending its differentiated outputs to CA3. Indeed, DG is ideally located to do this, receiving signals via the major projection from the entorhinal cortex (EC), the perforant path (PP), and sending signals to CA3. These results provide vigorous support for long-standing hypotheses attributing each hippocampal sub-region

© Springer Nature Switzerland AG 2021
N. Tsapatsoulis et al. (Eds.): CAIP 2021, LNCS 13052, pp. 393–402, 2021.
https://doi.org/10.1007/978-3-030-89128-2_38

with distinct roles in neural information processing and set the stage for exciting new research [25]. The deep learning models separate the main signal (e.g. images, sounds, text) in small signals using convolutional operation useful to improve the discrimination ability in the pattern recognition task. Recently, some works are considering forcing the pattern separation process using the semantic association (e.g. hierarchical structure) that comes from the hierarchy abstraction or by manual/automatic text annotation extracted for each image to achieve better performance of a deep learning model. Many of them apply sophisticated methods to extract specific crops on images in order to get more high discriminative features [13,36,38,42,43] instead to consider all manual annotations from a dataset to get them. In Computer Vision (e.g. Fine-Grained classification) implies a hierarchy organization structure composed by different levels of abstraction and it can be represented by a graph, in which all nodes closer to the root represents the abstract concept and as deep as we go far from the node root we find finer-grained abstraction. Also, humans use hierarchical information to recognize a specific object when it is unknown, therefore the categories hierarchy provides a rich semantic correlation among different categories across many levels of abstraction. In the learning process, this guidance can have a regulating effect on semantic space and can lead an algorithm to get better discriminative features for the fine-grained recognition task. In [4], the authors designed a model which considers different granularity levels and proves the usefulness to consider this information to enhance the capability of the main model. Again, in [17], the authors use hierarchical annotation taken by Word-Net to build an end-to-end model to focus on final classification jointly with the hierarchical classification task. They use a simple multi-layer perceptron considering 3 levels of abstraction demonstrating the capability of a model to solve both recognition tasks. The idea to feature fusion considering a multi-scale model was introduced in [20]. Recent works as [12,21] brings to light new interesting architecture to feature fusion at different levels of a deep model. These approaches use the lateral connections of a deep model to carry out fusion operations and combine them widely. In our approach, we use the semantic association as a hierarchical supervised signal to improve the ability of pattern recognition. In Fine-Grained classification, the focus of the most recent deep models is to generate an attention map that contains high discriminative feature such that they can outperform the results in the classification. However, the spatial information (e.g. all regions that contain the environment of the object itself) can also contain useful features to help the pattern recognition ability by models. In [24], observations of five species of Warbler proves that species divide up the resources of a community in such a way that each species is limited by a different factor, such for example the tree partition. Authors show the tree partitioning where at a certain percentage it is possible to find a species in a specific location of the trees. The environment (spatial information) in which the objects can be found is very important and must be considered by modern deep learning models. In our proposal, we do not avoid spatial information using cropping technique, but we consider all information without using any specific region location. In this paper, leveraging by the action makes by DG in our brain we simulate the pattern separation ability of DG neurons using a supervised approach through the semantic association extracted from the hierarchical information of the datasets. We force the pattern separation in a deep model in order to get discriminative features useful to recognize the hierarchy of the objects and distinguish very similar objects. The scientific contribution of this work is concluded as follows:

1. We introduce a Multiple Granularity Branch Network with Fine-Coarse graft grained for Fine-Grained classification task. Our model termed as EnGraf-Net, uses the hierarchical semantic associations from the datasets to force the pattern separation and improve the discrimination capability of a deep learning model.
2. We conduct experiments on Cifar-100, CUB200-2011 and FGVC-Aircraft datasets and demonstrating the effectiveness of our proposal over the baselines and proves to compete with the most recent algorithms compared. We investigating also in the contribution of each components using the Resnet family models conducting ablative studies. We released the code and all experimental reports at [16].

1.1 Related Work

NTS-Net [38] introduces a self-supervised mechanism to locate informative regions without using the bounding box and part annotations. Many works as [11,15,33,37], take advantage of fine-grained human annotations, like the location of some details of images. However, human annotations are expensive and far away from the deep learning concept where every single concept has to be automatic. NTS-Net [38] uses a mechanism to localize informative regions automatically (Navigator) and a Teacher module that evaluates the probability to belong to the ground-truth class using these regions extracted by the Navigator module. Finally, a Scrutinizer module uses these regions to make fine-grained classifications. The model takes the top-M informative regions with the highest score got and these last can represent a weakness of NTS-Net because of the fixed number of regions taken. In [14], authors developed a localization module integrated into an end-to-end setup that generates an attention map and then is used to predict the bounding box of the discriminative regions. The main model is composed of three main modules. The first two modules termed as *AttNet* and *AffNet* has the goal to perform the localization using a combined max-pooling method that merges the vertical/horizontal transformation. Finally, the last model represents the baseline useful to the make classification. In [13], similar to *Affnet*, a method was proposed to search relevant images regions introducing a module trained to build an attention map and a Global K-Max pooling function useful to find a single feature vector that describes the image. The final model requires multiple separate training runs instead to have an end-to-end model. In [43], authors proposed an *attentive pairwise interaction network* for Fine-grained classification based on the idea that humans often compare pairs of images jointly to recognize subtle differences between similar objects. Their method uses two paired images as input and cross-entropy (CE) loss function with a score ranking regularization. In this paper, we compare us with the most recent models who obtain excellent performance in Fine-Grained classification task and we conduct extensive ablation studies analyzing the performance of our proposal.

2   Methodology

The hippocampus and related structures have the capability to minimizes the sets overlap between similar patterns (pattern separation) and to reconstruct complete stored

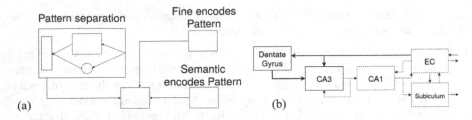

Fig. 1. a) Schematic diagram of our EnGraf-Net b) Schematic diagram of the regions of the hippocampus. The figure shows the feedforward pathway from the entorhinal cortex to the DG and the CA3 neurons. The EC, DG and CA3 blocks are very similar to ours blocks. We simulate the process of the pattern separation by DG and the others connection (EC, DG, CA3) with our proposed approach.

representations from partial patterns that are part of the stored representation (pattern completion). In Fig. 1(b) we show the main pathway diagram of the hippocampus regions and the structure similarity than our proposed approach (see Fig. 1(a)). Observing the nature of this process, we try to simulate the pattern separation/completion as a module engrafted into a branch of a convolutional neural network and analyze the performance model in Fine-Grained classification task. We force a branch through a graft to obtain two supervised patterns that come from the truth of the fine labels and the coarse labels (pattern separation) and finally, we concatenate these patterns into one going towards the next steps of EnGraft-Net (pattern completion). Instead to use manual/automatic annotation comes from images as supervised truth, we extract the semantic association that comes from the hierarchy of the entire datasets used. A semantic association is a process that quantifies the strength of the semantic connection between textual units, taking into account different kinds of relationships and it is an indispensable section of various applications having a spot with a huge number of fields, for instance, Cognitive Psychology and Computer Science. When a semantic association is organized as hierarchical structure, it is called *Taxonomy*. We use the *Taxonomy* of datasets and use the semantic association (class, superclass) as supervised signals useful to compute the loss functions used and we use a combination of different patterns comes from different branches to enhance the discriminative power and increase the main performance of the model. More precisely, given y^K be the fine-grained label from a dataset, we build upon y^K label the superclasses label y^{K-1}. Each image x is annotated using different granularity y^{K-1}, y^K and C_{K-1}, C_K is the number of the class categories considered. Our goal is to correct classify images x across two different types of granularity using an end-to-end model and CE loss functions.

2.1   Network Architecture

EnGraf-Net is based on Resnet family networks. We use a multi-branch approach (see Fig. 1) where the first two branches have the goal to find discriminative features using two types of supervised signals: all labeled classes of fine grained and all labeled superclass of coarse grained extracted by the semantic annotation of a the dataset. The third

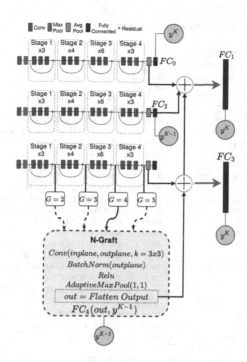

Fig. 2. An overview of our proposed EnGraf-Net model. It employs two branches to extract features at different grain and a third branch network where we engraft a sub-network useful to apply the pattern separation process.

branch is responsible to make the pattern separation/completion through both supervised signals (fine/coarse grained labels). We can choose different type of grafting. The graft block is composed by a convolutional layer, batch normalization and *relu* activation function. Then, we use an adaptive max pooling with output 1×1 and finally after a flatten operation of the output we have a fully connected layer, where y^{K-1} represents the hierarchy class labels used. We concatenate all signals from different branches and use fully connect layers where the last loss function is applied. Depending on which model of Resnet we select the total numbers of parameters of EnGraf-Net is increased.

2.2 Loss Functions

In the training process we use CE as loss function in the form:

$$\mathcal{L}_{xent} = -\frac{1}{m} \sum_{i=1}^{m} \log \frac{e^{W_{y_i}^T x_i + b_{y_i}}}{\sum_{j=1}^{n} e^{W_j^T x_i + b_j}}, \tag{1}$$

where W_{y_i} is the weight associated to class y of i-th instance, x_i are the deep feature of i-th instance and b is the bias term to class y of i-th instance.

Table 1. Experimental results

(a) CUB-200-2011

Method	Top-1
Prior Work	
Resnet-50	84.5
PN-DCN [1](BMVA 14)	85.4
DT-RAM [19](ICCV 17)	86.0
MC-Loss [3](Trans. Img Proc. 20)	87.3
MaxEnt [10](NeurIPS 18)	86.5
MA-CNN [39](ICCV 17)	86.5
KERL [6](IJCAI 18)	87.0
AP-CNN 1 st. [9](Trans. Img Proc. 21)	87.2
NTS-Net [38](ICCV 18)	87.5
DBTNet-50 [40](NeurIPS 19)	87.5
Cross-X [22](ICCV 19)	87.7
TASN [42](CVPR 19)	87.9
HSE [5](ACM-MM 18)	88.1
DBTNet-101 [40](NeurIPS 19)	88.1
CDL [36](ACM-MM 19)	88.4
AP-CNN 2 st. [9](Trans. Img Proc. 21)	88.4
Elope [13](WACV 20)	88.5
API-Net [43](AAAI 20)	88.6
Our Results	
EnGraf-Net50 (G=4, H=1)	87.94
EnGraf-Net101 (G=4, H=1)	88.00
EnGraf-Net152 (G=4, H=1)	88.31

(b) FGVC-Aircraft

Method	Top-1
Prior Work	
Kernel-Act [2](ICCV 17)	88.3
MaxEnt [10](NeurIPS 18)	89.8
MA-CNN [39](ICCV 17)	89.9
PA-CNN [41](Trans. Img Proc. 19)	91.0
DBTNet-50 [40](NeurIPS 19)	91.2
NTS-Net [38](ICCV 18)	91.4
iSQRT-COV [18](CVPR 18)	91.4
DBTNet-101 [40](NeurIPS 19)	91.6
DFL-CNN [35](CVPR 18)	92.0
SEF [23](IEEE Sign. Proc. Lett. 20)	92.1
AP-CNN 1 st [9](Trans. Img Proc. 21)	92.2
Cross-X [22](ICCV 19)	92.7
S3Ns [8](ICCV 19)	92.8
MC-Loss [3](Trans. Img Proc. 20)	92.9
EfficientNet-B7 [31](ICML 19)	92.9
API-Net [43](AAAI 20)	93.4
Elope [13](WACV 20)	93.5
AP-CNN 2 st [9](Trans. Img Proc. 21)	94.1
Our Results	
EnGraf-Net50 (G=4, H=1)	92.14
EnGraf-Net101 (G=4, H=1)	93.34

(c) Hierarchy classification

Method	CUB		AIR	
	acc coarse-fine		acc coarse-fine	
EnGraf-Net50	92.32-87.94		95.44-92.14	
EnGraf-Net101	92.70-88.00		96.10-93.34	

(d) Cifar-100

Method	top-1
Resnet-18	72.43
Two-Branch	72.95
Graft	73.85
EnGraf-net18 (G=2, H=1)	75.52
EnGraf-net18 (G=3, H=1)	75.13
EnGraf-net18 (G=4, H=1)	**75.85**
EnGraf-net18 (G=5, H=1)	75.41

(e) Cifar-100

Method	top-1	Ours	top-1
Resnet-18	72.43	EnGraf-net18	75.85
Resnet-50	75.42	EnGraf-net50	77.27
Resnet-101	75.49	EnGraf-net101	77.13

Considering the network proposed (see Fig. 2) we compute multiple CE loss in a different part of our proposal ($FC_0, FC_1, FC_2, FC_3, FC_4$) where each of them jointly with supervised signals is used in the learning process with Stochastic Gradient Descendent method to achieve the global minima (or a good approximation of it). To summarize our total loss function, we use the following formulation:

$$\mathcal{L} = \mathcal{L}_{xent}(FC_0, y^K) + \mathcal{L}_{xent}(FC_1, y^K) \qquad (2)$$
$$+ \mathcal{L}_{xent}(FC_2, y^{K-1}) + \mathcal{L}_{xent}(FC_3, y^K) + \mathcal{L}_{xent}(FC_4, y^{K-1})$$

The cardinality of the classes y considered in *EnGraf-Net* is different in FC_2, FC_4 than FC_0, FC_1, FC_3 due to the supervised signals selected (it depends on the datasets and by how many hierarchy annotations we consider).

3 Experiments

We conduct experiments on three well-known datasets: Cifar-100, CUB-200-2011 and FGVC-Aircraft and we investigate our performance model using the Resnet family comparing our proposal with the relative baselines and with some most recent architectures proposed in the literature (Table (1a) and Table (1b)). We conduct an ablation study on Resnet-18 using different type of *graft* and with some variations of it (Table (1d) and Table (1e)). We use Cifar100 dataset as a toy dataset to analyze the behaviour of our proposal. It contains 50,000 images 32×32 of training and 10,000 test images, labelled over 100 fine-grained classes. We use 20 coarse-grained classes as y^{K-1} semantic association in our hierarchical extraction. All other experiments have been performed on challenging Fine-Grained image classification benchmark datasets. **CUB-200-2011** [32] contains 11788 images of 200 species of birds split in 5994 and 5794 images for train and test respectively. In addition, we use 122 class labelled as *genera* of the species as supervised signals. **FGVC-Aircraft** [26] contains $10,000$ images of airplanes annotated with the model, specifically splitted in 6667 and 3333 for train and test set. This dataset is organised in four-level hierarchy. In addition to 100 classes (fine-labels) we use 70 classes (family) as superclass labelled. In all our experiments we use different pre-preprocessing data (see our code [16]). We report the upper-bound computational time of 19:43 h in CUB-200-2011 over 150 epochs using a learning rate optimizer (SGD in all our experiments) of 0.001 and batch-size 20 using an EnGraf-Net152.

3.1 Results

In Table (1a) and Table (1b) we report the comparison results between the proposed model and other existing models on the two widely used fine-grained classification benchmark. We measure the top-1 accuracy in each experiment demonstrating the improvements across both datasets used. We obtained 88.31% and 93.34% respectively on CUB-200-2011 and FGVC-Aircraft datasets overcoming the best models used of fine-grained task and being very competitive with the most recent algorithms designed for this specific task (e.g. API-Net). We investigated on Cifar-100 dataset analyzing the performance of our proposal using different graft (Table (1d)). In this last we achieve the best performance using a $graft = 4$ than the other type of graft applied in our experiments and overcome the performance than the baseline Resnet-18, two-branch or using only a branch with a simple graft. Starting to this assumption, we applied a $graft = 4$ for all experiment reported in Tables 1(a,b,c,e). In Table (1c) we report the accuracy from coarse and fine classes demonstrating the capability of our model to solve both tasks in a unique model end-to-end.

3.2 Visualization Analysis

In literature many techniques to visualize the class activation map has been proposed [30,34]. Gradient-weighted Class Activation mapping (Grad-CAM) is widely used because it can be applied in the pretrained models highlighting the discriminative regions of the images. These approaches are useful to analyze the behaviour of the main

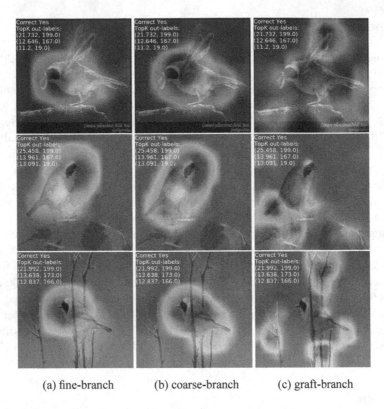

(a) fine-branch   (b) coarse-branch   (c) graft-branch

Fig. 3. Visualization of the attentions regions captured by EnGraf-Net50 in 3 types of layers (columns) and 3 different images (rows) of CUB. Using semantic association of the taxonomy, our method has the capability to detect subtle differences and spatial discriminative information without using part annotations. The third column show the effectiveness to focus the attention to other regions usually not considered in fine-grained models.

model and make it more transparent and understandable. In Fig. 3, we use Grad-CAM approach to visualize the attention map build by our model applying it on three different types of layers of our model. We emphasize that the three features obtained from these layers are combined using a concatenation function and feed into a fully connected layer where we make the final classification. In Fig. 3, we show the activation map build by the branch guided by the supervised signal that represents the class label (1st column), the branch guided by the supervised signal that represents the y^{K-1} class (2nd column) and at last, the branch responsible to make the graft using both supervised signals (3rd column). The discriminative regions usually considered in a fine-grained model belong in the object (1st column), however we force the model to find other discriminative regions from different area of images (3rd column) as the environment information or other useful details. It is extremely curious to observe the different highlighted regions from the graft branch (3rd column) than the others. The exploration of new discriminative regions (spatial information) using our approach is detected and combined with the regions from the others branch to increase the performance of the baselines without using any annotations (e.g. bounding box, location parts).

4 Conclusions

In this paper, we simulate the pattern separation/completion process follow the behaviour of the hippocampus brain circuit. We explore a way to fine-grained classification using only semantic association without the requirement to use bounding-box/part annotations or sophisticated cropping techniques. We conduct experiments along the Resnet models demonstrating that our proposed model can easily be integrated into recent convolutional neural networks. Experiments in CUB-200-2011, FGVC-Aircraft and Cifar-100 have demonstrated the effectiveness of proposed model across many models designed for fine-grained task overcoming the performance of them and to be competitive with the most recent models.

References

1. Branson, S., et al.: Bird species categorization using pose normalized deep convolutional nets (2014)
2. Cai, S., et al.: Higher-order integration of hierarchical convolutional activations for fine-grained visual categorization. In: ICCV (2017)
3. Chang, D., et al.: The devil is in the channels: mutual-channel loss for fine-grained image classification. IEEE Trans. Image Process. **29**, 4683–4695 (2020)
4. Chang, D., et al.: Your "labrador" is my "dog": fine-grained, or not. arXiv preprint arXiv:2011.09040 (2020)
5. Chen, T., et al.: Fine-grained representation learning and recognition by exploiting hierarchical semantic embedding. In: ACM-MM (2018)
6. Chen, T., et al.: Knowledge-embedded representation learning for fine-grained image recognition. In: IJCAI (2018)
7. Deshmukh, S.S., Knierim, J.J.: Representation of non-spatial and spatial information in the lateral entorhinal cortex. Front. Behav. Neurosci. **5**, 69 (2011)
8. Ding, Y., et al.: Selective sparse sampling for fine-grained image recognition. In: ICCV (2019)
9. Ding, Y., et al.: AP-CNN: weakly supervised attention pyramid convolutional neural network for fine-grained visual classification. IEEE Trans. Image Process. **30**, 2826–2836 (2021)
10. Dubey, A., et al.: Maximum-entropy fine grained classification. In: NIPS (2018)
11. Fu, J., et al.: Look closer to see better: recurrent attention convolutional neural network for fine-grained image recognition. In: CVPR (2017)
12. Ghiasi, G., et al.: NAS-FPN: learning scalable feature pyramid architecture for object detection. In: CVPR (2019)
13. Hanselmann, H., et al.: ELoPE: fine-grained visual classification with efficient localization, pooling and embedding. In: WACV (2020)
14. Hanselmann, H., et al.: Fine-grained visual classification with efficient end-to-end localization. arXiv (2020)
15. Jaderberg, M., et al.: Spatial transformer networks. arXiv preprint arXiv:1506.02025 (2015)
16. La Grassa, R., Gallo, I., Landro, N.: EnGraf-Net: multiple granularity branch network with fine-coarse graft grained for classification task. https://gitlab.com/artelabsuper/engraf-net

17. La Grassa, R., Gallo, I., Landro, N.: Learn class hierarchy using convolutional neural networks. Appl. Intell. **51**, 6622–6632 (2021). https://doi.org/10.1007/s10489-020-02103-6
18. Li, P., et al.: Towards faster training of global covariance pooling networks by iterative matrix square root normalization. In: CVPR (2018)
19. Li, Z., et al.: Dynamic computational time for visual attention. In: ICCV (2017)
20. Lin, T.Y., et al.: Feature pyramid networks for object detection. In: CVPR (2017)
21. Liu, S., et al.: Path aggregation network for instance segmentation. In: CVPR (2018)
22. Luo, W., et al.: Cross-x learning for fine-grained visual categorization. In: ICCV (2019)
23. Luo, W., et al.: Learning semantically enhanced feature for fine-grained image classification. IEEE Signal Process. Lett. **27**, 1545–1549 (2020)
24. MacArthur, R.H.: Population ecology of some warblers of northeastern coniferous forests. Ecology **39**, 599–619 (1958)
25. Madar, A.D., et al.: Pattern separation of spiketrains in hippocampal neurons. Sci. Rep. **9**, 1–20 (2019)
26. Maji, S., et al.: Fine-grained visual classification of aircraft. Tech. rep. (2013)
27. Neunuebel, J.P., et al.: Conflicts between local and global spatial frameworks dissociate neural representations of the lateral and medial entorhinal cortex. Neuroscience **33**, 9246–9258 (2013)
28. Neunuebel, J.P., et al.: Ca3 retrieves coherent representations from degraded input: direct evidence for ca3 pattern completion and dentate gyrus pattern separation. Neuron **81**, 416–427 (2014)
29. Newman, E.L., et al.: Ca3 sees the big picture while dentate gyrus splits hairs. Neuron **81**, 226–228 (2014)
30. Selvaraju, R.R., et al.: Grad-cam: visual explanations from deep networks via gradient-based localization. In: ICCV (2017)
31. Tan, M., et al.: EfficientNet: rethinking model scaling for convolutional neural networks. In: ICML (2019)
32. Wah, C., et al.: The Caltech-UCSD Birds-200-2011 Dataset (2011)
33. Wang, D., et al.: Multiple granularity descriptors for fine-grained categorization. In: ICCV (2015)
34. Wang, H., et al.: Score-cam: score-weighted visual explanations for convolutional neural networks. In: CVPR (2020)
35. Wang, Y., et al.: Learning a discriminative filter bank within a CNN for fine-grained recognition. In: CVPR (2018)
36. Wang, Z., et al.: Weakly supervised fine-grained image classification via correlation-guided discriminative learning. In: ACM-MM (2019)
37. Xie, L., et al.: Hierarchical part matching for fine-grained visual categorization. In: ICCV (2013)
38. Yang, Z., Luo, T., Wang, D., Hu, Z., Gao, J., Wang, L.: Learning to navigate for fine-grained classification. In: Ferrari, V., Hebert, M., Sminchisescu, C., Weiss, Y. (eds.) Computer Vision – ECCV 2018. LNCS, vol. 11218, pp. 438–454. Springer, Cham (2018). https://doi.org/10.1007/978-3-030-01264-9_26
39. Zheng, H., et al.: Learning multi-attention convolutional neural network for fine-grained image recognition. In: ICCV (2017)
40. Zheng, H., et al.: Learning deep bilinear transformation for fine-grained image representation. In: NIPS (2019)
41. Zheng, H., et al.: Learning rich part hierarchies with progressive attention networks for fine-grained image recognition. IEEE Trans. Image Process. **29**, 476–488 (2019)
42. Zheng, H., et al.: Looking for the devil in the details: learning trilinear attention sampling network for fine-grained image recognition. In: CVPR (2019)
43. Zhuang, P., et al.: Learning attentive pairwise interaction for fine-grained classification. In: Proceedings of the AAAI Conference on Artificial Intelligence (2020)

When Deep Learners Change Their Mind: Learning Dynamics for Active Learning

Javad Zolfaghari Bengar[1,2]([✉]) [iD], Bogdan Raducanu[1,2] [iD],
and Joost van de Weijer[1,2] [iD]

[1] Computer Vision Center (CVC), Barcelona, Spain
{jzolfaghari,bogdan,joost}@cvc.uab.es
[2] Univ. Autònoma of Barcelona (UAB), Barcelona, Spain

Abstract. Active learning aims to select samples to be annotated that yield the largest performance improvement for the learning algorithm. Many methods approach this problem by measuring the informativeness of samples and do this based on the certainty of the network predictions for samples. However, it is well-known that neural networks are overly confident about their prediction and are therefore an untrustworthy source to assess sample informativeness. In this paper, we propose a new informativeness-based active learning method. Our measure is derived from the learning dynamics of a neural network. More precisely we track the label assignment of the unlabeled data pool during the training of the algorithm. We capture the learning dynamics with a metric called label-dispersion, which is low when the network consistently assigns the same label to the sample during the training of the network and high when the assigned label changes frequently. We show that label-dispersion is a promising predictor of the uncertainty of the network, and show on two benchmark datasets that an active learning algorithm based on label-dispersion obtains excellent results.

Keywords: Active learning · Deep learning · Image classification

1 Introduction

Deep learning methods obtain excellent results for many tasks where large annotated dataset are available [14]. However, collecting annotations is both time and labor expensive. Active Learning(AL) methods [22] aim to tackle this problem by reducing the required annotation effort. The key idea behind active learning is that a machine learning model can achieve a satisfactory performance with a subset of the training samples if it is allowed to choose which samples to label. In AL, the model is trained on a small initial set of labeled data called initial label pool. An acquisition function selects the samples to be annotated by an external oracle. The newly labeled samples are added to the labeled pool and

We acknowledge the support of the Spanish Ministry of Science and Innovation for funding projects PID2019-104174GB-I00.

N. Tsapatsoulis et al. (Eds.): CAIP 2021, LNCS 13052, pp. 403–413, 2021.
https://doi.org/10.1007/978-3-030-89128-2_39

the model is retrained on the updated training set. This process is repeated until the labeling budget is exhausted.

One of the main groups of approaches for active learning use the network uncertainty, as contained in its prediction, to select data for labelling [5,22, 25]. However, it is known that neural networks are overly confident about their predictions; making wrong predictions with high certainty [19]. In this paper, we present a new approach to active learning. Our method is based on recent work of Toneva et al. [24], who study the learning dynamics during the training process of a neural network. They track for each training sample the transitions from being classified correctly to incorrectly (or vice-versa) over the course of learning. Based on these learning dynamics, they characterize a sample of being 'forgettable' (if its class label changes from subsequent presentation) or 'unforgettable' (if the class label assigned is consistent during subsequent presentations). Their method is only applicable for labeled data (and therefore not applicable to active learning) and was applied to show that redundant (forgettable) training data could be removed without hurting network performance.

Inspired by this work, we propose a new uncertainty-based active learning method which is based on the learning dynamics of a neural network. With learning dynamics, we refer to the variations in the predictions of the neural network during training. Specifically, we keep track of the model predictions on every unlabeled sample during the various epochs of training. Based on the variations of the predicted label of samples, we propose a new active learning metric called *label-dispersion*. This way, we can indirectly estimate the uncertainty of the model based on the unlabeled samples. We will directly use this metric as the acquisition function to select the samples to be labeled in the active learning cycles. Other than the forgetfulness measure proposed in [24], we do not require any label information.

Experimental results show that label-dispersion better resemble the true uncertainty of the neural networks, i.e. samples with low dispersion were found to have a correct label prediction, whereas those with high dispersion often had a wrong prediction. Furthermore, in experiments on two standard datasets (CIFAR 10 and CIFAR 100) we show that our method outperforms the state-of-the-art methods in active learning.

2   Related Work

The most important aspect for an active learner is the strategy used to query the next sample to be annotated. These strategies have been successfully applied to a series of traditional computer vision tasks, such as image classification [9,11], object detection [1,2], image retrieval [31], remote sensing [6], and regression [13].

Pool based methods are grouped into three main query strategies relying mostly on heuristics: informativeness [4,10,29], representativeness [21], and hybrid [12,28], a comprehensive survey of these frameworks and a detailed discussion can be found in [22].

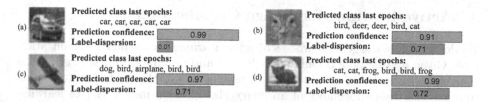

Fig. 1. Comparison between the dispersion and confidence scores. We show four examples images together with the predicted label for the last five epochs of training. The last predicted label is the network prediction when training is finished. We also report the prediction confidence and our label-dispersion measure. (a) Shows an example which is consistently and correctly classified as *car*. The confidence of model is 0.99 and the consistent predictions every epoch result in low dispersion score of 0.01. (b–d) present examples on which the model is highly confident despite a wrong final prediction and constant changes of predictions across the last epochs. This network uncertainty is much better reflected by the high label-dispersion scores.

Informativeness-Based Methods: Among all the aforementioned strategies, the informativeness-based approaches are the most successful ones, with uncertainty being the most used selection criteria in both bayesian [10] and non-bayesian frameworks [29]. In [15,30], the authors employed a loss module to learn the loss of a target model and select the images based on their output loss. More recently, query-synthesizing approaches have used generative models to generate informative samples [17,18,32].

Representativeness-Based Methods: In [23] the authors rely on selecting few examples by increasing diversity in a given batch. The Core-set technique was shown to be an effective representation learning method for large scale image classification tasks [21] and was theoretically proven to work best when the number of classes is small. However, as the number of classes grows, its performance deteriorates. Moreover, for high-dimensional data, using distance-based representation methods, like Core-set, is ineffective because in high-dimensions p-norms suffer from the curse of dimensionality which is referred to as the distance concentration phenomenon in the computational learning literature [7].

Hybrid Methods: Methods that aim to combine uncertainty and representativeness use a two-step process to select the points with high uncertainty as of the most representative points in a batch [16]. A weakly supervised learning strategy was introduced in [25] that trains the model with pseudo labels obtained for instances with high confidence in predictions. While most of the hybrid approaches are based on a two-step process, in [26] they propose a method to select the samples in a single step, based on a generative adversarial framework. An image selector acts as an acquisition function to find a subset of representative samples which also have high uncertainty.

3   Active Learning for Image Classification

We describe here the general process of active learning for the image classification task. Given a large pool of unlabeled data U and an annotation budget B, the goal of active learning is to select a subset of B samples to be annotated as to maximize the performance of an image classification model. Active learning methods generally proceed sequentially by splitting the budget in several cycles. Here we consider the batch-mode variant [21], which annotates multiple samples per cycle, since this is the only feasible option for CNN training. At the beginning of each cycle, the model is trained on the initial labeled set of samples. After training, the model is used to select a new set of samples to be annotated at the end of the cycle via an acquisition function. The selected samples are added to the labeled set $\mathcal{D}_L$ for the next cycle and the process is repeated until the total annotation budget is spent.

3.1   Label-Dispersion Acquisition Function

In this section, we present a new acquisition function for active learning. The acquisition function is the most crucial component and the main difference between active learning methods in the literature. In general, an acquisition function receives a sample and outputs a score indicating how valuable the sample is for training the current model. Most of informativeness-based active learning approaches consider to assess the certainty of the network on the unlabeled data pool which is obtained after training on the labeled data [5,22,25].

In contrast, we propose to track the labels of the unlabeled samples during the course of training. We hypothesize that if the network frequently changes the assigned label, it is unsure about the sample, therefore the sample is an appropriate candidate to be labeled. In Fig. 1 we depict the main idea behind our method and compare it to network confidence. While the confidence score is used to assign the label based on the certainty of the last epoch, the dispersion uses the prediction over all epochs in order to assess the certainty. The first example shows the case of a correct label prediction when both confidence score and dispersion agree. However, in the other three examples, we depict situations where the system predicts the wrong label with high certainty. However, a large dispersion value (i.e. high uncertainty) is the indication of an erroneous prediction.

This idea is based on the concept of *forgettable samples* recently introduced by [24]. [24] states that there exist a large number of unforgettable samples that are never forgotten once learnt. It is shown that they can be omitted from the training set while the generalization performance is maintained. Therefore it suffices to learn the forgettable samples in the train set. However to identify forgettable samples the ground-truth labels is needed. Since we do not have access to the labels in active learning, we propose to use a measure called the *label-dispersion*. The dispersion of a nominal variable is calculated as the proportion of predicted class labels that are not the modal class prediction [8]. It estimates the uncertainty of the model by measuring the changes in the predicted class as

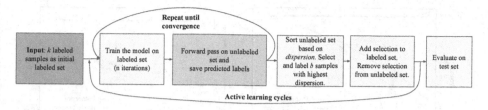

Fig. 2. Active learning framework using Dispersion. Active learning cycles start with initial labeled pool. The model trained on labeled pool is used to output the predictions and compute dispersion for each sample. The samples with highest dispersion are queried for labeling and added to labeled set. This cycle repeats until the annotation budget is exhausted.

following:

$$Dispersion(x) := 1 - \frac{f_x}{T}, \tag{1}$$

with

$$f_x = \sum_t 1[y^t = c^*],$$

$$c^* = \underset{c=1,\dots,C}{\arg\max} \sum_t 1[y^t = c], \tag{2}$$

where f_x is the number of predictions falling into the modal class for sample x and C is the number of classes. Larger values for dispersion means more uncertainty in model outputs. Similar to forgettable samples, we are interested in samples for which the model doesn't persistently output the same class.

Figure 2 presents the active learning framework with our acquisition function. During the training of a network at regular intervals we will save the label predictions for all samples in the unlabeled pool (green block in Fig. 2). In practice, we will perform this operation at every epoch. These saved label predictions allow us to compute the label-dispersion with Eq. 1. We then select the samples with highest dispersion to be annotated and continue to the next active learning cycle until the total label budget is used.

3.2   Informativeness Analysis

To assess the informativeness of methods, we compute the scores assigned to the unlabeled samples and sort the samples accordingly. Then we select several portions of the most informative samples (according to their score) and run the model to infer their labels. We argue that annotating the correctly classified samples would not provide much information for the model because the model already knows their label. In contrast, the model can learn from misclassified samples if labeled. We use the accuracy to implicitly measure the informativeness of unlabeled samples. The lower the accuracy, the more informative the samples will be if labeled. Figure 3 shows the accuracy of model on the unlabeled samples queried by each method. The model used in this analysis is trained on the initial

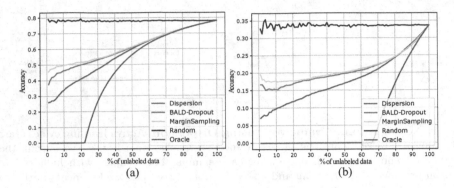

Fig. 3. Informativeness analysis of AL methods on CIFAR10(a) and CIFAR100(b) datasets. The model is used to infer the label of samples selected by AL methods before labeling and the accuracy is measured. For any amount of unlabeled samples, dispersion offers samples with lower accuracy and hence more informative for the model.

labeled set. The accuracy of samples selected randomly remains almost constant regardless of the amount of unlabeled samples. In this analysis, the oracle method by definition uses groundtruth and queries samples that the model misclassified and therefore the accuracy of the model is zero. Among the active learning methods, on both CIFAR10 and CIFAR100 datasets, and for any amount of unlabeled samples, dispersion queries misclassified samples the most, showing that high dispersion correlates well with network uncertainty. These samples can potentially increase the performance of the model if labeled.

4   Experimental Results

4.1   Experimental Setup

We start with model trained on initial labeled set from scratch and employ Resnet-18 as the model architecture. The initial labeled set consists of 10% of train dataset that is selected randomly once for all the methods. At each cycle, we use the model with the corresponding acquisition function to select b samples, which are then labeled and added to $\mathcal{D}_L$. We continue for 4 cycles until the total budget is completely exhausted. In all experiments, the budget per cycle is 5% and total budget is 30% of the entire dataset. Eventually for each cycle, we evaluate the model on the test set. To evaluate our method, we use CIFAR10 and CIFAR100 [14] datasets with 50K images for training and 10K for test. CIFAR10 and CIFAR100 have 10 and 100 object categories respectively and image size of 32×32. During training, we apply a standard augmentation scheme including random crop from zero-padded images, random horizontal flip, and image normalization using the channel mean and standard deviation estimated over the training set.

Dispersion is computed from the most probable class in the output of the model. During training we do an inference on the unlabeled pool at every epoch and save the model predictions. Based on these predictions we compute the label-dispersion for each sample specifically.

Implementation Details. Our method is implemented in PyTorch [20]. We trained all models with the momentum optimizer with value 0.9 and the initial learning rates 0.02. We train for 100 epochs and reduce the learning rate by a factor of 5 once after 60 epochs and again at 80 epochs. Finally, to obtain more stable results we repeat the experiments 3 times and report the mean and standard deviation in our results.

Baselines. We compare our method with several informative and representative-based approaches. *Random sampling:* selects an arbitrary subset of samples from all unlabeled samples. *BALD* [10]: method chooses samples that are expected to maximise the information gained about the model parameters. In particular, it select samples that maximise the mutual information between predictions and model posterior via dropout technique. *Margin sampling* [3]: uses the difference between the two classes with the highest scores as a measure of proximity to the decision boundary. *KCenterGreedy* [21]: is a greedy approximation of KCenter problem also known as min-max facility location problem [27]. Samples having maximum distance from their nearest labeled samples in the embedding space are queried for labeling. *CoreSet* [21]: finds samples as the 2-Opt greedy solution of Kcenter problem in the form of Mixed Integer Programming (MIP) problem. *VAAL* [23]: learns a latent space using a Variational Autoencoder (VAE) and an adversarial network trained to discriminate between unlabeled and labeled data. The unlabeled samples which the discriminator classifies with lowest certainty as belonging to the labeled pool are considered to be the most representative and queried for labeling. *Oracle method:* An acquisition function using ground-truth that selects samples that the model miss-classified. In order to study the potential of active learning, we evaluate oracle-based acquisition function. Note this is not a useful active learning function in practice, as we would not have access to the ground-truth annotations in a real scenario. In order to make a fair comparison with the baselines, we used their official code and adapted them into our code to ensure an identical setting.

4.2 Results

Results on CIFAR10: A comparison with several active learning methods, including both informativeness and representativeness, is provided in Fig. 4. As can be seen in Fig. 4(a) dispersion outperforms the other methods across all the cycles on CIFAR10, only the BALD-Dropout method obtains similar results at 30%. The active learning gain of dispersion against Random sampling is

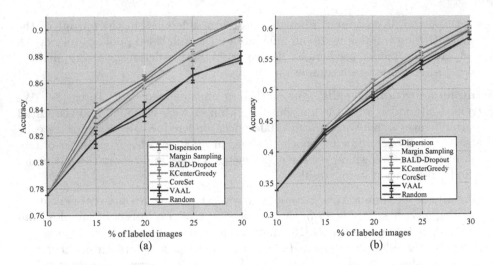

Fig. 4. Performance evaluation. Results for several active learning methods on CIFAR10 (a) and CIFAR100 (b) datasets. All curves are average of 3 runs.

around 7.5% at cycle 4, equivalent to annotating 4000 samples less. The informative methods such as Margin Sampling and BALD lie above the representative methods including KCenterGreedy, CoreSet, VAAL and Random highlighting the importance of informativeness on CIFAR10 where the number of classes is limited and each class is well-represented by many samples.

Results on CIFAR100: Figure 4(b) shows the performance of active learning methods on CIFAR100. As can be seen, the methods are closer and the overall performance of Dispersion, Margin sampling and CoreSet are comparable. However, the addition of labeled samples at cycle 3 and 4 makes the dispersion superior in performance to others. The smaller gap between the informative based methods and Random emphasizes the importance of representativeness on CIFAR100 dataset which has more diverse classes that are underrepresented with few samples in small budget size.

Additionally, Table 1 illustrates the full performance of models that are trained on the entire datasets. Dispersion manages to attain almost 97% and 82% of full performance on CIFAR10 and CIFAR100 respectively by using only 30% of the data, which is a significant reduction in the labeling effort.

Table 1. Active learning results. Performance of AL methods using 30% of dataset both in absolute performance and relative to using all data.

Methods	CIFAR 10		CIFAR 100	
	Acc.	Rel.	Acc.	Rel.
All data	93.61	100%	74.61	100%
Dispersion	**90.74**	96.93%	**60.66**	81.97%
Margin sampling [3]	90.44	96.61%	59.78	80.78%
BALD [10]	90.66	96.85%	59.54	80.46%
KCenterGreedy [21]	89.57	95.69%	59.64	80.59%
CoreSet [21]	89.45	95.56%	59.87	80.91%
VAAL [23]	87.88	93.88%	58.42	78.95%
Random sampling	87.65	93.63%	58.47	79.02%

5   Conclusion

We proposed an informativeness-based active learning algorithm based on the learning dynamics of neural networks. We introduced the label-dispersion metric, which measures label-consistency during the training process. We showed that this measure obtains excellent results when used for active learning on a variety of benchmark datasets. For future work, we are interested in exploring label-dispersion for other research fields such as out-of-distribution detection and within the context of lifelong learning.

References

1. Aghdam, H.H., Gonzalez-Garcia, A., Van de Weijer, J., López, A.M.: Active learning for deep detection neural networks. In: ICCV, pp. 3672–3680 (2019)
2. Bengar, J.Z., et al.: Temporal coherence for active learning in videos. In: ICCV-W, pp. 914–923 (2019)
3. Brust, C.A., Käding, C., Denzler, J.: Active learning for deep object detection. In: VISAPP (2019)
4. Cai, W., Zhang, Y., Zhou, S., Wang, W., Ding, C., Gu, X.: Active learning for support vector machines with maximum model change. In: Calders, T., Esposito, F., Hüllermeier, E., Meo, R. (eds.) ECML PKDD 2014. LNCS (LNAI), vol. 8724, pp. 211–226. Springer, Heidelberg (2014). https://doi.org/10.1007/978-3-662-44848-9_14
5. Chitta, K., Alvarez, J.M., Lesnikowski, A.: Large-scale visual active learning with deep probabilistic ensembles. arXiv preprint arXiv:1811.03575v3 (2019)
6. Deng, C., Liu, X., Li, C., Tao, D.: Active multi-kernel domain adaptation for hyperspectral image classification. Pattern Recogn. **77**, 306–315 (2018)
7. Donoho, D.L., et al.: High-dimensional data analysis: the curses and blessings of dimensionality. AMS Math Challenges Lecture **1**(2000), 32 (2000)
8. Freeman, L.: Elementary Applied Statistics: for Students in Behavioral Science. Wiley, New York (1965). https://books.google.es/books?id=r4VRAAAAMAAJ

9. Fu, W., Wang, M., Hao, S., Wu, X.: Scalable active learning by approximated error reduction. In: KDD, pp. 1396–1405 (2018)
10. Gal, Y., Islam, R., Ghahramani, Z.: Deep Bayesian active learning with image data. In: ICML, pp. 1183–1192 (2017)
11. Gavves, E., Mensink, T.E.J., Tommasi, T., Snoek, C.G.M., Tuytelaars, T.: Active transfer learning with zero-shot priors: Reusing past datasets for future tasks. In: ICCV, pp. 1–9 (2015)
12. Huang, S.J., Jin, R., Zhou, Z.H.: Active learning by querying informative and representative examples. IEEE Trans. PAMI **10**(36), 1936–1949 (2014)
13. Käding, C., Rodner, E., Freytag, A., Mothes, O., Barz, B., Denzler, J.: Active learning for regression tasks with expected model output changes. In: BMVC, pp. 1–15 (2018)
14. Krizhevsky, A.: Learning multiple layers of features from tiny images. Ph.D. thesis, University of Toronto (2012)
15. Li, M., Liu, X., van de Weijer, J., Raducanu, B.: Learning to rank for active learning: a listwise approach. In: ICPR, pp. 5587–5594 (2020)
16. Li, X., Guo, Y.: Adaptive active learning for image classification. In: CVPR, pp. 859–866 (2013)
17. Mahapatra, D., Bozorgtabar, B., Thiran, J.-P., Reyes, M.: Efficient active learning for image classification and segmentation using a sample selection and conditional generative adversarial network. In: Frangi, A.F., Schnabel, J.A., Davatzikos, C., Alberola-López, C., Fichtinger, G. (eds.) MICCAI 2018. LNCS, vol. 11071, pp. 580–588. Springer, Cham (2018). https://doi.org/10.1007/978-3-030-00934-2_65
18. Mayer, C., Timofte, R.: Adversarial sampling for active learning. In: WACV, pp. 3071–3079 (2020)
19. Ovadia, Y., et al.: Can you trust your model's uncertainty? evaluating predictive uncertainty under dataset shift. In: NeurIPS (2019)
20. Paszke, A., et al.: Automatic differentiation in Pytorch. In: NIPS-W (2017)
21. Sener, O., Savarese, S.: Active learning for convolutional neural networks: a coreset approach. In: International Conference on Learning Representations (2018). https://openreview.net/forum?id=H1aIuk-RW
22. Settles, B.: Active Learning. Morgan Claypool, New York (2012)
23. Sinha, S., Ebrahimi, S., Darrell, T.: Variational adversarial active learning. In: ICCV (2019)
24. Toneva, M., Sordoni, A., des Combes, R.T., Trischler, A., Bengio, Y., Gordon, G.J.: An empirical study of example forgetting during deep neural network learning. In: ICLR (2019)
25. Wang, K., Zhang, D., Li, Y., Zhang, R., Lin, L.: Cost-effective active learning for deep image classification. IEEE Trans. Circuits Syst. Video Technol. **27**(12), 2591–2600 (2016)
26. Wang, S., Li, Y., Ma, K., Ma, R., Guan, H., Zheng, Y.: Dual adversarial network for deep active learning. In: ECCV, pp. 1–17 (2020)
27. Wolf, G.W.: Facility location: concepts, models, algorithms and case studies. Int. J. Geogr. Inf. Sci. **25**(2), 331–333 (2011)
28. Yang, Y., Loog, M.: A variance maximization criterion for active learning. Pattern Recogn. **78**, 358–370 (2018)
29. Yang, Y., Ma, Z., Nie, F., Chang, X., Hauptmann, A.G.: Multi-class active learning by uncertainty sampling with diversity maximization. IJCV **113**(2), 113–127 (2015)
30. Yoo, D., Kweon, I.S.: Learning loss for active learning. In: CVPR, pp. 93–102 (2019)

31. Zhang, D., Wang, F., Shi, Z., Zhang, C.: Interactive localized content based image retrieval with multiple-instance active learning. Pattern Recogn. **43**(2), 478–484 (2010)
32. Zhu, J.J., Bento, J.: Generative adversarial active learning. arXiv preprint arXiv:1702.07956 (2017)

Learning to Navigate in the Gaussian Mixture Surface

Riccardo La Grassa[1]($\boxtimes$)(iD), Ignazio Gallo[1](iD), Calogero Vetro[2](iD),
and Nicola Landro[1](iD)

[1] University of Insubria, Department of Theoretical and Applied Sciences,
Varese, Italy
rlagrassa@uninsubria.it
[2] University of Palermo, Department of Mathematics and Computer Science,
Palermo, Italy

Abstract. In the last years, deep learning models have achieved remarkable generalization capability on computer vision tasks, obtaining excellent results in fine-grained classification problems. Sophisticated approaches based-on discriminative feature learning via patches have been proposed in the literature, boosting the model performances and achieving the state-of-the-art over well-known datasets. Cross-Entropy (CE) loss function is commonly used to enhance the discriminative power of the deep learned features, encouraging the separability between the classes. However, observing the activation map generated by these models in the hidden layer, we realize that many image regions with low discriminative content have a high activation response and this could lead to misclassifications. To address this problem, we propose a loss function called Gaussian Mixture Centers (GMC) loss, leveraging on the idea that data follow multiple unimodal distributions. We aim to reduce variances considering many centers per class, using the information from the hidden layers of a deep model, and decreasing the high response from the unnecessary areas of images detected along the baselines. Using jointly CE and GMC loss, we improve the learning generalization model overcoming the performance of the baselines in several use cases. We show the effectiveness of our approach by carrying out experiments over CUB-200-2011, FGVC-Aircraft, Stanford-Dogs benchmarks, and considering the most recent Convolutional Neural Network (CNN).

Keywords: Fine-grained image classification · Loss function

1 Introduction

Weakly supervised Fine-Grained Image Recognition (WFGIR) centers around discovering specific fine details under definite classes and granularity, starting only from class image labels. Firstly, due to the nature of the Fine-Grained dataset, the appearances of the classes can be the same, and how to distinguish the subtle differences between them is of imperative significance. Secondly, WFGIR task uses

© Springer Nature Switzerland AG 2021
N. Tsapatsoulis et al. (Eds.): CAIP 2021, LNCS 13052, pp. 414–423, 2021.
https://doi.org/10.1007/978-3-030-89128-2_40

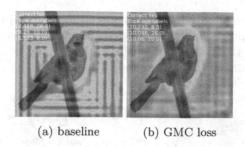

(a) baseline   (b) GMC loss

Fig. 1. Activation map generated by Grad-CAM algorithm show the high neurons response in unnecessary areas from the 3rd hidden layers (a). In (b) we show the attenuation of the neurons response in such regions. We highlight the captions inside the images corresponding to the class activation of the most three outputs from the last layer with the relative class index.

only image-level class annotations, becoming more complex to recognize the subtle changes between different classes. Two main approaches are present in Fine-Grained task, one of which aims to find the local regions in the instance using heuristic schemes as in [25]. The second tries to automatically find the discriminative regions using a sophisticated learning method [38,42]. The works mentioned before focus on the search regions containing subtle details of images, and they are useful to improve the recognition task. Using visualization techniques, we detect a high activation response by neuron groups into low discriminative location of the hidden layers (see Fig. 1 (a)). We decrease the high response of the activation in such regions (see Fig. 1 (b)) via appropriate loss function based on Gaussian Mixture Models (namely, GMC loss). Loss functions are considered the key during the training, where SGD moves in the error surface to find global minima through gradient evaluation and back-propagation procedure [18]. In literature, there exist many loss functions based-on CE loss applied in the last layers of a main Neural Network Model [6,8,9,21,27]. Recently, works as [11,12,26,41], leverage on Center Loss (CL) [37] approaches, achieve remarkable results using the second last layer as argument to their loss function, especially in computer vision task [32,33]. As the CL function is minimized, the variance of the data distribution is reduced to a single point (class center), see Fig. 2 (a). However, observing the features distribution space using the second last layer as an argument, the main effect is to increase the class separability and stretch the distribution by variance enlargement. This effect can bring overlaps for datasets containing many classes and this leads to misclassification. In [15,17], the authors propose a loss function where the error for each instance is adjusted by the values of the involved function. These approaches bring to light new interesting results overcoming the baseline CE loss performance models. In [31], the authors propose Large-margin Gaussian Mixture (L-GM) loss that leverages on the assumption that the deep features follow a Gaussian Mixture distribution. They use a single component to represent a class (one Gaussian peak), so they use a unimodal Gaussian approach. Here, we use multiple components shaped by GMM (see Fig. 2(b)) for each class of the batch, and use many centers of these Gaussian peaks as convergence points. Different by [31] we do not use density function to establish the main error, but compute Euclidean-type distance between multiple Gaussian components for each class. In addition, we use a *pointer* to extract the distribution of the features of the hidden layers pointed, and use them as argument for our proposal.

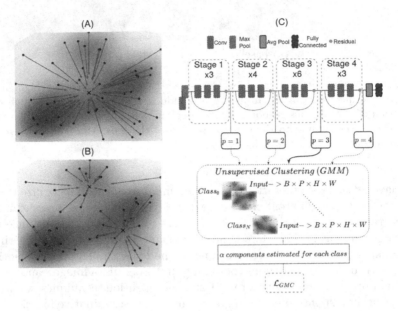

Fig. 2. (a), (b) Shows the differences between CL task and our proposal. Leveraging on the data distribution pointer p, we extract the Gaussian peak using an unsupervised clustering model (GMM). We compute suitable distance of each instance from the nearest centers. (c) Shows the contribution scheme in a Resnet-50 model.

2 Related Works

We recall three main categories of approaches for the Fine-Grained classification.

Manual Object Parts Localization. Bounding box/part annotations are used to enhance the discriminative power via useful details in the local regions of images. Although these methodologies increase the pattern recognition, they suffer from two issues: the manual annotations step is time-consuming, and the need for experts to create each dataset (e.g., Birds dataset).

Discriminative Feature Learning via Patches. Sophisticated approaches are used to extract discriminative patches from images in automatic step and unsupervised or weakly supervised manner [10,38]. A clear example is described by [38], where the model uses a Resnet as backbone jointly with the Navigator-Teacher-Scrutinizer whose goal is in localizing informative regions without the need of bounding-box/part annotations, and in isolating information regions with low information contents. However, the number of patches is fixed by a constraint. Works [7,35,40] produce attention-map based on a correlation between regions or considering the maximum response maps per classes to localize discriminative regions in images.

Pairwise Learning. These methods focus on the relationship between image pairs. Works as [9,12] use couple up to quadruplet instances [5] to improve the pattern recognition task. So, they get interesting results in different fields as person re-identification problems. A novel contribution is given in [42], where the features for couples of images are mixed via an interaction mechanism based on residual attentions. The motivation originates from the fact that humans are able to recognize an object by observing details of other images. [9,27] show beneficial for the discrimination in the final layer, that is obtained using the data distribution in the hidden layer as an argument for the Soft Nearest Neighbour loss. The main idea is to try to converge couples of instances, and as the loss is minimized the class features learned distribution are not necessarily collapsed to a single point as in CL function but can generate many clusters of the same class.

Discussion. All previous models aim to find discriminative details from high-level feature maps directly, ignoring other useful information from all not selected patches. To the best of our knowledge, we are the first to discover significant results for fine-grained classification through loss function, without adding new parameters or using the multi-branches methodology. Our contributions can be summarized as follow:

1. We introduce a novel loss function called GMC loss. We analyze and investigate its behaviour by highlighting its characteristics.
2. We prove the effectiveness of our proposal by using three well-known benchmarks: CUB-200-2011, FGVC-Aircraft and Stanford Dog datasets.
3. We investigate the generalization ability on Resnet family models and report remarkable results against the baseline comparisons. We release all pretrained models and experimental results.

3   Gaussian Mixture Models

A Mixture Model (MM) is a parametric probabilistic representation of data samples that combines (linearly) certain probabilistic components. A Gaussian Mixture Model (GMM) is a particular MM where the basic distributions are Gaussian functions. GMM is a useful key to catch the presence of sub-populations within an overall population, as it does not require the identification of the sub-population of interest. GMM allows to learn, unsupervised, the sub-populations automatically. We recall the basic formula of a GMM as follows:

$$f(x) = \sum_{i=1}^{N} \phi_i f_i(x) = \sum_{i=1}^{N} \phi_i \frac{1}{\sigma_i \sqrt{2\pi}} e^{-\frac{1}{2}\left(\frac{x-\mu_i}{\sigma_i}\right)^2}, \tag{1}$$

that is a weighted sum of N Gaussian distributions f_i. Here the weights ϕ_i ($i = 1, \cdots, N$) satisfy the convex combination constraint $\sum_{i=1}^{N} \phi_i = 1$, $\phi_i \geq 0$.

The GMM is determined by the weight set $\{\phi_i\}_{i=1,\cdots,N}$ and the parameter set $\{(\mu_i, \sigma_i^2)\}_{i=1,\cdots,N}$ (that is, the couple (mean, variance) of each Gaussian component). Whenever the parameter set is learned by a machine learning model, $f_i(x)$ $(i = 1, \cdots, N)$ gives us an a-posteriori estimate of the component probabilities over the studied data samples. We expect that a GMM has more than two components, to avoid that the combination in Eq. 1 reduces to a single Gaussian distribution. Indeed, if there is more than one "peak" in the distribution of the observed data sample (say, multimodal distribution), then a single Gaussian distribution (that is, a well-known example of unimodal distribution) leads generally to a poor fit. Then, to overcome this lack, GMM combines several unimodal Gaussian distributions to improve the fitting performance. For additional information on theoretical aspects and convergence properties of both MM and GMM, we refer to the book [29]. It is worth mentioning that even if the idea behind GMM arises naturally from the classical unimodal Gaussian model, it exhibits a source of mathematical difficulties. Precisely, we know that a crucial characteristic of a GMM is the number of its local maxima (that is generally different from the number of involved Gaussian functions, hence the number of above mentioned peaks). Indeed, it is not known if this number is (always) finite for a pure GMM, and this fact affects the convergence performances of the model (see, for example, [13]). This is an open problem, but there are some recent interesting bounds in the literature (see [1]).

4   Gaussian Mixture Centers Loss

We introduce our mathematical representation of generalized $\mathcal{L}_C$ as follows:

$$\mathcal{L}_{GMC} = \frac{1}{m} \sum_{i=1}^{m} \|x_i - \Omega_{c_{y_i}}^{\alpha}\|_2^2, \qquad (2)$$

with

$$\|x_i - \Omega_{c_{y_i}}^{\alpha}\|_2^2 = \min_{\mu \in \Omega_{c_{y_i}}^{\alpha}} \|\mu - x_i\|_2^2, \qquad (3)$$

where $\Omega_{c_{y_i}}^{\alpha}$ denotes the set of peaks given by a GMM for each batch i, and α means the number of Gaussian components (up to N, and greater than one). We recall that Eq. 3 means the distance between the instance x_i and the set of peaks $\Omega_{c_{y_i}}^{\alpha}$. That is, for each value α in the suitable range, we determine a corresponding decomposition into α Gaussian components, each one of them characterized by its own peak. Finally, via the minimization process (recall the projection method), we associate a single instance to the closer peak among all the peaks of unimodal Gaussian distributions. The above $\mathcal{L}_{GMC}$ is jontly used with a CE Loss of the form:

$$\mathcal{L}_{xent} = -\frac{1}{m} \sum_{i=1}^{m} \log \frac{e^{W_{y_i}^T x_i + b_{y_i}}}{\sum_{j=1}^{n} e^{W_j^T x_i + b_j}}, \qquad (4)$$

where W_{yi} is the weight associated to class y of i-th instance, x_i is the deep feature of i-th instance and b is the bias term to class y of i-th instance. Summarizing, our total loss function is given as follows:

$$\mathcal{L} = \lambda \mathcal{L}_{GMC} + \mathcal{L}_{xent},\qquad(5)$$

where the parameter λ is used to balance the effects of the two loss functions.

4.1 Scheme

In Fig. 2(c) we show an overview of our proposal loss function and the interaction with the used backbone Resnet-X. A variable p plays as a pointer to link the data distribution of the hidden layers. Each input contains $B \times P \times H \times W$ data, where B is the number of images selected per class, P are the planes extracted by the hidden layer pointed by p and H, W are the sizes of the planes. We apply a flatten operation to $H \times W$ to extract a vector, given as input to GMM. For each B images per class, we obtain a data distribution to $B \times P$. We also estimate the K components using GMM, and extract K Gaussian peaks for each class of the dataset. We compute the total error by Eq. 2. Potentially, one can use different values of p and apply the GMC loss using different data distribution.

5   Experiments

5.1   Datasets and Implementation Details

To evaluate our algorithm, we focus on the benchmark datasets: CUB [30], FGVC-Aircraft [24], Stanford Dogs [14]. CUB is composed of 200 classes and contains $11,788$ annotated images splitted into 5994 of train and 5794 of test sets. FGVC-Aircraft contains 6667 images of train and 3333 of test sets. Stanford dogs is composed by 200 classes and $20,580$ images splitted in $12,000$ of train and 8580 of test sets. All of them are subject to different images preprocessing as reported in our code [16]. We resized the images to 448×448 and use the Resnet family as backbone. We also use a Stochastic Gradient Descent (SGD) with momentum 0.9, learning rate 0.001 multiplied by 0.1 each 30 epochs until to 100 maximum epochs. We adopt a batch size of 12 and use a balancing value of $b = 3$ images per class and batch. We impose $\lambda = 0.001$ and change linearly the number of Gaussian Mixture Components over epochs from 20 to 30 per class.

5.2   Ablation Experiments

In Table 1(d) we provide a detailed ablation analysis using different pointer p. We conduct all ablation studies on CUB and FGVC-Aircraft with the ResNet-50 as backbone. In all experiments, we overcome the accuracy of the baseline with our proposal along with all pointers considered (1st column). In Tables 1 (a, b, c) we conduct experiments on Stanford-Dogs, CUB and FGVC-Aircraft with Resnet-50, Resnet-101, Resnet-152 as backbones and we compare our GMC loss function selecting prior work on the Fine-Grained recent classification models. In each table, we give the gain Δ as a percentage variation of our best accuracy and for each accuracy model reported (rows). We again emphasize that many models extend the used backbones, by introducing other parameters and applying techniques for attention map. Against, we achieve remarkable results overcoming many recent models or obtaining very competitive results via only loss function addition. We do not use any extra data augmentation and we use the backbone without variation. In Stanford-Dogs we overcome all prior papers that use best CNN. We make available our best models trained and all experiments ([16]).

Table 1. We compare the gain (Δ) of our GMC loss (best result) to other fine-grained models. We emphasize that some models extend the used backbone and apply some techniques to create attention map. We overcome or achieve competitive results without extending the model, but using only GMC + CE loss function.

(a) CUB-200-2011

Method	Top-1	Δ
Prior Work		
M-CNN [36]	84.2%	+3.82
Resnet-50 [20]	84.5%	+3.52
PN-DCN [2]	85.4%	+2.62
MaxEnt [8]	86.5%	+1.52
KERL [4]	87.0%	+1.02
ResNet-101	87.2%	+0.78
NTS-Net [38]	87.5%	+0.52
TASN [40]	87.9%	+0.12
CDL [35]	88.4%	-0.38
Our Results		
GMC (Resnet-50)	87.35%	
GMC (ResNet-101)	87.66%	
GMC (ResNet-152)	**88.02%**	

(b) FGVC-Aircraft

Method	Top-1	Δ
Prior Work		
Kernel-Act [3]	88.3%	+3.57
MaxEnt [8]	89.8%	+2.07
iSQRT-COV [19]	90.0%	+1.87
ResNet-50	90.3%	+1.57
PA-CNN [39]	91.0%	+0.87
NTS-Net [38]	91.4%	+0.47
DFL-CNN [34]	92.0%	-0.13
SEF [22]	92.1%	-0.23
S3Ns [7]	92.8%	-0.93
Our Results		
GMC (ResNet-50)	91.72%	
GMC (ResNet-101)	91.57%	
GMC (ResNet-152)	**91.87%**	

(c) Stanford Dogs

Method	Top-1	Δ
Prior Work		
MaxEnt [8]	83.6%	+7.0
DB [28]	87.7%	+2.9
API-Net (Resnet-50) [42]	88.3%	+2.3
SEF [22]	88.8%	+1.8
CrossX [23]	88.9%	+1.7
API-Net (Resnet-101) [42]	90.3%	+0.3
Our Results		
GMC loss (Resnet-50)	89.50%	
GMC loss (Resnet-101)	**90.60%**	

(d) Analysis

Method	Backbone	CUB	AIR
Baseline	Resnet-50	84.50%	90.30%
Our (p = 2)	Resnet-50	87.18%	91.57%
Our (p = 3)	Resnet-50	**87.35%**	91.33%
Our (p = 2, 3)	Resnet-50	87.12%	**91.72%**
Our (p = 4)	Resnet-50	87.21%	91.18%

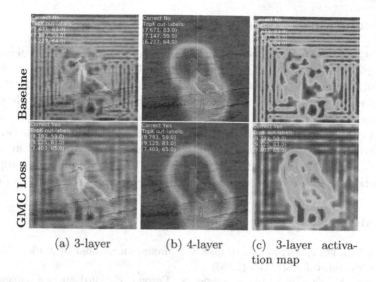

(a) 3-layer   (b) 4-layer   (c) 3-layer activation map

Fig. 3. Activation map using Grad-CAM algorithm shows the high neurons response in unnecessary areas from the 3rd and 2nd hidden layers (1st row). In 2nd row we show the attenuation of the neurons response in such regions. We highlight the captions inside the images corresponding to the class activation of the most three outputs from the last layer with the relative class index. In 3rd column we show only the activation map without the image in background.

5.3   Visualization Analysis

The minimization of the loss function in Eq. 5 allows us to improve the discriminative capability of the approach/model, and hence it plays a crucial role in the learning process. We use a well-known algorithm for data visualization termed Grad-CAM to explore and highlight all the regions with a high response by neurons responsible for these areas. In Fig. 3 we show 2 images from CUB dataset with their relative activation map generated by Grad-CAM in specific layers under consideration. In this experiment, we used a Resnet-152 with our loss function and a pointer $p = 3$ (after the 3rd layer from the baseline used). The 3rd column represents only the activation map created using the data distribution from the third layer and back-propagating the error in this layer following the step given by Grad-CAM algorithm. The different colour represents the importance relative by the activation of the neuron group focused on the various regions of the images (blue colour identify high attention and red dark colour very low attention). We emphasize the capability of the model trained using our loss function to decrease the attention in some regions that are not discriminative (see Fig. 3 1st column, in the 1st and 2nd rows). Instead, observing the attention map from the same baseline used without our approach, the model is not able to reduce the attention in many regions unnecessary to increase the discrimination power. So, GMC loss also serves to regularize the main attention of a model to decrease activation response by groups of neurons closely correlated on the image regions from the hidden layers, using the data distribution that comes from it.

6   Conclusions

Our new GMC loss together with CE loss, lead us to improve the generalization capability of a deep model over various datasets (CUB, FGVC-Aircraft, Stanford Dogs), with different data distribution from all Resnet family models. We prove by experiments, the effectiveness of our approach in using the data distribution of the hidden layers selected in the neural network, and leveraging on a set of Gaussian centers we exceed the performance of itself. GMC loss function could be applied in many neural classification models. Future work will analyze its behaviour, via a clustering dynamic approach to find the number of components of the distribution. We release the code, experiments and pretrained models [16].

References

1. Améndola, C., Engström, A., Haase, C.: Maximum number of modes of gaussian mixtures. Inf. Infer. J. IMA (2020)
2. Branson, S., Van Horn, G., Belongie, S., Perona, P.: Bird species categorization using pose normalized deep convolutional nets (2014)
3. Cai, S., et al.: Higher-order integration of hierarchical convolutional activations for fine-grained visual categorization. In: ICCV (2017)
4. Chen, T., et al.: Knowledge-embedded representation learning for fine-grained image recognition. In: Conference on Artificial Intelligence (2018)
5. Chen, W., et al.: Beyond triplet loss: a deep quadruplet network for person re-identification. In: CVPR (2017)
6. Deng, J., et al.: Arcface: Additive angular margin loss for deep face recognition. In: CVPR (2019)
7. Ding, Y., et al.: Selective sparse sampling for fine-grained image recognition. In: IEEE/CVF (2019)
8. Dubey, A., et al.: Maximum-entropy fine grained classification. In: NIPS (2018)
9. Frosst, N., Papernot, N., Hinton, G.E.: Analyzing and improving representations with the soft nearest neighbor loss. In: ICML (2019)
10. Fu, J., Zheng, H., Mei, T.: Look closer to see better: recurrent attention convolutional neural network for fine-grained image recognition. In: CVPR (2017)
11. Ghosh, P., Davis, L.S.: Understanding center loss based network for image retrieval with few training data. In: ECCV (2018)
12. He, X., et al.: Triplet-center loss for multi-view 3d object retrieval. In: CVPR (2018)
13. Hennig, C., Meila, M., Murtagh, F., Rocci, R.: Handbook of cluster analysis (2015)
14. Khosla, A., et al.: Novel dataset for fine-grained image categorization: stanford dogs. In: CVPR
15. Kulesza, A., Jiang, N., Singh, S.: Low-rank spectral learning with weighted loss functions. In: AISTATS (2015)
16. La Grassa, R., et al.: Learning to navigate in the gaussian mixturesurface. https://gitlab.com/artelabsuper/gmc_loss
17. La Grassa, R., et al.: σ^2 r loss: a weighted loss by multiplicative factors using sigmoidal functions. arXiv preprint arXiv:2009.08796 (2020)
18. LeCun, Y.A., et al.: Efficient backprop. In: Neural networks (2012)
19. Li, P., Xie, J., Wang, Q., Gao, Z.: Towards faster training of global covariance pooling networks by iterative matrix square root normalization. In: CVPR (2018)

20. Li, Z., et al.: Dynamic computational time for visual attention. In: ICCV (2017)
21. Liu, W., et al.: Sphereface: deep hypersphere embedding for face recognition. In: CVPR (2017)
22. Luo, W., Zhang, H., Li, J., Wei, X.S.: Learning semantically enhanced feature for fine-grained image classification. IEEE Signal Proc. Lett. **27**, 1545-9 (2020)
23. Luo, W., et al.: Cross-x learning for fine-grained visual categorization. In: ICCV (2019)
24. Maji, S., et al.: Fine-grained visual classification of aircraft Tech Rep (2013)
25. Peng, Y., He, X., Zhao, J.: Object-part attention model for fine-grained image classification. IEEE Transactions on Image Processing (2017)
26. Qi, C., Su, F.: Contrastive-center loss for deep neural networks. In: ICIP (2017)
27. Salakhutdinov, R., Hinton, G.: Learning a nonlinear embedding by preserving class neighbourhood structure. In: AISTATS (2007)
28. Sun, G., et al.: Fine-grained recognition: accounting for subtle differences between similar classes. In: Conference on Artificial Intelligence (2020)
29. Theodoridis, S., et al.: Pattern recognition. IEEE Trans. Neural Netw. (2008)
30. Wah, C., et al.: The Caltech-UCSD birds-200-2011 Dataset tech rep (2011)
31. Wan, W., Zhong, Y., Li, T., Chen, J.: Rethinking feature distribution for loss functions in image classification. In: CVPR (2018)
32. Wang, M., Deng, W.: Deep face recognition. Neurocomputing **393**, 1-14 (2020)
33. Wang, Q., et al.: A comprehensive survey of loss functions in machine learning. Annals of Data Sci. (2020)
34. Wang, Y., et al.: Learning a discriminative filter bank within a cnn for fine-grained recognition. In: CVPR (2018)
35. Wang, Z., et al.: Weakly supervised fine-grained image classification via correlation-guided discriminative learning. In: ACM-MM (2019)
36. Wei, X.S., et al.: Mask-cnn: localizing parts and selecting descriptors for fine-grained bird species categorization. Pattern Recogn. (2018)
37. Wen, Y., et al.: A discriminative feature learning approach for deep face recognition. In: ECCV (2016)
38. Yang, Z., et al.: Learning to navigate for fine-grained classification. In: ECCV (2018)
39. Zheng, H., et al.: Learning rich part hierarchies with progressive attention networks for fine-grained image recognition. IEEE Trans. Image Process. **29**,476-488 (2019)
40. Zheng, H., et al.: Looking for the devil in the details: learning trilinear attention sampling network for fine-grained image recognition. In: CVPR (2019)
41. Zhu, Y., et al.: Hetero-center loss for cross-modality person re-identification. Neurocomputing **386**, 97-109 (2020)
42. Zhuang, P., Wang, Y., Qiao, Y.: Learning attentive pairwise interaction for fine-grained classification. In: AAAI (2020)

A Deep Hybrid Approach for Hate Speech Analysis

Vipul Shah[1]($\boxtimes$), Sandeep S. Udmale[1] (ORCID), Vijay Sambhe[1], and Amey Bhole[2]

[1] Department of Computer Engineering and Information Technology,
Veermata Jijabai Technological Institute (VJTI), Mumbai 400019, Maharashtra, India
vpshah_m19@ce.vjti.ac.in, {ssudmale,vksambhe}@it.vjti.ac.in
[2] Faculty of Science and Engineering, University of Groningen, Bernoulliborg,
Nijenborgh, 9747AG Groningen, The Netherlands

Abstract. Hate speech is about making insults or stereotypes towards a person or a group of people based on its characteristics such as origin, race, gender, religion, and more. Thus, hate speech can be classified using machine learning and deep learning methods, and it gives a distinguished output from one class to another. Also, every day tons of data are getting accumulated from social media. However, the single deep learning model cannot provide the diversified feature for text classification due to data characteristics. Therefore, this paper proposes two methods for hate speech classification. Initially, a majority voting classifier with three deep learning hybrid models is presented. Finally, a multi-channel convolutional neural network with a bi-directional gated recurrent unit and capsule network is introduced. The proposed approach helps in improving the classification accuracy and ground truth information by reducing ambiguity. The proposed models are verified using six different data sets. The experimental outcomes demonstrate that the proposed methods achieve adequate results for hate speech classification.

Keywords: Bi-directional gated recurrent unit · Capsule network · Convolutional neural network · COVID-19 · Hate speech · Lockdown · COVID-19 vaccination · Farmer bill · US election

1 Introduction

Social media has been growing exponentially, and it is gaining popularity every day. Social networking sites offer a platform for every user to express their emotions and share comments that are visible on the Internet [6,12]. There are several advantages of social media where people can connect and help business to grows faster and rapidly. But, it has disadvantages as well; this technology increases cyberhate and hateful content, which are published and propagated very fast. The use of social media has increased recently, and consequently, cyberhate incidents have risen. Therefore, it has drawn the attention of researchers in hate speech analysis [6,11,12]. Literature informs that the individuals with negative

© Springer Nature Switzerland AG 2021
N. Tsapatsoulis et al. (Eds.): CAIP 2021, LNCS 13052, pp. 424–433, 2021.
https://doi.org/10.1007/978-3-030-89128-2_41

thoughts are biased towards some community, minority groups, women, or selves [11]. Thus, these hateful messages directly impact respect or financial loss or both for the victim and their family [11]. Such cyberhate propaganda, racist messaging, and comments on social networking sites are usually activated when certain events are triggered [1, 6, 11, 12].

The motivation of hate speech analysis is to detect and identify the reaction of the people towards others. In the past decade, the use of hate speech on social media has increased, and, as an effect, it becomes a popular research area for the internet domain. In 2019, Alorainy et al. have proposed a feature set algorithm to detect hate speech effectively. It focuses on feature set selection in which necessary features are getting selected for the embedding approach [1]. After that, Basak et al. have created and developed a web-based application (block shame) to identify and mitigate online public shaming. The application can silence and block the spammer, which defined shaming in six types: abusive, comparison, passing judgments to a user, sarcasm/jokes, and whataboutery. The pre-trained support vector machine (SVM) concept has been introduced to attain better results and simultaneously reduce computational time [2]. The multi-task mutual deep learning approach based on RNN has been proposed for the small datasets [8]. The use of complex attention mechanisms with multi-task learning has manifested the better classification of human sentiment. The MANDOLA, a web application, has several benefits for classifying hatred speech over social media [13]. It uses 3-layer stacked ensemble master-slave classifier. It has been constructed with two classifiers: a slave (CNN, deep neural network, Hybrid CNN-RNN) and one as a master (logistic regression and Linear SVM). It gives better results but requires high computational time [13]. Sequeira et al. in [14] have classified drug abuse tweets using sentence embedding with long short-term memory (LSTM), RNN, region-based CNN, and TextCNN. The proposed paradigm identifies the ambiguity in the sentence for more reliable classification. Recently, researchers and developers are employing the autoencoder (AE) for the detection of hate speech. Thus, deep learning AE-based hate speech detection has been performed to handle ambiguous data to enhance performance [17].

Many countries have laws towards hate speech, and the world's largest democracy India also has rules and regulations towards hate speech[1]. As per Sects. 153(A) and 295(A) of Indian law, hate speech by a word spoken or written is punishable with imprisonment. It may cause the detention of three years, or a fine, or both [3]. Thus, it motivates us to detect and block hate speech over social media. Besides, hate speech leads to violent riots, protests and anti-social events [11].

Based on the above discussion, detection and classification problems are addressed using intelligent approaches extensively through machine learning methods, becoming a powerful approach [6, 11, 12]. The deep learning models also perform well to understand the semantics meaning using recurrent neural network (RNN). It improves the accuracy and predicts well better than the machine learning model when data is significant [9]. Thus, majority voting-based multiple

---

[1] https://lawcommissionofindia.nic.in/reports/Report267.pdf.

Table 1. Political and COVID-19 datasets description for hate speech analysis.

Dataset	US election 2020	India farmer bill	COVID-19	Lockdown 2020	Lockdown 2021	COVID-19 vaccination
	DS1	DS2	DS3	DS4	DS5	DS6
Total tweets	13000	46317	142163	16045	61757	76753
Hate	5000	2049	4898	412	6757	15062
Non-hate	8000	44268	137265	15633	55000	61691
% of hate	38.46	4.42	3.44	2.56	10.94	19.62

hybrid ensembles of deep learning models and multi-channel convolutional neural network (CNN) with a bi-directional gated recurrent unit (Bi-GRU) and capsule network (CapsNet) is proposed in this paper for predicting and improving accuracy. The majority voting approach typically aims to perform voting on models which detect the ground truth of words. Also, it helps in identifying the actual meanings of sentences when a single model could not predict desired results. The multiple models predict the different outcomes, and based on voting, it assists in evaluating more efficiently. The second proposed method is a multi-channel CNN-Bi-GRU with CapsNet (MCCB-CapsNet) that elicits the features from MC-CNN and Bi-GRU model layers. The subsequent CapsNet layer performs dynamic feature routing for a better understanding of features. It identifies the actual ground truth of the word, which helps the model for effective prediction (Table 1)

2   Materials and Methods

2.1   Dataset Description and Construction

This section explains the dataset collection process and the information to detect and classify hate speech under various circumstances. In addition, the latest trending issues in the world are considered for the development of the dataset.

It is essential to understand that collecting the dataset based on an event is the most crucial part of hate speech analysis [15]. Therefore, we have analyzed six datasets, which comprise two political datasets; one US election 2020 and the second Indian Farms Bill 2020. It has collected from Twitter[2]. The Indian farmer bill data set is collected based on hashtag farmer protest and farmer bill from Twitter. The other datasets belong to COVID-19. The COVID-19 data set is an open-source dataset[3]. The fourth data set is the lockdown data set for the year 2020. It has been developed by separating lockdown tweets from the Indian COVID-19 data set. Next, the dataset was gathered from Twitter for the lockdown 2021 from the world. The final dataset is the COVID-19 vaccination

---

[2] https://developer.twitter.com/en.
[3] https://www.kaggle.com/abhaydhiman/covid19-sentiments.

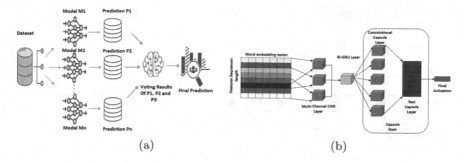

Fig. 1. (a) Proposed method 1: deep voting classifier and (b) proposed method 2: multi-channel CNN-Bi-GRU with CapsNet for hate speech classification.

dataset and developed from Twitter and an open-source dataset[4,5,6] to identify the hate speech and rumors spread over social media. It is worth noting that these raw data sets contain lots of unwanted noise, and it is not required to detect and classify hate speech.

Initially, data is cleaned and processed. After that, lemmatization is applied to the dataset for labeling the dataset into hate and non-hate-related tweets. Next, the hate-related profanity and swear words are collected from PyPI[7]. Then, the identified high-intensity words considered to be hatred words, we applied them to the unlabelled dataset to get a labeled dataset. Finally, we have developed the labeled dataset for hate speech classification.

2.2 Proposed Model Architecture and Experimental Setup

This section shares a detailed explanation of the proposed method for hate speech analysis. This paper presents two novel approaches, The first approach deep voting classifier based on an ensemble of hybrid models, as shown in Fig. 1(b). The second proposed method is a multi-channel CNN-Bi-GRU with CapsNet (MCCB-CapsNet), as shown in Fig. 1(a). The two approaches are explored to identify the distinct feature set for classification. Based on this motive, various classifiers are adopted to distinctly express the input data, which creates the novel feature set. The proposed deep voting approach is beneficial when a single method shows bias towards a particular class. Thus, the proposed method is utilized to derive a generalized fit from the individual models. Hence, a voting classifier attains a decent performance in comparison to a particular approach.

The text classification builds a large dimensional corpus because each unique word is considered a feature. Thus, the proposed ensemble deep voting model is designed on hybrid deep learning models and performs majority class selection

---

[4] https://www.kaggle.com/kaushiksuresh147/covidvaccine-tweets.
[5] https://www.kaggle.com/gpreda/all-covid19-vaccines-tweets.
[6] https://www.kaggle.com/neonian/vaccination-tweets.
[7] https://pypi.org/.

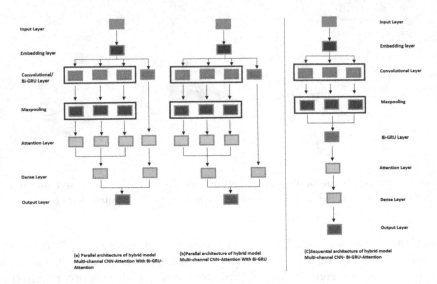

Fig. 2. Base hybrid model architectures used for proposed ensemble deep voting classifier.

voting on those model's results. We have implemented a deep voting classifier for the binary classification dataset. We have used three hybrid models in the proposed work and operated a majority voting for a classification. The hybrid model consists of one sequential and two parallel architectures. They are encapsulated with multi-channel CNN, Bi-GRU, and an attention mechanism. Single-channel CNN for text data classification [4] elicits unique feature information. To obtain multiple features, we have introduced multi-channel CNN. Multi-channel CNN algorithms produce localized word features. Also, we have defined the various hyperparameters for three discrete convolutional channels to obtain the different granularities of the sentence [4,7]. These three base architectures are consist of an input layer that extracts the input features from the datasets and then transmits them to the next subsequent embedding layer. Each text has converted into input features using the keras tokenizer and the One-hot encoding method. This process generates input features where each possible value for that feature has mapped to a new column. The embedding layer consists of a word vector that transforms each word into a word vector matrix. We defined each feature's vector length as 40. This word vector output is forwarded to three multi-channel convolution layers with kernel window sizes of 3, 4, and 5. Also, these channels consist of 100 filters with a dropout ratio of 0.4, which eventually removes the irrelevant features. As a result, these various channels obtain the diversified feature information from the sentence. Additionally, we have used the 'same' padding method to have the output shape similar to inputs. The ReLU activation function is employed in the base CNN architecture, which considers the only activated neurons for faster prediction.

Moreover, we have used GRU, which was developed by Cho et al. in [5]. In contrast to LSTM, GRU requires only two gates, i.e., update and reset gate [4,16]. These two gates accumulate the information and simultaneously determine the hidden ground truth of the words. The update gate operated to pass the relevant information at a particular time t. The reset gate drops the irrelevant information and sends only those required for the training phase [7]. This function of GRU accumulated the past information, which is necessary to understand for actual semantic meaning of the sentence in text classification. We have used GRU with 100 connected units and the attention layer. The sigmoid activation function is utilized as the final activation function. Additionally, we have used the word attention mechanism in each deep learning hybrid model to select the necessary features by assigning weights to actual word features operated to categorize hate speech. The attention mechanism works on the concept of encoder and decoder with query and key values [4].

Furthermore, we have implemented a deep voting classifier for the binary classification dataset. One of the proposed voting model models is a sequential model multi-channel CNN-Bi-GRU-Attention (MCCBA) in which multi-channel CNN is connected to Bi-GRU in sequential, and after that, the attention model is employed to classify hate speech. Also, two parallel models are proposed in the voting model. They are multi-channel CNN-Attention + Bi-GRU (MCCAB), multi-channel CNN-Attention + Bi-GRU-Attention (MCCABA) as shown in Fig. 2. We have selected the three deep learning models, voting based on each tweet of the three hybrid models and calculating the accuracy.

The second proposed MCCB-CapsNet method helps when a single channel CNN extracts the single feature. The CapsNet for text classification uses single-channel CNN and performs dynamic routing depending on the features selects from the embedding layer. So, the CapsNet is legged for large and complex datasets and unable to understand the ground truth of the word when a single-CNN and large feature matrix are applied. So we have used the multi-channel CNN and Bi-GRU for vital feature selection and, those features pass to the CapsNet for dynamic routing for better performance. In this dynamic routing, the output of the child capsule is weighted. Besides, the inner product of the results of the parent capsule is taken. Then, the output is passed to the parent who has the same capsule. Once we get the ground truth of each feature word, it is easy for the CapsNet to perform dynamic routing to overcome the issue of a large-scale feature set, where a single channel does not identify the relationship between words in a sentence. We have implemented an MCCB-CapsNet method for all six datasets. The embedding layer consists of a word vector that converts each word into a word vector matrix, Twitter for glove[8] has been used for word embedding method this converts 100 vector-matrix. We defined the vector length of each feature as 100. This word vector forwarded to multi-channel convolution layers contains three channels with a kernel window size of 3, 4, and 5. A CapsNet detects the characteristics and also identifies the contextual significance of each feature and how they relate to each other [10]. We have utilized ten numbers of

---

[8] https://nlp.stanford.edu/projects/glove/.

capsules, and each capsule determines lower-level features, activation parameters and it calculates the likelihood of that features. Also, the dimension of each capsule is set as 16. Furthermore, each capsule performs the five times dynamic routing, replaces the neuron scalar output with a vector output. The vector output calculated as follow

$$v_j = \frac{||s_j||^2}{1 + ||s_j||^2} \frac{s_j}{||s_j||} \tag{1}$$

Where v_j is the vector output of capsule j and s_j is a total input. For all except the first layer of capsules, the total input to a capsule s_j is a weighted sum over all "prediction vectors" $\hat{u}_{j|i}$. Each $\hat{u}_{j|i}$ of capsules in the layer produces the output u_i of a capsule by multiplying a weight matrix W_{ij} and coupling coefficient c_{ij}. This iterative dynamic routing process assists in enhancing the prediction.

$$s_j = \sum_i c_{ij} \hat{u}_{j|i} , \qquad \hat{u}_{j|i} = W_{ij} u_i \tag{2}$$

The coupling coefficients between capsule i and all the capsules in the layer above sum to 1. It is determined by a "routing softmax" whose initial logits b_{ij} are the log prior probabilities that capsule i should be coupled to capsule j, as shown in Eq. 2. Where $j \in [1, k]$, and k is the number of classes. We have performed our experiment with 50 Epochs and a batch size of 64.

3   Results and Discussion

The results are discussed in this section for the proposed ensemble hybrid deep learning voting classifier and MCCB-CapsNet. It is essential to verify the capability of the proposed system under various datasets with different conditions. Thus, the effectiveness of both the voting classifier and MCCB-CapsNet proposed models is evaluated using six datasets for hate speech analysis. Also, the proposed method is compared with the different state-of-art methods and fuzzy logic. It has been observed that the proposed methods perform better than machine learning models and fuzzy logic when classifying hate speech. The results are comprised of training and testing performance along with F-Score. Also, the proposed method is compared with the different machine learning and deep learning-based paradigms and fuzzy logic [11].

Table 2 indicates that all fuzzy classifiers' training and the testing accuracy is around 80.0% for DS1. But, for the individual deep learning model, performance ranges from 88.0% to 94.0%. This deviation in the result is observed due to the deep learning model's different capability to express the input data. The CNN attains a decent performance of 94.35% among the other competitive deep learning model. Also, the training accuracy of the proposed model and individual deep learning model is around 98.0%. But, the proposed model on the DS1 provides a performance between 94.0% to 95.0%. The most favorable result is achieved for the ensemble voting classifier, and it is 94.23%, and MCCB-CapsNet achieved

Table 2. Comparison of hate speech classifiers performance using political and COVID-19 dataset.

Methods	Tr (%) DS1	Te (%)	F1	Tr (%) DS2	Te (%)	F1	Tr (%) DS3	Te (%)	F1	Tr (%) DS4	Te (%)	F1	Tr (%) DS5	Te (%)	F1	Tr (%) DS6	Te (%)	F1
SVM	77.76	77.02	0.82	95.57	95.57	0.97	96.55	96.55	0.98	97.49	97.27	0.98	89.29	88.5	0.93	84.93	85.19	0.91
Naive Bayes	70.09	70.89	0.75	82.36	82.10	0.89	83.94	83.80	0.91	86.41	85.62	0.92	77.89	77.53	0.86	76.32	74.90	0.84
Decision tree	100.0	65.46	0.71	100.0	90.96	0.95	100.0	93.27	0.96	100.0	93.80	0.96	100.0	82.7	0.90	100.0	75.56	0.84
Gradient boosting tree	83.25	76.33	0.81	95.98	95.61	0.97	96.74	96.55	0.98	98.37	96.88	0.98	90.88	89.36	0.94	85.30	84.19	0.90
DNN	80.39	77.94	0.82	95.76	95.74	0.97	96.76	96.68	0.98	97.49	97.27	0.98	92.12	91.06	0.95	89.42	88.93	0.93
Fuzzy min norm	80.03	79.95	0.81	60.22	60.81	0.74	79.67	79.74	0.88	85.43	84.36	0.91	81.99	81.81	0.88	88.59	88.55	0.92
Fuzzy Prod norm	80.03	79.95	0.81	60.22	60.81	0.74	79.67	79.74	0.88	85.43	84.36	0.91	81.99	81.81	0.88	88.59	88.55	0.92
Fuzzy lukasiewicz norm	80.03	79.95	0.81	60.22	60.81	0.74	79.67	79.74	0.88	85.43	84.36	0.91	81.99	81.81	0.88	88.59	88.55	0.92
Fuzzy Yager norm	80.03	79.95	0.81	60.22	60.81	0.74	79.67	79.74	0.88	85.43	84.36	0.91	81.99	81.81	0.88	88.59	88.55	0.92
Fuzzy fusion	80.03	79.95	0.81	60.22	60.81	0.74	79.67	79.74	0.88	85.43	84.36	0.91	81.99	81.81	0.88	88.59	88.55	0.92
Fusion fuzzy KNN	87.30	83.84	0.90	95.05	99.01	0.99	95.05	97.42	0.98	96.47	95.53	0.98	90.22	90.05	0.94	80.58	79.54	0.88
LSTM	98.02	88.71	0.91	99.88	99.35	0.99	99.97	99.80	0.99	98.26	97.34	0.98	99.39	99.13	0.99	99.89	99.30	0.99
Bi-LSTM	99.78	88.51	0.90	99.84	98.90	0.99	99.96	99.76	0.99	97.49	97.27	0.98	99.89	99.82	0.99	99.93	99.31	0.99
CNN	99.31	94.35	0.95	99.88	98.95	0.99	99.97	99.72	0.99	97.94	97.27	0.98	99.92	99.10	0.99	99.90	99.42	0.99
MulC-CNN	99.38	93.35	0.94	99.68	97.09	0.98	99.96	99.61	0.99	99.63	97.34	0.98	99.92	99.01	0.99	99.90	99.37	0.99
GRU	99.23	91.69	0.93	99.92	99.31	0.99	99.97	99.73	0.99	99.88	98.52	0.99	99.93	99.09	0.99	99.91	99.33	0.99
Bi-GRU	99.14	91.41	0.92	99.93	99.41	0.99	99.98	99.84	0.99	99.88	98.56	0.99	99.92	99.12	0.99	99.89	99.37	0.99
MCCBA (1)	99.42	94.25	0.94	99.94	99.39	0.99	99.98	99.88	0.99	99.84	98.15	0.99	99.94	99.47	0.99	99.93	99.56	0.99
MCCAB (2)	99.43	93.64	0.94	99.94	99.36	0.99	99.99	99.88	0.99	99.80	98.23	0.99	99.93	99.23	0.99	99.90	99.25	0.99
MCCABA (3)	99.47	93.82	0.94	99.94	99.22	0.99	99.98	99.87	0.99	99.88	98.33	0.99	99.95	99.29	0.99	99.91	99.47	0.99
Ensemble voting (1+2+3)	99.52	94.23	0.95	99.94	99.46	0.99	99.99	99.89	0.99	99.88	98.33	0.99	99.95	99.93	0.99	99.92	99.55	0.99
MCCB-CapsNet	99.54	94.46	0.95	99.89	99.11	0.99	99.84	99.34	0.99	99.82	98.73	0.99	99.74	97.79	0.98	99.71	98.49	0.99

Tr: Training accuracy, Te: Testing accuarcy, F1: F1-score

94.46% with an F-score of 0.95. Further, the difference in the accuracy compared to the state-of-art models is scaled approximately from 0.01% to 5.0%. While testing accuracy, the difference of the state-of-art models with the proposed voting classifier is around 0.02% in the US election dataset. Similarly, nearby 99.0% and 98.0% training and testing performance is accomplished by individual deep learning models. However, a decent performance of 98.33% is demonstrated by the proposed ensemble voting classifier, and MCCB-CapsNet achieved 99.73% performance. The training accuracy is more than 99.5% for various the proposed architecture, and the testing accuracy is more than 98.1%. Figure 3 shows the confusion matrix for the proposed deep voting classifier and MCCB-CapsNet, respectively. Similar, observations are noted for other datasets.

The proposed method discovered the performance of more than 99.0% for the various proposed architecture. However, the best accuracy of 99.44% and 99.88% is attained by the ensemble voting classifier on the DS2 and DS3 dataset, respectively. The Fuzzy logic classifiers report the worst performance of around 60.0% and 79.0% on the DS2 and DS3 dataset. The best performance of 99.93% and 99.55% is achieved for the ensemble voting classifier with an F-score of 0.99 on the DS5 and DS6 datasets. Similarly, MCCB-CapsNet gains 97.79% and 98.49%, with an F-score of 0.98 and 0.99 for DS5 and DS6 datasets. For both the proposed architectures, the training accuracy is more than 99.5%. It is also observed that the voting classifier provides satisfactory results compared to the individual fuzzy and deep learning models. In addition, nearby 99.0% training and testing performance is accomplished by individual deep learning models. However, Fuzzy logic performance ranges between 79.0% and 90.0% for both the DS5 and DS6 datasets. Table 2 shows that the machine learning and fuzzy logic norms are not performing well than the deep learning state-of-art models

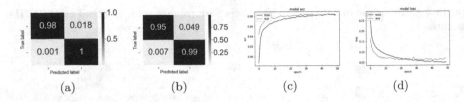

Fig. 3. Confusion matrix for DS6 (a) voting model and (b) MCCB-CapsNet model. MCCB-CapsNet model on DS6 with (c) accuracy and (d) loss function.

on both categories of datasets, i.e., small and large immense datasets. Figures 3 notify the training and validation learning curve graph for loss and accuracy, which helped better performance improvement judgments. Similar, observations are noted for other datasets. Overall, the outcomes show that the proposed deep voting strategy can efficiently deal with the text uncertainty effect, defeating the constraints of using a single deep learning method.

4   Conclusion

This research paper presents two methods, the deep voting classifier method and multi-channel CNN-Bi-GRU with capsule network for hate speech text classification. A voting method is proposed for different sizes of hate speech datasets with multiple hybrid deep learning models. An introduction of the attention mechanism in the proposed work focuses on vital words that increase accuracy. The attention mechanism helps to perform better than other baseline models. The second proposed multi-channel CNN-Bi-GRU with capsule network model focuses on the features and performs the dynamic routing between capsules, presenting better results than other baseline models. Both models reduce the problem of overfitting and work well with small-scale as well as large datasets. Dynamic routing has excellent capabilities while working with multi-channel CNN with a large-scale feature set compares to traditional, cutting-edge state-of-art methods. This research only used text tweets written in English, but it could be further extended to include tweets written in other languages such as Hindi, German, Japanese, Spanish, and others. Besides, this research can be expanded by hate speech detection through multimedia like image and video hate speech detection.

Our future work will focus on multiple hybrid models with small and large-scale data sets to better understand feature heterogeneity. The optimization algorithms using different techniques with different data sets can be explored for hate speech detection work, check with the other hybrid state-of-art models with ensemble techniques, and examine accuracy.

Acknowledgement. We thank the Department of Computer Engineering and Information Technology of the Veermata Jijabai Technological Institute (VJTI), Mumbai-19, for their support and for providing access to the high-performance computing resources developed under TEQIP.

References

1. Alorainy, W., Burnap, P., Liu, H., Williams, M.L.: "The enemy among us": detecting cyber hate speech with threats-based othering language embeddings. ACM Trans. Web **13**(3), 1–26 (2019)
2. Basak, R., Sural, S., Ganguly, N., Ghosh, S.K.: Online public shaming on Twitter: detection, analysis, and mitigation. IEEE Trans. Comput. Soc. Syst. **6**(2), 208–220 (2019)
3. Chaudhari, A., Parseja, A., Patyal, A.: CNN based hate-o-meter: a hate speech detecting tool. In: 2020 Third International Conference on Smart Systems and Inventive Technology (ICSSIT), pp. 940–944 (2020)
4. Cheng, Y., Yao, L., Xiang, G., Zhang, G., Tang, T., Zhong, L.: Text sentiment orientation analysis based on multi-channel CNN and bidirectional GRU with attention mechanism. IEEE Access **8**, 134964–134975 (2020)
5. Cho, K., et al.: Learning phrase representations using RNN encoder-decoder for statistical machine translation. arXiv:1406.1078 (2014)
6. Fortuna, P., Nunes, S.: A survey on automatic detection of hate speech in text. ACM Comput. Surv. **51**(4), 1–30 (2018)
7. Goodfellow, I., Bengio, Y., Courville, A.: Deep Learning, vol. 1. MIT Press, Cambridge (2016)
8. Gui, L., Jia, L., Zhou, J., Xu, R., He, Y.: Multi-task learning with mutual learning for joint sentiment classification and topic detection. IEEE Transactions on Knowledge and Data Engineering pp. 1–13 (2020)
9. Hingu, D., Shah, D., Udmale, S.S.: Automatic text summarization of wikipedia articles. In: 2015 International Conference on Communication, Information Computing Technology (ICCICT), pp. 1–4 (2015)
10. Kim, J., Jang, S., Park, E., Choi, S.: Text classification using capsules. Neurocomputing **376**, 214–221 (2020)
11. Liu, H., Burnap, P., Alorainy, W., Williams, M.L.: A fuzzy approach to text classification with two-stage training for ambiguous instances. IEEE Trans Comput. Soc. Syst. **6**(2), 227–240 (2019)
12. Naseem, U., Razzak, I., Eklund, P.W.: A survey of pre-processing techniques to improve short-text quality: a case study on hate speech detection on twitter. Multimedia Tools Appl. 1–28 (2020). https://doi.org/10.1007/s11042-020-10082-6
13. Paschalides, D., et al.: MANDOLA: a big-data processing and visualization platform for monitoring and detecting online hate speech. ACM Trans. Internet Technol. **20**(2), 1–21 (2020)
14. Sequeira, R., Gayen, A., Ganguly, N., Dandapat, S.K., Chandra, J.: A large-scale study of the Twitter follower network to characterize the spread of prescription drug abuse tweets. IEEE Trans. Comput. Soc. Syst. **6**(6), 1232–1244 (2019)
15. Tang, D., Qin, B., Wei, F., Dong, L., Liu, T., Zhou, M.: A joint segmentation and classification framework for sentence level sentiment classification. IEEE/ACM Trans. Audio Speech Lang. Process. **23**(11), 1750–1761 (2015)
16. Udmale, S.S., Singh, S.K., Bhirud, S.G.: A bearing data analysis based on kurtogram and deep learning sequence models. Measurement **145**, 665–677 (2019)
17. Zhao, R., Mao, K.: Cyberbullying detection based on semantic-enhanced marginalized denoising auto-encoder. IEEE Trans. Affect. Comput. **8**(3), 328–339 (2017)

On Improving Generalization of CNN-Based Image Classification with Delineation Maps Using the CORF Push-Pull Inhibition Operator

Guru Swaroop Bennabhaktula$^{(\boxtimes)}$ ⓘ, Joey Antonisse ⓘ,
and George Azzopardi ⓘ

University of Groningen, Groningen, The Netherlands
g.s.bennabhaktula@rug.nl

Abstract. Deployed image classification pipelines are typically dependent on the images captured in real-world environments. This means that images might be affected by different sources of perturbations (e.g. sensor noise in low-light environments). The main challenge arises by the fact that image quality directly impacts the reliability and consistency of classification tasks. This challenge has, hence, attracted wide interest within the computer vision communities. We propose a transformation step that attempts to enhance the generalization ability of CNN models in the presence of unseen noise in the test set. Concretely, the delineation maps of given images are determined using the CORF push-pull inhibition operator. Such an operation transforms an input image into a space that is more robust to noise before being processed by a CNN. We evaluated our approach on the Fashion MNIST data set with an AlexNet model. It turned out that the proposed *CORF-augmented* pipeline achieved comparable results on noise-free images to those of a conventional AlexNet classification model without CORF delineation maps, but it consistently achieved significantly superior performance on test images perturbed with different levels of Gaussian and uniform noise.

Keywords: CORF · Push-pull · Inhibition · Robustness · Perturbations · Noise suppression · CNN

1 Introduction

In most real-world image classification tasks, there is no control over the environment within which the images are captured. This means that such images might be affected by different types and severity of perturbations (e.g. sensor noise in low-light environments), which may differ from what was present in the training data. Noise is often dynamic and can change over time. Depending on the conditions, noise can suddenly increase due to events in the visual field, which may lead to perturbations in the image affecting the image quality. Examples include adversarial attacks that with very subtle changes to the input images may, for instance, confuse neural networks to classify a panda as a gibbon [8].

© Springer Nature Switzerland AG 2021
N. Tsapatsoulis et al. (Eds.): CAIP 2021, LNCS 13052, pp. 434–444, 2021.
https://doi.org/10.1007/978-3-030-89128-2_42

Image quality directly impacts the reliability and consistency of classification tasks [15]. This challenge has attracted wide interest within the image processing and computer vision communities [5]. Image quality can be affected by a host of factors, such as image compression, during encoding and decoding of images into different formats, resizing, and recoloring, among others. Such methods can also be used as an attack to fool the trained classifier [18]. A common approach to make models more robust to such attacks involves data augmentation during model learning. While data augmentation is effective, its robustness becomes limited when the trained models are deployed into environments where the test images contain noise different than what was present during training.

We hypothesize that giving more importance to the global perceptual contours of a scene will contribute to an image classification solution that is more robust to different types of image noise. To test this hypothesis we use the CORF contour delineation operator with push-pull inhibition, which has been shown to effectively suppress texture and high-frequency noise while delineating the salient contours [3]. We evaluate this transformation tool with respect to different levels of additive perturbations on the Fashion MNIST data set [24] when coupled with the breakthrough network AlexNet [13].

The details of the proposed transformation are presented in Sect. 3.2. Here, we compare two pipelines; a) one that uses the original and noise-free images for training; and b) one that first processes the images with the CORF operator before being fed to the CNN. In order to mimic the real-world scenario where noisy images can be given at the time of model deployment, we evaluate the two pipelines with images consisting of different types and severity of additive noise.

The rest of the paper is organized as follows. In Sect. 2 we present the related works followed by our proposed method in Sect. 3. Experiments and results are reported in Sect. 4, and in Sect. 5 we discuss certain aspects of our work. Finally, we draw our conclusions in Sect. 6.

2   Related Works

Dodge and Karma [7] analyzed how image quality affects the performance of state-of-the-art deep learning models. They trained a network on noise-free images to classify noisy, blurred, and compressed images. From their results, they concluded that image classification is directly proportional to image quality.

In machine/deep learning, this problem can be viewed from the distributions of the training and test data. Ideally, the distributions of the training and test data must be similar for a fair evaluation of the models. In practice, however, the distribution of the test data often deviates from that of the training. In order to account for this unpredictability and make the models more robust, augmented versions of the input data are added to the training set. Data augmentation is among the several techniques used to enhance generalization. Some other popularly used techniques are dropout [20], parameter weight regularization [17], and batch normalization [12], among others. While these techniques are effective, they may not be able to handle deviation in the test set distribution caused due to noise. In order to address this limitation, we propose a transformation step that attempts to enhance the generalization ability of the CNN models in the presence of unseen noise in the test set.

Our hypothesis states that training a model with contour maps of the salient objects instead of the original content results in a classification model that is more robust to unseen noise. This hypothesis requires a robust contour delineation operator that suppresses image noise as much as possible. For this purpose, we use the CORF (Combination of Receptive Fields) operator with push-pull inhibition [3]. It is inspired by the early stages of the mammalian visual system [11], and consists of a system of difference-of-Gaussians (DoG) operators with linearly aligned center-surround areas of support. The output of the operator is an AND-type aggregation of the involved DoG responses. This arrangement is based on the speculation of Hubel and Wiesel [11] that an orientation-selective simple cell is activated when all the afferent LGN cells with center-surround receptive fields are triggered. By means of experiments, it was demonstrated that the CORF model shares more properties with simple cells than the Gabor function model [1]. It is also more effective in contour detection. This operator has later been augmented with two types of inhibition phenomena, namely push-pull [3,21] and surround suppression [14]. It turns out that such inhibition is very effective in suppressing image noise, essentially random strokes and texture that do not belong to the perceptual objects in a given scene.

3   Methods

3.1   Overview

The overall idea is to transform the given images with the CORF contour operator before classification by a CNN model as depicted in Fig. 1.

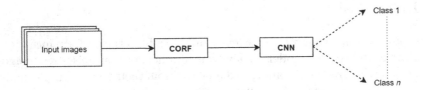

Fig. 1. The proposed application pipeline.

We evaluate the impact on the generalization that the CORF contour operator has on the concerned classification model. Therefore, we compare two pipelines, namely *CORF-free* and *CORF-augmented*. The former is the conventional pipeline that uses the given images as input to the CNN. The latter first delineates the salient contours from the given images by the CORF operator and then uses the resulting contour maps as input to the CNN. Figure 2 illustrates the training and test pipelines of the two approaches.

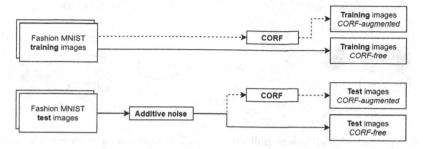

Fig. 2. (Top) Training and (bottom) test pipelines. The solid and dashed arrows indicate the *CORF-free* and the *CORF-augmented* approaches, respectively.

3.2 CORF Operator with Push-Pull Inhibition

The CORF operator is a computational model of orientation-selective simple cells of the mammalian brain [1]. In comparison to the linear Gabor function model, CORF is nonlinear and it achieves more properties of real simple cells; contrast invariant orientation tuning and cross-orientation suppression [3]. The nonlinearity and these two properties result in a CORF operator that is more effective in contour detection than the Gabor function model. The configuration of a model is trainable and its implementation has been found effective in other computer vision [2,9] and signal processing [16] applications.

Figure 3 depicts the structure of a CORF model that is selective for horizontal edges. The circles represent center-on and center-off DoG functions whose output is combined by geometric mean. The standard deviations of the DoG functions and the spacing between their areas of support are hyperparameters of the CORF model used to tune its selectivity. A CORF operator selective for a different orientation can be configured by rotating the alignment of the areas of support of the DoG functions. A rotation-tolerant response can then be achieved by taking the maximum response across all CORF operators selective for different orientations.

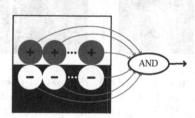

Fig. 3. A CORF computational model of a simple cell that is selective for horizontal edges of the type shown with the white-to-black stimulus behind the circles. The circles indicate the afferent center-on and center-off DoG functions.

(Push) CORF   (Pull) CORF

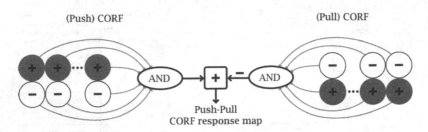

Fig. 4. CORF model with push-pull inhibition. It consists of two sub models, namely push and pull, with the same topology but of opposite selectivity. Their output is then combined with a linear function.

Later, Azzopardi et al. [3] proposed a push-pull CORF model of a simple cell with anti-phase inhibition, which takes as input the response of two CORF models of the type proposed in [1] but with opposing selectivity of luminance contrast. The output of a push-pull CORF model is then the difference between the response of an excitatory (push) CORF model that is stimulated by the pattern of interest and a (weighted) response of the inhibitory (pull) CORF model that is stimulated by the same pattern of interest but of opposite luminance contrast. Figure 4 illustrates the structure of the CORF model augmented with push-pull inhibition. For further technical details, we refer the reader to [3].

In Fig. 5 we illustrate the response maps of the push-pull CORF operator to examples of noise-free and noisy Fashion MNIST images. They demonstrate the operator is very little affected even with high Gaussian noise.

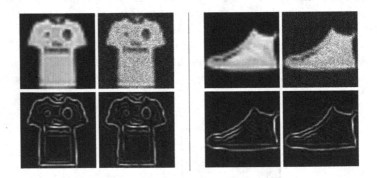

Fig. 5. Robustness of the push-pull CORF delineation operator to Gaussian noise. (Top) Two examples from the Fashion MNIST data set with and without additive Gaussian noise ($\sigma = 0.05$). (Bottom) The corresponding CORF contour maps. The Fashion MNIST images of size 28×28 pixels are resized to 227×227 pixels before the addition of noise.

3.3   AlexNet

We use the AlexNet architecture for our experiments, which was the winning entry in ILSVRC 2012, and was inspired by the Le-Net-5 model introduced in 1998 [26]. AlexNet consists of 8 layers including 5 convolutional layers, 3 fully connected layers, where the final one is the output layer. In order to process grayscale images, the input dimensions of the network are modified to $227 \times 227 \times 1$ pixels from the actual size of $227 \times 227 \times 3$ pixels. The convolutional layers are followed by batch normalization. In our work, batch normalization was used after every convolutional layer which is different from [13], where batch normalization was used only after the first two convolutional layers. The first, second, and the final convolutional layers are followed by a MaxPool layer of size 3×3 pixels with a stride of 2. The first two fully-connected layers consist of 4096 units, each of which is followed by a dropout layer with a factor of 0.5. The number of units in the final fully connected layer is lowered from the original 1000 to 10 classes, the class size of the Fashion MNIST data set. ReLU activations are used in all the convolutional and the fully connected layers, which make training faster in comparison to *tanh* units [13]. The architecture of the AlexNet is depicted in Fig. 6 and for a detailed overview, we refer the reader to [13].

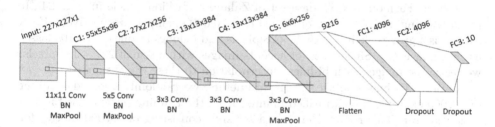

Fig. 6. An illustration of the AlexNet architecture with 5 convolutional (Conv) and 3 fully connected (FC) layers where all the Conv layers are followed by batch normalization (BN).

Sophisticated networks, such as VGG-16 [19] and ResNet-151 [10] have shown to improve the accuracy on ImageNet [6], when compared to AlexNet. However, when it comes to choosing a convolutional network for the relatively simple Fashion MNIST data set [24], we prefer to use AlexNet. This decision is motivated by the fact that the design of AlexNet is simple and it is efficient in terms of time complexity. Due to its simple architectural design and relatively fewer parameters, AlexNet was also found to generalize better when compared to more sophisticated networks [10,19]. Although we use AlexNet in our experiments, in principle, the proposed approach is can be augmented to any CNN.

3.4   Image Perturbations

In the evaluation phase, we experiment with two types of additive noise: Gaussian and uniform. Additive Gaussian noise is part of almost any signal [4], which makes it ideal for mimicking real-life scenarios. An image perturbed with Gaussian noise $\hat{I}_g$ is generated by adding a random value to each pixel (x, y), drawn from a normal distribution $\mathcal{N}$, with a zero mean and a given standard deviation σ to a given image I:

$$\hat{I}_g(x, y) = I(x, y) + \mathcal{N}(0, \sigma) \tag{1}$$

An image perturbed with additive uniform noise $\hat{I}_u$ is created by adding random values drawn from a uniform distribution $\mathcal{U}$, with values between 0 and 1, multiplied by a given weighting value η:

$$\hat{I}_u(x, y) = I(x, y) + \eta \cdot \mathcal{U}(0, 1) \tag{2}$$

4   Experiments and Results

4.1   Data Set

We use the Fashion MNIST data set of Zalando's fashion article images [24]. It has a training and a test set of 60,000 and 10,000 examples, respectively. Each sample is a gray-scale image of 28×28 pixels and belongs to one of the 10 classes as shown in Fig. 7. Since AlexNet accepts images with a size of 227×227 pixels, we resize the images with bi-linear interpolation to these dimensions.

In order to fine-tune the hyper-parameters, we randomly selected a subset of 10,000 examples in a stratified manner from the training set and used it as a validation set. This resulted in a data set split consisting of 50,000 images for training, 10,000 for validation, and 10,000 for testing.

Fig. 7. An example of each of the 10 classes in the Fashion MNIST data set.

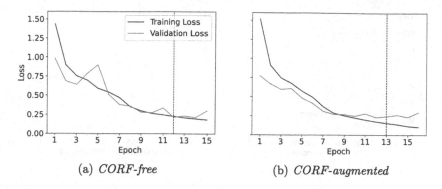

(a) *CORF-free*   (b) *CORF-augmented*

Fig. 8. Training and validation loss without perturbations.

4.2   Experiments

In our experiments, we compare a *CORF-free* pipeline against the proposed *CORF-augmented* pipeline. The hyperparameters of AlexNet used in both pipelines are the same. We used the categorical cross-entropy loss along with the Adam optimizer to train the models, and a batch size of 32. We used an initial learning rate of 0.001 that was decayed at the end of every epoch using an exponential method with a decay rate of 0.96. In order to avoid overfitting, we use a stopping criterium that stops training when the validation accuracy does not improve for three consecutive epochs. This criterium is met at the 12th and 13th epoch for the *CORF-free* and *CORF-augmented* approaches, respectively. From a training point of view, the two pipelines have very similar convergence patterns, depicted in Fig. 8.

The aforementioned pipelines are implemented as follows: *CORF-free* uses the original Fashion MNIST grayscale images and is used as the baseline. Whereas the *CORF-augmented* pipeline first transforms the given images into CORF contour maps[1] before processing them for classification purposes. Both pipelines are trained with noise-free images and are evaluated with the given test set perturbed by Gaussian and uniform noise of increasing severity.

Figure 9 shows the performance of the *CORF-free* and the *CORF-augmented* models to different levels of Gaussian and uniform noise.

5   Discussion

The results of our experiments confirm our hypothesis on the Fashion MNIST data set, in that an AlexNet trained with CORF contour maps is more robust

---

[1] The CORF parameters are set as follows. The afferent DoG functions have a standard deviation $\sigma = 5$. As suggested in [1] for $\sigma = 5$, we use two parallel sets of ten center-on and ten center-off collinear DoG functions, whose distances from the center are 34, 18, 9, 5, and 3 pixels. The two parallel sets of center-on and center-off DoG functions are separated by $\beta = 4.0$ pixels and the inhibition factor $\alpha = 5$.

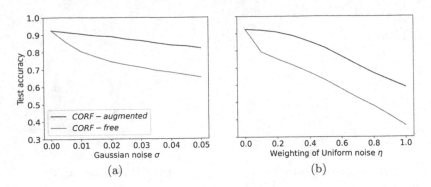

Fig. 9. Test accuracy for different levels of Gaussian and uniform additive noise.

to unseen additive noise than its counterpart trained with the original grayscale images. As a matter of fact, the proposed *CORF-augmented* approach outperforms that of the conventional *CORF-free* pipeline consistently for different levels of noise. For the maximum severity of noise that we test with, the *CORF-augmented* approach achieves an accuracy of 0.8209 and 0.5874 and reduces the error rate by 48.36 and 35.09 percent for Gaussian and uniform noise, respectively.

Notable is the fact that for noise-free images both pipelines achieve comparable results. The improved robustness, therefore, does not influence the baseline data. To the best of our knowledge, our work is the first to investigate the robustness of CNN-based image classification with CORF push-pull contour maps. In [22] the authors investigated an embedded approach of the push-pull mechanism in CNN models, but it does not involve contour maps as we propose here.

In this preliminary study we investigate only two types of additive noise. Our method, however, has the potential to work on a variety of image corruptions, which we will investigate in future. The conducted experiments only use AlexNet and the Fashion MNIST data set. In future, this work may be extended by investigating other CNNs, such as ZFNet [25], Inception [23] or ResNet [10], other data sets, as well as other types of noise and adversarial attacks. Furthermore, we speculate that using a multi-channel approach to train CNNs can further improve the robustness. The channels may, for instance, include the original color channels of a given image along with CORF response maps with different inhibition strengths.

6   Conclusion

In this work, we show that the proposed pipeline that uses the CORF delineation operator with push-pull inhibition is a promising approach to increase the generalization ability of CNNs. Our experiments included an AlexNet architecture and the Fashion MNIST data set. The proposed *CORF-augmented* pipeline exhibits substantially higher generalization ability for additive Gaussian and uniform noise than a conventional AlexNet without the CORF transformation step.

References

1. Azzopardi, G., Petkov, N.: A CORF computational model of a simple cell that relies on LGN input outperforms the Gabor function model. Biol. Cybern. **106**(3), 177–189 (2012)
2. Azzopardi, G., Petkov, N.: Trainable cosfire filters for keypoint detection and pattern recognition. IEEE Trans. Patt. Anal. Mach. Intell. **35**(2), 490–503 (2013). https://doi.org/10.1109/TPAMI.2012.106
3. Azzopardi, G., Rodríguez-Sánchez, A., Piater, J., Petkov, N.: A push-pull CORF model of a simple cell with antiphase inhibition improves SNR and contour detection. PLoS One **9**(7), e98424 (2014)
4. Boncelet, C.: Image noise models. In: The Essential Guide to Image Processing, pp. 143–167. Elsevier (2009)
5. Da Costa, G.B.P., Contato, W.A., Nazare, T.S., Neto, J.E., Ponti, M.: An empirical study on the effects of different types of noise in image classification tasks. arXiv preprint arXiv:1609.02781 (2016)
6. Deng, J., Dong, W., Socher, R., Li, L.J., Li, K., Fei-Fei, L.: Imagenet: a large-scale hierarchical image database. In: 2009 IEEE Conference on Computer Vision and Pattern Recognition, pp. 248–255. IEEE (2009)
7. Dodge, S., Karam, L.: Understanding how image quality affects deep neural networks. In: 2016 Eighth International Conference on Quality of Multimedia Experience (QoMEX), pp. 1–6. IEEE (2016)
8. Goodfellow, I.J., Shlens, J., Szegedy, C.: Explaining and harnessing adversarial examples. arXiv preprint arXiv:1412.6572 (2014)
9. Guo, J., Shi, C., Azzopardi, G., Petkov, N.: Recognition of architectural and electrical symbols by COSFIRE filters with inhibition. In: Azzopardi, G., Petkov, N. (eds.) CAIP 2015. LNCS, vol. 9257, pp. 348–358. Springer, Cham (2015). https://doi.org/10.1007/978-3-319-23117-4_30
10. He, K., Zhang, X., Ren, S., Sun, J.: Deep residual learning for image recognition. In: Proceedings of the IEEE Conference on Computer Vision and Pattern Recognition, pp. 770–778 (2016)
11. Hubel, D.H., Wiesel, T.N.: 8. Receptive fields of single neurones in the cat's striate cortex. In: Brain Physiology and Psychology, pp. 129–150. University of California Press (2020)
12. Ioffe, S., Szegedy, C.: Batch normalization: accelerating deep network training by reducing internal covariate shift. In: International Conference on Machine Learning, pp. 448–456. PMLR (2015)
13. Krizhevsky, A., Sutskever, I., Hinton, G.E.: Imagenet classification with deep convolutional neural networks. Adv. Neural Inf. Process. Syst. **25**, 1097–1105 (2012)
14. Melotti, D., Heimbach, K., Rodríguez-Sánchez, A., Strisciuglio, N., Azzopardi, G.: A robust contour detection operator with combined push-pull inhibition and surround suppression. Inf. Sci. **524**, 229–240 (2020)
15. Nazaré, Tiago, S., Da Costa, G.B.P., Contato, W.A.., Ponti, M.: Deep convolutional neural networks and noisy images. In: Mendoza, M., Velastín, S. (eds.) CIARP 2017. LNCS, vol. 10657, pp. 416–424. Springer, Cham (2018). https://doi.org/10.1007/978-3-319-75193-1_50
16. Neocleous, A., Azzopardi, G., Schizas, C.N., Petkov, N.: Filter-based approach for ornamentation detection and recognition in singing folk music. In: Azzopardi, G., Petkov, N. (eds.) CAIP 2015. LNCS, vol. 9256, pp. 558–569. Springer, Cham (2015). https://doi.org/10.1007/978-3-319-23192-1_47

17. Ng, A.Y.: Feature selection, l 1 vs. l 2 regularization, and rotational invariance. In: Proceedings of the Twenty-First International Conference on Machine Learning, p. 78 (2004)
18. Shamsabadi, A.S., Sanchez-Matilla, R., Cavallaro, A.: Colorfool: Semantic adversarial colorization. In: Proceedings of the IEEE/CVF Conference on Computer Vision and Pattern Recognition (CVPR) (2020)
19. Simonyan, K., Zisserman, A.: Very deep convolutional networks for large-scale image recognition. arXiv preprint arXiv:1409.1556 (2014)
20. Srivastava, N., Hinton, G., Krizhevsky, A., Sutskever, I., Salakhutdinov, R.: Dropout: a simple way to prevent neural networks from overfitting. J. Mach. Learn. Res. **15**(1), 1929–1958 (2014)
21. Strisciuglio, N., Azzopardi, G., Petkov, N.: Robust inhibition-augmented operator for delineation of curvilinear structures. IEEE Trans. Image Process. **28**(12), 5852–5866 (2019). https://doi.org/10.1109/TIP.2019.2922096
22. Strisciuglio, N., Lopez-Antequera, M., Petkov, N.: Enhanced robustness of convolutional networks with a push-pull inhibition layer. Neural Comput. Appl, pp. 1–15 (2020)
23. Szegedy, C., et al.: Going deeper with convolutions. In: Proceedings of the IEEE Conference on Computer Vision and Pattern Recognition, pp. 1–9 (2015)
24. Xiao, H., Rasul, K., Vollgraf, R.: Fashion-mnist: a novel image dataset for benchmarking machine learning algorithms. arXiv preprint arXiv:1708.07747 (2017)
25. Zeiler, M.D., Fergus, R.: Visualizing and understanding convolutional networks. In: Fleet, D., Pajdla, T., Schiele, B., Tuytelaars, T. (eds.) ECCV 2014. LNCS, vol. 8689, pp. 818–833. Springer, Cham (2014). https://doi.org/10.1007/978-3-319-10590-1_53
26. Zhai, J., Shen, W., Singh, I., Wanyama, T., Gao, Z.: A review of the evolution of deep learning architectures and comparison of their performances for histopathologic cancer detection. Proc. Manuf. **46**, 683–689 (2020)

Fast Hand Detection in Collaborative Learning Environments

Sravani Teeparthi[1]([✉]), Venkatesh Jatla[1], Marios S. Pattichis[1],
Sylvia Celedón-Pattichis[2], and Carlos LópezLeiva[2]

[1] The University of New Mexico, Albuquerque, NM, USA
{steeparthi,venkatesh369,pattichi}@unm.edu
[2] Department of Language, Literacy, and Sociocultural Studies,
Albuquerque, NM, USA
{sceledon,callopez}@unm.edu

Abstract. Long-term object detection requires the integration of frame-based results over several seconds. For non-deformable objects, long-term detection is often addressed using object detection followed by video tracking. Unfortunately, tracking is inapplicable to objects that undergo dramatic changes in appearance from frame to frame. As a related example, we study hand detection over long video recordings in collaborative learning environments. More specifically, we develop long-term hand detection methods that can deal with partial occlusions and dramatic changes in appearance.

Our approach integrates object-detection, followed by time projections, clustering, and small region removal to provide effective hand detection over long videos. The hand detector achieved average precision (AP) of 72% at 0.5 intersection over union (IoU). The detection results were improved to 81% by using our optimized approach for data augmentation. The method runs at 4.7× the real-time with AP of 81% at 0.5 intersection over the union. Our method reduced the number of false-positive hand detections by 80% by improving IoU ratios from 0.2 to 0.5. The overall hand detection system runs at 4× real-time.

Keywords: Hand detection · Video analysis · Data augmentation

1 Introduction

We study the problem of developing a robust method for detecting student hands in collaborative learning environment [3]. Here, we define a collaborative learning environment as a small group of students working together in a single table as shown in Fig. 1. Our goal is to recognize writing and typing activities over the detected hand regions. We will then use the writing and typing activities to assess student participation.

This material is based upon work supported by the National Science Foundation under the AOLME project (Grant No. 1613637), the AOLME Video Analysis project (Grant No. 1842220), and the ESTRELLA project (Grant No. 1949230). Any opinions or findings of this paper reflect the views of the authors. They do not necessarily reflect the views of NSF.

ⓒ Springer Nature Switzerland AG 2021
N. Tsapatsoulis et al. (Eds.): CAIP 2021, LNCS 13052, pp. 445–454, 2021.
https://doi.org/10.1007/978-3-030-89128-2_43

(a) Sample video frame showing fully visible hands, occluded hands, and hands belonging to other groups.

(b) Sample video frame occuring 2 seconds after the frame in (a). On the lower-left, a new set of hands appears.

Fig. 1. Hand detection in collaborative learning environments. The problem is restricted to detecting student hands that are nearer to the camera. We use green bounding boxes to identify unobstructed hands that need to be detected. We use yellow bounding boxes to identify occluded hands that need to be detected through projection methods. We use red bounding boxes to identify hands that belong to groups that are associated with hands outside our group of interest. We use a white bounding box in (b) to highlight the appearance of a hand that was fully occluded in (a). (Color figure online)

For robust detection, we require that our hand detection results are consistent throughout the video, implying that we need to deal with occlusions. Furthermore, we need to reject hands that belong to students that belong to other groups, as opposed to the collaborative group that is closer to the camera (see Fig. 1). Since our ultimate goal is to apply our methods to about 1,000 h of digital videos, we also require that our methods are fast.

We also recognize the dynamic aspects of the hand detection problem. First, it is clear that we need to associate hands with different people and that there is a need to deal with the fact that hands can disappear from view due to occlusion (see Fig. 1). Second, we note that the same hands assume very different appearances throughout the video and that there is a need to associate their variations with a single instance.

We summarize some earlier research on the same problem in the M.Sc. thesis by C.J. Darsey [4]. In her thesis, the author studied the problem of accurate hand segmentation over a limited dataset. The dataset consisted of 15 video clips of a maximum duration of 99 seconds. While the methods were successful over a limited video dataset, it is important to note that we are dramatically extending this prior research to long-term detection of hand regions over long video segments. Thus, unlike [4], the current paper also deals with occlusion, rejecting hands outside the group, and associating hand regions with different students. We also have an earlier attempt to detect hands using deep learning in [6]. The current paper dramatically extends this prior research that was focused on very short video datasets without considering occlusion, appearance issues, and associating hands with different people. We also note that head detection and person recognition has been studied in [14,15,19] and [17]. Human activity classification over cropped regions was studied in [5], [16] and [8]. In addition, we note that speech recognition using speaker geometry is studied in [18].

The current paper uses transfer learning from deep learning methods to provide initial hand detection results. For this initial step, we tested several well-known methods. We tested Faster R-CNN [13], YOLO [11], and SSD [9]. We then decided to adopt Faster R-CNN as our baseline model due to the fact that it is more widely supported within human activity recognition systems. We then build our system by post-processing the results from Faster R-CNN. More specifically, we project the results over short video segments to address occlusion and then develop a clustering approach and small area removal to identify the students within the current collaborative group, which are not addressed by traditional hand tracking methods (e.g., [12]). Our approach yields significant improvements over the standard use of Faster R-CNN.

The rest of the paper is organized into three additional sections. We summarize the methodology in Sect. 2. We then present results in Sect. 3 and provide concluding remarks in Sect. 4.

2   Methodology

We summarize our methodology into two sections. First, we present a summary of our hand detection method. Second, we present an optimal data augmentation approach to extend our ground truth dataset.

2.1   Hand Detection Method

We present a block diagram and the corresponding pseudo-code of our approach in Fig. 2. We begin with a deep-learning method that detects hands at the rate of one frame per second. The output of the hand detection method is assumed to be 1 over pixel regions that represent hand regions, and 0 over other regions. Then, we take the projection of the detected regions every 12 s. The projected images $\{PI_1, PI_2, ..., PI_{\lfloor n/12 \rfloor}\}$ can hold a maximum of 12 that represents hand detection over all images, and a minimum of 0 that represents the lack of any hands detected over any image.

To account for occlusion, appearance, and disappearance, we apply a clustering method over the projected image. Several other standard clustering were investigated during the training process (e.g., Otsu, Li, mean, min, etc. [10]). We found that ISODATA [1] performed best. ISODATA is an iterative method that uses Euclidean distance to determine the clusters.

We illustrate the proposed approach in Fig. 3. We show hand detections, obtained after non-maximum suppression [2], and clusters using time projections respectively in Figs. 3a and 3b. Following this, we were able to reject out of group hand clusters with high confidence based on a cluster area constraint [7] as shown in Fig. 3d. The final clusters are then shown in Fig. 3e.

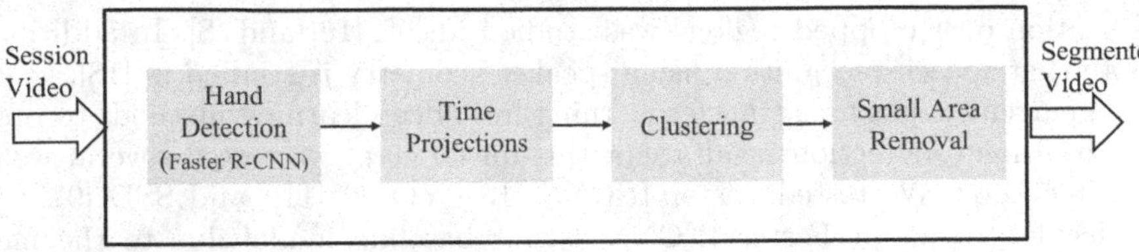

```
function DETECTHANDS(w_*, V, a_th)
▷ Input:
▷     w_* represents a pre-trained single-frame hand detector.
▷     V represents a short Video segment of fixed n seconds duration.
▷     a_th represents a minimum area requirement.
▷ Output:
▷     H contains the detected hand regions for each 12-second video segment.
     BI ← w_*(V) ▷ detect hands at the rate of one frame per second.
     H ← {} ▷ initialize H to store hand detections.
     for each 12-second video segment i: do
         Project the detected hand regions using:
             PI_i ← Σ_s BI_s
         Cluster the projected hand regions using:
             CI_i ← Cluster(PI_i)
         Remove small hand regions of far-away groups:
             H_i ← AreaThreshold(CI_i, a_th)
         H ← Append(H, H_i)
     end for
     return H
end function
```

Fig. 2. Proposed hand detection method using time-projections, clustering, and small region removal

2.2   Optimal Data Augmentation

For robust detection, developed an optimization method for augmenting the dataset. Our goal here is to significantly extend the hand dataset for different scenarios.

The hand detection dataset was created by extracting frames from 44 different collaborative learning sessions. These sessions were selected across 3 years providing a diverse dataset. We labeled every hand instance for a total of 4,548 instances. We partition the dataset into training, validation, and testing samples as given in Table 1.

The ground truth images span multiple video sessions. For training, we sampled hands from 33 video sessions. For validation, we sampled hands from another four video sessions. For testing, we used another set of 7 complete video sessions. Video sessions were collected over three years. Video sessions were forty-five to one hour and fifteen minutes long. The training dataset described in Table 1 was carefully selected to have diversity with 350 samples.

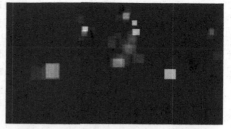

(a) Hand detections using Faster R-CNN.

(b) Hand detection projections for 12 seconds.

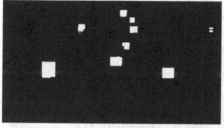

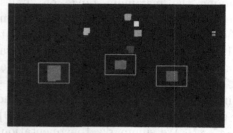

(c) Binary image showing clusters

(d) Green boxes showing valid clusters after removing small clusters.

(e) Hand detections using our method.

Fig. 3. Hand detection images that demonstrate the proposed approach. (Color figure online)

Table 1. Dataset for training, validation, and testing. The training, validation, and testing examples come from different video sessions.

	# Sessions	# Images	# Hand instances
Training	33	305	1803
Validation	4	100	714
Testing	7	313	2031
Total	44	718	4,548

Table 2. Optimal augmentation parameter value ranges.

Method	Optimal range
Shear	$[-3°, 3°]$
Rotate	$[-7°, 7°]$
Translate	$[-20, 20]$

We develop a separable optimization approach that starts with determining the maximum range of angles for shear, rotation, and pixels to be translated. To establish the maximum range of values to consider for shear and rotation, we calculate validation accuracy at multiple angles: $\theta \in \{1°, 2°, 4°, 8°, 16°, 32°\}$. The maximum range is determined based on the largest angle that results in a significant decrease in validation accuracy. Let $[-\theta_r^*, \theta_r^*]$, $[-\theta_s^*, \theta_s^*]$ denote the optimal ranges for rotation and shear, respectively. Similarly, we evaluate validation accuracy at multiple horizontal translations: $\tau \in \{1, 2, 4, 8, 16, 32, 64, 128, 256, 512, 800\}$, and compute the maximum interval: $[-\tau^*, \tau^*]$. We summarized augmentation methods, along with their respective optimal ranges in Table 2.

In addition to determining the best parameter values for each augmentation method, we also optimize the probability, p, for applying data augmentation. For example, for p $= 1$, data augmentation is always applied. We compute the optimal data augmentation probability p* as described in Fig. 4.

3 Results

We present the results in two sections. We first present improvement in hand detection by using optimal data augmentation method described in Sect. 2.2.

```
 1: for each p ∈ {0, 0.25, 0.5, 0.75, 1} do
 2:     for each image in training do
 3:         Apply random horizontal flips with p probability.
 4:         Apply random scaling of {0.8,1.2} with p probability.
 5:         Apply random shear angle sampled from {−θ_s^*, ..., θ_s^*} with p probability.
 6:         Apply random rotation angle sampled from {−θ_r^*, ..., θ_r^*}
 7:             with prbability p.
 8:         Apply random horizontal translation with pixels
 9:             uniformly sampled from {−τ^*, ..., τ^*} with p probability.
10:     end for
11:     Train the model with the augmented data.
12:     Record validation accuracies.
13: end for
14: Select optimal probability (p^*) that has the highest validation accuracy
```

Fig. 4. Pseudocode for finding the optimal probability for data augmentation.

Table 3. Hand detection validation and testing average precision. From the table, it is clear that p of 0.5 gave the best performance.

Data split	Model	Probability of applying each data augmentation				
		0	0.25	0.5	0.75	1.0
Val	Best	0.77	0.86	**0.86**	0.85	0.84
	Last	0.76	0.85	**0.86**	0.84	0.82
Test	Best	0.75	0.80	**0.80**	0.79	0.78
	Last	0.71	0.80	**0.81**	0.78	0.76

Table 4. Reduction in number of hand detections for each test session.

Session	# Hand detections		Median IoU		
	Faster RCNN	Ours	Faster RCNN	Ours	Reduction
C1L1P-C, Mar30	55,914	9,804	0.22	0.38	82.5%
C1L1P-C, Apr13	34,665	8,028	0.18	0.45	76.8%
C1L1P-E, Mar02	50,312	9,968	0.15	0.46	80.0%
C2L1P-B, Feb23	48,073	9,924	0.22	0.47	79.3%
C2L1P-D, Mar08	31,875	7,724	0.27	0.40	75.7%
C3L1P-C, Apr11	36,757	9,536	0.23	0.43	74.0%
C3L1P-D, Mar19	57,319	9,536	0.23	0.54	83.3%

We then present the final detention results that demonstrate that our method reduced the number of false positive regions by 78.8% without sacrificing any true positive detections.

We used an Intel Xeon 4208 CPU @ 2.10 GHz server, having 128 GB DDR4 RAM and an NVIDIA RTX 5000 GPU for all the experiments. For training Faster R-CNN, we used the recommended learning rate of 0.001 for 12 epochs with a mini-batch size of 2 images. We can train the model in less than 13 min.

3.1 Results for Optimal Data Augmentation

Table 2 provides the optimal maximum range angles for shear, rotation, and pixels to be translated for hand detection. We applied the optimal augmentation values at different probabilities as summarized in Table 3. From this table, it is clear that 0.5 probability provided the best performance.

3.2 Hand Detection Results

We summarize our results in Table 4. Compared to Faster R-CNN, our approach reduced the number of false positives by 80% while improving IoU ratios from 0.2 to 0.5. Overall, our hand detector achieved average precision (AP) of 72% at 0.5 intersection over union (IoU). The detection results were improved to 81%

by using our optimized approach for data augmentation. Our method runs at 4.7× the real-time.

We present results against Faster R-CNN in Fig. 5. Overall, we can see that our approach results in a significant reduction in the number of detected hand regions. In some instances, our approach produces two overlapping hand regions that are associated with the same student.

(a) Initial hand regions detected using Faster RCNN.

(b) Ours.

(c) Initial hand regions detected using Faster RCNN.

(d) Ours.

(e) Initial hand regions detected using Faster RCNN with significant hand movements.

(f) Ours.

Fig. 5. Comparison between Faster RCNN (left column) and our proposed approach (right column).

4   Conclusion

We presented a fast and robust method for detecting hands in collaborative learning environments. Our method performed significantly better than the standard use of Faster R-CNN. In future work, the detected proposal regions will be used for the accurate detection of writing and typing activities which can inform educational researchers identify moments of interest in collaborative learning environments.

References

1. Ball, G.H., Hall, D.J.: Isodata, a novel method of data analysis and pattern classification. Technical Report, Stanford Research Institute, Menlo Park, CA (1965)
2. Bodla, N., Singh, B., Chellappa, R., Davis, L.S.: Soft-NMS - improving object detection with one line of code. In: Proceedings of the IEEE International Conference on Computer Vision (ICCV), October 2017
3. Celedón-Pattichis, S., LópezLeiva, C.A., Pattichis, M.S., Llamocca, D.: An interdisciplinary collaboration between computer engineering and mathematics/bilingual education to develop a curriculum for underrepresented middle school students. Cult. Stud. Sci. Educ. **8**(4), 873–887 (2013). https://doi.org/10.1007/s11422-013-9516-5
4. Darsey, C.J.: Hand movement detection in collaborative learning environment videos (2018)
5. Eilar, C.W., Jatla, V., Pattichis, M.S., LópezLeiva, C., Celedón-Pattichis, S.: Distributed video analysis for the advancing out of school learning in mathematics and engineering project. In: 2016 50th Asilomar Conference on Signals, Systems and Computers, pp. 571–575. IEEE (2016)
6. Jacoby, A.R., Pattichis, M.S., Celedón-Pattichis, S., LópezLeiva, C.: Context-sensitive human activity classification in collaborative learning environments. In: 2018 IEEE Southwest Symposium on Image Analysis and Interpretation (SSIAI), pp. 1–4. IEEE (2018)
7. Jatla, V., Pattichis, M.S., Arge, C.N.: Image processing methods for coronal hole segmentation, matching, and map classification. IEEE Trans. Image Process. **29**, 1641–1653 (2019)
8. Jatla, V., Teeparthi, S., Pattichis, M.S., Celedón-Pattichis, S., Leiva, C.L.: Long-term human video activity quantification of student participation. In: 2021 55th Asilomar Conference on Signals, Systems, and Computers. IEEE (2021)
9. Liu, W., et al.: SSD: single shot MultiBox detector. In: Leibe, B., Matas, J., Sebe, N., Welling, M. (eds.) ECCV 2016. LNCS, vol. 9905, pp. 21–37. Springer, Cham (2016). https://doi.org/10.1007/978-3-319-46448-0_2
10. Pedregosa, F., et al.: Scikit-learn: machine learning in Python. J. Mach. Learn. Res. **12**, 2825–2830 (2011)
11. Redmon, J., Divvala, S., Girshick, R., Farhadi, A.: You only look once: unified, real-time object detection. In: Proceedings of the IEEE Conference on Computer Vision and Pattern Recognition, pp. 779–788 (2016)
12. Rehg, J.M., Kanade, T.: Visual tracking of high DOF articulated structures: an application to human hand tracking. In: Eklundh, J.-O. (ed.) ECCV 1994. LNCS, vol. 801, pp. 35–46. Springer, Heidelberg (1994). https://doi.org/10.1007/BFb0028333
13. Ren, S., He, K., Girshick, R., Sun, J.: Faster R-CNN: towards real-time object detection with region proposal networks. IEEE Trans. Pattern Anal. Mach. Intell. **39**(6), 1137–1149 (2016)

14. Shi, W., Pattichis, M.S., Celedón-Pattichis, S., LópezLeiva, C.: Dynamic group interactions in collaborative learning videos. In: 2018 52nd Asilomar Conference on Signals, Systems, and Computers, pp. 1528–1531, October 2018
15. Shi, W., Pattichis, M.S., Celedón-Pattichis, S., LópezLeiva, C.: Robust head detec- . tion in collaborative learning environments using AM-FM representations. In: 2018 IEEE Southwest Symposium on Image Analysis and Interpretation (SSIAI), pp. 1–4, April 2018. https://doi.org/10.1109/SSIAI.2018.8470355
16. Shi, W., Pattichis, M.S., Celedón-Pattichis, S., Leiva, C.L.: Person detection in collaborative group learning environments using multiple representations. In: 2021 55th Asilomar Conference on Signals, Systems, and Computers. IEEE (2021)
17. Shi, W., Pattichis, M.S., Celedón-Pattichis, S., Leiva, C.L.: Talking detection in collaborative learning environments. In: 19th International Conference CAIP. Springer, Cham(2021)
18. Tapia, L.S., Pattichis, M.S., Celedón-Pattichis, S., Leiva, C.L.: Bilingual speech recognition by estimating speaker geometry from video data. In: 19th International Conference CAIP. Springer, Cham (2021)
19. Tran, P., Pattichis, M.S., Celedón-Pattichis, S., Leiva, C.L.: Facial recognition in collaborative learning videos. In: 19th International Conference CAIP. Springer, Cham (2021)

Assessing the Role of Boundary-Level Objectives in Indoor Semantic Segmentation

Roberto Amoroso$^{(\boxtimes)}$ (iD), Lorenzo Baraldi (iD), and Rita Cucchiara (iD)

University of Modena and Reggio Emilia, Modena, Italy
{roberto.amoroso,lorenzo.baraldi,rita.cucchiara}@unimore.it

Abstract. Providing fine-grained and accurate segmentation maps of indoor scenes is a challenging task with relevant applications in the fields of augmented reality, image retrieval, and personalized robotics. While most of the recent literature on semantic segmentation has focused on outdoor scenarios, the generation of accurate indoor segmentation maps has been partially under-investigated. With the goal of increasing the accuracy of semantic segmentation in indoor scenarios, we focus on the analysis of boundary-level objectives, which foster the generation of fine-grained boundaries between different semantic classes and which have never been explored in the case of indoor segmentation. In particular, we test and devise variants of both the Boundary and Active Boundary losses, two recent proposals which deal with the prediction of semantic boundaries. Through experiments on the NYUDv2 dataset, we quantify the role of such losses in terms of accuracy and quality of boundary prediction and demonstrate the accuracy gain of the proposed variants.

Keywords: Indoor scene understanding · Segmentation · Boundary losses

1 Introduction

The automatic understanding of indoor scenes is a core Computer Vision task that aims at providing detailed information about the objects in a scene, such as their type and how they interact with each other [27]. Such a level of understanding can have a high impact in many applications, ranging from augmented reality to image retrieval and the navigation of mobile robots in indoor spaces.

One of the core subtask of indoor scene parsing is performing a semantic segmentation over the input image, which can be either a plain RGB image or include depth information depending on the sensor of choice. While most of the recent indoor understanding literature has focused on the usage of RGBD data [6,8,14], and while most of the semantic segmentation literature has adopted outdoor datasets [9,17,21,26], some applications require to focus

© Springer Nature Switzerland AG 2021
N. Tsapatsoulis et al. (Eds.): CAIP 2021, LNCS 13052, pp. 455–465, 2021.
https://doi.org/10.1007/978-3-030-89128-2_44

on RGB data in indoor scenarios. Examples include the understanding of photos taken from users for augmented reality applications, the processing of pictures taken from the web and social networks, and every application in which employing a depth camera is unfeasible.

In such contexts, providing accurate and fine-grained pixel-wise classification is of great importance. Recently, the research on semantic segmentation models has focused on the introduction of fully convolutional networks [4,15] which leverage convolutional layers and downsampling operations to achieve a large receptive field, while upsampling operations are employed to increase the output resolution. Although this protocol is necessary to encode contextual information and deal with objects at large scales, it inevitably leads to feature smoothing across object boundaries. Thus, the segmentation results might be blurred and lack fine object boundary details – a defect that can be particularly detrimental in the case of indoor applications.

With the aim of improving the quality of semantic segmentation in indoor scenarios, in this paper we investigate the incorporation of boundary-aware losses when training semantic segmentation architectures. Starting from two recently proposed classes of loss functions, namely the Boundary loss [11] and the Active Boundary loss [20], we analyze their role and propose variations that can increase the overall quality of the segmentation. Experimentally, we assess the role of boundary-aware losses and of the proposed strategies on the NYUDv2 dataset for indoor semantic segmentation, employing an RGB-only setting. We quantify and show, through quantitative and qualitative experiments, the role of both losses in the case of indoor scene segmentation and the appropriateness of the proposed variants.

2   Related Work

Localizing semantic boundaries or exploiting boundary information to improve the semantic segmentation has been the focus of several previous studies [1,23]. Gated-SCNN [19], for instance, designs a two-stream network to exploit the duality between the segmentation predictions and the boundary predictions, integrating shape information. Other works [3,5,10], instead, learn pairwise pixel-level affinity and monitor information flow across boundaries to preserve feature disparity for semantic boundaries and feature similarity for interior pixels.

While most of these methods [5,10,19] depend on the segmentation model and require re-training, extensive studies [13,25] have proposed post-processing techniques to improve boundary details of segmentation results. DenseCRF [13] considers fully connected CRF models defined at the pixel level to improve segmentation accuracy around boundaries. SegFix [25], instead, proposes a model-agnostic method to refine segmentation maps, by training a separate network to transfer the label of interior pixels to boundary pixels. PointRend [12] presents a rendering approach to refine boundary information by performing point-based predictions at selected locations based on an iterative subdivision algorithm.

Boundary loss (BL) [11] and Active Boundary loss (ABL) [20], finally, propose a model-agnostic end-to-end trainable approach to tackle the problem of

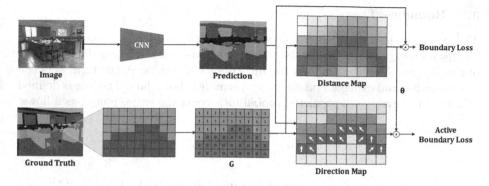

Fig. 1. We analyze two loss functions for improving boundary-level predictions in semantic segmentation: (a) a *Boundary loss* which weights pixels predictions according to their distance to semantic boundaries; (b) an *Active Boundary* loss which promotes the alignment between predicted and ground truth boundaries. Best seen in color. (Color figure online)

semantic segmentation at boundaries. BL promotes the refinement of the semantic boundaries by optimizing the sum of the linear combinations of the regional probability predictions and their distance transforms. ABL monitors the changes in the boundaries of the segmentation predictions and encourages the alignment between predicted boundaries and ground-truth boundaries, leveraging the distance transform of the prediction maps to regularize the network behavior.

Despite the empirical success of boundary-aware approaches in improving segmentation precision, there are still substantial segmentation errors at object boundaries. In this work, we investigate the reciprocal dependency between semantic segmentation and boundary-level objectives to increase the accuracy of semantic segmentation performance.

3   Method

Motivated by the need of providing a more precise segmentation along boundaries in the case of indoor scene segmentation, we investigate the usage of loss functions that explicitly model the prediction of semantic boundaries. In particular, we draw inspiration from the Boundary loss [11] and the Active Boundary loss [20], two loss functions that mitigate the problem of boundary segmentation error. We propose novel variations of both of them, with the aim of improving their results, and compare the results obtained (Fig. 1).

Hereafter, we consider a segmentation setting characterized by C classes and input image resolution $H \times W$. $\boldsymbol{P} \in \mathbb{R}^{C \times H \times W}$, instead, will be used to indicate the class probability map predicted by the network. Thorough the rest of the section, given a tensor with spatial support $\boldsymbol{Z}$, the notation $\boldsymbol{Z}_i$ will be employed to denote the value(s) stored at the i-th spatial location of $\boldsymbol{Z}$, thus employing a "flattened" indexing of the two spatial dimensions.

3.1 Boundary Loss

The Boundary loss, originally proposed by Kervadec *et al.* [11], conceptually calculates an integral over the points between regions which capture the proximity of two shapes. As such, it allows the incorporation of a weighting term between the estimated and expected pixels along a semantic boundary. The loss is defined as a weighted average of predicted probabilities over the entire image, as follows:

$$\mathbf{BL} = \frac{1}{N} \sum_{i}^{N} P_i D_i^{\mathsf{T}},\tag{1}$$

where N is the number of pixels of the input image and $\boldsymbol{D} \in \mathbb{R}^{C \times H \times W}$ is a distance map that does the actual probability weighting. Noticeably, negative values in $\boldsymbol{D}_i \in \mathbb{R}^C$ will increase the probability of predicting a given class in a pixel, while positive values will discourage the network from predicting a given class in a spatial location.

Given a one-hot ground-truth tensor $\boldsymbol{G} \in \{0,1\}^{C \times H \times W}$, the distance map can be calculated by means of the distance transform operator, which computes for each positive pixel its distance to the closest zero-valued pixel on the same channel, *i.e.* the closest pixel which does not belong to a given class. In the original formulation [11], the distance map was defined as follows:

$$\boldsymbol{D}_i = -\mathrm{Dist}(\boldsymbol{G}_i) \odot \boldsymbol{G}_i + \mathrm{Dist}(1 - \boldsymbol{G}_i) \odot (1 - \boldsymbol{G}_i)\tag{2}$$

where $\odot$ indicates the Hadamard element-wise product and $\mathrm{Dist}(\cdot)$ is the distance transform[1]. As it can be observed, pixels that belong to a class are given a negative weight, thus promoting the prediction of high probability values for that class – while pixels that do not belong to a class are given a positive weight, thus discouraging the network from predicting the same class. When considering the magnitude of the weights, instead, it can be seen that pixels far from the boundaries, for which the $\mathrm{Dist}(\cdot)$ function produces high values, play a larger role in determining the loss in this formulation – while pixels close to the boundary are given less importance. In other words, the network is encouraged to give correct predictions in regions that do not lie close to the boundaries between classes and is allowed to be less precise in boundary regions.

With the aim of increasing the quality of predictions at the boundary level, we propose and investigate two variations of the Boundary loss, which correspond to the following distance maps:

$$\boldsymbol{D}_i^+ = -\boldsymbol{G}_i + \mathrm{Dist}(1 - \boldsymbol{G}_i) \odot (1 - \boldsymbol{G}_i),\tag{3}$$

$$\boldsymbol{D}_i^- = -\mathrm{Dist}(\boldsymbol{G}_i) \odot \boldsymbol{G}_i + (1 - \boldsymbol{G}_i).\tag{4}$$

As it can be observed, in the two proposed variants the distance map values are replaced with constant values which are independent of the distance from the

---

[1] In our implementation, we employ the `ndimage.morphology.distance_transform_edt` function from the `scipy` Python package.

boundary. This is done in the case of pixels that do not belong to the target class (*i.e.*, negative pixels) for D_i^-, and in the case of pixels that belong to the target class (*i.e.*, positive pixels) for D_i^+, respectively. In this manner, greater importance is given to boundary pixels, compared to the original formulation.

3.2   Active Boundary Loss

The Active Boundary loss is instead formulated as a differentiable direction vector prediction problem, which gradually promotes the alignment between predicted boundaries (which in the following will be named, for brevity, PBs) and ground truth boundaries (for brevity again, GTBs). The pipeline for computing the loss can be conceptually divided into two phases.

Phase 1. During this phase, we compute the PBs starting from the probability map predicted by the network and devise a target direction map D^g which will be employed to align PBs with GTBs.

Specifically, boundary pixels of the predicted boundary map are recovered through the computation of the Kullback–Leibler (KL) divergence between the probabilities predicted for adjacent pixels. The i-th pixel of the PB is defined as

$$PB_i = \begin{cases} 1 \text{ if } \exists \mathbb{KL}(P_i \| P_j) > \epsilon, j \in \mathcal{N}_2(i); \\ 0 \qquad \text{otherwise}, \end{cases} \tag{5}$$

where $\mathcal{N}_2(\cdot)$ indicates the 2-neighborhood of a pixel, corresponding to the offset $\{\{1,0\},\{0,1\}\}$ (*i.e.*, the pixels to the right and below the current pixel). The threshold value ϵ is calculated dynamically to ensure that the number of boundary pixels in PB is less than $1/100$ of the area of the input image.

The pixels of GTBs are, accordingly, determined by applying Eq. 5 to the one-hot ground-truth tensor and replacing the KL divergence with a simpler equality condition on the class labels between the pixels in $\mathcal{N}_2(\cdot)$.

As a second point, we compute a target direction map containing offset vectors which will encourage pixels on the PBs to move towards pixels of the GTBs. In the original version of the Active Boundary loss, the offset was encoded as a one-hot vector. In our version, we encode the coordinate of the offset vector as a progressive index indicating its position within the 8-neighborhood of a pixel, ranging from 0 (*i.e.* offset $\{-1,-1\}$ or top-left corner) to 8 (*i.e.* offset $\{1,1\}$ or bottom-right corner) following the row-major order, and excluding index 4 which is associated with the central pixel itself.

Formally, the target direction map $D^g \in \mathbb{R}^{H \times W}$ is computed by considering the offset direction which would move a pixel closer to a GTB, *i.e.*:

$$D_i^g = \arg\min_j M_{i+\Delta_j}, \ j \in \{0, 1, ..., 7\}, \tag{6}$$

where $M = \text{Dist}(\text{GTBs})$ is the result of the distance transform applied to GTBs and Δ_j represents the j-th element in the set of directions $\Delta = \{\{-1,-1\}, \{0,-1\}, \{1,-1\}, \{-1,0\}, \{1,0\}, \{-1,1\}, \{0,1\}, \{1,1\}\}$.

Phase 2. By using the KL divergence between the predictions for a pixel i and those for one of its neighbor pixels j as logits in a cross-entropy loss, the predicted boundary at pixel i is pushed towards the pixel j in a probabilistic way. The purpose is to increase the KL divergence between the class probability distribution of i and j while reducing the KL divergence between i and its 8-neighborhood pixels. To this aim, a predicted direction map $\boldsymbol{D}^p \in \mathbb{R}^{8 \times H \times W}$ is computed as follows:

$$\boldsymbol{D}_i^p = \left\{ \frac{e^{\mathrm{KL}(P_i, P_{i+\Delta_k})}}{\sum_{h=0}^7 e^{\mathrm{KL}(P_i, P_{i+\Delta_h})}}, k \in \{0, 1, ..., 7\} \right\}, \tag{7}$$

Employing the predicted and the target direction map, the Active Boundary loss can be defined as a weighted cross-entropy (**CE**) loss, as follows:

$$\mathbf{ABL} = \left(\sum_i \Lambda(M_i) \cdot \mathbf{CE}(D_i^p, D_i^g) \cdot \boldsymbol{PB}_i \right) \cdot \frac{1}{\sum_i \boldsymbol{PB}_i} \tag{8}$$

Through the weight function $\Lambda(x) = \frac{\min(x, \theta)}{\theta}$, the distance of the pixel i from the nearest boundary of GTBs is used as weight to penalize its divergence from the GTBs. The hyper-parameter θ is empirically set to 20.

Managing Collisions. Noticeably, collisions between offset vectors of neighboring pixels are possible, especially in the case of complex boundary shapes. To address this problem, the original formulation of the Active Boundary loss [20] suggests detaching the gradient flow for all non-boundary pixels. As a result, the gradient is calculated only for the pixels on the predicted boundaries, ignoring all the other pixels.

To overcome any conflicts, we adopted an equivalent strategy. In our implementation, we multiply the result of the weighted cross-entropy loss by the predicted boundary map $\boldsymbol{PB}$, so that the only pixels that contribute to the loss calculation are the boundary pixels. The final value is the average calculated by dividing the sum of the weighted and masked values of the cross-entropy by the number of predicted boundary pixels.

Finally, the Active Boundary loss is regularized through label smoothing [18], to prevent the network from taking over-confident decisions. During label smoothing, the highest probability of the one-hot target distribution is set at 0.8, while the rest of the distribution is set to 0.2/7. Both values have been empirically determined during our preliminary experiments.

Alternative Distances. In the original formulation of the Active Boundary loss, the KL divergence is used to measure the distance between two probability distributions when identifying the predicted boundaries and when computing the predicted direction map.

As the KL divergence has the disadvantage of being asymmetrical, we propose and explore two variants of Active Boundary loss in which we replace the KL divergence with the Jensen-Shannon (JS) divergence and Bhattacharyya (BC) distance, both of which are also symmetrical. In particular, the Jensen-Shannon divergence provides a smoothed and normalized version of KL divergence.

4 Experiments

4.1 Dataset

We conduct our analyses on the image segmentation dataset NYU-Depth V2 [16], which provides densely annotated images of indoor environments. Specifically, the NYU-Depth V2 dataset consists of 1449 RGB-D frames showing interior scenes, acquired through the Microsoft Kinect sensor and with a size of 640×480. Since the distortion of the images has been corrected, they showcase a thin white border which we remove by cropping the original images to a size of 608×448 pixels. We use the segmentation labels provided in [6], in which all labels were mapped to 40 classes. We employ the standard training/test split with 795 and 654 images, respectively, and train our models on RGB images only.

In NYU-Depth V2, ground-truth labels are given as semantic regions, rather than pixel-level segmentation. This occasionally results in thin strips of unlabeled pixels between two adjacent regions and creates an issue when evaluating segmentation results at boundary level. To remedy the issue, we pre-processed the ground truth to remove small unlabeled regions through the median filtering strategy proposed in [22]. Overall, the NYUDv2 is a challenging dataset due to difficult lighting conditions and cluttered scenes.

4.2 Implementation Details and Evaluation Protocol

We train our semantic segmentation models using two loss functions $\mathcal{L}_{bl}$ and $\mathcal{L}_{abl}$, both consisting of the traditional cross-entropy and IoU losses, which are paired with the considered boundary-level losses:

$$\mathcal{L}_{bl} = \mathbf{CE} + \mathbf{IoU} + w_a \mathbf{BL},$$
$$\mathcal{L}_{abl} = \mathbf{CE} + \mathbf{IoU} + w_b \mathbf{ABL}. \tag{9}$$

Here, $\mathbf{CE}$ is the cross-entropy loss and $\mathbf{IoU}$ refers to the lovász-softmax loss [2], a surrogate IoU loss. While the $\mathbf{CE}$ loss focuses on per-pixel classification, the lovász-softmax loss prevents small objects from being ignored. The weights w_a and w_b regulate the contribution of $\mathbf{BL}$ and $\mathbf{ABL}$ to the final loss, respectively. In particular, our experimental results are obtained by setting w_a to 1 both for the original version of BL and its proposed variants, while w_b is set to 1.5 in the case of the original version of ABL and to 2 for the two proposed variants implementing the JS divergence and BC distance.

In all experiments, we employ a DeepLabV3 [4] with ResNet-50 [7] as our default backbone architecture. Following the training protocol of [24], we use random scaling, crop, left-right flipping, and brightness jittering during data augmentation. We use a plain SGD optimizer, with an initial learning rate of

Table 1. Quantitative results on the NYUDv2 dataset, when training with the Boundary loss and the proposed variations.

Loss function	Pixel accuracy	Mean accuracy	Mean IoU
CE	65.02	53.01	38.84
CE + IoU	65.33	54.43	39.60
CE + IoU + BL	65.20	**55.14**	39.39
CE + IoU + BL$^-$	65.36	54.47	39.79
CE + IoU + BL$^+$	**65.53**	55.09	**39.90**

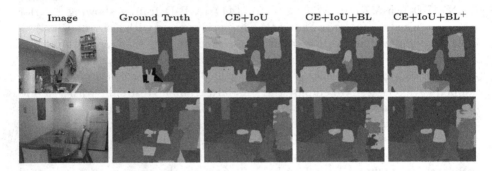

Fig. 2. Qualitative comparison between boundary loss functions

0.005 and weight decay equal to 0.0005. Training is performed with a mini-batch size of 4 and conducted for 200 training epochs. The learning rate is divided by 10 after 60, 80, 100, and 150 epochs.

4.3 Quantitative Evaluation

Table 1 reports the results obtained on the NYUDv2 dataset when training with the Boundary loss, and with the proposed variations, in terms of mean intersection-over-union, pixel accuracy, and mean accuracy [15]. As it can be seen, the combination of cross-entropy loss and IoU loss leads to improved results in terms of all metrics, proving that this combination is useful in the domain of indoor segmentation.

When turning to the evaluation of the losses based on BL, we first notice that the combination of cross-entropy, IoU, and Boundary loss leads to an improvement in terms of mean accuracy and to a decrease in pixel accuracy and mean IoU, highlighting that the original loss struggles to improve the results. The usage of the two proposed variations of the distance maps (D^+ and D^- – indicated in Table 1, respectively, as BL$^+$ and BL$^-$), helps to recover this quantitative loss, leading to improved results in terms of accuracy and mean IoU. Figure 2 reports some qualitative samples, comparing the predictions obtained with CE + IoU and those with CE + IoU + BL and CE + IoU + BL$^+$.

Table 2. Quantitative results on the NYUDv2 dataset, when training with the Active Boundary loss and the proposed variations.

Loss function	Pixel accuracy	Mean accuracy	Mean IoU
CE	65.02	53.01	38.84
CE + IoU	65.33	54.43	39.60
CE + IoU + ABL_{KL}	**65.54**	54.45	**39.87**
CE + IoU + ABL_{JS}	65.31	**54.66**	39.72
CE + IoU + ABL_{BC}	65.40	54.39	39.67

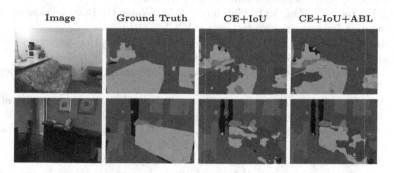

Fig. 3. Qualitative comparison between active boundary loss functions

In Table 2, instead, we turn to the evaluation of the Active Boundary loss, and its proposed variations. Firstly, we notice that in this case the ABL, in its original formulation, does not show a loss in performance when compared with the CE + IoU baseline. Indeed, a CE + IoU + ABL setting, with the original KL-divergence as a distance measure, leads to an improvement in terms of pixel accuracy, mean accuracy, and mean IoU. Further, applying the JS divergence in place of the KD divergence increases the performance in terms of mean accuracy. Overall, this highlights that the Active Boundary loss can be a reasonable choice in indoor settings to improve the quality of predictions at boundary level. Finally, in Fig. 3 we show qualitative samples comparing the results obtained when employing the CE + IoU baselines, in comparison with the ABL loss.

5   Conclusion

We considered the usage of boundary loss functions when training segmentation models in indoor scenarios. To this end, we have considered two recently proposed boundary-level objectives, *i.e.* the Boundary loss, and the Active Boundary loss, quantified their role, and proposed variants for both of them. Through quantitative and qualitative experiments on the NYUDv2 dataset, we have shown that ABL and our proposed variation of the Boundary loss can improve segmentation results at the boundary level.

References

1. Acuna, D., Kar, A., Fidler, S.: Devil is in the edges: learning semantic boundaries from noisy annotations. In: CVPR (2019)
2. Berman, M., Triki, A.R., Blaschko, M.B.: The lovász-softmax loss: a tractable surrogate for the optimization of the intersection-over-union measure in neural networks. In: CVPR (2018)
3. Bertasius, G., Torresani, L., Yu, S.X., Shi, J.: Convolutional random walk networks for semantic image segmentation. In: CVPR (2017)
4. Chen, L.C., Papandreou, G., Schroff, F., Adam, H.: Rethinking atrous convolution for semantic image segmentation. In: CVPR (2017)
5. Ding, H., Jiang, X., Liu, A.Q., Thalmann, N.M., Wang, G.: Boundary-aware feature propagation for scene segmentation. In: ICCV (2019)
6. Gupta, S., Arbelaez, P., Malik, J.: Perceptual organization and recognition of indoor scenes from RGB-D images. In: CVPR (2013)
7. He, K., Zhang, X., Ren, S., Sun, J.: Deep residual learning for image recognition. In: CVPR (2016)
8. Hu, W., Zhao, H., Jiang, L., Jia, J., Wong, T.T.: Bidirectional projection network for cross dimension scene understanding. In: CVPR (2021)
9. Huang, Z., Wang, X., Huang, L., Huang, C., Wei, Y., Liu, W.: Ccnet: criss-cross attention for semantic segmentation. In: ICCV (2019)
10. Ke, T.W., Hwang, J.J., Liu, Z., Yu, S.X.: Adaptive affinity fields for semantic segmentation. In: Ferrari, V., Hebert, M., Sminchisescu, C., Weiss, Y. (eds.) ECCV 2018. LNCS, vol. 11205, pp. 605–621. Springer, Cham (2018). https://doi.org/10.1007/978-3-030-01246-5_36
11. Kervadec, H., Bouchtiba, J., Desrosiers, C., Granger, E., Dolz, J., Ayed, I.B.: Boundary loss for highly unbalanced segmentation. In: MIDL (2019)
12. Kirillov, A., Wu, Y., He, K., Girshick, R.: Pointrend: image segmentation as rendering. In: CVPR (2020)
13. Krähenbühl, P., Koltun, V.: Efficient inference in fully connected crfs with gaussian edge potentials. NeurIPS 24, 109–117 (2011)
14. Kundu, A.: Virtual multi-view fusion for 3D semantic segmentation. In: V, A., B, H., Brox, T., Frahm, J.M. (eds.) ECCV 2020. LNCS, vol. 12369, pp. 518–535. Springer, Cham (2020). https://doi.org/10.1007/978-3-030-58586-0_31
15. Long, J., Shelhamer, E., Darrell, T.: Fully convolutional networks for semantic segmentation. In: CVPR (2015)
16. Silberman, N., Hoiem, D., Kohli, P., Fergus, R.: Indoor segmentation and support inference from RGBD images. In: Fitzgibbon, A., Lazebnik, S., Perona, P., Sato, Y., Schmid, C. (eds.) ECCV 2012. LNCS, vol. 7576, pp. 746–760. Springer, Heidelberg (2012). https://doi.org/10.1007/978-3-642-33715-4_54
17. Pang, Y., Li, Y., Shen, J., Shao, L.: Towards bridging semantic gap to improve semantic segmentation. In: CVPR (2019)
18. Szegedy, C., Vanhoucke, V., Ioffe, S., Shlens, J., Wojna, Z.: Rethinking the inception architecture for computer vision. In: CVPR (2016)
19. Takikawa, T., Acuna, D., Jampani, V., Fidler, S.: Gated-scnn: Gated shape cnns for semantic segmentation. In: ICCV (2019)
20. Wang, C., et al.: Active boundary loss for semantic segmentation. arXiv preprint arXiv:2102.02696 (2021)
21. Wang, L., Li, D., Zhu, Y., Tian, L., Shan, Y.: Dual super-resolution learning for semantic segmentation. In: CVPR (2020)

22. Xiaofeng, R., Bo, L.: Discriminatively trained sparse code gradients for contour detection. In: NeurIPS (2012)
23. Yu, Z., Feng, C., Liu, M.Y., Ramalingam, S.: Casenet: deep category-aware semantic edge detection. In: CVPR (2017)
24. Yuan, Y., Chen, X., Wang, J.: Object-contextual representations for semantic segmentation. In: Vedaldi, A., Bischof, H., Brox, T., Frahm, J.M. (eds.) ECCV 2020. LNCS, vol. 12351, pp. 173–190. Springer, Cham (2020). https://doi.org/10.1007/978-3-030-58539-6_11
25. Yuan, Y., Xie, J., Chen, X., Wang, J.: Segfix: model-agnostic boundary refinement for segmentation. In: ECCV (2020)
26. Zhao, H., Shi, J., Qi, X., Wang, X., Jia, J.: Pyramid scene parsing network. In: CVPR (2017)
27. Zhuo, W., Salzmann, M., He, X., Liu, M.: Indoor scene parsing with instance segmentation, semantic labeling and support relationship inference. In: CVPR (2017)

Skin Lesion Classification Using Convolutional Neural Networks Based on Multi-Features Extraction

Samia Benyahia[1], Boudjelal Meftah[2]([✉]) [iD], and Olivier Lézoray[3] [iD]

[1] Department of Computer Science, Faculty of Exact Sciences,
University of Mascara, Mascara, Algeria
[2] LRSBG Laboratory, University of Mascara, Mascara, Algeria
boudjelal.meftah@univ-mascara.dz
[3] Normandie Univ, UNICAEN, ENSICAEN, CNRS, GREYC, Caen, France

Abstract. In the recent era, deep learning has become a crucial technique for the detection of various forms of skin lesions. Indeed, Convolutional neural networks (CNN) have became the state-of-the-art choice for feature extraction. In this paper, we investigate the efficiency of three state-of-the-art pre-trained convolutional neural networks (CNN) architectures as feature extractors along with four machine learning classifiers to perform the classification of skin lesions on the PH2 dataset. In this research, we find out that a DenseNet201 combined with Cubic SVM achieved the best results in accuracy: 99% and 95% for 2 and 3 classes, respectively. The results also show that the suggested method is competitive with other approaches on the PH2 dataset.

Keywords: Feature extraction · Classification · Skin lesion · Convolutional neural networks

1 Introduction

Cancer is considered one of the leading causes of death in the world. It is estimated that the number of people diagnosed with cancer will double in the next few decades. Fortunately, some cancers have a high chance of cure through early detection, appropriate treatment, and care in the early stages. Skin lesions are an abnormal change in the tissue either on the surface of the skin or under the skin. Skin lesion usually grows in an irregular way beyond their usual boundaries as compared to the surrounding tissue. Is is primarily caused by excessive exposure to ultraviolet radiation. Skin lesions can be classified into two categories: benign skin tumors such as nevus, or malignant tumors such as melanoma that is the least common but the most harmful form of skin cancer. Until the last few years, computer-aided diagnosis was used for the early detection of skin cancer from dermoscopy images. It was based on handcrafted features extraction, such as statistical pixel-level features, shape features, texture features, and relational features from the images in order to train classical machine learning models for

N. Tsapatsoulis et al. (Eds.): CAIP 2021, LNCS 13052, pp. 466–475, 2021.
https://doi.org/10.1007/978-3-030-89128-2_45

distinguishing the lesion from the surrounding healthy skin. Currently, such a computer-aided diagnosis is still a challenging task. Most of researchers are now investigating deep learning techniques. It is expected that they will in the next future be able to reach a performance similar to those of dermatologists, directly from dermoscopic images. Indeed, in recent years there have been many successful applications of machine learning techniques (ANN, SVM, KNN) as well as deep learning approaches and in particular of Convolutional Neural Networks (CNNs) such as AlexNet, VGGNet, ResNet, DenseNet, GoogleNet, Inception, EffcientNets. Several contributions have been proposed recently in terms of new models that have made significant improvements in the detection and the classifications of skin lesions. Hopefully, such new techniques will help to improve patient survival rates.

In this paper, we investigate the efficiency of various commonly pre-trained convolutional neural networks architectures as feature extractors with various machine learning classifiers to perform the classification for dermoscopic images from the PH2 dataset.

The rest of the paper is structured as follows. Section 2 presents some studies on the application of CNN for melanoma diagnosis on the PH2 dataset. Section 3 describes the dataset used, the various pretrained CNN used as feature extractor, the machine learning classifiers, and the evaluation metrics. Section 4 describes expriments and discusses of the obtained results. Last Section provides the conclusion of this study.

2   Related Works

Several automated recognition methods for skin lesion images have been proposed in the last decade. In particular, recent classification approaches of skin lesions have been dominated by CNN approaches. In this section, we review some studies on the application of CNN for melanoma diagnosis on the PH2 dataset.

Ozkan et al. [1] proposed a study on skin lesions classification based on dermoscopic images to classify images of the PH2 datasets into three classes: normal, abnormal, and melanoma. They used four different machine learning classifiers: ANN, SVM, KNN, and Decision Tree. The achieved accuracies were 92.50%, 89.50%, 82.00%, and 90.00% for ANN, SVM, KNN, and DT, respectively.

Ghasem et al. [2] proposed two-hybrid approaches to combine four heterogeneous classifiers KNN, SVM, ENN, and MLP. Their first approach was the Structure-Based on Stacking (SBS), while the other was the Hierarchical Structure Based on Stacking (HSBS). The authors have considered a preprocessing step, a segmentation step for analyzing the lesion area and have applied different feature extraction methods based on the shape, color, and texture. Finally, a classification step is used either SBS or HSBS by combining the different classifiers. The evaluation was based on the PH2 dermoscopic images with different selected features. The achieved accuracy was 96.7% for the HSBS method and 98.5% for SBS.

Ann et al. [3] used a deep learning approach to classify skin lesions in dermoscopic images. First, the authors preprocess the images to remove unwanted artifacts, such as hairs, using morphological operators and an inpainting algorithm. They classify the images using a CNN AlexNet architecture, and finally, tested the classifier using both preprocessed and unprocessed images from the PH2 dataset. The obtained accuracy for the two classes was 93%, while the accuracy for the three classes was 67.5% with preprocessed images.

Singh et al. [4] proposed an approach for the skin lesions classification using a segmentation step by a thresholding method and the ABCD rule for feature extraction. Finally, a SVM classifier is used to make a decision. The proposed system achieved an accuracy of 92.5% on the PH2 data set.

Filali et al. [5] proposed a skin lesion classification system based on a fusion of handcrafted features (shape, skeleton, color, and texture) and features extracted from deep learning architectures (VggNet16, ResNet18, AlexNet, and GoogLeNet). They used a Support Vector Machine (SVM) classifier to make a final decision. Their approach achieved an accuracy of 98% on the PH2 dataset.

Sanket et al. [6] proposed a skin lesion diagnosis system. The authors begin with a processing step using median and Wiener filters to remove noise, followed by a segmentation step using a watershed and morphological filters. After that, a feature extracting step is performed using the Grey Level Co-occurrence Matrix (GLCM), color and geometrical features. Finally, the classification is performed using KNN, SVM, and an ensemble method. They achieved an accuracy of 92% on the PH2 dataset.

Khalid et al. [7] proposed a skin lesions classification framework based on fine-tuning a pre-trained deep learning network AlexNet. They replaced the last layer with a softmax to classify the lesion on three different classes from the PH2 dataset and achieved an accuracy of 98.61%.

3   Methodology and Materials

The process we propose is made up of four successive parts. The first component is the input of dermoscopic images. The second is the feature extraction part with three commonly pre-trained convolutional neural networks (CNN) architectures. From the obtained features, two databases are created: learning and test databases. The third component of the system is the learning step by a set of classifiers where each one is performed individually. Finally, the last is the validation step from the test dataset.

In addition, as shown in Fig. 1, there are two kinds of extracted features with the skin lesions dataset:

1. Extracted features using the original dataset.
2. Extracted features with an augmented dataset using data augmentation techniques (will be detailed later).

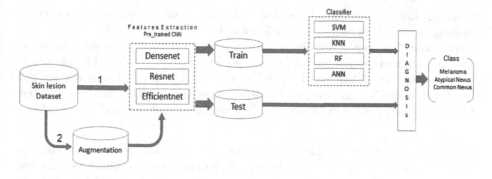

Fig. 1. Proposed method's flowchart.

3.1 PH2 Dataset

The Portuguese dermatological service of the Pedro Hispano Hospital and the University of Porto have collaborated to establish the PH2 database [15]. There are 200 dermoscopic images in the PH2 database, with 80 common nevus, 80 atypical nevus, and 40 melanomas. These are 8-bit RGB color images with a 768 × 560 pixels. The images can be obtained online from Hospital Pedro Hispano (https://www.fc.up.pt/addi/ph2%20database.html) and some are shown in Fig. 2.

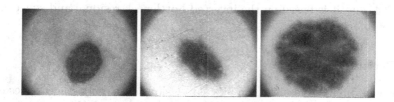

Fig. 2. Example of PH2 skin images.

3.2 Convolutional Neural Networks

The convolutional neural networks (CNN) are a special type of neural networks, that possesses several convolutional layers to extract learned features directly from images in a hierarchical manner, and they can be used for classification tasks. Recently, CNN have been used to improve the performance in many different applications with various architectures such as LeNet, AlexNet, VGGNet, ResNet, DenseNet, GoogLeNet among others. Such CNN architectures can be used as pre-trained architectures that were initially trained on large datasets. In this section, we present a brief overview of such pre-trained CNNs that can be used as extractors of features for skin lesion images.

ResNet. ResNet [12] introduces residual blocks that include skip-connection between layers where each layer feeds into the next layer and directly into the layers about 2–3 hops away. ResNet architecture contains a 3 × 3 convolution filter, global average pooling, and max-pooling layers, residual blocks, batch normalization layers followed by a fully connected layer, and softmax for the classification. ResNet architecture has several variations with a different number of layers such as ResNet18, ResNet34, ResNet50, ResNet101, and ResNet152.

DenseNet. DenseNet architecture [13] uses shortcut connections composed of a dense block linked by transition layers. Each dense block contains convolution layers where each layer is connected to all preceding other layers. All the feature maps from previous layers are passed to all subsequent layers of the same block. DenseNet architecture contains a convolution filter, global average pooling, max-pooling layers, transition layers, dense blocks followed by fully connected layers, and softmax for the classification. DenseNet architecture proposes several variations with a different number of layers such as DensNet-121, DensNet-169, DensNet-201, and DensNet-246.

EfficientNet. EfficientNet architecture [14] is based on the principle of scaling the different dimensions such as depth, width, and image resolution of the network at the same time uniformly by using a fixed compound coefficient. EfficientNet is considered as a family of eight different CNN models: EfficientNet-B0 to B7. The EfficientNet-B0 represents the baseline version with an input size of 224 × 224, and it is based on the inverted bottleneck residual and squeeze-and-excitation blocks.

3.3   Classifiers

Accurate automated classification of a skin lesion in its early stages saves effort, time, and human life. The purpose of our study will be to find the most accurate classifier from several machines learning classifiers for skin lesion classification. This section provides a summary of four machine learning classifiers that will be employed for the classification of skin lesions: Artificial neural network, Support Vector Machines, K-Nearest Neighbor, and Random Forest.

Artificial Neural Network. ANN are described as compound systems made up of at least two layers of neurons, an input layer, and an output layer, usually including hidden layers [8]. Each layer contains a large number of artificial neurons that constitute an interconnected network in a weighted way. Each neuron in the network receives digital information as signals from neighboring neurons, and each of these values is assigned a particular "weight" representative of the strength of the connection.

Support Vector Machines. SVMs algorithm plots the feature's value at a point in a high-dimensional space [9]. Then, it performs classification by finding the hyperplane, that determines a an hyperplane that can classify the data and differentiates the space into two zones. The set of points near the hyperplane is referred to as the Support Vectors.

K-Nearest Neighbor. KNN is a nonparametric method associated with only one parameter that represents the number K of nearest neighbors and a training data set [10]. The principle of the KNN model consists of choosing the K data points closest to the point under study in order to predict its class.

Random Forest. RF [11] consists in constructing and training multitude of decision trees in parallel on slightly different data subsets with random variables.

3.4 Evaluation Metrics

In order to evaluate the performance of our classifiers and to compare the different results of different scenarios, we used four evaluation metrics. These measures are accuracy, sensitivity, specificity, and precision.

$$Accuracy = \frac{T_P + T_N}{T_P + F_P + T_N + F_N} \tag{1}$$

$$Sensitivity = \frac{T_P}{T_P + F_N} \tag{2}$$

$$Specificity = \frac{T_N}{T_N + F_P} \tag{3}$$

$$Precision = \frac{T_P}{T_P + F_P} \tag{4}$$

Where T_P, T_N, F_P, F_N, and refer to true positive, true negative, false positive and false negative respectively.

4   Experimental Results and Discussion

In our study for a melanoma diagnosis, two scenarios of experiments are proposed using the PH2 dataset:

- Classification of skin lesions into two types of lesions: melanoma or non-melanoma
- Classification of skin lesions into three types: melanoma, atypical nevus, or common nevus

Table 1. Summary of results for melanoma or non-melanoma

	DenseNet201				ResNet50				EfficientB0			
	Acc. %	Sens. %	Spec. %	Prec. %	Acc. %	Sens. %	Spec. %	Prec. %	Acc. %	Sens. %	Spec. %	Prec. %
SVM	99	99	99	99	97.5	83.33	83.33	98.68	92.5	65.31	65.31	72.36
KNN	97.5	98.64	98.64	87.5	95	81.98	91.98	91.98	97.5	98.64	98.64	87.5
ANN	95	96.96	96.96	88.88	95	97.36	97.36	75.00	87.50	78.12	78.12	81.16
RF	99	99	99	99	97.5	83.33	83.33	98.68	95	97.29	97.29	80

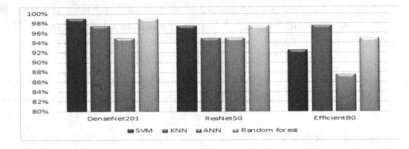

Fig. 3. Accuracy according to DenseNet201, ResnNet50, and EfficientB0 with four classifiers for melanoma or non-melanoma .

For both scenarios DenseNet201, ResNet50, EfficientB0 pre-trained CNN models are used as feature extractors, and ANN, SVM, KNN, RF classifiers used for classification. The latter are compared one to the other on the PH2 dataset. As previously mentioned, the PH2 dataset contains a total of 200 images that we divided into two parts.

- 80% of the dataset (160 images) is used for training
- 20% of the dataset (40 images) is used for testing the effectiveness of the models.

For the second scenario concerning the classification into three classes, a data augmentation techniques are included with the following augmentations: rotations, zooming, shearing, flips (top-bottom, left right), skew-left-right, contrast enhancement. All experiments were performed with the same training dataset and tested with the same test set.

Table 1 and Fig. 3 depict the performance results obtained for classifying PH2 dataset in two classes: melanoma or non-melanoma with DenseNet201, ResNet50, EfficientB0 CNN extractor and ANN, SVM, KNN, RF classifiers.

Table 2 and Fig. 4 depict the performance results obtained for classifying PH2 dataset in 3 classes: melanoma, atypical nevus or common nevus without augmentation with DenseNet201, ResNet50, EfficientB0 CNN extractor and ANN, SVM, KNN, RF classifiers.

Table 3 and Fig. 5 depicts the performance results obtained for classifying PH2 dataset in 3 classes: melanoma, atypical nevus or common nevus with augmentation with DenseNet201, ResNet50, EfficientB0 CNN extractor and ANN, SVM, KNN, RF classifiers.

Table 2. Summary of results for melanoma, atypical nevus or common nevus without augmentation

	DenseNet201				ResNet50				EfficientB0			
	Acc. %	Sens. %	Spec. %	Prec. %	Acc. %	Sens. %	Spec. %	Prec. %	Acc. %	Sens. %	Spec. %	Prec. %
SVM	62.50	62.77	78.38	73.19	60.00	60.20	77.16	71.44	62.50	40.09	76.82	60.45
KNN	70.00	70.37	81.48	75.91	60.00	57.06	75.59	61.11	62.50	41.14	76.20	62.19
ANN	70	**72.80**	**83.33**	**72.87**	60.00	28.57	74.04	34.43	50.00	30.51	72.17	57.44
RF	57.50	58.69	75.30	69.59	70.00	44.75	82.09	76.50	70.00	47.78	82.38	67.35

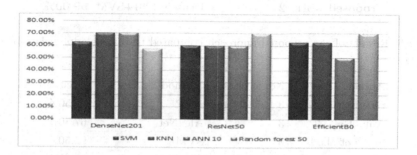

Fig. 4. Accuracy according to DenseNet201, ResnNet50, and EfficientB0 with four classifiers for melanoma, atypical nevus or common nevus without augmentation.

Table 3. Summary of results for melanoma, atypical nevus or common nevus with augmentation.

	DenseNet201				ResNet50				EfficientB0			
	Acc. %	Sens. %	Spec. %	Prec. %	Acc. %	Sens. %	Spec. %	Prec. %	Acc. %	Sens. %	Spec. %	Prec. %
SVM	**95**	**96.67**	**97.22**	**96.29**	87.50	91.25	92.77	91.02	90.00	92.91	94.16	92.59
KNN	87.50	91.25	93.51	86.31	82.50	80	90.37	77.14	82.50	87.50	89.72	87.30
ANN	92.50	87.77	93.88	92.50	95.00	96.29	96.00	96.66	85.00	88.88	90.90	89.16
RF	85.00	88.75	91.57	83.57	82.50	87.50	90.92	79.39	82.50	87.50	90.04	82.47

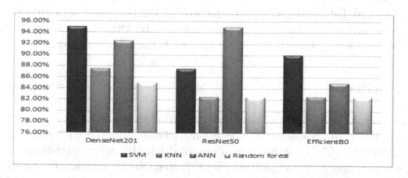

Fig. 5. Accuracy according to DenseNet201, ResnNet50, and EfficientB0 with four classifiers for melanoma, atypical nevus or common nevus with augmentation.

Table 4. Results of the proposed approach compared to various approaches for PH2 dataset

Authors	No. of classes	Method	Accuracy
Ghasem et al. [2]	2	SBS	98.50%
Ghasem et al. [2]	2	HSBS	96.70%
Ann et al. [3]	2	AlexNet	93.00%
Filali et al. [5]	2	SVM	98.00%
Sanket et al. [6]	2	SVM	92.00%
Proposed work	2	DenseNet201+SVM	**99.00%**

Table 5. Results of the proposed approach compared to various approaches for PH2 dataset

Authors	No. of classes	Method	Accuracy
Ozkan et al. [1]	3	MLP	92.50%
Ann et al. [3]	3	AlexNet	67.50%
Singh et al. [4]	3	SVM	92.50%
Khalid et al. [7]	3	AlexNet	98.61%
Proposed work	3	DenseNet201+SVM	95.00%

Through the evaluation and results obtained from Tables 1, 2 and 3, we find that the DenseNet201 model achieved the best results as compared to other CNN models. On the other side, the cubic SVM classifier achieved the best results as compared to others classifiers ANN, KNN, RF. Through Figs. 3, 4 and 5, we notice that data augmentation considerably improves the results of classification for three classes. From Tables 1, Table 2 and Table 3, the statistics show that SVM classifiers perform the highest accuracy of 99% for melanoma and non-melanoma detection while 95% for melanoma, atypical nevus, or common nevus with augmentation combined with DenseNet architecture.

Table 4 and Table 5 depict the comparison of the proposed approach with the state-of-the-art models in terms of accuracy for the PH2 dataset. As it can be seen our proposed approach is competitive with the state-of-the-art.

5   Conclusion

In order to take advantage of deep learning models' ability to extract features from skin lesion images, several experiments were conducted with convolutional neural networks (CNN) architectures combined with various classifiers. In this work, we have proposed to use DenseNet201, ResnNet50, EfficientB0 pre-trained CNN architectures as feature extractors and the Artificial neural network (ANN), Support Vector Machines (SVM), K-Nearest Neighbor (KNN), Random Forest (RF) as classifiers to evaluate the classification of skin lesions from PH2 datasets with 2 or 3 classes. The results found show that the DenseNet201 model combined with the SVM classifier gives the better score for two classes with an accuracy of 99%, moreover for three classes with 95%, which is competitive with the actual state-of-the-art.

Acknowledgement. This work was completed as part of the Hubert Curien Partnership (PHC) TASSILI cooperation program between France and Algeria under the project code 19MDU212.

References

1. Ozkan, I., Koklu, M.: Skin Lesion classification using machine learning algorithms. Int. J. Intell. Syst. Appl. Eng. **5**, 285–289 (2017)
2. Ghasem Shakourian, G., Kordy, H.M., Ebrahimi, F.: A hierarchical structure based on stacking approach for skin lesion classification. Expert Syst. Appl. **145**, 113–127 (2020)
3. Salido, J.A., Ruiz, C.R.: Using deep learning to detect melanoma in dermoscopy Images. Int. J. Mach. Learn. Comput. **8**(1), 61–68 (2018)
4. Singh, L., Janghel, R.R., Sahu, S.: Designing a retrieval-based diagnostic aid using effective features to classify skin Lesion in dermoscopic images. Procedia Comput. Sci. **167**, 2172–2180 (2020)
5. Filali, Y., El Khoukhi, H., Sabri, M., Aarab, A.: Efficient fusion of handcrafted and pre-trained CNNs features to classify melanoma skin cancer. Multimedia Tools Appl. **79**, 31219–31238 (2020)
6. Sanket, K., Chandra, J.: Skin Cancer Classification using Machine Learning for Dermoscopy Image 1457 (2019)
7. Khalid, M.H., Kassem, M.A., Foaud, M.M.: Skin cancer classification using deep learning and transfer learning. In: 2018 9th Cairo International Biomedical Engineering Conference (CIBEC), pp. 90–93 (2018)
8. Livingstone, D.J.: Artificial Neural Networks: Methods and Applications. Humana Press, Totowa, USA (2011)
9. VapniK, V.: Statistical learning theory (1998)
10. Larose, D.T., Larose, C.D.: Discovering Knowledge in Data: An Introduction to Data Mining. Wiley, Hoboken, USA (2014)
11. Breiman, L.: Random Forests. Machine Learning **4**, 5–32 (2001)
12. He, K., Zhang, X., Ren, S., Sun, J.: Deep residual learning for image recognition. In: IEEE Conference on Computer Vision and Pattern Recognition (CVPR), pp. 770–778 (2016)
13. Huang, G., Liu, Z., Weinberger, K. Q.: Densely connected convolutional networks. In: IEEE Conference on Computer Vision and Pattern Recognition (CVPR), pp. 2261–2269 (2017)
14. Tan, M., Le, Q.V.: EfficientNet: Rethinking Model Scaling for Convolutional Neural Networks, ArXiv (2019)
15. Mendonçan, T., Ferreira, P., Marques, J., Marçal, A., Rozeira, J.: PH2 - a dermoscopic image database for research and benchmarking. In: 35th Annual International Conference of the IEEE Engineering in Medicine and Biology Society (EMBC), pp. 5437–5440 (2013)

Recursively Refined R-CNN: Instance Segmentation with Self-RoI Rebalancing

Leonardo Rossi$^{(\boxtimes)}$ (iD), Akbar Karimi$^{(\boxtimes)}$ (iD), and Andrea Prati$^{(\boxtimes)}$ (iD)

IMP Lab - D.I.A. - University of Parma, Parma, Italy
{leonardo.rossi,akbar.karimi,andrea.prati}@unipr.it
http://implab.ce.unipr.it/

Abstract. Within the field of instance segmentation, most of the state-of-the-art deep learning networks rely nowadays on cascade architectures [1], where multiple object detectors are trained sequentially, re-sampling the ground truth at each step. This offers a solution to the problem of exponentially vanishing positive samples. However, it also translates into an increase in network complexity in terms of the number of parameters. To address this issue, we propose Recursively Refined R-CNN (R^3-CNN) which avoids duplicates by introducing a loop mechanism instead. At the same time, it achieves a quality boost using a recursive re-sampling technique, where a specific IoU quality is utilized in each recursion to eventually equally cover the positive spectrum. Our experiments highlight the specific encoding of the loop mechanism in the weights, requiring its usage at inference time. The R^3-CNN architecture is able to surpass the recently proposed HTC [4] model, while reducing the number of parameters significantly. Experiments on COCO *minival* 2017 dataset show performance boost independently from the utilized baseline model. The code is available online at https://github.com/IMPLabUniPr/mmdetection/tree/r3_cnn.

Keywords: Instance segmentation · Object detection · Roi rebalancing · Deep learning

1 Introduction

Computer vision is a field of continuous experimentation, where new and better performing algorithms are developed every day and are able to operate in environments with increasingly extreme conditions. In particular, object detection, and instance segmentation as its narrower extension, offers complex challenges which are utilized in various applications, including medical diagnostics [3], autonomous driving [11], visual product search [14], and many others. All these applications demand high-performing systems in terms of prediction quality, as well as low memory usage. Therefore, a desirable architecture is as light as possible regarding the parameter count since it reduces the search space and enhances generalization, while retaining high-quality detection and segmentation performance.

© Springer Nature Switzerland AG 2021
N. Tsapatsoulis et al. (Eds.): CAIP 2021, LNCS 13052, pp. 476–486, 2021.
https://doi.org/10.1007/978-3-030-89128-2_46

However, often these two goals are conflicting. The R^3-CNN architecture and the corresponding training mechanism that we propose present a trade-off between these two conflicting goals. We show that our model is able to obtain the same performance of complex networks (such as HTC [4]) with a network as light as Mask R-CNN [9].

The accuracy of instance segmentation systems is strongly based on the concept of intersection over union (IoU), which is used to identify the detection precision with respect to the ground truth. The higher this value is, the more accurate and the less noisy the predictions are. However, by increasing the IoU threshold, a problem called *exponentially vanishing positive samples* [1] (EVPS) is also introduced, meaning that it can give rise to the problem of good proposals scarcity compared to low-quality ones. This usually leads to a training that is excessively biased towards low-quality predictions. In order to solve this issue, Cascade R-CNN [1] first, and its descendant HTC later, introduced a cascade mechanism where multiple object detectors are trained sequentially in order to take advantage of the previous one and to increase the prediction quality gradually. This means that each stage performs two tasks: first, the detector is training itself, and, then, it is also devoted to identifying the region proposals for subsequent stages. Unfortunately, this also translates into an increase in network complexity in terms of the number of parameters.

In this work, we propose a new way to balance positive samples by exploiting the re-sampling technique, introduced by the cascade models. Our proposed technique generates new proposals with a pre-selected IoU quality in order to equally cover all IoU values. We carry out an extensive ablation study and compare our results with the state of the art in order to demonstrate the advantages of the proposed solution and its applicability to different existing architectures.

The main contributions of this paper are the following:

- An effective solution to deal with the EVPS problem with a single-detector model, rebalancing the proposals with respect to the IoU thresholds through a recursive re-sampling mechanism. This mechanism has the goal of eventually feeding the network with an equal distribution of samples.
- An exhaustive ablation study on all the components of our R^3-CNN architecture in order to evaluate how the performance is affected by each component.
- Our R^3-CNN is introduced into major state-of-the-art models to demonstrate that it boosts the performance independently from the baseline model used.

2   Related Works

Multi-stage Detection/Instance Segmentation. The early works on object detection and instance segmentation were based on the assumption that single-stage end-to-end networks are sufficient to recognize and segment the objects. For instance, YOLO network [19] optimizes localization and classification in one step. Starting with the R-CNN network [8], the idea of a two-stage architecture was introduced, where, in the first stage, a network called RPN (Region Proposal Network) analyzes the whole image and identifies the regions where the probability of finding an object is high. In the second stage, another network performs a more refined analysis on each single region. After this seminal work, others have further refined this idea. The Cascade R-CNN architecture [1] uses multiple bounding-box heads connected sequentially, where each one refines the proposals produced by the previous one. The minimum IoU required for positive examples is increased at each stage, taking into account a different set of proposals. Other studies [22, 23, 25] introduced a similar cascade concept, but applied to the RPN network, where multiple RPNs are sequenced and the results from the previous stage are fed into the next stage. Our work is inspired by HTC network [4], which introduces a particular cascade operation also on the mask extraction modules. However, all these multi-stage networks are quite complex in terms of the number of parameters.

IoU distribution imbalance. Authors in [17] describe the problem as a skewed IoU distribution observed in bounding boxes used in training and evaluation. In [21], the authors highlight the significant imbalance between background and foreground RoI examples and present a hard example mining algorithm to easily select the most significant ones. While in their case the aim is balancing the background (negative) and the foreground (positive) RoIs, in our work the primary goal is to balance RoIs across the entire positive spectrum of the IoU. In [18], an IoU-balanced sampling technique is proposed to mine hard examples. However, the sampling always takes place on the results of the RPN which, as we will see, is not very optimized to provide high-quality RoIs. In our case, we apply re-sampling to the detector itself, which has, on average, a much higher probability of returning more significant RoIs. In [6], the sources of false positives are analyzed and an extra independent classification head is introduced to be plugged into the original architecture to reduce hard false positives. In [26], the authors introduce a new IoU-prediction branch which supports classification and localization. They propose to manually generate samples around ground truths instead of using RPN for localization and IoU prediction branches in training. In [1, 4], overfitting due to EVPS problem for large thresholds is addressed using multiple detectors connected sequentially. They re-sample the ground truth in a sequential manner to progressively improve hypothesis quality. Unlike them, we tackle the problem with a single detector and a single segmentation head. In [16], they offer an interpretation similar to ours about the fact that IoU imbalance has an adverse effect on performance. However, while they use an algorithm to systematically generate the RoIs with the chosen quality, we rely only on the capabilities of the detector itself.

3   Recursively Refined R-CNN

In this section, first we briefly introduce the idea behind multi-stage processing. Then we describe our R^3-CNN architecture with its evolution from a sequential to a recursive pipeline, which offers a change of perspective on training.

As shown in Fig. 1 (a), the HTC (Hybrid Task Cascade) multi-stage architecture [4] mainly follows the idea that a single detector is unlikely to be able to

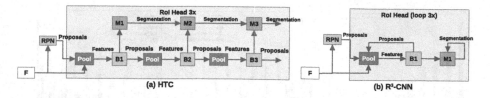

Fig. 1. Network design. (a) HTC: a multi-stage network which trains each head in a cascade fashion. (b) R^3-CNN: our architecture which introduces a loop mechanism to self-train the heads.

train uniformly on all quality levels of IoU. The cascade architecture tries to solve the EVPS problem by training multiple regressors connected sequentially, each of which is specialized in a predetermined and growing IoU minimum threshold. Each regressor performs a conversion of its localization results into a new list of proposals for the following regressor. Although this type of architecture clearly improves the overall performance, it also introduces a considerable number of new parameters into the network. In fact, with respect to its predecessor Mask R-CNN, the number of detection and segmentation modules triples. To reduce the complexity of cascade networks and to address the EVPS problem, we design a lighter architecture with single detection and mask heads uniformly trained on all IoU levels. Authors in [1] underlined the cost-sensitive learning problem [7,15], where the optimization of different IoU thresholds requires various loss functions. Inspired by this study, we address the problem using multiple selective training, which focuses on a specific IoU quality in each step and recursively feeds them into the detector. The intuition is that the detector training and its ability to return an adequate number of proposals of a certain quality level will happen at the same time.

In Fig. 1 (b), the new R^3-CNN architecture along with our training paradigm are shown. In this loop (recursive) architecture, the detector and the RoI pooling modules are connected in a cycle. As in HTC, the first set of RoI proposals is provided by the RPN. After that, the RoI pooler crops and converts them to fixed-size feature maps, which are used to train the B1 block. Then, with an appropriate IoU threshold, the ground truth re-sampling takes place by the B1 block to generate a new proposal set. The result is then used both in the segmentation module M1 and as the new input for the pooler which closes the loop. By the IoU threshold manipulation, the network can force the detection to extract those RoIs with IoU quality levels which are typically missed. The cycle continues three times (3x loop) to guarantee the rebalancing of RoI levels.

Figure 2 (a) shows the generated RoI distribution for each IoU level in Mask R-CNN as well as the EVPS problem. The distribution of the rebalanced samples by our model, on the other hand, can be seen in Fig. 2 (b) and (c). For the latter, it is worth emphasizing some important details emerging from these graphs: (i) Considering only the first loop trend, R^3-CNN looks quite similar to Mask R-CNN; (ii) Conversely, considering the sum of the first two loops, our distribution looks much more balanced; (iii) The third loop significantly increases the number

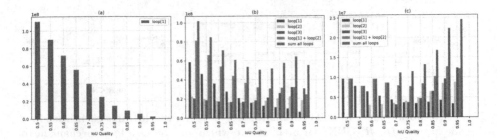

Fig. 2. The IoU histogram of training samples for Mask R-CNN with a 3x schedule (36 epochs) (a), and R^3-CNN where each loop uses different IoU thresholds [0.5, 0.6, 0.7], decreasingly (b) and increasingly (c). Better seen in color. (Color figure online)

of high-quality RoIs. Despite the fact that our architecture contains a single detector, its behavior shows a unique and well-defined trend in terms of RoI distribution within different loops. We believe this is the reason why R^3-CNN outperforms Mask R-CNN. It is able to mimic the Cascade R-CNN behavior in the RoI distribution (as also shown in [1]), achieved by HTC, but using only a single detector and significantly fewer parameters.

For a given loop t, let us define h as the sole classifier and f as the sole regressor which is trained for a selected IoU threshold u^t, with $u^t > u^{t-1}$, by minimizing the loss function of Cascade R-CNN [1]:

$$L(x^t, g) = L_{cls}\left(h\left(x^t\right), y^t\right) + \lambda\left[y^t \geqslant 1\right] L_{loc}\left(f\left(x^t, b^t\right), g\right) \tag{1}$$

where x^t represents the input features of the t-th loop, $b^t = f\left(x^{t-1}, b^{t-1}\right)$ is the new sampled set of proposals coming from the previous loop (with b^0 coming from the RPN), g is the ground truth, and λ is a positive coefficient. y^t represents the label of x^t given the IoU threshold u^t, the proposals b^t and the ground truth label g_y with the following equation:

$$y^t = \begin{cases} g_y & \text{if } IoU\left(b^t, g\right) \geqslant u^t \\ 0 & \text{otherwise} \end{cases} \tag{2}$$

At inference time, the same loop procedure is applied and all the predictions are merged together by computing the mean of the classification values. As it will be shown in the experiments, using loops also at inference (or evaluation) time is not optional, meaning that the loop mechanism is intrinsic to the weights of the network.

4   Experiments

4.1   Dataset and Evaluation Metrics

Dataset. As the majority of recent literature on instance segmentation, we perform our tests on the MS COCO 2017 dataset [13]. The training dataset consists of more than 117,000 images and 80 different classes of objects.

Evaluation Metrics. We used the same evaluation functions offered by the python *pycocotools* software package. All the evaluation phases have been performed on the COCO minival 2017 validation dataset, which contains 5000 images. We report the Average Precision (AP) with different IoU thresholds for both bounding box and segmentation tasks. The main metric (AP) is computed with IoUs from 0.5 to 0.95. Others include AP_{50} and AP_{75} with 0.5 and 0.75 minimum IoU thresholds, and AP_s, AP_m and AP_l for small, medium and large objects, respectively.

4.2 Implementation Details

To perform a fair comparison, we obtain all the reported results by training the networks with the same hardware and, when possible, the same software configuration. When available, the original code released by the authors or the corresponding implementation in MMDetection [5] framework were used. Our code is also developed within this framework. In the case of HTC, we did not consider the semantic segmentation branch.

We performed a distributed training on 2 servers, each equipped with 2 x 16 IBM POWER9 cores, 256 GB of memory and 4 x NVIDIA Volta V100 GPUs with Nvlink 2.0 and 16 GB of memory. Each training consists of 12 epochs with Stochastic Gradient Descent (SGD) optimization algorithm, an initial learning rate of 0.02, a weight decay of 0.0001, and a momentum of 0.9. The learning rate decays at epochs 8 and 11. We used batch size of 2 for each GPU. We fixed the long edge and short edge of the images to 1333 and 800, maintaining the aspect ratio. ResNet 50 [10] was used as the backbone. If not specified differently, the number of loops in training and evaluation are the same.

4.3 Analysis of R^3-CNN

Description. In this part, we demonstrate the potentiality offered by a naive three-stage loop compared to Mask R-CNN and the original three-stage cascade HTC. To have a fair comparison, we select the optimal configuration for the HTC network as baseline and also apply it to training our R^3-CNN. In the *advanced* version, we replace fully-connected layers from detection head with lightweight convolutions with kernel 7×7 and a Non-Local block [24] with incremented kernel size of 7×7 to better exploit information. We also build a brand new branch using only convolutions and Non-Local blocks to include a new learning task to improve segmentation as described in [12]. Since our naive version has slightly fewer parameters than Mask R-CNN, it is also insightful to compare it with our model. Finally, we also want to demonstrate the following important claim: it does not matter the way or order with which the IoU thresholds are changed (either incrementally or decrementally), since in both cases a more balanced IoU distribution is achieved (see Fig. 2).

Results. In Table 1, we report speed in evaluation and memory usage in training, distinguishing between *memory usage* of the entire training process and *model*

Table 1. Comparing trainable parameters with the bounding box and segmentation average precision. K: thousand. Column L_t: number of stages. Speed is image per second. TS: Training strategies. Inc: progressively increasing and Dec: decreasing IoU quality through loops.

#	Model	# Params	TS	L_t	H	B_{AP}	S_{AP}	Speed	Mem. usage	Model size
1	Mask (1x)	44,170 K	–	1	1	38.2	34.7	11.5	4.4 GB	339 MB
2	Mask (3x)	44,170 K	–	1	1	39.2	35.5	5.4	4.4 GB	339 MB
3	HTC	77,230 K	Inc	3	3	41.7	36.9	5.4	6.8 GB	591 MB
4	R^3-CNN (naive)	43,912 K	Inc	3	1	40.9	36.8	5.5	5.9 GB	337 MB
5	R^3-CNN (naive)	43,912 K	Dec	3	1	40.4	36.7	5.5	5.9 GB	337 MB
6	R^3-CNN (advanced)	50,072 K	Inc	3	1	42.0	38.2	1.0	6.8 GB	384 MB

size (proportional to the number of parameters). Comparing the naive version (row #4) with HTC (row #3), it can be seen that our model has significantly fewer parameters and is more memory efficient. While the segmentation precision (S_{AP}) is practically the same, there is a slight loss in B_{AP}. Also, the speed of naive is slightly better than HTC. Regarding the advanced version (row #6), it surpasses the HTC accuracy in both tasks, while saving a significant number of parameters and using the same amount of memory in training. The only disadvantage is the reduced speed due to Non-Local blocks.

Compared to Mask R-CNN (row #1), the naive R^3-CNN has the same complexity, but achieves a much higher precision in both tasks. To further investigate how well our recursive mechanism works, we also compare it with Mask R-CNN trained with triple number of epochs (row #2). While more training helps Mask R-CNN produce a higher precision, it is still outperformed by naive R^3-CNN. This demonstrates that our loop mechanism is not simply another way of training the network for more epochs, but that it represents a different and more effective training strategy. This can be explained by the fact that while in Mask R-CNN the RoI proposals are always provided by the RPN, in our case they are provided by the detection head which generates higher quality and more balanced RoIs (see Fig. 2).

Finally, to show that the order of changes in IoU threshold is not crucial to performance, in rows #4 and #5, we report a comparison between increasing and decreasing IoU thresholds through loops. Although there is a slight degradation of precision using the decreasing training strategy, it is almost negligible due to a more balanced IoU distribution achieved in both cases, but skewed to high-quality RoIs in the first case and low-quality in the latter.

4.4   Ablation Study on the Evaluation Phase

Description. In this subsection we focus on how the results are affected by the number of cycles in the evaluation phase. We consider the naive version mentioned above as pre-trained model, which consists of three loops in the training and evaluation phases.

Table 2. Impact of evaluation loops L_e in a 3-loop and one-head-per-type R^3-CNN model. Row #4 is the naive R^3-CNN in Table 1.

#	Model	L_t	H	L_e	B_{AP}	S_{AP}
1	Mask	1	1	1	38.2	34.7
2	R^3-CNN	3	1	1	37.9	35.4
3		3	1	2	40.5	36.6
4		3	1	3	40.9	36.8
5		3	1	4	40.9	36.7
6		3	1	5	40.9	36.7

Table 3. Impact of the number of training loops in a one-head-per-type R^3-CNN model. Row #4 is the naive R^3-CNN in Table 1.

#	Model	L_t	H	B_{AP}	S_{AP}
1	Mask	1	1	38.2	34.7
2	R^3-CNN	1	1	37.7	34.7
3		2	1	40.4	36.4
4		3	1	40.9	36.8
5		4	1	40.9	36.8

Results. From Table 2, it is evident that the loop mechanism is of paramount importance for the evaluation phase too. In fact, when we train the network with the 3x loops and then evaluate it with one loop, it performs even worse than Mask R-CNN (row #2). On the contrary, with two loops, the result is significantly better (row #3). It also underperforms the three-loop evaluation only slightly (row #4). Therefore, this version could be considered a good compromise between execution time and detection quality. From four loops onward, the performance tends to remain almost stable. This is consistent with our initial hypothesis of a link between evaluation and the loop mechanism, and confirms that in order to have higher performance with more than three loops in the evaluation, we also need to increase the number of loops in the training phase.

4.5 Ablation Study on the Training Phase

Description. In this experiment, the network is trained with a number of loops varying from 1 to 4. The number of loops for the evaluation changes accordingly.

Results. The comparative results are reported in Table 3. The precision of the single loop (row #2) is comparable to Mask R-CNN, and not much different from the above-mentioned model with one-loop evaluation (row #2 of Table 2). This connection with the previous experiment suggests that the detector is strictly optimized on the corresponding sample distribution. The performance is improved by the training strategies with two and three loops, though significantly by the former and only slightly over that by the latter. Regarding more than 3 loops (row #5), the improvement is negligible.

4.6 Extensions on R^3-CNN

Description. Our final experiments show that R^3-CNN model can be plugged in seamlessly to several state-of-the-art architectures for instance segmentation, consistently improving their performance, which demonstrates its generalizability. In this experiment, we select our best-performing version previously called

Table 4. Performance of the state-of-the-art models with and without R^3-CNN model. Bold values are best results, red ones are second-best values.

#	Method	Bounding box (object detection)						Mask (instance segmentation)					
		AP	AP_{50}	AP_{75}	AP_s	AP_m	AP_l	AP	AP_{50}	AP_{75}	AP_s	AP_m	AP_l
1	Mask	37.3	58.9	40.4	21.7	41.1	48.2	34.1	55.5	36.1	18.0	37.6	46.7
2	HTC	41.7	60.4	45.2	24.0	44.8	54.7	36.9	57.6	39.9	19.8	39.8	50.1
3	R^3-CNN-L	42.0	61.0	46.3	24.5	45.2	55.7	38.2	58.0	41.4	20.4	41.0	52.8
4	GRoIE	38.6	59.4	42.1	22.5	42.0	50.5	35.8	56.5	38.4	19.2	39.0	48.7
5	R^3-CNN-L+GRoIE	42.0	61.2	45.6	24.4	45.2	55.7	39.1	58.8	42.3	20.7	42.1	54.3
6	GC-Net	40.5	62.0	44.0	23.8	44.4	52.7	36.4	58.7	38.5	19.7	40.2	49.1
7	HTC+GC-Net	43.9	63.1	47.7	26.2	47.7	57.6	38.7	60.4	41.7	21.6	42.2	52.5
8	R^3-CNN-L+GC-Net	44.3	64.1	48.4	27.0	47.1	58.9	40.2	61.1	43.5	22.6	42.8	56.0
9	DCN	41.9	62.9	45.9	24.2	45.5	55.5	37.6	60.0	40.0	20.2	40.8	51.6
10	HTC+DCN	44.7	63.8	48.6	26.5	48.2	60.2	39.4	61.2	42.3	21.9	42.7	54.9
11	R^3-CNN-L+DCN	44.8	64.3	48.9	26.6	48.3	59.6	40.4	61.3	44.0	22.3	43.6	56.1

advanced (see Table 1 row #6) and renamed R^3-CNN-L model. The experiment tested different state-of-the-art models, namely GRoIE [20], GC-net [2], DCN [27], with and without the R^3-CNN-L version. When compatible, we also merged the original model with HTC as baseline. For example, *HTC+GC-Net* is composed of both HTC and GC-Net merged together.

Results. The results are summarized in Table 4, where best results for each comparison are reported in bold, while the second best is in red. From the table we can see that almost all the best and second-best scores belong to R^3-CNN architectures, both in object detection and instance segmentation. However, there are two cases in which this does not happen. The first one is the combination of GC-Net and HTC which outperforms R^3-CNN-L in AP_m by 0.6% (row #7), and the other one is the combination of DCN and HTC which outperforms R^3-CNN-L in AP_l by the same amount (row #10). Apart from these rare cases, these experiments confirm that the proposed R^3-CNN consistently brings benefits to existing object detection and instance segmentation models, in terms of both precision and reduced number of parameters.

5 Conclusions

In this paper, we introduced the R^3-CNN architecture to address the issue of exponentially vanishing positive samples in training by rebalancing the training proposals with respect to the IoU thresholds, through a recursive re-sampling mechanism in a single detector architecture. We demonstrated that a good training needs to take into account the diversity of IoU quality of the RoIs used to learn, more than aiming to have only high quality RoIs. Our extensive set of experiments and ablation studies provide a comprehensive understanding of the benefits and limitations of the proposed models. R^3-CNN offers a good flexibility to use intermediate versions between the naive version and HTC, permitting to play with the number of loops, depending if we privilege precision, number of parameters or speed. Overall, the proposed R^3-CNN architecture demonstrates its usefulness when used in conjunction with several state-of-the-art models, achieving considerable improvements over the existing models.

References

1. Cai, Z., Vasconcelos, N.: Cascade R-CNN: delving into high quality object detection. In: Proceedings of the IEEE Conference on Computer Vision and Pattern Recognition (CVPR), June 2018
2. Cao, Y., Xu, J., Lin, S., Wei, F., Hu, H.: Gcnet: non-local networks meet squeeze-excitation networks and beyond. In: Proceedings of the IEEE International Conference on Computer Vision Workshops (2019)
3. Chen, H., Qi, X., Yu, L., Dou, Q., Qin, J., Heng, P.A.: Dcan: deep contour-aware networks for object instance segmentation from histology images. Med. Image Anal. **36**, 135–146 (2017)
4. Chen, K., et al.: Hybrid task cascade for instance segmentation. In: Proceedings of the IEEE Conference on Computer Vision and Pattern Recognition, pp. 4974–4983 (2019)
5. Chen, K., et al.: MMDetection: open mmlab detection toolbox and benchmark. arXiv preprint arXiv:1906.07155 (2019)
6. Cheng, B., Wei, Y., Shi, H., Feris, R., Xiong, J., Huang, T.: Revisiting rcnn: On awakening the classification power of faster rcnn. In: Proceedings of the European Conference on Computer Vision (ECCV), pp. 453–468 (2018)
7. Elkan, C.: The foundations of cost-sensitive learning. In: In Proceedings of the Seventeenth International Joint Conference on Artificial Intelligence, pp. 973–978 (2001)
8. Girshick, R., Donahue, J., Darrell, T., Malik, J.: Rich feature hierarchies for accurate object detection and semantic segmentation. In: Proceedings of the IEEE Conference on Computer Vision and Pattern Recognition, pp. 580–587 (2014)
9. He, K., Gkioxari, G., Dollár, P., Girshick, R.: Mask r-cnn. In: Proceedings of the IEEE International Conference on Computer Vision, pp. 2961–2969 (2017)
10. He, K., Zhang, X., Ren, S., Sun, J.: Deep residual learning for image recognition. In: Proceedings of the IEEE Conference on Computer Vision and Pattern Recognition, pp. 770–778 (2016)
11. Huang, L., Zhe, T., Wu, J., Wu, Q., Pei, C., Chen, D.: Robust inter-vehicle distance estimation method based on monocular vision. IEEE Access **7**, 46059–46070 (2019)
12. Huang, Z., Huang, L., Gong, Y., Huang, C., Wang, X.: Mask scoring r-cnn. In: Proceedings of the IEEE Conference on Computer Vision and Pattern Recognition, pp. 6409–6418 (2019)
13. Lin, T.Y., et al.: Microsoft COCO: common objects in context. In: Fleet, D., Pajdla, T., Schiele, B., Tuytelaars, T. (eds.) ECCV 2014. LNCS, vol. 8693, pp. 740–755. Springer, Cham (2014). https://doi.org/10.1007/978-3-319-10602-1_48
14. Liu, S., et al.: Matching-cnn meets knn: quasi-parametric human parsing. In: Proceedings of the IEEE Conference on Computer Vision and Pattern Recognition, pp. 1419–1427 (2015)
15. Masnadi-Shirazi, H., Vasconcelos, N.: Cost-sensitive boosting. IEEE Trans. Pattern Anal. Mach. Intell. **33**(2), 294–309 (2010)

16. Oksuz, K., Cam, B.C., Akbas, E., Kalkan, S.: Generating positive bounding boxes for balanced training of object detectors. In: The IEEE Winter Conference on Applications of Computer Vision, pp. 894–903 (2020)

17. Oksuz, K., Cam, B.C., Kalkan, S., Akbas, E.: Imbalance problems in object detection: a review. IEEE Trans. Pattern Anal. Mach. Intell. (2020)

18. Pang, J., Chen, K., Shi, J., Feng, H., Ouyang, W., Lin, D.: Libra r-cnn: towards balanced learning for object detection. In: Proceedings of the IEEE Conference on Computer Vision and Pattern Recognition, pp. 821–830 (2019)

19. Redmon, J., Divvala, S., Girshick, R., Farhadi, A.: You only look once: unified, real-time object detection. In: Proceedings of the IEEE Conference on Computer Vision and Pattern Recognition, pp. 779–788 (2016)

20. Rossi, L., Karimi, A., Prati, A.: A novel region of interest extraction layer for instance segmentation. arXiv preprint arXiv:2004.13665 (2020)

21. Shrivastava, A., Gupta, A., Girshick, R.: Training region-based object detectors with online hard example mining. In: Proceedings of the IEEE Conference on Computer Vision and Pattern Recognition, pp. 761–769 (2016)

22. Vu, T., Jang, H., Pham, T.X., Yoo, C.: Cascade rpn: Delving into high-quality region proposal network with adaptive convolution. In: Advances in Neural Information Processing Systems, pp. 1432–1442 (2019)

23. Wang, J., Chen, K., Yang, S., Loy, C.C., Lin, D.: Region proposal by guided anchoring. In: Proceedings of the IEEE/CVF Conference on Computer Vision and Pattern Recognition, pp. 2965–2974 (2019)

24. Wang, X., Girshick, R., Gupta, A., He, K.: Non-local neural networks. In: Proceedings of the IEEE Conference on Computer Vision and Pattern Recognition, pp. 7794–7803 (2018)

25. Zhong, Q., Li, C., Zhang, Y., Xie, D., Yang, S., Pu, S.: Cascade region proposal and global context for deep object detection. Neurocomputing 395, 170–177 (2020)

26. Zhu, L., Xie, Z., Liu, L., Tao, B., Tao, W.: Iou-uniform r-cnn: breaking through the limitations of rpn. arXiv preprint arXiv:1912.05190 (2019)

27. Zhu, X., Hu, H., Lin, S., Dai, J.: Deformable convnets v2: more deformable, better results. In: Proceedings of the IEEE Conference on Computer Vision and Pattern Recognition, pp. 9308–9316 (2019)

Layer-Wise Relevance Propagation Based Sample Condensation for Kernel Machines

Daniel Winter[1], Ang Bian[2(✉)] ⓘ, and Xiaoyi Jiang[1] ⓘ

[1] Faculty of Mathematics and Computer Science, University of Münster,
Münster, Germany
[2] College of Computer Science, Sichuan University, Chengdu, China
bian@scu.edu.cn

Abstract. Kernel machines are a powerful class of methods for classification and regression. Making kernel machines fast and scalable to large data, however, is still a challenging problem due to the need of storing and operating on the Gram matrix. In this paper we propose a novel approach to sample condensation for kernel machines, preferably without impairing the classification performance. To our best knowledge, there is no previous work with the same goal reported in the literature. For this purpose we make use of the neural network interpretation of kernel machines. Explainable AI techniques, in particular the Layer-wise Relevance Propagation method, are used to measure the relevance (importance) of training samples. Given this relevance measure, a decremental strategy is proposed for sample condensation. Experimental results on three data sets show that our approach is able to achieve the goal of substantial reduction of the number of training samples.

1 Introduction

A fundamental result of learning theory is the family of representer theorems [7], which lead to the powerful kernel machines. Although trained to have zero classification error, kernel machines generalize well to unseen test data [4]. Compared to deep neural networks (DNN), they can be interpreted as two-layer NNs. Despite the simplicity, however, kernel machines turned out to be a good alternative to DNNs, capable of matching and even surpassing their performance while utilizing less computational resources in training [8,9].

Making kernel machines fast and scalable to large data is still a challenging problem. A major limiting factor is the need of saving all training samples, computing the corresponding Gram matrix, and solving the related linear equation system (see Sect. 2). In this paper we thus consider the problem of condensing the training samples, preferably without impairing the classification performance. Based on the interpretation of kernel machines as two-layer neural networks, we make use of explainable AI techniques [15], in particular Layer-wise Relevance Propagation (LRP) [14], as a means to measure the relevance (importance) of training samples. A decremental strategy is proposed to use this measure for sample condensation.

© Springer Nature Switzerland AG 2021
N. Tsapatsoulis et al. (Eds.): CAIP 2021, LNCS 13052, pp. 487–496, 2021.
https://doi.org/10.1007/978-3-030-89128-2_47

Sample condensation has been studied in other contexts, where the whole training set has to be saved and used for classification. Starting from the pioneer work [6], more advanced techniques have been proposed to boost the performance of nearest neighbor based classifiers [2,12]. In addition, nearest neighbor condensation has been applied to speed up the training of support vector machines [1] and convolutional neural networks [12].

The remainder of the paper is organized as follows. In Sect. 2 we introduce the fundamentals of kernel machines and discuss the need of sample condensation, thus motivating our work. Our sample condensation method is described in Sect. 3. Experimental results are reported in Sect. 4. Finally, Sect. 5 concludes the paper.

2   Kernel Machines

Kernels are an efficient way to compute the similarity of two samples in a higher dimensional space. In this section we introduce a technique to fully interpolate the training data using kernel functions, known as kernel machines. Let $X = \{x_1, x_2, \ldots, x_n\} \subset \Omega^n$ be a set of n training samples with their corresponding targets $Y = \{y_1, y_2, \ldots, y_n\} \subset T^n$ in the target space. A function $f : \Omega \to T$ interpolates this data iif

$$f(x_i) = y_i, \quad \forall i \in 1, \ldots, n \tag{1}$$

Representer Theorem [7]. Let $k : \Omega \times \Omega \to \mathbb{R}$ be a positive definite kernel, X and Y a set of training samples and targets as defined above, and $g : [0, \infty) \to \mathbb{R}$ a strictly monotonically increasing function for regulation. We define E as an error function that calculates the loss l of f on the whole sample set with

$$E(X,Y) = E((x_1, y_1), \ldots, (x_n, y_n)) = \frac{1}{n} \sum_{i=1}^{n} l(f(x_i), y_i) + g(\|f\|) \tag{2}$$

Then, the function f^* that minimizes E, $f^* = \operatorname{argmin}_f\{E(X,Y)\}$, has the form

$$f^*(z) = \sum_{i=1}^{n} \alpha_i k(z, x_i) \quad \text{with } \alpha_i \in \mathbb{R} \tag{3}$$

We now can use f^* from Eq. (3) to interpolate our training data. Note that the only learnable parameters are $\alpha = (\alpha_1, \ldots, \alpha_n)$. Learning α is equivalent to solving the system of linear equations

$$K(\alpha_1^*, \ldots, \alpha_n^*)^T = (y_1, \ldots, y_n)^T \tag{4}$$

where $K \in \mathbb{R}^{n \times n}$ is the Gram matrix with elements $K_{ij} = k(x_i, x_j)$. Since the kernel function k is assumed to be positive definite, the Gram matrix K is invertible. Therefore, we can find the optimal α^* to construct f^* by

$$(\alpha_1^*, \ldots, \alpha_n^*)^T = K^{-1}(y_1, \ldots, y_n)^T \tag{5}$$

After learning, the kernel machine then uses the interpolating function from Eq. (3) to make prediction for test samples. In this work we focus on classification problems. In this case $f(z)$ is encoded as a one-hot vector $f(z) = (f_1(z), \ldots f_t(z))$ with $t \in \mathbb{N}$ being the number of output classes. When predicting a test sample z, the output vector $f(z)$ is not a one-hot vector, in fact not even a probability vector, in general. The class which gets the highest output value is considered as the predicted class. If needed, e.g. for the purpose of classifier combination, the output vector $f(z)$ can also be converted into a probability vector by applying the softmax function.

The practical usability of kernel machines strongly depends on the size n of training set. Solving the optimal α^* in (5) in a naive manner requires computation of order $\mathcal{O}(n^3)$ and is thus not feasible for many applications. Recently, a highly efficient solver EigenPro has been developed [13] to enable significant speedup for training on GPUs.

Sample condensation is another way of efficiency boosting, which is required even when using high-performance solvers like EigenPro. After training, the testing using (3) still needs the whole set of training samples, which is similar to the situation with nearest neighbor based classifiers. In complex domains like strings and graphs the kernel computation may be costly [3, 11, 18] so that the need of considerably reducing the number of samples remains. Even in case of easy-to-compute kernel functions, it can be typically expected that not all training samples are relevant to the classification. This observation has been made before, e.g. when working with nearest neighbor based classifiers [2, 12]. Thus, there is a general need of sample condensation for kernel machines. In this work we propose a novel approach tailored to sample condensation for kernel machines. To our best knowledge, there is no previous work with the same goal reported in the literature.

3   Sample Condensation Method

We make use of the neural network interpretation of kernel machines and apply the Layer-wise Relevance Propagation method to measure the relevance of training samples. Given the relevance estimation of training samples, a decremental strategy is then applied to select the most relevant samples out of a training set.

3.1   LRP for Relevance Measure of Kernel Machine

The kernel machine (3) can be seen as a network with one hidden layer. Let z be the test sample to which the target $f(z)$ should be computed. Given a training sample x_i out of the training set $X \subset \Omega^n$, we denote α_{it} as the trained weight between x_i and the value in the output $f_t(z)$. Figure 1 shows the network architecture of a kernel machine. The input z is represented by a single input neuron. Each training sample is represented by a single neuron in the hidden layer and connected to the input by a special connection applying the kernel function.

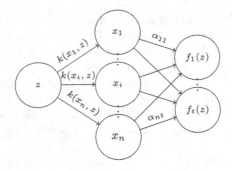

Fig. 1. Neural network representation of a kernel machine.

Each output class is represented by a neuron in the output layer, connected by the individually learned weight α_{it}.

The recent research on explainable AI has spawned many techniques, e.g. for studying the influence of hyper-parameters on training deep neural networks [5] and interpreting the behavior of neural networks [15]. In particular, it is possible to estimate the relevance of features (also hidden neurons) to the network decision. We apply such relevance estimation, concretely Layer-wise Relevance Propagation (LRP) [14], to determine the relevance of training samples.

Overall, the relevance estimation in our proposed approach consists of two steps. In the first phase the optimal α^* in (3) is computed, by means of a highly efficient solver like EigenPro [13] if needed. In the second phase a set of validation samples is used to estimate the relevance of training samples, which builds the foundation for sample condensation described in Sect. 3.

We apply LRP to propagate the relevance back from the output layer to the hidden layer and so assign each training sample a relevance measure. The formula to compute the relevance of a neuron x_i with connection to neurons x_t on a given validation sample z is given by

$$R(x_i, z) = \sum_t \frac{x_i w_{it}^+}{\sum_{i'} x_{i'} w_{i't}^+} \cdot R(x_t) \tag{6}$$

where $w^+ = \max(w, 0)$. Since x_t is in the output layer, its relevance is the target itself $R(x_t) = f_t(z)$. The weight w_{it} between the neurons representing x_i and x_t is given by the weight of the kernel machine $w_{it} = \alpha_{it}$. The activation on the neuron representing x_i is given by the kernel function $k(z, x_i)$. We thus can express the relevance $R(x_i, z)$ of a training sample x_i to the output $f(z)$ on a given validation sample z by

$$R(x_i, z) = \sum_t \frac{k(x_i, z)\alpha_{it}^+}{\sum_{i'} k(x_{i'}, z)\alpha_{i't}^+} \cdot f_t(z) \tag{7}$$

The target vector $f(z)$ is a one-hot vector in our case, i.e. the value in the vector corresponding to the correct target class t^* is 1 and all others are 0. Therefore, we do not need to apply the outer sum but only calculate it for t^* since all other elements in the sum would be 0, which leads to

$$R(x_i, z) = \frac{k(x_i, z)\alpha^+_{it^*}}{\sum_{i'} k(x_{i'}, z)\alpha^+_{i't^*}} \tag{8}$$

Equation (8) only focuses on the relevance of one validation sample z. To get a good estimation of the general relevance of a training sample x_i, we split the training set $X \subset \Omega^n$ in two distinct subsets X_{train} and X_{val} with $X = X_{train} \cup X_{val}$, $X_{train} \cap X_{val} = \emptyset$. We train a kernel machine only using the set X_{train}. For each training sample $x_i \in X_{train}$ we then add all relevances on validation samples $x_j \in X_{val}$

$$R(x_i) = \sum_{x_j \in X_{val}} R(x_i, x_j) = \sum_{x_j \in X_{val}} \frac{k(x_i, x_j)\alpha^+_{it^*}}{\sum_{i'} k(x_{i'}, x_j)\alpha^+_{i't^*}} \tag{9}$$

3.2 Relevance-Based Sample Condensation

Given the relevance estimation of training samples, we first sort the training samples by the relevance measure. A decremental strategy is then applied to select the most relevant samples out of a training set X_{train} by slowly eliminating the least relevant samples until only m ($< n$) samples are left.

The idea is to select the m samples with the highest relevance scores. A problem with this simple approach is a proper choice of the parameter m. If m is chosen too small, the selected samples will not suffice to reach a model of good accuracy. On the other hand, if m is chosen too big, the selection is sub-optimal since the same accuracy could be reached with fewer samples. Therefore, we define a parameter μ that expresses the minimum share of the original score (on the whole training set) that we like to retain. This means that for the whole training set X_{train}, the selected samples $X_{selected_m} \subset X_{train}$, and a score measure s, e.g. the accuracy, the following should apply

$$s(X_{selected_m}) \geq s(X_{train}) \cdot \mu \tag{10}$$

We assume that the evolution of the score is approximately monotonously rising, i.e. the score is in general higher for greater m, but may have small local noise, which is however small enough to be ignorable. Later in Sect. 4 we will show that the score indeed is of such a form.

A possible way to find m samples that represent the data best is to drop the Δ least relevant samples in each step. We train a new kernel machine in each step with the remaining samples and re-calculate the relevances with this machine. In general, we hope that in each step there is less redundancy. For example, a medium relevant sample can become more relevant in the next iterations, when other samples that are similar to it are dropped out. In each iteration, we thus train a kernel machine and drop the least relevant Δ samples regarding to the validation set. The algorithm is depicted in Algorithm 1. Due to the fact that a

---

Algorithm 1. Decremental sample condensation

1: **procedure** DECREMENTAL _ SELECTION(
$$X_{train}, Y_{train}, X_{val}, Y_{val}, X_{test}, Y_{test}, k, \Delta, \mu)$$
2:     $model = \text{TRAIN\_KM}(X_{train}, Y_{train}, k)$
3:     $score = \text{TEST\_KM}(model, X_{test}, Y_{test}, k)$
4:
5:     $score_i = score$
6:     **while** $score_i > score * \mu$ **do**
7:         $relevances = \text{GET\_RELEVANCES}($
$$model.\alpha, X_{train}, X_{val}, Y_{val}, k)$$
8:         $X_{train} = \text{ARGSORT}(X_{train}, relevances)$ ▷ Sort by the relevances
9:         $Y_{train} = \text{ARGSORT}(Y_{train}, relevances)$
10:
11:         $X_{train} = X_{train}[1 : -\Delta]$ ▷ drop the last m elements
12:         $Y_{train} = Y_{train}[1 : -\Delta]$
13:
14:         $model = \text{TRAIN\_KM}(X_{train}, Y_{train}, k)$
15:         $score_i = \text{TEST\_KM}(model, X_{test}, Y_{test}, k)$
16:     **end while**
17:     **return** $X_{train}, Y_{train}, score_i$
18: **end procedure**

---

new kernel machine is computed in each iteration and its weight vector is used in the next iteration, the runtime therefore is always of order $\mathcal{O}(\frac{n_{train}}{\Delta})$.

Another way to find m samples that represent the data best is to start at the other side of the set, i.e. add the Δ most relevant samples in each step. This incremental strategy, however, turns put not to be competitive against the decremental strategy [17] and is thus not further discussed in this paper.

4   Experimental Validation

4.1   Data Sets

For our purposes, we have chosen three data sets for image classification that are broadly used and well studied. The **MNIST** data set contains 60,000 handwritten digits (graylevel images are of 28×28 pixels) for training and 10,000 handwritten digits for testing, written by 250 different people. The **MNIST-Fashion** data set has the same structure as the original MNIST data set (i.e. 60,000 training images and 10,000 test images, all of size 28×28). It contains images of clothes of 10 different classes (t-shirt/top, trouser, pullover, dress, coat, sandal, shirt, sneaker, bag, ankle boot). The **CIFAR-10** data set is formed by selecting and labeling proper images out of the *80 million tiny images* data set. It contains 50,000 training images plus 10,000 designated test images, each being a 32×32 RGB-image and labeled with one of the ten classes (airplane, automobile, bird, cat, deer, dog, frog, horse, ship, truck). The objects in the images were captured from different view points and from different distances, which leads to

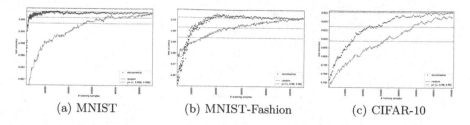

(a) MNIST   (b) MNIST-Fashion   (c) CIFAR-10

Fig. 2. Accuracy with selected training samples for the three data sets.

Table 1. Required number of samples to reach a certain accuracy level.

Data set	μ	Approach	
		Decremental	Random
MNIST	0.998	2,500	26,000
	0.999	3,000	35,200
MNIST-Fashion	0.98	8,400	14,200
	0.99	9,700	22,900
CIFAR-10	0.98	16,600	25,600
	0.99	21,600	33,900

more variety in the data set compared to the other two data sets. For all three data sets, 90% is of the training data is really used as training data while the remainder 10% serves as validation data for relevance estimation.

The state-of-the-art classification results on these data sets can be found in [10, 16, 20], respectively. It is important to emphasize that it is not our goal to beat these results. Instead, we use them to study the ability of our approach to sample condensation without impairing the classification performance of kernel machines. The power of kernel machines themselves as classifier has already been demonstrated in the literature [4, 9].

4.2 Results

Since convolutional neural networks (CNN) are powerful in feature learning, we train a CNN with all the samples and resort to using the learned features from the convolutional layers as input to a kernel machine. In all our experiments we use the Laplacian kernel $k(x_1, x_2) = \exp\left(-\frac{\|x_1 - x_2\|}{2\sigma}\right)$ with a bandwidth $\sigma = 7$. We chose the bandwidth $\sigma = 7$ since the experiments only show minor improvements with larger bandwidths.

In Fig. 2 we show the performance of our approach on the three test sets (the step size is set to $\Delta = 100$). For comparison purpose, we also show the performance of the same number of randomly selected samples. We marked the original accuracy with the whole test set and $\mu = 99\%$ and $\mu = 98\%$ of this accuracy as stop criterion described in the algorithm. Note that for the MNIST

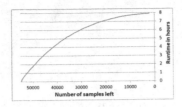

Fig. 3. Runtime with decremental approach on MNIST.

Table 2. Comparison of the accuracies of CNN trained using the selected samples.

Data set	Number of samples	Selected samples	Random samples
MNIST	2,500	0.9799	0.9580
	3,000	0.9831	0.9654
MNIST-Fashion	8,400	0.8067	0.8698
	9,700	0.8449	0.8779
CIFAR-10	16,600	0.7191	0.7436
	21,600	0.8146	0.8108

set we alternately chose $\mu = 99.9\%$ and $\mu = 99.8\%$ since the accuracy for this set was still high even with only 2,500 of 60,000 samples left. The required number of samples to reach a certain accuracy is shown in Table 1.

When decreasing the number of samples (i.e. reading the figures from right to left), the accuracy of the randomly selected samples considerably decreases, while the accuracy of the samples selected with our approach only decreased slowly (CIFAR-10) or stays static (MNIST and MNIST-Fashion). For MNIST-Fashion, reducing the samples even increases the accuracy slightly.

To produce the data for the decremental approach in Fig. 2 we reduced the training set down to 2,000 samples. This took 7.9 h for the two MNIST data sets and 11.6 h for CIFAR-10 because the feature vectors contain more elements here. Note that the runtime is recorded on a computer with 16 GB memory and Nvidia GTX 760. As an example, Fig. 3 shows the accumulated runtime in dependency to the number of samples left on the MNIST data set.

We now investigate if the selected samples have a higher expressiveness in general or if it is limited only to our special experiment with kernel machines. Therefore, we only use the individual selected samples to train the original CNN and compare the accuracy of the resulting model to the accuracy with a model trained of the same number of randomly selected features. Table 2 shows the result of this comparison for the samples selected with our approach. We can see that the selected samples do not seem to have a higher expressiveness in general. Only for the MNIST data set, where we managed to select very few samples, the accuracy of the selected samples is higher. On the other data sets, the accuracy is mostly the same or slightly worse.

Overall, our LRP-based sample condensation technique can reduce the number of training samples while still preserving the high accuracy of kernel machines.

As input for these studies we have chosen the output of the convolutional (feature learning) part of a CNN, which was trained once with the whole data set. In the comparison shown in Table 2, the CNN, especially its convolutional part, was only trained with the remaining, selected samples of the previous experiments. Since we could not show that the selected samples do lead to greater accuracy than randomly selected ones, we come to the following conclusion: The convolutional (feature learning) part of a CNN really benefits from a large base of training samples, whereas for the fully connected (classification) part a smaller base on training samples is sufficient.

It is important to mention that achieving a general higher expressiveness of the selected training samples is not the goal of this work. In fact, it cannot necessarily be expected since our approach is tailored to kernel machines. Our goal is to reduce the number of training samples to store for model inference and classification of unseen patterns, which is clearly achieved. We could reduce both the size of the model and the complexity of computation for kernel machines. To maintain 99% (99.9% for MNIST) of the original accuracy, we could reduce the number of training samples to 5% of the original training set for an easier task like MNIST and 43% for a more complex task like CIFAR-10.

5 Conclusion

This work intends to achieve substantial reduction of training data to store for model inference and classification of unseen patterns for kernel machines. Based on the neural network interpretation of kernel machines, we apply explainable AI techniques, in particular the Layer-wise Relevance Propagation method, to measure the relevance (importance) of training samples. A decremental strategy has been proposed for sample condensation. Our experimental results demonstrated the ability of our approach to considerably condense the training set without impairing the classification performance. Currently, we apply a rather straightforward decremental strategy for the condensation purpose. More sophisticated techniques can be studied in future. For instance, the concept of sparse representations [19] may be an option to model the importance of training samples.

To our best knowledge, our work is the first contribution to sample condensation tailored to kernel machines. As such it contributes to making kernel machines fast and scalable to large data. In addition, it also represents a novel application of explainable AI techniques.

Acknowledgment. This work was partly supported by the EU Horizon 2020 RISE Project ULTRACEPT under Grant 778062.

References

1. Angiulli, F., Astorino, A.: Scaling up support vector machines using nearest neighbor condensation. IEEE Trans. Neural Netw. **21**(2), 351–357 (2010)
2. Batchanaboyina, M.R., Devarakonda, N.: Design and evaluation of outlier detection based on semantic condensed nearest neighbor. J. Intell. Syst. **29**(1), 1416–1424 (2020)

3. Belazzougui, D., Cunial, F.: A framework for space-efficient string kernels. Algorithmica **79**(3), 857–883 (2017). https://doi.org/10.1007/s00453-017-0286-4
4. Belkin, M., Ma, S., Mandal, S.: To understand deep learning we need to understand kernel learning. In: 35th International Conference on Machine Learning, pp. 540–548 (2018)
5. Hamid, S., Derstroff, A., Klemm, S., Ngo, Q.Q., Jiang, X., Linsen, L.: Visual ensemble analysis to study the influence of hyper-parameters on training deep neural networks. In: Proceedings of the of 2nd Workshop on Machine Learning Methods in Visualisation for Big Data (MLVis@EuroVis), pp. 19–23 (2019)
6. Hart, P.E.: The condensed nearest neighbor rule. IEEE Trans. Inf. Theory **14**(3), 515–516 (1968)
7. Herbrich, R.: Learning Kernel Classifiers: Theory and Algorithms. MIT Press, Cambridge (2002)
8. Huang, P., Avron, H., Sainath, T.N., Sindhwani, V., Ramabhadran, B.: Kernel methods match deep neural networks on TIMIT. In: IEEE International Conference on Acoustics, Speech and Signal Processing, pp. 205–209 (2014)
9. Hui, L., Ma, S., Belkin, M.: Kernel machines beat deep neural networks on mask-based single-channel speech enhancement. In: Proceedings of 20th Annual Conference of International Speech Communication Association, pp. 2748–2752 (2019)
10. Kowsari, K., Heidarysafa, M., Brown, D.E., Meimandi, K.J., Barnes, L.E.: RMDL: random multimodel deep learning for classification. In: Proceedings of the 2nd International Conference on Information System and Data Mining, pp. 19–28 (2018)
11. Kriege, N.M., Johansson, F.D., Morris, C.: A survey on graph kernels. Appl. Netw. Sci. **5**(1), 1–42 (2019). https://doi.org/10.1007/s41109-019-0195-3
12. Liang, T., Xu, X., Xiao, P.: A new image classification method based on modified condensed nearest neighbor and convolutional neural networks. Pattern Recogn. Lett. **94**, 105–111 (2017)
13. Ma, S., Belkin, M.: Kernel machines that adapt to GPUs for effective large batch training. In: Proceedings of Conference on Machine Learning and Systems (2019)
14. Montavon, G., Binder, A., Lapuschkin, S., Samek, W., Müller, K.-R.: Layer-wise relevance propagation: an overview. In: Samek, W., Montavon, G., Vedaldi, A., Hansen, L.K., Müller, K.-R. (eds.) Explainable AI: Interpreting, Explaining and Visualizing Deep Learning. LNCS (LNAI), vol. 11700, pp. 193–209. Springer, Cham (2019). https://doi.org/10.1007/978-3-030-28954-6_10
15. Samek, W., Montavon, G., Vedaldi, A., Hansen, L.K., Müller, K.-R. (eds.): Explainable AI: Interpreting, Explaining and Visualizing Deep Learning. LNCS (LNAI), vol. 11700. Springer, Cham (2019). https://doi.org/10.1007/978-3-030-28954-6
16. Springenberg, J.T., Dosovitskiy, A., Brox, T., Riedmiller, M.A.: Striving for simplicity: the all convolutional net. In: Proceedings of 3rd International Conference on Learning Representations (2015)
17. Winter, D.: Sample Condensing for Kernel Machine based Classification. Master's thesis, University of Münster (2020)
18. Xu, L., Bai, L., Jiang, X., Tan, M., Zhang, D., Luo, B.: Deep Rényi entropy graph kernel. Pattern Recognit. **111**, 107668 (2021)
19. Zhang, Z., Xu, Y., Yang, J., Li, X., Zhang, D.: A survey of sparse representation: Algorithms and applications. IEEE Access **3**, 490–530 (2015)
20. Zhong, Z., Zheng, L., Kang, G., Li, S., Yang, Y.: Random erasing data augmentation. In: Proceedings of The Thirty-Fourth AAAI Conference on Artificial Intelligence, pp. 13001–13008 (2020)

Author Index

Printed in the United States
by Baker & Taylor Publisher Services